Short Contents

Contents

Preface

By writing this book we hope to share with you some of our wonder at the complexity of nature, but we must all also be aware that there is a darker side: the fear that we are destroying our natural environments and the services they provide. All of us need to be ecologically literate so that we can take part in political debate and contribute to solving the ecological problems that we carry with us in this new millennium. We hope our book will contribute to this objective.

The genesis of this book can be found in the more comprehensive treatment of ecology in our big book *Ecology: from Individuals to Ecosystems* (Begon, Townsend & Harper, 4th edn, 2006). This is used as an advanced university text around the world, but many of our colleagues have called for a more succinct treatment of the essence of the subject. Thus, we were spurred into action to produce a distinctively different book, written with clear objectives for a different audience—those taking a semester-long beginners course in the essentials of ecology. We hope that at least some readers will be excited enough to go on to sample the big book and the rich literature of ecology that it can lead into.

In this fourth edition of *Essentials of Ecology* we have continued to make the text, including mathematical topics, accessible while updating the material and expanding our coverage of ecosystem science and biogeochemistry. The fourth edition extensively covers both terrestrial and aquatic ecology, and we have strived to demonstrate how ecological principles apply equally to both types of environments. While we have expanded coverage on some topic areas in the fourth edition, we worked hard to not expand the size of the book. We want this text to be a readily accessible read.

Ecology is a vibrant subject and this is reflected by our inclusion of literally hundreds of new studies. Some readers will be engaged most by the fundamental principles of how ecological systems work. Others will be impatient to focus on the ecological problems caused by human activities. We place heavy emphasis on both fundamental and applied aspects of ecology: there is no clear boundary between the two. However, we have chosen to deal first in a systematic way with the fundamental side of the subject, and we have done this for a particular reason. An understanding of the scope of the problems facing us (the unsustainable use of ecological resources, pollution, extinctions and the erosion of natural biodiversity) and the means to

counter and solve these problems depend absolutely on a proper grasp of ecological fundamentals.

The book is divided into five sections. In the introduction we deal with two foundations for the subject that are often neglected in texts. Chapter 1 aims to show not only what ecology is but also how ecologists do it—how ecological understanding is achieved, what we understand (and, just as important, what we do not yet understand) and how our understanding helps us predict and manage. We then introduce 'Ecology's evolutionary backdrop' and show that ecologists need a full understanding of the evolutionary biologist's discipline in order to make sense of patterns and processes in nature (Chapter 2).

What makes an environment habitable for particular species is that they can tolerate the physicochemical conditions there and find in it their essential resources. In the second section we deal with conditions and resources, both as they influence individual species (Chapter 3) and in terms of their consequences for the composition and distribution of multispecies communities and ecosystems, for example in deserts, rain forests, rivers, lakes and oceans (Chapter 4).

The third section (Chapters 5–8) deals systematically with the ecology of individual organisms and populations, with chapters on 'birth, death and movement' (Chapter 5), 'interspecific competition' (Chapter 6), and 'predation, grazing, and disease' (Chapter 7). This section also includes a chapter on 'Molecular and evolutionary ecology', added originally in the third edition and responding to the feelings of some readers that, although evolutionary ideas pervade the book, there was still not sufficient evolution for a book at this level.

In the fourth section (Chapters 9–11), we move up the hierarchical scale of ecology to consider communities consisting of many populations, and ecosystems, where we focus on the fluxes of energy and matter between and within systems.

Finally, armed with knowledge and understanding of the fundamentals, the book turns to the application of ecological science to some of the major environmental challenges of our time. Our goal in these final chapters is not to provide encyclopedic coverage to these environmental problems, but rather to illustrate how ecology contributes to understanding the problems, and can potentially help with their solution. In Chapter 12, we focus on global biogeo-chemical cycles, such as the global carbon dioxide cycle and how this has been dramatically changed by burning fossil fuels and other human activities. In 'conservation ecology' (Chapter 13), we develop an armory of approaches that may help us to save endangered species from extinction and conserve some of the biodiversity of nature for our descendants. The final chapter, 'the ecology of human population growth, disease, and food supply,' takes an ecological approach to examining the issues of the population problem, of human health, and of the sustainability of agriculture and fisheries.

A number of pedagogical features have been included to help you.

- Each chapter begins with a set of key concepts that you should understand before proceeding to the next chapter.

- Marginal headings provide signposts of where you are on your journey through each chapter—these will also be useful revision aids.

- Each chapter concludes with a summary and a set of review questions, some of which are designated challenge questions.

- You will also find three categories of boxed text:

  - 'Historical landmarks' boxes emphasize some landmarks in the development of ecology.

  - 'Quantitative aspects' boxes set aside mathematical and quantitative aspects of ecology so they do not unduly interfere with the flow of the text and so you can consider them at leisure.

  - 'ECOncerns' boxes highlight some of the applied problems in ecology, particularly those where there is a social or political dimension (as there often is). In these, you will be challenged to consider some ethical questions related to the knowledge you are gaining.

An important further feature of the book is the companion web site, accessed through Wiley at www.wiley.com/college/begon. This provides an easy-to-use range of resources to aid study and enhance the content of the book. Features include self-assessment multiple choice questions for each chapter in the book, an interactive tutorial to help students to understand the use of mathematical modeling in ecology, and high-quality images of the figures in the book that teachers can use in preparing their lectures or lessons, as well as access to a Glossary of terms for use with this book and for ecology generally.

Acknowledgments

It is a pleasure to record our gratitude to the people who helped with the planning and writing of this book. Going back to the first edition, we thank Bob Campbell and Simon Rallison for getting the original enterprise off the ground and Nancy Whilton and Irene Herlihy for ably managing the project; and for the second edition, Nathan Brown (Blackwell, US) and Rosie Hayden (Blackwell, UK) for making it so easy for us to take this book from manuscript into print. For the third edition, we especially thank Nancy Whilton and Elizabeth Frank in Boston for persuading us to pick up our pens again (not literally) and Rosie Hayden, again, and Jane Andrew and Ward Cooper for seeing us through production. For this fourth edition, we thank Rachel Falk (Wiley, USA) for getting the ball rolling and for bringing in one of us (RWH) as a new author, Elisa Adams for her superb assistance with text editing, Chloe Moffett, Elizabeth Baird, MaryAnn Price and Lisa Torri (Precision Graphics) for their excellent overseeing of the final production, and the entire Wiley team for their dedicated efforts and cheerful "can-do" attitude.

We note with sadness the passing in 2009 of our long-time mentor and collaborator John Harper, author on the first three editions of this book. We owe him a special debt of gratitude that extends far beyond the past co-authorship of this book into all aspects of our lives as ecologists. He is sorely missed.

Colin Townsend, the lead author on the first three editions of *Essentials of Ecology*, has stepped from the treadmill of revisions and let us take the lead on this fourth edition. His imprint on the book remains strong, and we gratefully acknowledge his tremendous contribution to the series.

We are also grateful to the following colleagues who provided insightful reviews of early drafts of one or more chapters in this or earlier editions, or who gave us important advice and leads: William Ambrose (Bates College), Vickie Backus (Middlebury College), James Cahill (University of Alberta), Liane Cochrane-Stafira (Saint Xavier University), Mark Davis (Macalester College), Tim Crews (The Land Institute), Kevin Dixon (Arizona State University, West), Stephen Ellner (Cornell University), Alex Flecker (Cornell University), Bruce Grant (Widener University), Christy Goodale (Cornell University), Don Hall (Michigan State University), Jenny Hodgson, Greg Hurst (both University of Liverpool), William Kirk (Keele University, UK), Hans deKroon (University of Nijmegen), Zen Lewis (University of Liverpool), Sara Lindsay (Scripps Institute of Oceanography), James Maki (Marquette University), George Middendorf (Howard University), Paul Mitchell (Staffordshire University, UK), Tim Mousseau (University of South Carolina), Katie O'Reilly (University of Portland), Clayton Penniman (Central Connecticut State University), Tom Price (Univeristy of Liverpool), Jed Sparks (Cornell University), Catherine Toft (UC Davis), David Tonkyn (Clemson University), Saran Twombly (University of Rhode Island), Jake Weltzin (University of Tennessee at Knoxville), and Alan Wilmot (University of Derby, UK).

Last, and perhaps most, we are glad to thank our wives and families for continuing to support us, listen to us, and ignore us, precisely as required—thanks to Linda, and to Roxanne and Marina.

Michael Begon, Liverpool, UK and
Robert Howarth, Ithaca, NY USA

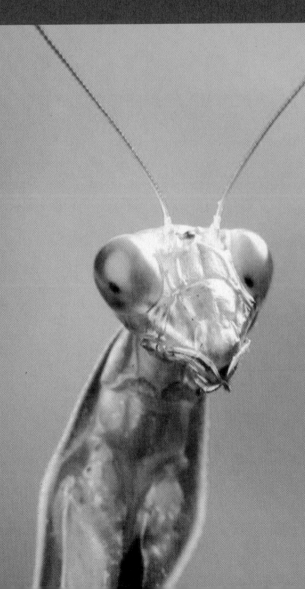

Part 1

Introduction

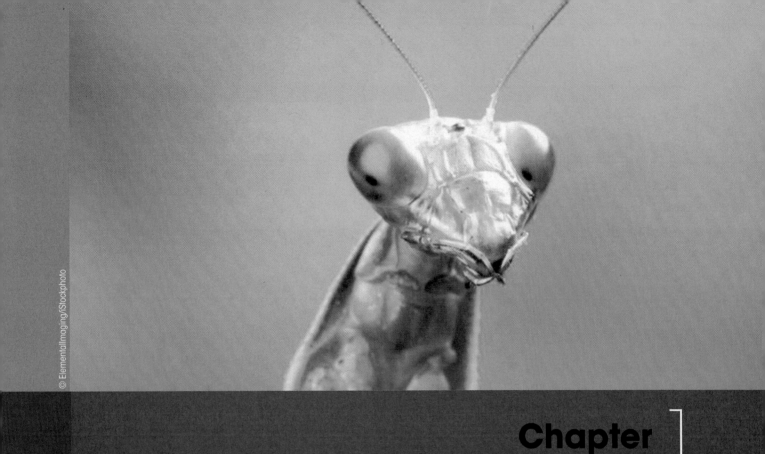

Chapter 1

Ecology and how to do it

CHAPTER CONTENTS

KEY CONCEPTS

After reading this chapter you will be able to:

- explain how ecologists seek to describe and understand, and on the basis of their understanding, to predict, manage, mitigate, and control

- describe the variety of spatial and temporal scales on which ecological phenomena occur

- describe how ecologists use observations, field and laboratory experiments, and mathematical models to collect scientific evidence

E cology today is a subject about which almost everyone has heard and most people consider to be important—even when they are unsure about the exact meaning of the term. There can be no doubt that it is important, but this makes it all the more critical that we understand what ecology is and how to do it.

1.1 WHAT IS ECOLOGY?

We could answer the question 'What is ecology?' by examining various definitions that have been proposed and choosing one as the best (Box 1.1). But while definitions have conciseness and precision, and they

the earliest ecologists

are good at preparing you for an examination, they are not so good at capturing the flavor and excitement of ecology. There is a lot to be gained by replacing that single question about a definition with a series of more provocative ones: 'What do ecologists *do*?' 'What are ecologists *interested* in?' and 'Where did ecology emerge from in the first place?'

1.1 Historical Landmarks

Definitions of ecology

Ecology (originally in German, *Öekologie*) was first defined in 1866 by Ernst Haeckel, an enthusiastic and influential disciple of Charles Darwin. To him, ecology was 'the comprehensive science of the relationship of the organism to the environment.' The spirit of this definition is very clear in an early discussion of biological subdisciplines by Burdon-Sanderson (1893), in which ecology is 'the science which concerns itself with the external relations of plants and animals to each other and to the past and present conditions of their existence,' to be contrasted with physiology (internal relations) and morphology (structure).

In the years after Haeckel, plant ecology and animal ecology drifted apart. Influential works defined ecology as 'those relations of *plants*, with their surroundings and with one another, which depend directly upon differences of habitat among plants' (Tansley, 1904), or as the science 'chiefly concerned with what may be called the sociology and economics of *animals*, rather than with the structural and other adaptations possessed by them' (Elton, 1927). The plant ecologists and animal ecologist, though, have long since agreed that they belong together, and more recent definitions of ecology include all organisms, including bacteria, archaea, algae, and fungi in addition to plants and animals. Most modern definitions stress the relationships between and among organisms. For example, two textbooks from the 1970s defined ecology as 'the study of the natural environment, particularly the interrelationships between organisms and their surroundings' (Ricklefs, 1973) and as 'the scientific study of the *interactions* that determine the distribution and abundance of organisms' (Krebs, 1972).

Ecology certainly includes the investigation of organisms and their interactions, but to many ecologists, definitions that focus only on these interactions and on the distribution and abundance of organisms are too narrow. Ecologists also examine the interaction between life and the physical environment, for instance studying how organisms affect material fluxes in nature. The sequestration of carbon dioxide by a forest would be one example of this. Beginning in the mid-20th century, the American ecologist E. P. Odum (1953) pushed for a broader definition of ecology: 'the study of the structure and function of nature, which includes the living world.' Many have thought this definition overly broad, as geologists and meteorologists also study aspects of the structure and function of nature. In 1992, G. E. Likens stressed the need for the definition of ecology to include 'the interactions between organisms and the transformation and flux of energy and matter.' We agree, and in this text define ecology as:

the scientific study of the distribution and abundance of organisms, the interactions that determine that distribution and abundance, and the relationships between organisms and the transformation and flux of energy and matter.

Ecology can lay claim to being the oldest science, as the most primitive humans must have been ecologists of sorts, driven by the need to understand where and when their food and their (nonhuman) enemies were to be found. The earliest agriculturalists needed to be even more sophisticated, with knowledge of how to manage their domesticated sources of food. These early ecologists, then, were *applied* ecologists, seeking to understand the distribution, abundance, and productivity of organisms in order to apply that knowledge for their own benefit. Applied ecologists today still have many of the same interests: how to optimize the rate at which food is collected from natural environments in a sustainable way; how domesticated plants and animals can best be managed so as to maximize rates of return; how food organisms can be protected from their own natural enemies; and how to control the populations of pathogens and parasites that live on us.

In the last century or so, however, since ecologists have been self-conscious enough to give them-

a pure and applied science

selves a name, ecology has consistently covered not only applied but also fundamental, 'pure' science. A. G. Tansley was one of the founding fathers of ecology. He was concerned especially to understand, for understanding's sake, the processes responsible for determining the structure and composition of different plant communities. When, in 1904, he wrote from Britain about 'The problems of ecology' he was particularly worried by a tendency for too much ecology to remain at the descriptive and unsystematic stage (such as accumulating descriptions of communities without knowing whether they were typical, temporary, or whatever), too rarely moving on to experimental or systematically planned, or what we might call a *scientific* analysis.

Tansley's worries were echoed in the United States by another of ecology's founders, F. E. Clements, who in 1905 in his *Research Methods in Ecology* complained:

The bane of the recent development popularly known as ecology has been a widespread feeling that anyone can do ecological work, regardless of

preparation. There is nothing . . . more erroneous than this feeling.

On the other hand, the need for *applied* ecology to be based on its *pure* counterpart was clear in the introduction to Charles Elton's (1927) *Animal Ecology* (Figure 1.1):

Ecology is destined for a great future . . . The tropical entomologist or mycologist or weed-controller will only be fulfilling his functions properly if he is first and foremost an ecologist.

In the intervening years, the coexistence of these pure and applied threads has been maintained and built upon. Many applied sciences such as forestry, agronomy, and fisheries biology have contributed to the development of ecology and have seen their own

courtesy Robert Elton

FIGURE 1.1 One of the great founders of ecology: Charles Elton (1900–1991). *Animal Ecology* (1927) was his first book but *The Ecology of Invasions by Animals and Plants* (1958) was equally influential. (After Breznak, 1975.)

development enhanced by ecological ideas and approaches. All aspects of food and fiber gathering, production, and protection have been involved. The biological control of pests (the use of pests' natural enemies to control them) has a history going back at least to the ancient Chinese but has seen a resurgence of ecological interest since the shortcomings of chemical pesticides began to be widely apparent in the 1950s. The ecology of pollution has been a growing concern from around the same time and expanded further in the 1980s and 1990s from local to regional and global issues. The last few decades have also seen expansions in both public interest and ecological input into the conservation of endangered species and the biodiversity of whole areas, the control of disease in humans as well as many other species, and the potential consequences of profound human-caused changes to the global environment.

And yet, at the same time, many fundamental problems of ecology remain unanswered. To what extent does competition for food determine which species can coexist in a habitat? What role does disease play in the dynamics of populations? Why are there more species in the tropics than at the poles? What is the relationship between soil productivity and plant community structure? Why are some species more vulnerable to extinction than others? Are wetlands net sources or sinks of greenhouse gas emission to the atmosphere? And so on. Of course, unanswered questions—if they are *focused* questions—are a symptom of the health, not the weakness, of any science. But ecology is not an easy science, and it has particular subtlety and complexity, in part because ecology is peculiarly confronted by 'uniqueness': millions of different species, countless billions of genetically distinct individuals, all living and interacting in a varied and ever-changing world. The beauty of ecology is that it challenges us to develop an understanding of very basic and apparent problems—in a way that recognizes the uniqueness and complexity of all aspects of nature – but seeks patterns and predictions within this complexity rather than being swamped by it.

| unanswered questions |

Let's come back to the question of what ecologists do. First and foremost ecology is a science, and ecologists therefore try to *explain* and *understand*. Explanation can be either 'proximate' or 'ultimate,' and ecologists are interested in both. For example, the present distribution and abundance of a particular species of bird may be 'explained' in terms of the physical environment that the bird tolerates, the food that it eats, and the

| understanding, description, prediction, and control |

parasites and predators that attack it. This is a *proximate* explanation – an explanation in terms of what is going on 'here and now.' We can also ask how this bird came to have these properties that now govern its life. This question has to be answered by an explanation in evolutionary terms; the *ultimate* explanation of the present distribution and abundance of this bird lies in the ecological experiences of its ancestors (see Chapter 2).

In order to understand something, of course, we must first have a description of whatever it is we wish to understand. Ecologists must therefore *describe* before they explain. On the other hand, the most valuable descriptions are those carried out with a particular problem or 'need for understanding' in mind. Undirected description, carried out merely for its own sake, is often later found to have selected the wrong things and has little place in ecology—or any other science.

Ecologists also often try to *predict*. For example, how will global warming affect the sequestration (storage) of carbon in natural ecosystems? Will warming reduce this storage, and therefore result in even more global warming since less carbon dioxide will be removed from the atmosphere? Often, ecologists are interested in what will happen to a population of organisms under a particular set of circumstances, and on the basis of these predictions to control, exploit or conserve the population. We try to minimize the effects of locust plagues by predicting when they are likely to occur and taking appropriate action. We try to exploit crops most effectively by predicting when conditions will be favorable to the crop and unfavorable to its enemies. We try to preserve rare species by predicting the conservation policy that will enable us to do so. Some prediction and control can be carried out without deep explanation or understanding: it is not difficult to predict that the destruction of a woodland will eliminate woodland birds. But what if the woodland is not destroyed, but rather fragmented into distinct parts with suburbs or agricultural fields between them? What effect may this have on the woodland birds? Insightful predictions, precise predictions, and predictions of what will happen in unusual circumstances can be made only when we can also explain and understand what is going on.

This book is therefore about:

1 How ecological understanding is achieved.

2 What we do understand, and what we do not.

3 How ecological understanding can help us predict, manage, mitigate, and control.

1.2   SCALES, DIVERSITY OF APPROACHES, AND RIGOR

Ecology is a diverse discipline, and ecologists use a vast array or tools and approaches. Later in this chapter, we briefly give some examples of this diversity, but first we elaborate on three general points:

- ecological phenomena occur at a variety of scales;

- ecological evidence comes from a variety of different sources;

- ecology relies on truly scientific evidence.

Questions of scale

Ecology operates at a range of scales: time scales, spatial scales, and 'biological' scales. It is important to appreciate the breadth of these and how they relate to one another.

Life is studied at a variety of hierarchical levels, with much of biology focused on levels from molecules, | the 'biological' scale

to organelles, cells, tissues, organs, and whole organisms. Ecologists study levels from individual organisms, to populations, communities, ecosystems, and the global biosphere (Figure 1.2).

- Populations are functioning groups of individual organisms of the same species in a defined location.

- Communities consist of all the species populations present in a defined location.

- Ecosystems include both the community of organisms and the physical environment in which they exist.

- The biosphere is the totality of all of life interacting with the physical environment at the scale of the entire planet.

At the level of the organism, ecology deals primarily with how individuals are affected by their environment and with their physiological and behavioral responses to the environment. **Population ecology** stresses the trends and fluctuations in the number of individual of a particular species at a particular time and place, as determined by the interactions of birth and death rates and the interactions between the populations themselves (such as predators and prey). **Community ecology** focuses on questions such as what controls the diversity of species of in a given area. **Ecosystem ecology** strives to understand the functioning of entire lakes, forests, wetlands, or other portions of the Earth in terms of energy and material inputs and outputs. Across all scales of biological hierarchy—including these ecological ones—three generalities emerge.

1   The properties observed at a particular level arise out of the functioning of parts at the level below. For example, how a tissue functions is the result of the functioning of the cells in that tissue, and how an ecosystem functions is the result of the functioning of the communities within it interacting with the physical environment.

2   In order to understand the mechanistic reasons that a particular property is observed at any level of biological organization, a scientist needs to look at the next lowest level of organization. To understand dysfunction in an individual organism, we must look at the functioning of the organs in that organism; and to understand the controls on birth rate in a population, we must look at reproduction in individual organisms.

3   However, properties observed at a given level of organization may be predicted without fully understanding the functioning at lower levels. This third generality may seem to contradict the other two, but it does not. Consider an analogy from the physical sciences. As early as 1662, Boyle knew that when the pressure of a gas is doubled, its volume is halved, if temperature remains constant. This behavior of the gas as a whole is the result of the interactions of the gas molecules, yet Boyle's law provided valuable predictive power for centuries, long before the concept of the molecule was developed. Today, physical chemists can indeed explain gas behavior based on understanding of the behavior of individual molecules, but the explanation is complex, and not even taught to most undergraduate college students. Similarly, ecologists can predict patterns in ecosystems without understanding all of the details of the dynamics of constituent populations, and can predict patterns in populations without understanding all of the details of the responses of individual organisms.

Within the living world, there is no arena too small nor one so large that it does not have an ecol- | a range of spatial scales

ogy. Even the popular press talk increasingly about the 'global ecosystem', and there is no question that several ecological problems can be examined only at this very large scale. These include the relationships between ocean currents and fisheries, or between climate patterns and the distribution of deserts and tropical rain forests, or between elevated carbon dioxide in the atmosphere (from burning fossil fuels) and global climate change.

At the opposite extreme, an individual cell may be the stage on which two populations of pathogens

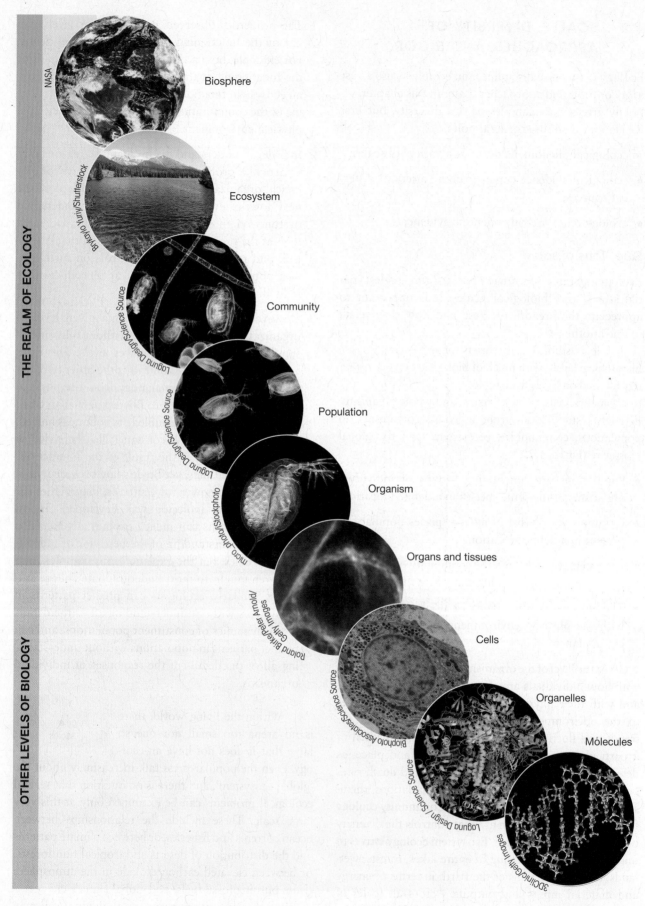

Biosphere

Ecosystem

Community

Population

Organism

Organs and tissues

Cells

Organelles

Molecules

FIGURE 1.2 Ecology is studied at many hierarchical levels.

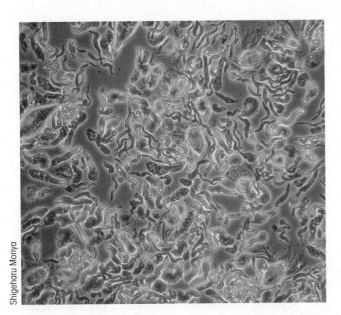

Shigeharu Moriya

FIGURE 1.3 The diverse community of a termite's gut. Termites can break down lignin and cellulose from wood because of their mutualistic relationships (see Chapter 8) with a diversity of microbes that live in their guts.

compete with one another for the resources that the cell provides. At a slightly larger spatial scale, a termite's gut is the habitat for bacteria, protozoans, and other species (Figure 1.3) – a community whose diversity is comparable to that of a tropical rain forest in terms of the richness of organisms living there, the variety of interactions in which they take part, and indeed the extent to which we remain ignorant about the species identity of many of the participants. Between these extremes, different ecologists, or the same ecologist at different times, may study the inhabitants of pools that form in small tree-holes, the temporary watering holes of the savannas, or the great lakes and oceans; others may examine the diversity of fleas on different species of birds, the diversity of birds in different sized patches of woodland, or the diversity of woodlands at different altitudes.

To some extent related to this range of spatial scales, and to the levels in the biological hierarchy, | a range of time scales

ecologists also work on a variety of time scales. **Ecological succession** – the successive and continuous colonization of a site by certain species populations, accompanied by the local extinction of others – may be studied over a period from the deposition of a lump of sheep dung to its decomposition (a matter of weeks), from the abandonment of a patch of tropical rain forest cleared for slash-and-burn agriculture (years to decades), or from the development of a new forest on land wiped clean to bedrock by the retreat of a glacier in the arctic or high mountains (centuries). Migration

may be studied in butterflies over the course of days, or in the forest trees that are still (slowly) migrating into deglaciated areas following the last ice age.

The appropriate time scale for ecological investigation varies with the question to be answered. | the need for long-term studies

However, many ecological studies end up being shorter than appropriate for the question, due to human frailties. Longer studies cost more and require greater dedication and stamina. The often short-term nature of funding, an impatient scientific community, and the requirement for concrete evidence of activity for career progression all put pressure on ecologists (and all scientists) to publish their work sooner rather than later. Why are long-term studies potentially of such value? The reduction over a few years in the numbers of a particular species of wild flower, or bird, or butterfly might be a cause for conservation concern—but one or more decades of study may be needed to be sure that the decline is more than just an expression of the random ups and downs of 'normal' population dynamics. One of the longest, continuously run ecological studies is at the Hubbard Brook Experiment Forest in the White Mountains of New Hampshire. Among other measures, Gene Liken and other scientists there have monitored the acidity of rain since the early 1960s. In the 1960s, the rain was quite acidic (low pH: high hydrogen ion concentrations), and this was in fact one of the earliest discoveries anywhere of the phenomenon of acid rain. The long-term trend, though, has been for precipitation to become less acidic over subsequent decades (Figure 1.4); but we can observe this only

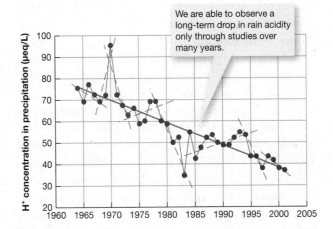

FIGURE 1.4 Hydrogen ion concentration in precipitation at the Hubbard Brook Experimental Forest over time. Note the long-term trend of decreasing concentration, indicating that the pH has been rising, and the rain has become less acidic over time. However, analysis of periods of only a few years in duration can show sharp increases or decreases in the hydrogen ion concentration, and are quite misleading with regard to the long-term trend (After Likens 2004).

because we have that long-term record. Observations over periods of even 5 to 10 years at a time would be very misleading.

This does not mean that all ecological studies need to last for 20 years – nor that every time an ecological study is extended the answer changes. But it does emphasize the great value to ecology of the small number of long-term investigations that have been carried out or are ongoing.

The diversity of ecological evidence

Ecological evidence comes from a variety of different sources and approaches. The principal tools are:

- *Observations*, often of changes in abundance or system functioning over either time or space, and often involving comparisons across and between different areas or systems.

- *Experiments*, including both those in the lab and in the field.

- *Mathematical models* that capture some component of ecological interactions, function, and structure.

Ecologists often combine two or more of these approaches. For instance, they may use models or inferences from comparative observations across systems to inform experiments, or they may use experiments and observations to calibrate models.

Many ecological studies include careful observation and monitoring in the natural environment, for

> observations and field experiments

instance, of the changing abundance of one or more species over time, or over space, or both. In this way, ecologists may establish patterns, for example, that red grouse (birds shot for 'sport') exhibit regular cycles in abundance peaking every 4 or 5 years. Documenting the pattern does not provide explanation for the cause of the cycle, but it is a start toward understanding. A next step might be to development one or more hypotheses to explain the pattern: for instance, perhaps the 4- to 5-year cycle is caused by a gradual accumulation of parasitic worms in the grouse populations over this period of time. A **manipulative field experiment** is one approach to test such a hypothesis, in this case by ridding the grouse of the parasites and monitoring whether the 4- to 5-year population cycle persists. Treating the grouse for their parasites strongly dampens the population cycle, giving strong support to this hypothesis (Hudson, Dobson, & Newborn, 1998).

We can also use **comparative field observations** to test hypotheses. That is, we explicitly compare the same sort of data from many different sites. Consider the question whether the amount of nitrogen

> observations can test hypotheses

pollution deposited onto the landscape in rain, snow, dust, and gases affects the biodiversity of grassland communities (nitrogen is a major component of acid rain). The extent of this nitrogen pollution varies greatly over the landscape of Europe, so we can test the hypothesis that more nitrogen pollution lowers biodiversity by comparing diversity in different grasslands receiving different inputs of nitrogen pollution from the atmosphere (Figure 1.5). The diversity of forbs (broad-leaved herbs) is indeed lower when nitrogen pollution is greater, but the diversity of grasses increases with increasing nitrogen pollution. The scatter in the relationships is great, but the relationships are nonetheless significant. (We will discuss what we mean by "significant" later in this chapter.) The scatter is undoubtedly the result of other factors across the landscape – in addition to the nitrogen pollution – that might also affect diversity, such as types of soil, differences in precipitation and other climate variables, and other types of pollution and disturbance.

Rather than relying on this comparative observational approach, we could conduct a manipulative field experiment to test the hypothesis that nitrogen pollution affects biodiversity. The result of one such experiment shows a very tight and pronounced effect of increasing nitrogen supply on biodiversity (expressed as species richness, the number of species) after just 4 years of nitrogen addition (Figures 1.6). This experiment has an advantage over the comparative observational study, in that other confounding variables – soil type, climate, disturbance history – are held constant between treatments and hence eliminated, and this probably explains the tighter relationship between nitrogen and diversity. On the other hand, the experiment is conducted on just one type of soil, with one type of climate and disturbance history, and over a fairly limited period of time (4 years in this case), and so the result may not fully apply to other areas and to longer time scales. Both manipulative experiments and observations are critical to ecology, and ecologists gain confidence in their understanding of nature when the two approaches lead to similar conclusions.

Why might nitrogen affect plant biodiversity? One hypothesis is that the effect is one of fertilization, with plants growing more as the nitrogen supply increases, and this leading to less diversity as species that grow particularly well come to dominate the community and shade out other plants. A creative experiment gave some

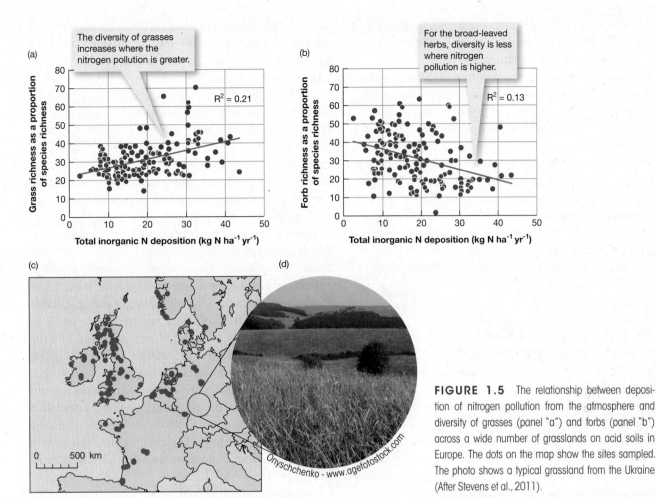

FIGURE 1.5 The relationship between deposition of nitrogen pollution from the atmosphere and diversity of grasses (panel "a") and forbs (panel "b") across a wide number of grasslands on acid soils in Europe. The dots on the map show the sites sampled. The photo shows a typical grassland from the Ukraine (After Stevens et al., 2011).

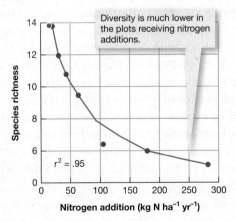

FIGURE 1.6 Experimental addition of nitrogen to a grassland in Minnesota reduces species diversity. Reprinted from Huston (1997), based on data in Tilman (1996).

Perhaps less obviously, ecologists also often turn to laboratory systems or to mathematical models designed to capture ecological processes. These have played a crucial role in the development of ecology, and they are certain to continue to do so. Field experiments are almost inevitably costly and difficult to carry out. Moreover, even if time and expense were not issues, natural field systems may simply be too complex to allow us to tease apart the consequences of the many different processes that may be going on. Are the intestinal worms actually capable of having an effect on reproduction or mortality of individual grouse? How do light and nitrogen interact to regulate the growth rate of the various species in a grassland ecosystem? Controlled laboratory experiments are often the best way to provide answers to specific questions that are key parts of an overall explanation of the complex situation in the field.

Of course, the complexity of natural ecological communities may simply make it inappropriate for an

laboratory experiments and mathematical models

preliminary support for this hypothesis: when artificial light was supplied to the shaded plants of the understory in nitrogen fertilized plots, biodiversity was maintained despite the higher nitrogen (Hautier et al., 2009).

ecologist to dive straight into them in search of understanding. We may wish to explain the structure and dynamics of a particular community of 20 animal and plant species comprising various competitors, predators, parasites, and so on (relatively speaking, a community of remarkable simplicity). But we have little hope of doing so unless we already have some basic understanding of even simpler communities of just one predator and one prey species, or two competitors, or (especially ambitious) two competitors that also share a common predator. For this, it is usually most appropriate to construct, for our own convenience, simple laboratory systems that can act as benchmarks or jumping-off points in our search for understanding.

simple laboratory systems . . .

What is more, you have only to ask anyone who has tried to rear caterpillar eggs, or take a cohort of shrub cuttings through to maturity, to discover that even the simplest ecological communities may not be easy to maintain or keep free of unwanted pathogens, predators, or competitors. Nor is it necessarily possible to construct precisely the particular, simple, artificial community that interests you; nor to subject it to precisely the conditions or the perturbation of interest. In many cases, therefore, there is much to be gained from the analysis of mathematical models of ecological communities: constructed and manipulated according to the ecologist's design.

. . . and mathematical models

On the other hand, although a major aim of science is to simplify, and thereby make it easier to understand the complexity of the real world, ultimately it is the real world that we are interested in. The worth of models and simple laboratory experiments must always be judged in terms of the light they throw on the working of more natural systems. They are a means to an end—never an end in themselves. Like all scientists, ecologists need to 'seek simplicity, but distrust it' (Whitehead, 1953).

Statistics and scientific rigor

For a scientist to take offense at some popular phrase or saying is to invite accusations of a lack of a sense of humor. But it is difficult to remain calm in the face of phrases such as 'There are lies, damn lies and statistics' or 'You can prove anything with statistics.' Statistics *are* regularly misused, but more often in the public media and by those who may seek to manipulate popular opinion, and rarely if ever in the scientific literature. You should not mistrust statistics. Rather, you should understand their strengths and limitations. An essential point: you cannot *prove* anything with statistics. Rather, statistical analysis allows us to attach a

level of confidence to our conclusions. Ecology, like all science, is a search not for statements that have been 'proved to be true' but for conclusions in which we can be confident.

What distinguishes science – what makes science rigorous – is that it is based not simply on assertions, but rather on conclusions resulting from investigations that test specific hypotheses, and to which we can attach a level of confidence, measured on an agreed-upon scale.

ecology: a search for conclusions in which we can be confident

Statistical analyses are carried out after data have been collected, and they help us to interpret those data. Really good science, though, requires forethought. Ecologists, like all scientists, must know what they are doing, and why they are doing it, *while* they are doing it. Ecologists must plan, so as to be confident that they will collect the right kind of data, and a sufficient amount of data, to address the question they hope to answer. As discussed in Box 1.2, more data are required to obtain statistically significant results when the relationship being tested is a weak one, or when the relationship is confounded by other factors, as is likely the case for the relationship between nitrogen deposition and diversity illustrated in Figure 1.5.

ecologists must think ahead

Many ecological field experiments rely on a large number of replicates for each treatment, and this increases the likelihood of obtaining statistically significant results. For example, ecologists experimentally testing the effect of nitrogen deposition on plant diversity in grasslands might have 8 different levels of nitrogen inputs, with 10 different plots for each treatment (a total of 80 plots). However, replication can be expensive and time consuming, particularly if the ecologists include in the responses they monitor processes that are difficult to measure. Determining the biomass of the plants at the end of the experiment is relatively simple (cutting, drying, and weighing); characterizing the diversity of the community is more difficult, particularly if the diversity is high with many species potentially present; measuring the rate at which each species is assimilating nitrogen is far, far more difficult and time consuming. Most experimentalists feel a constant tug between having a large number of replicates and keeping their experiments doable.

replication in experiments

As noted by David Schindler (1998), experiments can often involve a trade-off between realism and replication. Smaller scale experiments – such as small plots of grassland, or

replication in whole-ecosystem experiments

1.2   Quantitative Aspects

Interpreting probabilities

Ecologists need to know, as do any scientists dealing with sets of data, what conclusions can be drawn from those data. Imagine we are interested in determining whether high abundances of a pest insect in summer are associated with high temperatures the previous spring, and imagine we have data on summer insect abundances and mean spring temperatures for each of a number of years. How do we use statistical analysis to conclude, with a stated degree of confidence, either than there is or is not a relationship between the spring temperature and summer insect numbers?

Null hypotheses and *P*-values

To carry out a statistical test we first need a *null hypothesis*, which simply means in this case that there is *no* association; that is, no association between insect abundance and temperature. The statistical test (stated simply) then generates a probability (a *P*-value) of getting a data set like ours if the null hypothesis is correct.

Suppose the data were like those in Figure 1.7a. The probability generated by a statistical test of association on these data is $P = 0.5$ (equivalently 50%). This means that, if the null hypothesis really was correct (no association), then 50% of studies like ours should generate just such a data set, or one even further from the null hypothesis. We therefore could have no confidence in any claim that there *was* an association.

Suppose, however, that the data were like those in Figure 1.7b, where the *P*-value is 0.001 (0.1%). This would mean that such a data set (or one even further from the null hypothesis) could be expected in only 0.1% of similar studies if there was really no association. In other words, either something very improbable has occurred, or there *was* an association between insect abundance and spring temperature. Thus, since we do not expect highly improbable events to occur, we can have a high degree of confidence in the claim that there *was* an association between abundance and temperature.

Significance testing

Both 50% and 0.01%, though, make things easy for us. Where, between the two, do we draw the line? There is no absolute answer to this, but scientists and statisticians have established a convention in *significance testing*, which says that if *P* is less than 0.05 (5%), written $P < 0.05$ (e.g., Figure 1.7d), then results are described as 'statistically significant' and confidence can be placed in the effect being examined; whereas if $P > 0.05$, then there is no statistical foundation for claiming the effect exists (e.g., Figure 1.7c). A further elaboration of the convention often describes results with $P < 0.01$ as 'highly significant.'

'Insignificant' results?

Some effects are naturally strong (there is a powerful association between people's weight and their height) and others are weak (the association between people's weight and their risk of heart disease is real but weak, since weight is only one of many important factors). More data are needed to establish support for a weak effect than for a strong one. Hence a *P*-value of greater than 0.05 (lack of statistical significance) may mean one of two things in an ecological study:

1   There really is no effect of ecological importance.

2   The data are simply not good enough, or there are not enough of them, to support the effect even though it exists, possibly because the effect itself is real but weak.

Throughout this book, then, studies of a wide range of types are described, and their results often have *P*-values attached to them. Remember that statements like $P < 0.05$ and $P < 0.01$ mean that these are studies where: (i) sufficient data have been collected to establish a conclusion in which we can be confident; (ii) that confidence has been established by agreed means (statistical testing); and (iii) confidence is being measured on an agreed and interpretable scale.

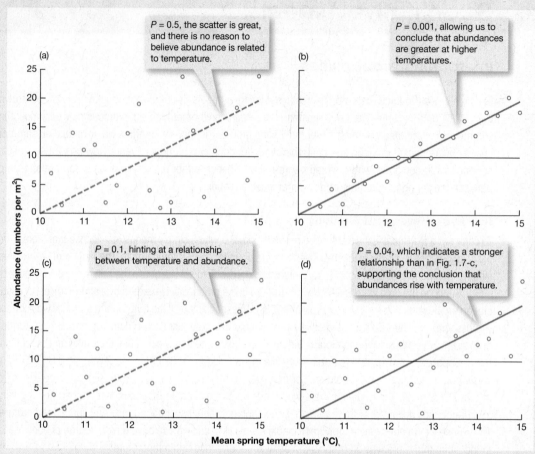

FIGURE 1.7 The results from four hypothetical studies of the relationship between insect pest abundance in summer and mean temperature the previous spring. In each case, the points are the data actually collected. Horizontal lines represent the *null hypothesis* – that there is no association between abundance and temperature, and thus the best estimate of expected insect abundance, irrespective of spring temperature, is the mean insect abundance overall. The second line is the *line of best fit* to the data, which in each case offers some suggestion that abundance rises as temperature rises. However, whether we can be confident in concluding that abundance does rise with temperature depends, as explained in the text, on statistical tests applied to the data sets. (a) The suggestion of a relationship is weak ($P = 0.5$). There are no good grounds for concluding that the true relationship differs from that supposed by the null hypothesis and no grounds for concluding that abundance is related to temperature. (b) The relationship is strong ($P = 0.001$) and we can be confident in concluding that abundance increases with temperature. (c) The results are suggestive ($P = 0.1$) but it would not be safe to conclude from them that abundance rises with temperature. (d) The results are not vastly different from those in (c) but are powerful enough ($P = 0.04$, i.e., $P < 0.05$) for the conclusion that abundance rises with temperature to be considered safe.

Standard errors and confidence intervals

Another way in which our confidence in results is assessed is through reference to 'standard errors,' which statistical tests often allow to be attached either to mean values calculated from a set of observations or to slopes of lines like those in Figure 1.7. These mean values and slopes can only ever be estimates of the 'true' mean value or slope, because they are calculated from data that are only a sample of all the imaginable items of data that could be collected. The standard error, then, sets a band around the estimated value within which the true value can be expected to lie, with a given, stated probability. In particular, there is a 95% probability that the true mean lies within roughly two standard errors (2 SE) of the estimated mean; we call this the *95% confidence interval*.

Large standard errors (little confidence in the estimated value) can arise when data are, for whatever reason, highly variable; but they may also be due to only a small data set having been collected. Standard errors are smaller, and confidence in estimates greater, *both* when data are more consistent (less variable) *and* when there are more data.

50-gallon containers of pond water – make replication relatively easy, but this scale cannot capture all of the complex interactions inherent in natural ecosystems. Schindler has devoted his career to whole-ecosystem experiments, for instance, adding nutrients, acids, or toxins to whole lakes and following the response of the lake ecosystem. At this scale, replication is difficult at best, and many whole-ecosystem experiments have not used replicate treatments. An excellent example is the work of Schindler and colleagues (1985), who added sulfuric acid to one lake (no replication) over a period of many years, to determine what the consequences of acid rain on lakes might be. To understand the effects of the acidification, they observed the lake for a period of years before acidification, and then observed the behavior of the acidified lake in comparison to other nearby reference lakes as acidification was gradually increased. They found remarkable effects at levels of acidity far less than lab studies of the time had indicated were likely to be a problem. For instance, the populations of crayfish and small minnows in the lake declined, and this led to deleterious consequences for the top predator, the lake trout (Figure 1.8), and to their eventual disappearance from the lake. Note also that while the experimental treatment (acidification) was not replicated, the scientists took a large number of replicate samples for the response variables of interest (for instance, numbers of crayfish) in both the acidified and reference lakes over time, making statistically significant conclusions possible. While at first glance it may seem distasteful to purposely pollute a lake, this whole-lake experiment clearly demonstrated more severe effects of acid rain on lakes than had previously been believed possible, and this led to pressure to reduce acid rain, protecting a far greater number of lakes than the one damaged in the experiment.

Ecologists often seek to draw conclusions about overall groups of organisms or overall trends in processes or fluxes: what is the | ecology relies on representative samples

trend in abundance over time of the crayfish in the lake experiment, described above? What is the birth rate of the bears in Yellowstone Park? What is the density of weeds in a wheat field? What is the rate of loss in streams of nitrogen from forests? In doing so, we can only very rarely examine every individual in a group, or in the entire sampling area, or all the nitrogen in all the water leaving a forest. We must therefore rely on what we hope will be a representative sample from the group or habitat or flux. Indeed, even if we examined a whole group (we might examine every fish in a small pond, say), we are likely to want to draw general conclusions from it: we might hope that the fish in 'our'

Lake trout in a reference lake receiving no acid are healthy.

Fisheries and Oceans Canada

Lake trout in the acidified lake are clearly underweight in the years before they are finally gone.

Fisheries and Oceans Canada

FIGURE 1.8 After several years of acidification, lake trout in the experimentally acidified lake (bottom) are obviously less healthy than those from a nearby reference lake (top). This resulted not from a direct toxicity effect of the acid on the fish, but rather on less prey organisms available for the fish. Reprinted from Schindler et al. (1985).

pond can tell us something about fish of that species in ponds of that type, generally. In short, ecology relies on obtaining *estimates* from representative samples. This is elaborated in Box. 1.3.

1.3 ECOLOGY IN PRACTICE

To discover the real problems faced by ecologists and how they try to solve them, we consider some real research programs in a little detail. Every chapter in this book will contain descriptions of similar studies, but in the context of a systematic survey of the driving forces in ecology (Chapters 2–11) or of the application of this knowledge to solve applied problems (Chapters 12–14). For now, we want to highlight some of the excitement and diversity of ecological studies, giving you the flavor of our field. We have chosen the following examples either because they are classic studies that pioneered new approaches, or because they exemplify the use of multiple approaches in addressing fundamental ecological questions.

Successions on old fields in Minnesota: a study in time and space

Ecological succession is a concept familiar to you if you have simply taken a walk in open country – the

Quantitative Aspects

Estimation: sampling, accuracy, and precision

In any sampling program, the aim is to;

1 obtain an estimate that is accurate and unbiased; that is, neither systematically too high nor too low as a result of some flaw in the program.

2 obtain data that have as little variation as possible.

3 use the time, money, and human effort invested in the program as effectively as possible, because these are always limited.

    To understand the application of these goals, consider another hypothetical example: the density of a particular weed (e.g., wild oat) in a wheat field. To prevent bias, each part of the field should have an equal chance of being selected for sampling, so sampling units should be selected at random. We might, for example, divide the field into a measured grid, pick points on the grid at random, and count the wild oat plants within a 50 cm radius of the selected grid point. This unbiased method can be contrasted with a plan to sample only weeds from between the rows of wheat plants, giving too high an estimate, or within the rows, giving too low an estimate (Figure 1.9a).

    What, though, if we suspect that the slope of a field affects weed density, and half the field slopes to the southeast and half to the southwest? The individual values from samples may then fall into two groups a substantial distance apart on the density scale: high from the southwest slope; low (mostly zero) from the southeast slope. The estimated mean weed density would be close to the true mean (it would be accurate and unbiased), but the variation among samples would be large and the estimate therefore imprecise (Figure 1.9b). If, however, we acknowledge the difference between the two slopes and treat them separately from the outset, then we obtain means for each that have much smaller confidence intervals.

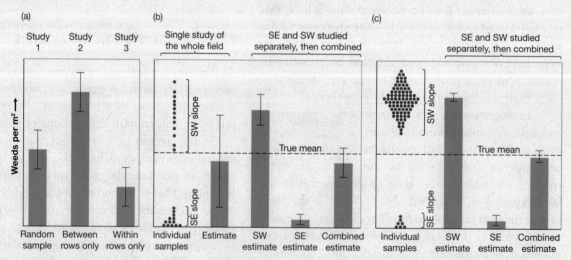

FIGURE 1.9 The results of hypothetical programs to estimate weed density in a wheat field. (a) The three studies have equal potential error or variation associated with their estimates (as shown by the "error bars"), but only the first (from a random sample) is accurate. (b) In the first study, individual samples from different parts of the field (southeast and southwest) fall into two groups (left); thus, the estimate, although accurate, is not precise (right). In the second study, separate estimates for southeast and southwest are both accurate and precise – as is the estimate for the whole field obtained by combining them. (c) Following on from (b), most sampling effort is directed to the southwest, reducing the confidence interval there, but with little effect on the confidence interval for the southeast. The overall interval is therefore reduced: precision has been improved.

What is more, if we average those means and combine their confidence intervals to obtain an estimate for the field as a whole, then that interval too is much smaller than previously.

But has our effort been directed sensibly, with equal numbers of samples from the southwest slope, where there are lots of weeds, and the southeast slope, where there are virtually none? The answer is no. If our efforts had been directed mostly at sampling the southwest slope, the increased amount of data would have noticeably decreased the variation, whereas less sampling of the southeast slope would have made very little difference (Figure 1.9c). Careful direction of a sampling program can clearly increase overall precision for a given investment in effort.

idea that a newly created habitat, or one in which a disturbance has created an opening, will be inhabited, in turn, by a variety of species appearing and disappearing in some recognizably repeatable sequence. Widespread familiarity with the idea, however, does not mean that we understand fully the processes that drive or fine-tune successions; yet developing such understanding is important not just because succession is one of the fundamental forces structuring ecological communities, but also because human disturbance of natural communities has become ever more frequent and profound. We need to know how communities may respond to, and hopefully recover from, such disturbance, and how we may aid that recovery.

One particular focus for the study of succession has been the old agricultural fields of the eastern and northern United States, abandoned as farmers moved west in search of 'fresh fields and pastures new.' One such site is now the Cedar Creek Natural History Area, roughly 50 km north of Minneapolis, Minnesota. The area was first settled by Europeans in 1856 and was initially subject to logging. Clearing for cultivation then began about 1885, and land was first cultivated between 1900 and 1910. Now there are agricultural fields that are still under cultivation and others that have been abandoned at various times since the mid-1920s. Cultivation led to depletion of nitrogen from soils that already were naturally poor in this important plant nutrient.

Studies at Cedar Creek illustrate the value of **natural experiments**. A natural experiment is not planned before hand, but rather ecologists take advantage of a situation where either natural events (the retreat of a glacier, to evaluate how plants recolonize soils) or human-controlled events set the stage to learn something about ecological processes. The Cedar Creek studies, begun only in the 1980s, take advantage of the abandonment of agricultural land beginning many decades earlier. To understand the

an unplanned experiment

successional sequence of plants that occur in fields in the years following abandonment, we *could* plan an artificial manipulation, under our control, in which a number of fields currently under cultivation were forcibly abandoned and the communities in them sampled repeatedly into the future. (We would need a number of fields because any single field might be atypical, whereas several would allow us to calculate mean values for, say, the number of new species per year, and place confidence intervals around those means.) But the results of this experiment would take decades to accumulate. The natural experiment alternative, therefore, was to use the fact that records already exist of when many of the old fields were abandoned. This is what David Tilman and his team did. Figure 1.10 illustrates data from a group of 22 old fields surveyed in 1983, having been abandoned at various times between 1927 and 1982 (i.e., between 1 and 56 years previously). These can be treated as 22 'snapshots' of the continuous process of succession in old fields at Cedar Creek in general, even though each field was itself only surveyed once.

A number of the shifting balances during succession are clear from the figure as statistically significant trends. Over the 56 years, the cover of invader species (mostly agricultural weeds) decreased (Figure 1.10a) while the cover of species from nearby prairies increased (Figure 1.10b): the natives reclaimed their land. Of more general applicability, the cover of annual species decreased over time, while the cover of perennial species increased (Figure 1.10c, d). **Annual species** – those that complete a whole generation from seed to adult through to seeds again within a year – tend to be good at increasing in abundance rapidly in relatively empty habitats (the early stages of succession); whereas **perennials** – those that live for several or many years and may not reproduce in their early years – are slower to establish but more persistent once they do. Over time, nitrogen accumulates in the soil (Figure 1.10e). Such data suggest that the change in nitrogen may drive some

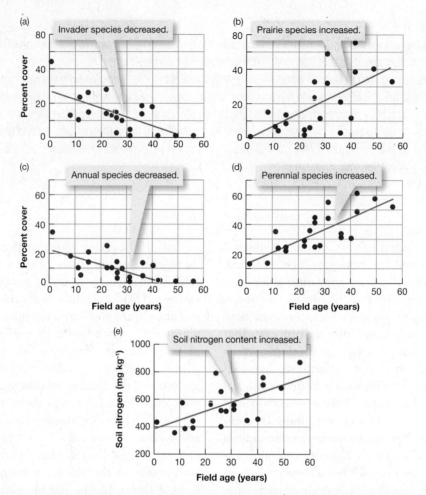

FIGURE 1.10 Twenty-two fields at different stages in an old-field succession were surveyed to generate the following trends with successional stage (field age): (a) invader species decreased, (b) native prairie species increased, (c) annual species decreased, (d) perennial species increased, and (e) soil nitrogen content increased. The best fit lines (see Box 1.2) are highly significant in every case ($P < 0.01$). (After Inouye et al., 1987.)

of the changes in plant community structure. Or conversely, perhaps the differences in nitrogen are the result of the differences in the plant community. Hypotheses based on these observations have encouraged many field manipulative experiments and sparked some lively debates over the past few decades, topics we explore in more detail later in this book.

Hubbard Brook: a long-term commitment to study at the ecosystem scale

The Cedar Creek study took advantage of space-for-time substitutions, inferring information on ecological processes that occur over decades (old field succession in Minnesota) from current spatial patterns. For longer time scales, this may be the only approach for understanding ecological processes and controls. At shorter time scales, though, actual observations are possible. As noted briefly above, a project started by Gene Likens together with many other colleagues, notably

Herb Bormann, has been doing precisely this in the Hubbard Brook Experimental Forest since the early 1960s. In part because of the noted success of having long-term continuous data demonstrated at Hubbard Brook, the U.S. National Science Foundation has now for several decades supported a network of Long-Term Ecological Research sites, in a variety of different types of ecosystems. Currently, 26 such sites are funded (http://www.lternet.edu/). The goal is to continue these efforts indefinitely into the future, providing a critical laboratory for studying ecological processes and how they change over time.

An early accomplishment of the Hubbard Brook team was to develop the small watershed technique to measure the input and output of chemicals from individual forest areas. Because most of the chemical losses from forests such as those at Hubbard Brook (with shallow soils, underlain by impermeable bedrock) are channeled through streams,

the catchment area as a unit of study

TABLE 1.1 Annual chemical budgets for forested catchment areas at Hubbard Brook (kg ha⁻¹ yr⁻¹). Inputs are for dissolved materials in precipitation or in dryfall (gases or associated with particles falling from the atmosphere). Outputs are losses in stream water as dissolved material plus particulate organic material in the streamflow. The source of the excess chemicals (where outputs exceeded inputs) was weathering of parent rock and soil. The exception was nitrogen (as ammonium or nitrate ions) – less was exported than arrived in precipitation because of nitrogen uptake in the forest (After Likens and Bormann 1994).

	NH_4^+	NO_3^-	SO_4^{2-}	K^+	Ca^{2+}	Mg^{2+}	Na^+
Input	2.7	16.3	38.3	1.1	2.6	0.7	1.5
Output	0.4	8.7	48.6	1.7	11.8	2.9	6.9
Net Change*	+ 2.3	+ 7.6	− 10.3	− 0.6	− 9.2	− 2.2	− 5.4

*Net change is positive when the catchment gains matter and negative when it loses it.

a comparison of the chemistry of stream water with that of incoming precipitation can reveal a lot about the differential uptake and cycling of chemical elements by the terrestrial biota. These small watersheds are often called **catchments**.

In their catchment studies, the Hubbard Brooks scientists measured (and continue to measure!) the inputs of acids, nutrients, and other materials from the atmosphere. They use a variety of tools, but the simplest is to measure total precipitation (rain and snow) and the concentrations of materials in the precipitation (measured in plastic buckets, which are only open to the air when it is raining), as well as what falls in other buckets at times of no rain (automatic sensors move covers depending upon whether it is raining or not). The scientists also measure what leaves the catchments in streamflow, using carefully gauged streams which provide a precise measure of water flow over time and periodic sampling of the chemicals dissolved in that stream water. Their early results (Table 1.1) were very surprising for the time, and of great interest: for many substances, the export from the forest exceeded inputs from the air (the result of weathering of the soil and bedrock material). But for nitrogen (both ammonium, NH_4^+, and nitrate, NO_3^-) the export from the ecosystem was less than the inputs: the forest was retaining a large amount of the nitrogen inputs.

Likens and colleagues also pioneered the idea of performing large-scale ecological experiments at the catchment scale. Hydrologists had done similar experiments before but had not explored the ecological ramifications. In the first of the ecologically oriented experiments, the Hubbard Brook team completely cut down all of the trees in one small catchment, as commercial

insights from a large-scale field experiment

FIGURE 1.11 The Hubbard Brook Experimental Forest. Note the experimental stream catchment from which all trees were removed, left of photo.

loggers sometimes do in what is called a "clear cut" (Figure 1.11), and compared the response of export of materials in stream-water with that in another 5 reference watersheds. The overall export of dissolved inorganic substances from the disturbed catchment rose to 13 times the normal rate (Figure 1.12). Two phenomena were responsible. First, the enormous reduction in transpiring surfaces (leaves) on live trees led to 40% more of the input precipitation being exported again as water in the streams, carrying with it more dissolved substances. Second, and more importantly, as microbes decomposed organic matter and made substances available in solution, the trees no longer were there to assimilate these materials; rather, they were flushed from the system with the exported water.

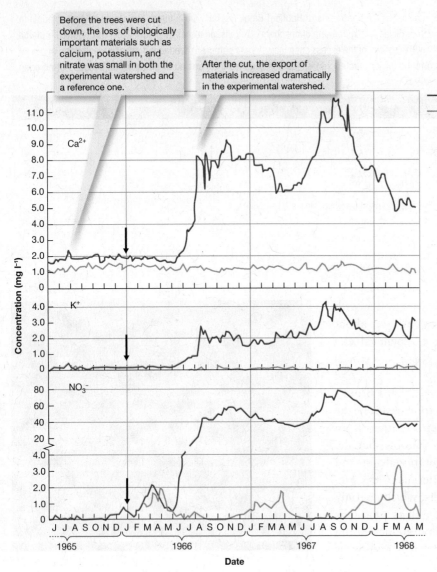

Before the trees were cut down, the loss of biologically important materials such as calcium, potassium, and nitrate was small in both the experimental watershed and a reference one.

After the cut, the export of materials increased dramatically in the experimental watershed.

— Deforested catchment
— Control catchment

FIGURE 1.12 Concentrations of ions in stream water from the experimentally deforested watershed 2 and the control (unmanipulated) watershed 6 at Hubbard Brook. The timing of deforestation is indicated by arrows. In each case, there was a dramatic increase in export of the ions after deforestation. Note that the 'nitrate' axis has a break in it. (After Likens & Bormann, 1975.)

The Hubbard Brook team has gone on to perform many more experiments, and their combination of careful long-term observation and experiments has informed much of our understanding of how forested catchments work, and in particular how they interact with acid rain.

Canada's Experimental Lakes Area: decades of exploring the consequences of human activities on lakes

Earlier in this chapter, we briefly noted the work by David Schindler and others on a whole-ecosystem experiment to understand the effects of acid rain on lakes. That experiment was one of many performed at the Experimental Lakes Area (ELA), a research reserve run by the Canadian government in a sparsely populated area of central Canada. ELA was originally established in 1968 to study the problem of **eutrophication** in lakes: excess growth of algae and cyanobacteria as a result of excessive inputs of nutrients.

During the 1950s and 1960s, water quality in lakes deteriorated in many areas across North America and Europe, with eutrophication being one of the biggest concerns. A considerable amount of research tried to discern the cause. Most of this consisted of laboratory experiments, often adding nutrients such as phosphorus or nitrogen, or adding carbon as inorganic bicarbonate, to bottles of water from ponds or lakes, and analyzing the response in terms of growth of phytoplankton over a period of a few days to a week. The results were conflicting, with many studies suggesting that algal growth and production was primarily limited by carbon dioxide and bicarbonate availability, while others indicated control by phosphorus or sometimes nitrogen. Without a clear cause, strong corrective policies could not be developed. For instance, if the problem were phosphorus, one partial solution would be to ban phosphorus from detergents that contributed to wastewater inputs. But without a clear consensus, agreement on a phosphorous ban or other policy actions was not possible. If algal growth were really controlled by the amount of carbon dioxide and bicarbonate in a lake, then controlling phosphorus would be of little value.

In response to concern over eutrophication in the Great Lakes, the Canadian government established the Freshwater Institute in 1966, hiring Schindler early in 1968 to run the whole-lake eutrophication experiment (Figure 1.13). Schindler and his team put in an access road to the remote area, set up a base camp, and surveyed many lakes, identifying over 50 that were suitable for whole-lake experiments. The ELA group hypothesized that eutrophication was caused by phosphorus, and they designed a set of experiments to test this. Their first experiment was a daring one: they chose a lake (Lake 227) that had extremely low levels of dissolved carbon dioxide, less than any other lake anywhere in the world studied to that time, and proposed to fertilize it with phosphorus. Lab experiments with water from Lake 227 had already "proven" that the low carbon dioxide of the lake strongly limited production by algae, but Schindler and his colleagues had confidence that a whole-lake experiment would yield a different result. As Schindler (2009) later argued, if they could prove that phosphorus was actually the main stimulant of production even in an ecosystem such as Lake 227, with extremely low carbon dioxide, it would disprove the carbon hypothesis generally.

Over the winter of 1968–1969, they hauled the phosphorus to the remote lake, using sleds pulled by snowmobiles. The team began adding phosphorus in June 1969, and Lake 227 rapidly became eutrophic. Despite the low carbon dioxide in the lake, enough became available to support algal growth, due to diffusion both from the atmosphere and from the carbon-rich sediments. The carbon dioxide limitation demonstrated in the laboratory experiments with water from Lake 227 had been an artifact: the rate of supply of new carbon dioxide to the water in the bottles of lab experiments was far slower than that to the waters of the lake itself. Phosphorus was the real control on algal growth and eutrophication in lakes. A subsequent experiment in another lake divided in half by a plastic barrier showed huge stimulation from fertilization with both phosphorus and nitrogen, compared to nitrogen alone (look again at Figure 1.13). The publication of the first results from these experiments led quickly to regulation across North America and Europe to reduce phosphorus pollution, and the Great Lakes and other lakes responded quickly in improved water quality. The problem of eutrophication in coastal marine ecosystems has unfortunately taken many decades longer to solve, since the cause there is primarily nitrogen and not phosphorus (Box 1.4).

Fisheries and Oceans Canada

FIGURE 1.13   In the Lake 226 experiment, one side of the lake received nitrogen additions alone, while the other side received both nitrogen and phosphorus. A plastic barrier prevented exchange of water between the two sides of the lake. The stimulation of algal growth by the phosphorus is clearly seen by the green color in that half of the lake. Reprinted from Schindler (1977).

An introduction of an exotic fish species to New Zealand: investigation on multiple biotic scales

Historically, studies that encompass more than one or two of the four levels in the ecological hierarchy (individuals, populations, communities, ecosystems) have been rare. For most of the 20th century, physiological and behavioral ecologists (studying individuals), population dynamicists, and community and ecosystem ecologists tended to follow separate paths, asking different questions in different ways. However, over the past decade or two, ecologists have increasingly recognized that, ultimately, our understanding will be enhanced considerably when the links between all these levels are made clear – a point illustrated by studies of an the introduction of an exotic fish to streams in New Zealand.

Prized for the challenge they provide to anglers, brown trout (*Salmo trutta*) have been transported from their native Europe all around the world for over a century. Introduced to New Zealand in 1867, they have formed self-sustaining populations in many streams and rivers, often at the apparent expense of native fish

The deterioration of coastal marine ecosystems

While lake-water quality has improved dramatically as a result of regulations to reduce phosphorus inputs, coastal marine ecosystems have become steadily more eutrophic since the early 1970s. In fact, by 2007, a majority of the coastal marine waters in the United States were moderately to severely degraded from eutrophication, despite the phosphorus controls. Worldwide, harmful algal blooms in coastal waters have become more common (Figure 1.14), as have areas depleted of oxygen ("dead zones") such as that in the northern Gulf of Mexico in the plume of the Mississippi River. Why? Eutrophication of these marine ecosystems is caused more by nitrogen than by phosphorus, and the regulation of nitrogen pollution has lagged behind phosphorus regulation by decades in most countries. That nitrogen is the major cause of coastal eutrophication was well demonstrated in a "mesocosm" experiment on the shore of Narragansett Bay, Rhode Island, USA. (Mesocosms are experimental arenas large enough to capture many features of natural systems, in contrast to smaller and less natural 'microcosms.') Pairs of large fiberglass tanks filled with seawater from the Bay were enriched with either phosphorus or nitrogen, or left unenriched as controls. The rate of phosphorus input was the same as for the Lake 227 experiment at ELA, described in the main text. Phosphorus caused no increased in production or amount of algae in the water, but nitrogen fertilization resulted in much higher quantities and growth of algae (Figure 1.14).

Many factors lead to the differences in the control of eutrophication in lakes and coastal marine ecosystems. One of the most important is the large supply of phosphorus relative to nitrogen that often enters coastal marine ecosystems from adjacent oceanic waters. Another factor is nitrogen fixation, the process whereby bacteria can convert atmospheric molecular nitrogen into biologically available nitrogen. When nitrogen is in short supply in lakes, cyanobacteria typically increase rapidly in abundance ('bloom') and fix

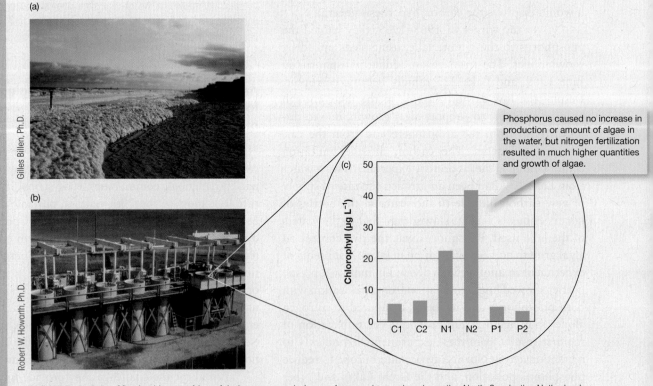

(a)

Gilles Billen, Ph.D.

(b)

Robert W. Howarth, Ph.D.

(c)

Phosphorus caused no increase in production or amount of algae in the water, but nitrogen fertilization resulted in much higher quantities and growth of algae.

FIGURE 1.14   Massive blooms of harmful algae cause windrows of scum along a beach on the North Sea in the Netherlands (a, top). Fertilization of 5-meter deep "mesocosms" filled with seawater from Narragansett Bay (b, bottom) clearly showed a large increase in algae, as measured by the amount of chlorophyll, from adding nitrogen but not phosphorus (c, right) (Oviatt et al., 1995).

nitrogen, alleviating the shortage and maintaining phosphorus as the limiting nutrient. This response by cyanobacteria simply does not occur in most coastal marine ecosystems, as we discuss further in Chapter 11.

1 For many years before the mesocosm experiment described above, many short-term lab experiments with bottles of water from coastal marine ecosystems had shown a response to nitrogen, and not to phosphorus. Water quality managers tended to discount this evidence, and assume that eutrophication in coastal waters was the result of the same forces as in lakes. Was this justified?

2 In many cases, the nitrogen that flows to coastal waters and causes eutrophication originates upstream in river basins at distances of hundreds to a thousand kilometers from the coast, often in different countries or states. The necessary actions to solve the problems occur upstream, but the damage that occurs without action is along the coast. How can societies solve coastal eutrophication in an equitable manner, given this spatial disjunction? Who should pay to reduce the nitrogen pollution?

species such as nonmigratory fish in the genus *Galaxias*. Today, many streams in New Zealand have brown trout, many still have only *Galaxias*, and some have both the introduced and native species of fish.

Microscopic algae growing on rocks compose the base of the food chain in most south New Zealand streams. Mayfly nymphs and other invertebrates graze on the microscopic algae, and they in turn are fed upon by fish. Strikingly, the behavior of nymphs differs depending on whether they are in *Galaxias* or brown-trout streams. In one experiment, nymphs collected from a trout stream and placed in small artificial laboratory channels were less active during the day than the night, whereas those collected from a *Galaxias* stream were active both day and night (Figure 1.15). Why might this difference have evolved? Trout rely principally on vision to capture prey, whereas *Galaxias* rely on mechanical cues. Thus, invertebrates in a trout stream are considerably more at risk of predation during daylight hours when the trout can see best and are most active.

> the individual level – consequences for invertebrate feeding behavior

That an exotic predator such as brown trout has direct effects on mayfly behavior is not surprising. Are there broader effects that cascade through the community to other species? An experiment involving artificial flow-through channels (several meters long, with mesh ends to prevent escape of fish but to allow invertebrates to colonize naturally) placed in a stream tested whether the exotic trout affect the stream food web differently from the displaced native *Galaxias*. Three treatments were established: no fish, *Galaxias* present, and trout present, at naturally occurring densities. Algae and invertebrates were allowed to colonize for 12 days before introducing the

> the community – brown trout cause a cascade of effects

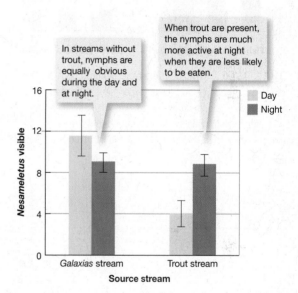

FIGURE 1.15 Mean number (± SE) of *Nesameletus ornatus* mayfly nymphs collected either from a trout stream or a *Galaxias* stream that were recorded by means of video as visible on the substrate surface in laboratory stream channels during the day and night (in the absence of fish). Mayflies from the trout stream are more nocturnal than their counterparts from the *Galaxias* stream. (After Mcintosh & Townsend, 1994.)

fish. After a further 12 days, invertebrates and algae were sampled (Figure 1.16). A significant effect of brown trout reducing invertebrate biomass was evident ($P = 0.026$), but the presence of *Galaxias* had no effect on invertebrate biomass compared to the no-fish control. Algal biomass achieved its highest values, too, in the trout treatment ($P = 0.02$), probably because there were fewer invertebrates grazing the algae.

What are the consequences to the ecosystem, in terms of overall photosynthesis by the algae and energy flow to the mayflies and other invertebrates, and ultimately to the fish? This

> the ecosystem – trout and energy flow

was investigated in two neighboring tributaries of the Taieri River, one where brown trout was the only fish species and one containing only *Galaxias*. Photosynthesis by the algae was six times greater in the trout stream than in the *Galaxias* stream. This resulted in more energy to support the entire food web, and the invertebrates that eat algae produced new biomass in the trout stream at about 1.5 times the rate found in the *Galaxias* stream. Trout themselves produced new biomass at roughly nine times the rate that *Galaxias* did (Figure 1.17). Thus, the algae, invertebrates and fish are all more productive in the trout stream, than in the *Galaxias* stream.

Does this mean that the introduction of the non-native species, brown trout, was beneficial for New Zealand streams? Yes, if the sole goal is to have a high level of fish production, and keep anglers happy. Of course, invasive species – whether introduced on purpose or otherwise – are a huge and increasing problem, and very often have a net negative influence on both basic ecological function and on a variety of issues directly connected to human well-being. This is a major topic for ecologists, and is explored further later in the text.

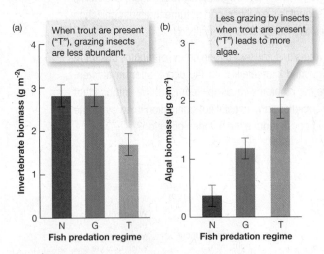

FIGURE 1.16 (a) Total invertebrate biomass and (b) algal biomass (chlorophyll *a*) (± SE) for an experiment performed in summer in a small New Zealand stream. In experimental replicates where trout are present, grazing invertebrates are rarer and graze less; thus, algal biomass is highest. G, *Galaxias* present; N, no fish; T, trout present. (After Flecker & Townsend, 1994.)

Why Asian vultures were heading for extinction: The value of a modeling study

In 1997, vultures in India and Pakistan began dropping from their perches. Local people were quick to notice dramatic declines in numbers of the oriental

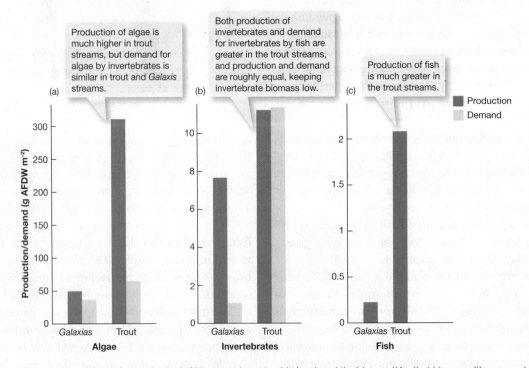

FIGURE 1.17 Annual estimates for 'production' of biomass at one trophic level, and the 'demand' for that biomass (the amount consumed) at the next trophic level, for (a) primary producers (algae), (b) invertebrates (which consume algae), and (c) fish (which consume invertebrates). Estimates are for a trout stream and a *Galaxias* stream. In the former, production at all trophic levels is higher, but because the trout consume essentially all of the annual invertebrate production (b), the invertebrates consume only 21% of primary production (a). In the *Galaxias* stream, these fish consume only 18% of invertebrate production, 'allowing' the invertebrates to consume the majority (75%) of annual primary production. AFDW stands for ash-free dry weight. (After Huryn, 1998.)

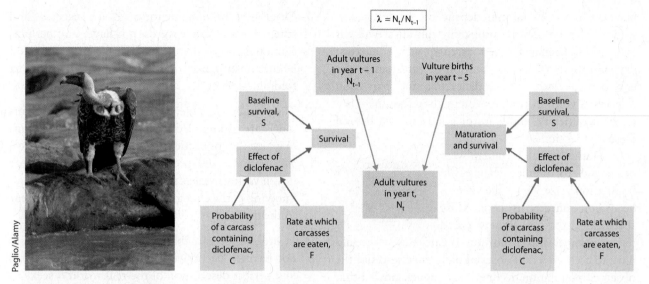

FIGURE 1.18 Flow diagram showing the elements of a model of how the number of adult vultures in the population changes from one year (N_{t-1}) to the next (N_t). The oriental white-backed vulture, whose populations have shown disastrous declines in India and Pakistan, is shown in the inset. The number of adult vultures in year t depends on the number present the previous year ($t-1$), some of which die from natural causes (baseline survival) and others because of diclofenac poisoning. The number of adults in year t also depends on the number of vultures born 5 years previously ($t-5$), because vultures do not mature until they are 5 years old. Again, some newborn vultures die before maturity from natural causes and others because of diclofenac poisoning. The reduction in survival due to diclofenac depends on two things: the probability that a carcass contains diclofenac (C) and the rate at which carcasses are eaten (F). (© Alamy Images AGJX38.)

white-backed vulture *Gyps bengalensis* (Figure 1.18) and the long-billed vulture *G. indicus*. Repeated population surveys from 2000 to 2003 confirmed alarming rates of decline: between 22% and 50% of the population was being lost each year. This was of very great concern because of the crucial role vultures play in everyday life for people in India and Pakistan, disposing of the dead bodies of large animals, both wild and domestic. The loss of vultures enhanced carrion availability to wild dogs and rats, allowing their populations to increase and raising the probability of diseases such as rabies and plague being transmitted to humans. Moreover, contamination of nearby wells and the spread of disease by flies became more likely now that dead animals were not quickly picked clean by vultures. One group of people, the Parsees, were even more intimately affected, because their religion calls for the dead to be taken in daylight to a special tower (dakhma) where the body is stripped clean by vultures within a few hours. It was crucial for ecologists to quickly determine the cause of vulture declines so that action could be taken.

It took scientists a few years to find a common element in the deaths of otherwise healthy birds – each had suffered from visceral gout (accumulation of uric acid in the body cavity) followed by kidney failure. Soon a crucial piece in the jigsaw became clear: vultures dying of visceral gout contained residues of the drug diclofenac (Oaks et al., 2004). Then it was confirmed that carcasses of domestic animals treated with diclofenac were lethal to captive vultures. Diclofenac, a non-steroidal anti-inflammatory drug developed for human use in the 1970s, had only recently come into common use as a veterinary medicine in Pakistan and India. Thus, a drug that benefited domestic mammals proved lethal to the vultures that fed on their bodies.

The circumstantial evidence was strong, but given the relatively small numbers of diclofenac-contaminated dead bodies available

... caused by drug-contaminated carcasses?

to wild vultures, was the associated vulture mortality sufficient explanation for the population crashes? Or might other factors also be at play? Were there other toxins? This was the question addressed by Rhys Green and his team by means of a simulation population model (Green et al. 2004). On the basis of their surveys of population declines and knowledge of birth, death and feeding rates, the researchers built a mathematical model to predict the behavior of the vulture populations. We show their model as a flow diagram (Figure 1.18).

The model was based on a series of simple and well-articulated assumptions. For example, they assumed that vultures do not breed until they are five years old, and then are capable of rearing only one juvenile per year, but only if both parents survive the breeding season of 160 days. Other model inputs specify normal birth and death rates for the population, but also how survival may depend on dicofenac poisoning. This reflects the probability an adult will eat from a diclofenac-affected carcass, which in turn depends partly on the proportion of carcasses in the environment that contain diclofenac and partly on how often vultures feed. As you might guess, there is a lot of complexity, and a lot of good biology, behind these assumptions.

Mathematical formulae were developed to predict changes in population size, but the details need not concern us here. More specifically, the researchers posed the question: what proportion of carcasses (C) would have to contain lethal doses of diclofenac to cause the observed population declines? The model showed that only 1 in 135 carcasses would have to contain diclofenac to cause the observed population decline. The proportions of vultures found dead or dying in the wild with signs of diclofenac poisoning were closely similar to the proportions of deaths expected from the model if the observed population decline was due *entirely* to diclofenac poisoning. The researchers concluded, therefore, that diclofenac poisoning was a sufficient cause for the dramatic decline of wild vultures. As a result of this research, governments have taken action to ban the use of diclofenac.

This example, then, has illustrated a number of important general points about mathematical models in ecology:

> diclofenac-contaminated cattle are sufficient to explain vulture decline

1 Models can be valuable for exploring scenarios and situations for which we do not have, and perhaps cannot expect to obtain, real data (e.g., what would be the consequences of different baseline survival or feeding rates?).

2 They can be valuable, too, for summarizing our current state of knowledge and generating predictions in which the connection between current knowledge, assumptions, and predictions is explicit and clear (given various values for S and F in Figure 1.18, and knowing λ, the change in population from one year to the next, what values of C do these imply?).

3 In order to be valuable in these ways, a model does not have to be (indeed, cannot possibly be) a full and perfect description of the real world it seeks to mimic – all models incorporate approximations (the vulture model was, of course, a very 'stripped down' version of its true life history).

4 Caution is therefore always necessary – all conclusions and predictions are provisional and can be no better than the knowledge and assumptions on which they are based – but applied cautiously they can be useful (the vulture model prompted changes in management practices and research into new drugs).

5 Nonetheless, a model is inevitably applied with much more confidence once it has received support from real sets of data.

SUMMARY

The science of ecology
Ecology is the scientific study of the distribution and abundance of organisms, the interactions that determine that distribution and abundance, and the relationships between organisms and the transformation and flux of energy and matter.

Scales, diversity of approaches, and rigor
Ecology deals with four levels of ecological organization: individual organisms, populations, communities, and ecosystems. Ecology is a rigorous science using standard scientific techniques.

Ecology in practice
Ecologists use a wide variety of approaches including comparative observations, experiments, and models. Experiments include natural experiments such as the observation of agricultural fields that were abandoned at various times over the past century, small scale experiments such as the addition of nitrogen to small plots, and whole-ecosystem experiments such as cutting all the trees from a catchment or adding phosphorus to a whole lake.

REVIEW QUESTIONS

1   Why do some temporal patterns in ecology need long runs of data to detect them, while other patterns need only short runs of data?

2   Discuss the pros and cons of descriptive studies as opposed to experiments of the same ecological phenomenon.

3   What is a 'natural field experiment'? Why are ecologists keen to take advantage of them?

4   Search the library for a variety of definitions of ecology; which do you think is most appropriate and why?

5   How might the results of the Cedar Creek study of old-field succession have been different if a single field had been monitored for 50 years, rather than simultaneously comparing fields abandoned at different times in the past?

6   When all the trees were felled in a Hubbard Brook catchment, there were dramatic differences in the chemistry of the stream water draining the catchment. How do you think stream chemistry would change in subsequent years as plants begin to grow again in the catchment area?

7   Why should we have confidence in whole-lake experiments, even though they are frequently performed without replication?

8   By what mechanism did the introduction of brown trout in New Zealand streams lead to changes in the production of algae?

9   What are the main factors affecting the confidence we can have in predictions of a mathematical model?

Challenge Questions

1   Discuss the different ways that ecological evidence can be gained. How would you go about trying to answer one of ecology's unanswered questions, namely, 'Why are there more species in the tropics than at the poles?'

2   The variety of microorganisms that live on your teeth have an ecology like any other community. What do you think might be the similarities in the forces determining species richness (the number of species present) in your oral community as opposed to a European grassland?

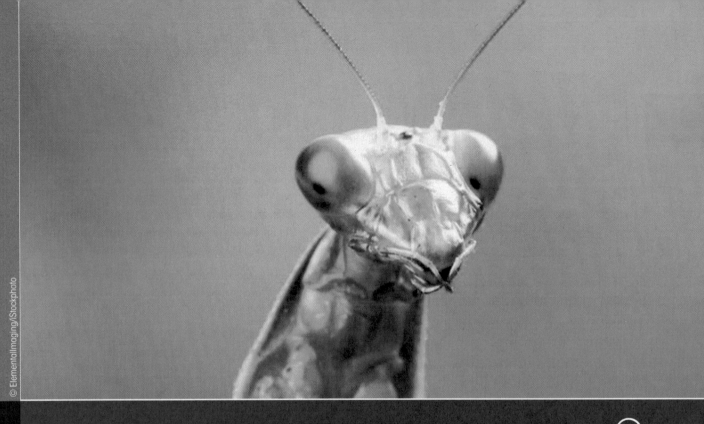

Chapter 2

Ecology's Evolutionary Backdrop

CHAPTER CONTENTS

KEY CONCEPTS

After reading this chapter you will be able to:

- explain that Darwin and Wallace, who were responsible for the theory of evolution by natural selection, were both, essentially, ecologists

- appreciate that natural selection can act very quickly on heritable variation, and that we can study it in action and control it in experiments

- identify what is required for the origin of new species

- explain that the evolutionary history of species constrains what future selection can achieve

- compare and contrast convergent and parallel evolution

- explain that natural selection fits organisms to their past, rather than anticipating their future

As the great Russian-American biologist Dobzhansky said, "Nothing in biology makes sense, except in the light of evolution." But equally, very little in evolution makes sense except in the light of ecology: ecology provides the stage directions through which the "evolutionary play" is performed. Ecologists and evolutionary biologists need a thorough understanding of each other's disciplines to make sense of key patterns and processes.

2.1   EVOLUTION BY NATURAL SELECTION

The Earth is inhabited by very many different types of organism. They are distributed neither randomly nor evenly over the surface of the globe. Any sampled area contains only a small subset of the variety of species present on Earth. Why are there so many types of organism? Why are the distributions of so many so restricted? Answering these ecological questions requires an understanding of the processes of evolution that have led to present-day patterns of diversity and distribution.

Until relatively recently, the emphasis with biological diversity was on using it (for example, for medicine), exhibiting it in zoos and botanic gardens, or cataloging it in museums (Box 2.1). Yet such catalogs are more like stamp collecting than science unless accompanied by

all species are so specialized that they are always absent from almost everywhere

---

2.1 | **Historical Landmarks**

A Brief History of the Study of Diversity

An awareness of the diversity of living organisms, and of what lives where, is part of the knowledge that the human species accumulates and hands down through the generations. Hunter–gatherers needed (and still need) detailed knowledge of the natural history of the plants and animals in their immediate environment in order to obtain food successfully and escape the hazards of being poisoned or eaten.

More than 4,000 years ago, the Chinese emperor Shen Nung compiled what was perhaps the first written 'herbal' of useful plants; and by around 2,000 years ago, the Greek, Dioscorides, had described 500 species of medicinal plants and illustrated many of them in his *De Materia Medica*, which remained in use through the following 1,500 years.

Collections of living specimens in zoos and botanic gardens also have a long history – certainly back to Greece almost 3,000 years ago. Then, in the 17th century, the urge to collect from the diversity of nature developed in the West, allowing some individuals to make their living by finding interesting specimens for other people's collections. For example, John Tradescant the father (died 1638) and John Tradescant the son (1608–1662) spent most of their lives collecting plants abroad for the gardens of the British aristocracy. The father was the first English botanist to visit Russia (1618), bringing back many living plants. His son made three visits to collect specimens in the American colonies.

Wealthy individuals built up vast collections into personal museums and traveled or sent travelers in search of novelties from new lands as they were discovered and colonized. Naturalists and artists (often the same people) were sent to accompany the major voyages of exploration, to report and take home, dead or alive, collections of the diversity of organisms and artifacts that they found. Taxonomy (giving names to the various types of organism) and systematics (organizing and classifying them) developed and flourished.

When big national museums were established (the British Museum in 1759 and the Smithsonian in Washington in 1846), they were largely compiled from the gifts of personal collections. Like zoos and botanic gardens, the museums' main role was to make a public display of the diversity of nature, especially the new and curious and rare.

There was no need to explain the diversity—the biblical theory of the seven-day creation of the world sufficed. However, the idea that the diversity of nature had 'evolved' over time, by progressive divergence from preexisting stocks, was beginning to be discussed around the end of the 18th century. Then, in 1844, Robert Chambers, a Scottish journalist, published his *Vestiges of the Natural History of Creation* (though it was published anonymously at the time)—a popular account of the idea that animal species had descended from other species. This paved the way for the wider acceptance of Darwin's *On the Origin of Species*, published 15 years later.

an understanding of how this diversity has developed, or how it affects the functioning of nature. In the 19th century, Charles Darwin and Alfred Russell Wallace provided ecologists with the scientific foundations needed for such understanding.

Darwin and Wallace (Figure 2.1) were both ecologists (although their seminal work was performed before the term was coined) who were exposed to the diversity of nature in the raw. Darwin sailed around the world as naturalist on the 5-year expedition of H.M.S. *Beagle* (1831–1836), recording and collecting specimens in the enormous variety of environments that he explored on the way. He gradually developed the view that the natural diversity of nature was the result of a process of evolution in which

> Darwin and Wallace were both ecologists

natural selection favored some variants within species through a 'struggle for existence.' He developed this theme over the next 20 years through detailed study and an enormous correspondence with his friends, as he prepared a major work for publication with all the evidence carefully marshalled. But he was in no hurry to publish.

In 1858, Wallace, who had read Darwin's journal of the voyage of the *Beagle*, wrote to Darwin, spelling out, in all its essentials, the same theory of evolution. Wallace was a passionate amateur naturalist. From 1847 to 1852, with his friend H. W. Bates, he explored and collected in the river basins of the Amazon and Rio Negro, and from 1854 to 1862 he made an extensive expedition in the Malay Archipelago. He recalled lying on his bed in 1858 'in the hot fit of intermittent fever,

(a)

National Library of Medicine/NIH

(b)

Hulton Archive/Getty Images

FIGURE 2.1 Photographs of Charles Darwin and Alfred Russell Wallace.

when the idea [of natural selection] suddenly came to me. I thought it all out before the fit was over, and . . . I believe I finished the first draft the next day.'

Today, competition for fame and financial support would no doubt lead to fierce conflict about priority—who had the idea first. Instead, in an outstanding example of selflessness in science, sketches of Darwin's and Wallace's ideas were presented together at a meeting of the Linnean Society in London (where portraits of both still hang today). Darwin's *On the Origin of Species* was then hastily prepared and published in 1859. It is arguably the first major textbook of ecology. Aspiring ecologists would do well to read at least the third chapter.

Both Darwin and Wallace had read *An Essay on the Principle of Population*, published by Thomas Malthus in 1798. Malthus's essay was concerned with human population growth, which, if it remained unchecked, could easily overrun the planet. He realized, though, that limited resources, as well as disease, wars, and other disasters, slowed the growth of populations—that only a small proportion of those born, or who might have been born, actually survived. As experienced field naturalists, Darwin and Wallace realized that the Malthusian argument applied with equal force to the whole of the plant and animal kingdoms.

> influence of Malthus's essay on Darwin and Wallace

Darwin noted the great fecundity of some species— a single individual of the sea slug *Doris* may produce 600,000 eggs; the parasitic roundworm *Ascaris* may produce 64 million. But he realized that every species 'must suffer destruction during some period of its life, and during some season or occasional year, otherwise, on the principle of geometrical increase, its numbers would quickly become so inordinately great that no country could support the product.' And he checked his ideas: in one of the earliest examples of population ecology, Darwin counted all the seedlings that emerged from a plot of cultivated ground three feet by two feet: 'Out of 357 no less than 295 were destroyed, chiefly by slugs and insects.' Darwin and Wallace, then, emphasized that most individuals die before they can reproduce and contribute nothing to future generations. Both, though, tended to ignore the important fact that those individuals that do survive in a population may leave different numbers of descendants.

The theory of **evolution by natural selection** that we owe to Darwin and Wallace rests on a series of established truths:

> fundamental truths of evolutionary theory

1 Individuals that form a population of a species are not identical.

2 Some of the variation between individuals is **heritable**—that is, it has a genetic basis and is therefore capable of being passed down to descendants.

3 All populations could grow at a rate that would overwhelm the environment; but in fact, most individuals die before reproducing and most (usually all) reproduce at less than their maximal rate. Hence, in each generation, the individuals in a population are only a subset of those that 'might' have arrived there from the previous generation.

4 Different ancestors leave different numbers of descendants (descendants, *not* just offspring): they do not all contribute equally to subsequent generations. Hence, those that contribute most have the greatest influence on the heritable characteristics of subsequent generations.

Evolution is the change, over time, in the heritable characteristics of a population or species. Given the above four truths, the heritable features that define a population will inevitably change. Evolution is inevitable.

But which individuals make the disproportionately large contributions to subsequent generations and hence determine the direction that evolution takes? The answer is: those that were best able to survive the risks and hazards of the environments in which they were born and grew; and those who, having survived, were most capable of successful reproduction. Thus, interactions between organisms and their environments—the stuff of ecology—lie at the heart of the process of evolution by natural selection.

> "the survival of the fittest"?

The philosopher Herbert Spencer described the process as "the survival of the fittest," and the phrase has entered everyday language—which is regrettable. First, we now know that survival is only part of the story: differential reproduction is often equally important. But more worryingly, even if we limit ourselves to survival, the phrase gets us nowhere. Who are the fittest?—those that survive. Who survives?—those that are fittest. Nonetheless, the term *fitness* is commonly used to describe the success of individuals in the process of natural selection. An individual will survive better, reproduce more, and leave more descendants—it will be fitter—in some environments than in others. And in a given environment, some individuals will be fitter than others.

Darwin had been greatly influenced by the achievements of plant and animal breeders—for example, the extraordinary variety of pigeons, dogs, and farm animals that had been deliberately bred by selecting individual parents with

> natural selection has no aim for the future

exaggerated traits. He and Wallace saw nature doing the same thing—'selecting' those individuals that survived from their excessively multiplying populations—hence the phrase 'natural selection.' But even this phrase can give the wrong impression. There is a great difference between human and natural selection. Human selection has an aim for the future—to breed a cereal with a higher yield, a more attractive pet dog, or a cow that will produce more milk. But nature has no aim. Evolution happens because some individuals have survived the death and destruction of the past, and have reproduced more successfully in the past: not because they were somehow chosen or selected as improvements for the future.

Hence, past environments may be said to have selected particular characteristics of individuals that we see in present-day populations. Those characteristics are suited to present-day environments only because environments tend to remain the same, or at least change only very slowly. We shall see later in this chapter that when environments do change more rapidly, often under human influence, organisms can find themselves, for a time, left 'high and dry' by the experiences of their ancestors.

2.2   EVOLUTION WITHIN SPECIES

The natural world is not composed of a continuum of types of organism each grading into the next: we recognize boundaries between one sort of organism and another. In

> to understand the evolution, of species we need to understand evolution within species

one of the great achievements of biological science, the Swede, Carl Linnaeus, devised in 1789 an orderly system for naming the different sorts. Part of his genius was to recognize that there were features of both plants and animals that were not easily modified by the organisms' immediate environment, and that these 'conservative' characteristics were especially useful for classifying organisms. In flowering plants, the form of the flowers is particularly stable. Nevertheless, within what we recognize as species, there is often considerable variation, and some of this is heritable. It is on such intraspecific variation, after all, that plant and animal breeders work. In nature, some of this intraspecific variation is clearly correlated with variations in the environment and represents local specialization.

Darwin called his book *On the Origin of Species by Means of Natural Selection*, but evolution by natural selection does far more than create new species. Natural selection and evolution occur *within* species, and we now know that we can study them in action

within our own lifetime. Moreover, we need to study the way that evolution occurs within species if we are to understand the origin of new species.

Geographical variation within species

Since the environments experienced by a species in different parts of its range are themselves different (to at least some extent), we might expect natural selection to have favored

> the characteristics of a species may vary over its geographical range

different variants of the species at different sites. But the characteristics of populations will diverge from each other only if there is sufficient heritable variation on which selection can act, and only if the forces of selection favoring divergence are strong enough to counteract the mixing and hybridization of individuals from different sites. Two populations will not diverge completely if their members (or, in the case of plants, their pollen) are continually migrating between them, mating and mixing their genes.

The sapphire rockcress, *Arabis fecunda*, is a rare perennial herb restricted to calcareous soil outcrops in western Montana—so rare, in fact, that there are just 19 existing populations separated into two groups ('high elevation' and 'low elevation') by a distance of around 100 km. Whether there is local adaptation here is of practical importance: four of the low-elevation populations are under threat from spreading urban areas and may require reintroduction from elsewhere if they are to be sustained. But the low-elevation sites were more prone to drought—both the air and the soil were warmer and drier there—and reintroduction may therefore fail if local adaptation is too marked. Observing plants in their own habitats and checking for differences between them would not tell us if there was local adaptation in the evolutionary sense. Differences may simply be the result of immediate responses to contrasting environments made by plants that are essentially the same. Hence, high- and low-elevation plants were grown together in a 'common garden,' eliminating any influence of contrasting immediate environments. The low-elevation plants in the common garden were indeed significantly more drought-tolerant (Figure 2.2). They had significantly better 'water use efficiency': their rate of water loss through the leaves was low compared to the rate at which carbon dioxide was taken in.

A second study using this 'common garden' approach illustrates the important phenomenon of countergradient variation. Common

> Scandinavian frogs: countergradient variation

frogs (*Rana temporaria*), monitored at sites along an

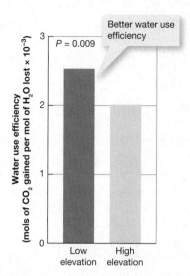

FIGURE 2.2 When plants of the rare sapphire rockcress from low-elevation (drought-prone) and high-elevation sites were grown together in a common garden, local adaptation was apparent: those from the low-elevation site had significantly better water use efficiency. (From McKay et al., 2001.)

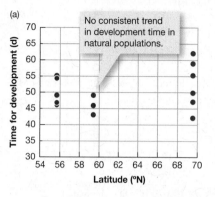

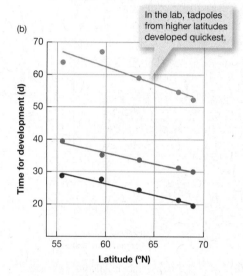

FIGURE 2.3 (a) Tadpoles (*Rana temporaria*) from ponds (separate dots) from two areas of Sweden, to the south, and from Finland, to the north, had varied development times but showed no consistent trend with latitude. (b) However, when tadpoles originating from these same sites and from intermediate latitudes in Sweden were reared in the laboratory, those from the highest latitudes consistently developed fastest whether reared at 14°C (orange dots), 18°C (blue dots) or 22°C (red dots) [population (latitude) effect: $F = 208.7$, $P < 0.001$]. (From Laugen et al., 2003.)

extended latitudinal gradient in Sweden and Finland, varied considerably in their development time (the time from complete gill absorption in the tadpole to emergence of the first foreleg), but there was no consistent trend with latitude (Figure 2.3a). However, when tadpoles from these and other sites along the gradient were reared in a common environment, at a range of temperatures, there was a clear and consistent trend, with those from the higher latitudes developing significantly faster (Figure 2.3b). In other words, there had indeed been local adaptation. Those consistently experiencing the coldest temperatures (higher latitudes) had evolved compensatory increases in development rate, hence **countergradient variation**. This was impossible to observe in nature, but the end result was that developmental times were similar at all latitudes.

Differentiation over a much smaller spatial scale has been demonstrated in the evolution of tolerance to zinc, at otherwise toxic concentrations, in sweet vernal grass, *Anthoxanthum odoratum*, growing along a 90-m transect covering the boundary between mine and pasture soils at the Trelogan zinc and lead mine in Wales. The work started in the 1960s with plants being taken from various points along the transect and grown for almost 2 years in a common garden before being tested. There was a most striking increase in tolerance over a distance of only 3 m at the transition from the pasture to the mine-derived plants (Figure 2.4a)—testimony to the power of the selective forces favoring tolerance in plants growing in contaminated

variation over very short distances

soil. But as we have already noted, such evolution can have occurred only if hybridization between plants on either side of the transition had failed to counteract it. **Hybridization**, the production of offspring sharing the characteristics of two parents, of course requires the exchange of genetic material between plants, which in turn requires that the plants be flowering at the same time. In fact, though, there was a pattern in flowering times that paralleled almost perfectly the pattern in tolerance, and this same pattern has been maintained for more than 40 years since the site was first investigated (Figure 2.4b). Plants growing on the mine soil flowered later than those in the pasture. Even if the plants had been thoroughly intermingled (not strung out along the transect), mating between tolerant and nontolerant

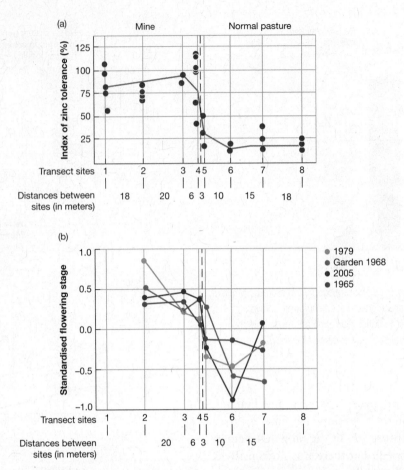

FIGURE 2.4 (a) Zinc tolerance in *Anthoxanthum odoratum* plants taken from various points along a transect at the Trelogan zinc and lead mine, as indicated, and grown in a common garden. The points are means of replicates from original sampled plants; the lines join the means of these means. Tolerance was measured as mean root length in soil with added zinc as a percentage of root length in a zinc-free control. (From McNeilly, Antonovics, & Bradshaw, 1970.) (b) Site-specific mean flowering stage across the middle of the transect (sites 2–7), measured in 2005 and compared with previous measurements made from site material in 1965, from common garden material in 1966 [see (a)], and from the field again in 1979. Flowering stage is given as a deviation from the mean flowering stage measured across the whole transect at the time, on a scale from 1 to 5, with a difference of 1 representing around 7–8 days difference in flowering time. (From Antonovics, 2006.)

individuals would have been more than 40% less than if flowering had been synchronous. This is not perfect reproductive isolation, clearly, but it is enough to allow the inherited variation in zinc tolerance within this species to become established and be maintained over remarkably short distances.

On the other hand, it would be quite wrong to imagine that local selection always overrides hybridization – that all species exhibit geographically distinct variants with a genetic basis. For example, in a study of *Chamaecrista fasciculata*, an annual legume from disturbed habitats in the central plains of North America, plants were also grown together in a common garden, having been taken either from the 'home' site or transplanted from distances of 0.1, 1, 10, 100, 1000, and 2000 km. Five characteristics were measured—germination, survival, vegetative biomass, fruit production, and the number of fruit—but for all characters in all replicates there was little or no evidence for differentiation, except at the very

farthest spatial scales (e.g., Figure 2.5). There is 'local adaptation'—but it's clearly not *that* local.

We can also test whether organisms have evolved to become specialized to life in their local environment in **reciprocal transplant experiments**: comparing their performance when they are grown 'at home' (i.e., in their original habitat) with their performance 'away' (i.e., in the habitat of others).

reciprocal transplants test the match between organisms and their environment— sea anemones transplanted into each other's habitats

It can be difficult to detect the local specialization of animals by transplanting them into each other's habitat: if they don't like it, most species will run away. But invertebrates like corals and sea anemones are sedentary, and some can be lifted from one place and established in another. The sea anemone *Actinia tenebrosa* is found in rock pools and on boulders around the coast of New South Wales, Australia. A reciprocal transplant

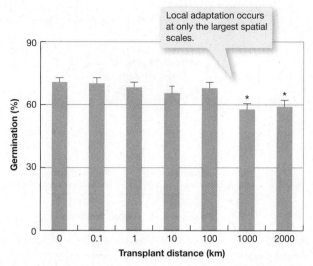

FIGURE 2.5 Percentage germination of local (transplant distance zero) and transplanted *Chamaecrista fasciculata* populations to test for local adaptation along a transect in Kansas. Data for 1995 and 1996 have been combined because they do not differ significantly. Populations that differ from the home population at $P < 0.05$ are indicated by an asterisk. (From Galloway & Fenster, 2000.)

experiment was carried out at two locations, Cape Banks and Bass Point, at each of which sea anemones were translocated between replicate sites in pool and boulder habitats. At each site, all anemones, between 50 and 85 individuals, were removed and then transplanted, all together, at one of the other sites at that location, chosen at random. Following this, then, at both locations, there were boulder sites with rock pool anemones, boulder sites with (translocated) boulder anemones, rock pool sites with boulder anemones and rock pool sites with (translocated) rock pool anemones. This kind of design is essential: only by translocating not only to a different but also to the same habitat is it possible to distinguish the effects of local adaptation from any effects of translocation in its own right. Individuals were then monitored at their new sites for two years and a number of measures of fitness examined: the proportion surviving, the proportion brooding juveniles, and the growth (change in size) of survivors (Figure 2.6). In almost every case, and in every case that was statistically significant, local

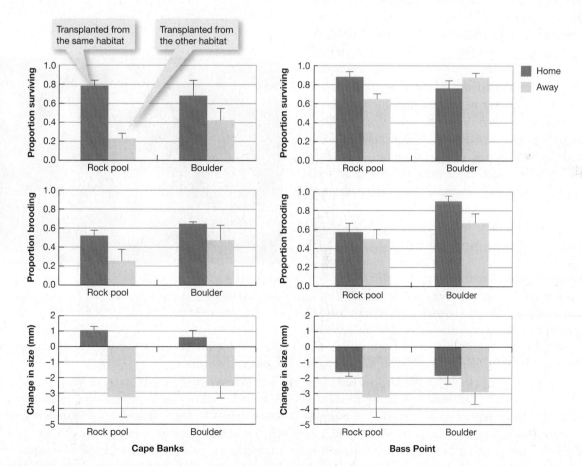

FIGURE 2.6 A reciprocal transplant experiment with sea anemones, *Actinia tenebrosa*, carried out at two locations in Australia, Cape Banks and Bass Point, and in each case between (light blue bars) and within (blue bars) two habitats, rock pool and boulder. Error bars represent ±SE. For all three fitness measures examined, 'home' anemones (blue bars) were significantly fitter than 'away' ones (light blue bars). In detail: proportion surviving ($P = 0.032$, with a significant effect of location, but none of habitat nor any interaction between habitat and treatment), proportion brooding juveniles ($P = 0.029$, with no significant effects of location, habitat nor any interaction between habitat and treatment), change in size ($P = 0.001$, with a significant effect of location, but none of habitat nor any interaction between habitat and treatment). (After Sherman & Ayre, 2008.)

adaptation was apparent: individuals did better in their own than in the alternative habitat. They were more likely to survive, more likely to be brooding juveniles of their own, and they grew more (or at least shrunk less).

Once again, however, it is important to recognize that while reciprocal transplant experiments are usually undertaken to investigate local adaptation, confirmation of local adaptation is certainly not a general rule. This can be seen from a survey carried out all studies of reciprocal transplant experiments performed up to 2005 (Hereford, 2009). Confining attention to those that were conducted in natural field environments, and that measured fitness directly through effects either on survival and/or fecundity, the survey yielded 74 studies, with 50 on plants and 21 on animals (and three on fungi or protists). These generated 360 cases where reciprocal local adaptation could be sought (because there were often several measures of survival and fecundity and because some populations were measured twice); but of these, only around half were able to confirm the pattern, and effect sizes were not always large (Figure 2.7). In almost as many cases, one of the populations was simply fitter than the other, both in its native and in its nonnative site. Reciprocal transplant experiments are a good way of detecting local adaptation. But we should not be tempted into thinking that all populations are perfectly adapted to their local environments. Lack of genetic variation on which natural selection can act, the forces of hybridization, and delays between alterations to the environment and evolved, adaptive change in the population, all combine to ensure that this is by no means invariably so.

> reciprocal transplants generally: local adaptation often but not always

In many of the examples so far, geographic variants of species have been identified, but the selective forces favoring them have not. This is not true of the next example, involving the guppy (*Poecilia reticulata*), a small freshwater fish living in Trinidad, where many rivers are subdivided by waterfalls that isolate fish populations above the falls from those below. Guppies are present in many of these water bodies, coexisting in the lower waters with various species of predatory fish that are absent higher up the rivers. The populations of guppies in Trinidad differ from each other in almost every feature that biologists have examined. Forty-seven of these traits tend to vary in step with each other (they *covary*) and with the intensity of the risk from predators. This correlation suggests that the guppy populations have been subject to natural selection from the predators. But the fact that two phenomena are correlated does not prove that one causes

> natural selection by predation—a controlled field experiment in fish evolution

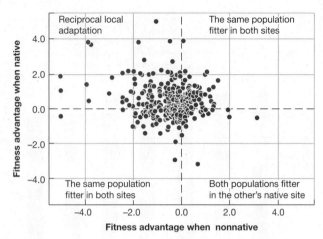

FIGURE 2.7 Summary of the results of a survey of 74 reciprocal transplant experiments published up to 2005, generating 360 estimates of local adaptation. In 48% of cases there was reciprocal local adaptation, where both populations were fitter in their native sites; these appear in the upper left quadrant. In 43% of cases, one of the populations was fitter both in its native site and in that of the other population; these appear in the upper left and lower right quadrants. In just 9% of cases, therefore, both populations were fitter in the native site of the other population. Fitnesses are relative to the mean fitness at each site. Hence, local adaptation was frequently but not invariably apparent. (After Hereford 2009.)

Courtesy Anne Magurran

FIGURE 2.8 Male and female guppies (*Poecilia reticulata*) showing two flamboyant males, courting a typical dull-colored female.

the other. Only controlled experiments can establish cause and effect.

Whenever we study natural selection in action, it becomes clear that compromises are involved. For every selective force that favors change, there is a counteracting force that resists the change. Color in male guppies is a good example. Where guppies have been free or relatively free from predators, the males are brightly decorated with different numbers and sizes of colored spots (Figure 2.8). Female guppies prefer to mate with the most gaudily decorated males, but these are more readily captured by predators because they are easier to see.

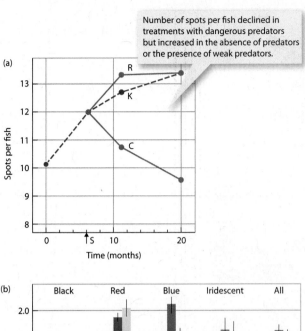

Number of spots per fish declined in treatments with dangerous predators but increased in the absence of predators or the presence of weak predators.

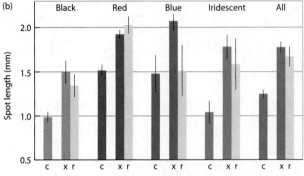

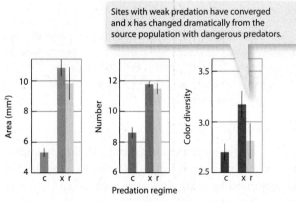

Sites with weak predation have converged and x has changed dramatically from the source population with dangerous predators.

FIGURE 2.9 (a) An experiment showing changes in populations of guppy, *Poecilia reticulata*, exposed to predators in experimental ponds. The graph shows changes in the number, size, and diversity of colored spots per fish in ponds with different populations of predatory fish. The initial population was deliberately collected from a variety of sites so as to display high variability and was introduced to the ponds at time 0. At time S weak predators (*Rivulus hartii*) were introduced to ponds R, a high intensity of predation by the dangerous predator *Crenicichila alta* was introduced into ponds C, while ponds K continued to contain no predators (the vertical lines show ± two standard errors). (b) Results of a field experiment. A population of guppies originating in a locality with dangerous predators (*c*) was transferred to a stream having only the weak predator (*Rivulus hartii*) and, until the introduction, no guppies (*x*). Another stream nearby with guppies and *R. hartii* served as a control (*r*). The results shown are from guppies collected at the three sites 2 years after the introductions. In the absence of strong predators, the size, number, and diversity of colored spots increased significantly within 2 years. (After Endler, 1980.)

This sets the stage for some revealing experiments on the ecology of evolution. Guppy populations were established in ponds in a greenhouse and exposed to different intensities of predation. The number of colored spots per guppy fell sharply and rapidly over the generations when the population suffered heavy predation (Figure 2.9a). Then, in a field experiment, 200 guppies were moved from a site far down the Aripo River where predators were common and introduced to a site high up the river where there were neither guppies *nor* predators. The transplanted guppies thrived in their new site, and within just two years the males had more and bigger spots of more varied color (Figure 2.9b). The females' choice of the more flamboyant males had dramatic effects on the gaudiness of their descendants, but this was only because predators were not present to reverse the direction of selection.

The speed of evolutionary change in this experiment in nature was as fast as that in artificial selection experiments in the laboratory. Many more fish were produced than would eventually survive (as many as 14 generations of fish occurred in the 23 months during which the experiment took place) and there was considerable genetic variation in the populations upon which natural selection could act.

Variation within a species with man-made selection pressures

It is not surprising that some of the most dramatic examples of natural selection in action have been driven by the forces of environmental pollution, since these can provide powerful selection pressures as a result of recent and rapid changes in the environment. We have already seen this illustrated for sweet vernal grass growing on an old zinc and lead mine (Figure 2.4). Pollution of the atmosphere during and after the Industrial Revolution has also left its evolutionary fingerprints. **Industrial melanism** is the phenomenon in which black or blackish forms of species of moths and other organisms have come to dominate populations in industrial areas. In the dark individuals, a dominant gene is responsible for producing an excess of the black pigment melanin. Industrial melanism is known in most industrialized countries, including some parts of the United States (e.g., Detroit, Pittsburgh), and more than a hundred species of moth have evolved forms of industrial melanism.

The earliest species known to have evolved in this way was the peppered moth, *Biston betularia*. The first black ('*carbonaria*') specimen was caught in Manchester (England) in 1848. By 1895, about 98% of the

natural selection by pollution—the evolution of a melanic moth

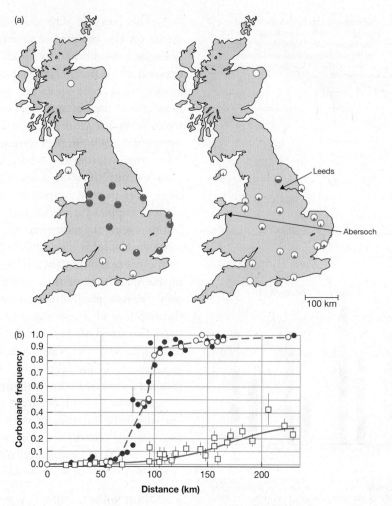

FIGURE 2.10 (a) The distribution of melanic, *carbonaria*, and pale forms (blue-and-white portions of the pie diagrams) of the peppered moth, *Biston betularia*, for 1952–56 (left) and 1996 (right), limited to sites where a reasonable comparison between the two periods could be made. The two arrows indicate the positions of Abersoch in the west and Leeds in the east, the two ends of the transect examined in (b). (After Grant et al., 1998.) (b) Clines in the frequency of the melanic form along a transect running WSW to NSE from Abersoch to Leeds [see (a)] for the periods 1964–69 (filled circles), 1970–75 (open circles), and 2002 (open squares); lines are standard errors. (After Saccheri et al., 2008.)

Manchester peppered moth population was melanic. Following many more years of pollution, a large-scale survey in the 1950s of pale and melanic forms in Britain recorded many thousands of specimens (Figure 2.10a). The winds in Britain are predominantly westerlies, spreading industrial pollutants (especially smoke and smog) toward the east. Melanic forms were concentrated toward the east and were completely absent from unpolluted western parts of England, Wales, and northern Scotland. It was apparent, moreover, along a west-to-east transect where the moths were most intensively monitored, that the transition from nonmelanic to melanic could be extremely sharp (Figure 2.10b). Adaptation was remarkably local, given that the moths can fly 2 km in a single night.

The moths are preyed upon by insectivorous birds that hunt by sight. In a field experiment, large numbers of melanic and pale ('typical') moths were reared and released in equal numbers in a rural and largely unpolluted area of southern England. Of the 190 moths that were captured by birds, 164 were melanic and 26 were typicals. An equivalent study was made in an industrial area near the city of Birmingham. Twice as many melanics as typicals were recaught. This showed that a significant selection pressure was exerted through bird predation, and that moths of the typical form were clearly at a disadvantage in the polluted industrial environment (where their light color stood out against a sooty background), whereas the melanic forms were at a disadvantage in the pollution-free countryside (Kettlewell, 1955).

In the 1960s, however, industrialized environments in Western Europe and the United States started to change as oil and electricity began to replace coal, and legislation was passed to impose greatly reduced air pollution (see Chapter 12). The frequency of melanic forms then fell back rapidly to near pre-Industrial levels, both nationwide and along the intensively studied transect (Figure 2.10a, b).

The forces of selection worked first in favor, and then against, the melanic forms, and this has clearly been related to industrial pollution. But the idea that melanic forms were favored simply because they were camouflaged against smoke-stained backgrounds may only be part of the story. The moths rest on tree trunks during the day, and nonmelanic moths are well hidden against a background of mosses and lichens. Industrial pollution had not just blackened the moths' background; atmospheric pollution, especially SO_2, had also destroyed most of the moss and lichen on the tree trunks. Indeed the distribution of melanic forms in Figure 2.10a closely fits the areas in which tree trunks were likely to have lost lichen cover as a result of SO_2 and so ceased to provide such effective camouflage for the nonmelanic moths. Thus, SO_2 pollution may have been as important as smoke in selecting melanic moths, and its amelioration may have been as important in favoring the return of the typicals.

Evolution and coevolution

It is easy to see that a population of plants faced with repeated drought is likely to evolve a tolerance of water shortage, and an animal repeatedly faced with cold winters is likely to evolve habits of hibernation or a thick protective coat. But droughts do not become any less severe as a result, nor winters milder. The situation is quite different when two species interact: predator on prey, parasite on host, competitive neighbor on neighbor. Natural selection may select from a population of parasites those that are more efficient at infecting their host. But this immediately sets in play forces of natural selection that favor more resistant hosts. As they evolve, they put further pressure on the ability of the parasite to infect. Host and parasite are then caught in never-ending reciprocating selection: they **coevolve**. In many other ecological interactions, the two parties are not antagonists but positively beneficial to one another: **mutualists**. Pollinators and their plants, and leguminous plants and their nitrogen-fixing bacteria are well-known examples. We consider coevolution in some detail when we return to more evolutionary aspects of ecology in Chapter 8.

2.3 THE ECOLOGY OF SPECIATION

We have seen that natural selection can force populations of plants and animals to change their character—to evolve. But none of the examples we have considered has involved the evolution of a new species. Indeed, Darwin's *On the Origin of Species* is about natural selection and evolution: not really about the origin of species. Black and typical peppered moths are forms within a species not different species. Likewise, the zinc-tolerant and zinc-sensitive forms of the grasses at Trelogan mine, and the dull and flamboyant races of guppies, are just local genetic classes. None qualifies for the status of distinct species. So: what criteria justify naming two populations as different species?

What do we mean by a 'species'?

Cynics have said, with some truth, that a species is what a competent taxonomist regards as a species.

biospecies do not exchange genes

Darwin himself regarded species, like many other taxonomic groupings, as 'merely artificial combinations made for convenience.' In the 1930s, however, two American biologists, Mayr and Dobzhansky, introduced a now broadly accepted empirical test that could be used to decide whether two populations were part of the same or of two different species. They recognized organisms as being members of a single **species** if they could, at least potentially, breed together in nature to produce fertile offspring. They called species tested and defined in this way **biospecies**. In the examples that we have used earlier in this chapter, we know that melanic and normal peppered moths can mate and that the offspring are fully fertile; this is also true of colored and dull guppies and of zinc-tolerant and zinc-sensitive grass plants. They are all variations within species—not separate species.

In practice, however, biologists do not apply the Mayr–Dobzhansky test before they recognize every species: there is simply not enough time and resources. What is more important is that the test recognizes a crucial element in the evolutionary process. Two parts of a population can evolve into distinct species only if some sort of barrier prevents gene flow between them. If the members of two populations are able to hybridize and their genes are combined and reassorted in their progeny, then natural selection can never make them truly distinct.

Thus, there are two crucial components of **ecological speciation**: that is, speciation where there

ecological speciation

is both an ecological source of divergent selection and a means of reproductive isolation (Rundle & Nosil, 2005). (Ecological speciation is thus distinct from speciation dominated by chance events, such as the sudden doubling of a chromosome number that occasionally occurs during abnormal cell division.) The source of divergent selection is likely to be either a difference in the environment experienced by different subpopulations or a difference in their interaction with other species. Hence, the most orthodox scenario for speciation comprises a number of stages (Figure 2.11). First, two subpopulations become geographically isolated and natural selection drives genetic adaptation to their local environments. Next, coincident with but not caused by this genetic differentiation, a degree of reproductive isolation builds up between the two. This may, for example, be a difference in courtship ritual, tending to prevent offspring production in the first place (hence, called 'prezygotic' isolation), or the hybrid offspring themselves may simply display a reduced fitness, being 'neither one thing nor the other' (post-zygotic isolation). Finally, in a phase of secondary contact, the two subpopulations remeet, and natural selection will then favor any feature in either subpopulation that *reinforces* reproductive isolation, especially pre-zygotic characteristics, preventing the production of low-fitness hybrid offspring. These breeding barriers then cement the distinction between what have now become separate species.

We should not, however, imagine that all examples of speciation conform to this orthodox picture. In Figure 2.11, there is a dividing line between the **allopatric phase** (when the subpopulations are in *different* places) and the **sympatric phase** (subpopulations in the *same* place). But that dividing line may sometimes be to the extreme right, so that there is never secondary contact. This purely *allopatric speciation* is especially likely for island species, which are examined next. Alternatively, the line may be to the extreme left, with populations diverging in the absence of physical separation: *sympatric speciation* (see below).

> allopatric and sympatric speciation

We must always remember, too, that the origin of a species, whether allopatric or sympatric, is a process, not an event. We have already seen, in the case of the grasses growing at Trelogan mine, that local adaptation there, while not associated with the genesis of new species, was accompanied by a difference in flowering time, and hence a degree of reproductive isolation, which facilitated the local adaptation. We shall see below that even after species have been formed, there may be some exchange of genes between them. For the formation of a new species, like the boiling of an egg, there is some freedom to argue about when it is completed.

Allopatric speciation

As we have already noted, it is when a population becomes split into completely isolated populations, especially when dispersed onto different islands, that they most readily diverge into distinct species. The most celebrated example of speciation on islands is the case of Darwin's finches in the Galapagos archipelago. The Galapagos are volcanic islands, isolated in the Pacific Ocean about 1000 km west of Equador and 750 km from the island of Cocos, which is itself 500 km from Central America. Above an altitude of 500 m, the vegetation is open grassland. Below this, there is a humid zone of forest and then a coastal strip of desert vegetation with some endemic species of prickly pear cactus (*Opuntia*). There is, therefore, a wide variety of habitats on the islands, and 14 different species of finch are found there (Figure 2.12).

> islands and speciation— Darwin's finches

Members of one group, including *Geospiza fuliginosa* and *G. fortis*, have strong bills and hop and scratch for seeds on the ground. *Geospiza scandens* has a narrower and slightly longer bill and feeds on the flowers and pulp of the prickly pears as well as on seeds. Finches of a third group have parrot-like bills and feed on leaves, buds, flowers, and fruits, and a fourth group with a parrot-like bill (*Camarhynchus*

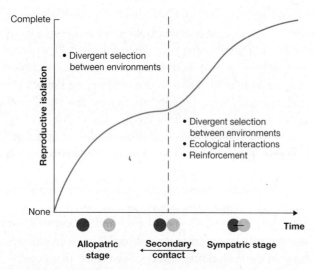

FIGURE 2.11 The process of ecological speciation. As time progresses, initially two populations of a species that are neither ecologically different nor reproductively isolated from one another become geographically separated: the start of the allopatric phase. In their now-separate environments, the populations diverge through independent natural selection (the different colored circles) and a degree of reproductive isolation develops. There is subsequently secondary contact between the two populations – the start of the sympatric stage—where strengthened reproductive isolation is favored by natural selection because of the much-reduced fitness of hybrid offspring. However, the dividing line between the allopatric and sympatric stages may be to the extreme right (no secondary contact: no sympatric stage) or, probably more rarely, to the extreme left (no initial allopatric stage but with incipient new species coexisting but living in different microenvironments or engaging in different ecological interactions). (After Rundle & Nosil, 2005.)

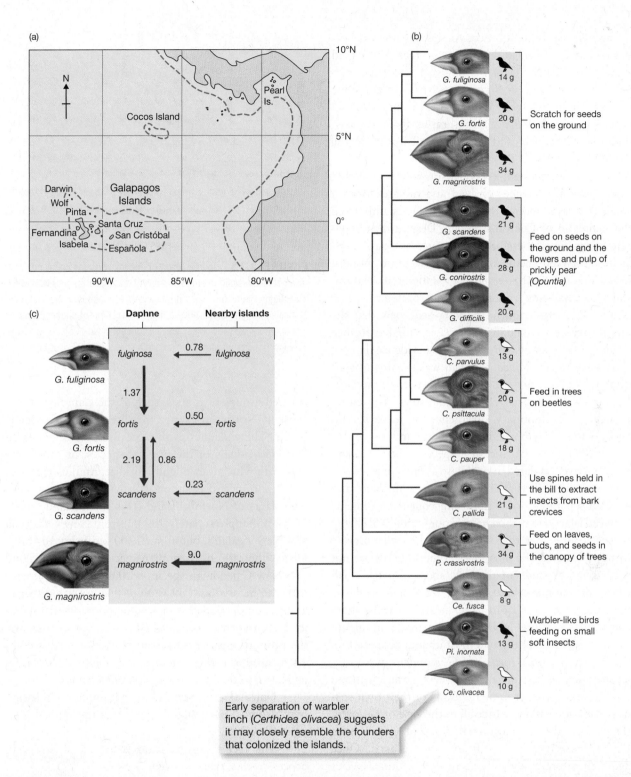

FIGURE 2.12 (a) Map of the Galapagos Islands showing their position relative to Central America; on the equator 5° equals approximately 560 km. (b) A reconstruction of the evolutionary history of the Galapagos finches based on variation in the length of microsatellite DNA. A measure of the genetic difference between species is shown by the length of the horizontal lines. The feeding habits of the various species are also shown. Drawings of the birds' heads are proportional to actual body size. The maximum amount of black coloring in male plumage and the average body mass are shown for each species. G = Geospiza, C = Camarhynchus, Ce = Certhidea. (Adapted from Petren et al., 1999.) (c) Gene flow for the four species on Daphne Major Island, through interbreeding with other species on the island and with immigrants of the same and other species from the nearby islands. Flow is measured as the effective number of individuals per generation. For genes to flow, the first-generation hybrid offspring must themselves mate with one of the parental species. Genes flow from G. fortis to G. scandens when the hybrid sings the G. scandens song (because its father did) and vice versa for genes flowing from G. scandens to G. fortis. The population of G. fulginosa on Daphne is very small, and hence the flow of genes into G. fortis comes from immigrants from other islands. (After Grant & Grant, 2010.)

psittacula) has become insectivorous, feeding on beetles and other insects in the canopy of trees. A so-called woodpecker finch, *Camarhynchus* (*Cactospiza*) *pallida*, extracts insects from crevices by holding a spine or a twig in its bill, while yet a further group includes a species (*Certhidea olivacea*) that, rather like a warbler, flits around actively and collects small insects in the forest canopy and in the air.

There is, though, every reason to suppose that these Galapagos finches evolved and radiated from a single ancestral species that invaded the islands from the mainland of Central America. Despite their separate adaptations, there are strong morphological similarities between them, and indeed, accurate modern tests (Figure 2.12b)—tracing evolutionary relationships by analyzing variation in 'microsatellite' DNA (see Chapter 8)—confirm the long-held view that the family tree of the Galapagos finches radiated from a single trunk. They also provide strong evidence that the warbler finch (*Certhidea olivacea*) was the first to split off from the founding group and is likely to be the most similar to the original colonist ancestors. The entire process of evolutionary divergence of these species appears to have happened in less than 3 million years.

Populations of ancestor species, then, became reproductively isolated, most likely after chance colonization of different islands within the archipelago, and evolved separately for a time. Subsequent movements between islands may then have brought nonhybridizing biospecies together that have evolved to fill different niches. However, the biospecies compartments are not water-tight. A study of the four species on the small island of Daphne Major, and of their possible interbreeding with birds from larger nearby islands, again using microsatellite DNA, is summarized in Figure 2.12c. It shows that the two most abundant species on the island, *Geospiza fortis* and *G. scandens*, were subject to a greater flow of genes between one another than they were to genes from immigrants of their own species from other islands. Indeed, in the case of *G. fortis*, there was also a substantial flow of genes from *G. fulginosa* immigrants from other islands. Once again, therefore, the 'ideal' of gene flow within a species but not between them is not borne out by the data. But the fact that there are 'gray areas' partway through the process does not diminish the importance of either the process of speciation or the concept of biospecies.

The flora and fauna of many archipelagos of islands show similar | island endemics | examples of great richness of species with many local **endemics**, species known only from one island or area. Lizards of the genus *Anolis* have evolved a kaleidoscopic diversity of species on the islands of the Caribbean; and

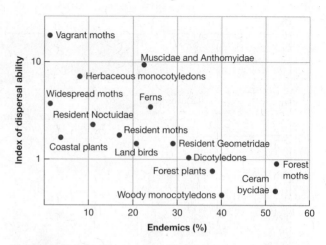

FIGURE 2.13 The evolution of endemic species on islands as a result of their isolation from individuals of an original species with which they might interbreed. Poorly dispersing (and therefore more 'isolated') groups on Norfolk Island have a higher proportion of endemic species and are more likely to contain species from either New Caledonia or New Zealand than from Australia, which is farther away. (After Holloway, 1977.)

isolated groups of islands, such as the Canaries off the coast of North Africa, are treasure troves of endemic plants. The endemics evolve because they are isolated from individuals of the original species, or other species, with which they might hybridize. An illustration of the importance of isolation is provided by the animals and plants of Norfolk Island. This small island (about 70 km²) is approximately 700 km from New Caledonia and New Zealand, but about 1200 km from Australia. Hence, the ratio of Australian species to New Zealand and New Caledonian species within a group of organisms can be used as a measure of that group's dispersal ability, and of course, groups with poorer dispersal ability are, in effect, more isolated. As Figure 2.13 shows, the proportion of endemics on Norfolk Island is highest in groups with poor dispersal ability. As isolation increases, so does the proportion of endemics.

Unusual and often rich communities of endemics may also pose particular problems for the applied ecologist (Box 2.2).

The fact that speciation is a process rather than an event is further illustrated by the extraordinary case of ring species. In these, races or subspecies of a species that fall | evolution in Californian salamanders: ring species | short of being full species themselves (i.e., distinct forms that are nonetheless capable of producing fertile hybrids) are arranged along a geographic gradient, but in such a way that the two ends of the gradient themselves meet, hence forming a ring, and where they do, they behave as good species despite being linked, back around the ring, by the series of interbreeding races.

2.2   ECOncerns

Deep-sea vent communities at risk?

Deep-sea volcanic vents are islands of high food availability, with the chemical energy of the vents trapped and passed up the food web by special free-living and symbiotic 'chemosynthetic' bacteria. As a consequence, they support unique communities, rich in endemic species. One of the latest controversies to pit environmentalists against industrialists concerns these deep sea vents, which are also now known to be sites potentially rich in minerals. The full version of this article by Alok Jha appeared on the website of *The Guardian* newspaper on December 28, 2011.

Exotic creatures discovered living at deep-sea vent in Indian Ocean

British scientists have found a remarkable array of creatures, some of them new to science, in one of the most inhospitable regions of the deep sea. In the first ever expedition to explore and take samples from the 'Dragon Vent' in the southwest Indian Ocean, remotely operated submarines spotted yeti crabs, sea cucumbers, and snails living around the boiling column of mineral-rich water that spews out of the seafloor.

Dr. Jon Copley, a marine biologist at the University of Southampton who led the exploration of the Dragon Vent, said: 'We found a new type of yeti crab [see Figure below 2.0]. Yeti crabs are known at vents in the eastern Pacific and there are two species described so far, but they have very long, hairy arms – ours have short arms and their undersides are covered in bristles. They're quite different to the ones that are known from the Pacific,' said Copley. 'This is the first time a Yeti crab has been seen in the Indian Ocean.'

Ifremer, A. Fifis/AP Images

His team also found sea cucumbers, vent shrimps, and scaly-foot snails. Sea cucumbers have previously only been seen at deep-sea vents in the eastern Pacific. 'This is the first time they've been seen at vents in the Indian Ocean and they're not known from the central Indian or mid-Atlantic vents so far,' he said.

Copley said that characterizing the life at the world's hydrothermal vents was a race against time. 'Earlier this year, China was granted a licence by the UN International Seabed Authority for exploratory mining at deep-sea vents on the southwest Indian ridge,' he said. 'The vent chimneys are very rich in copper, zinc, gold, and uranium. But we have no idea what's actually living there.'

'Just like the 19th century naturalists used to go to the Galápagos and other islands to find species there that are different to elsewhere and then use that to understand patterns of dispersal and evolution, we can use deep-sea vents to do the same things beneath the waves,' he said.

Consider the following options and debate their relative merits:

1 Allow the mining industry free access to all deep-sea vents, since the wealth created will benefit many people.

2 Ban mining and other disruption of all deep-sea vent communities, recognizing their unique biological and evolutionary characteristics.

3 Carry out biodiversity assessments of known vent communities and prioritize according to their conservation importance, permitting mining in cases that will minimize overall destruction of this category of community.

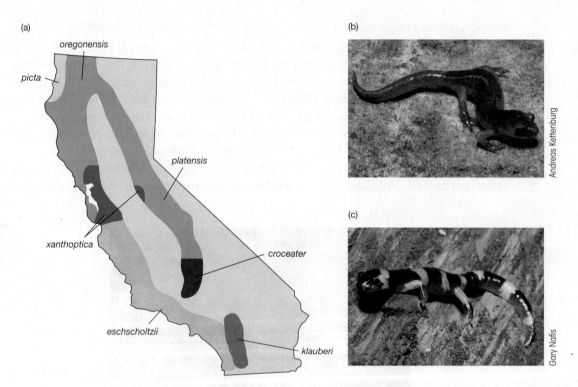

FIGURE 2.14 (a) The distributions of the subspecies of the salamander *Ensatina eschscholtzii* in California. They form a 'ring species' with adjacent subspecies hybridising readily, with the exception of *E. e. eschscholtzii* and *E. e. klauberi*, which behave as good species. (After Irwin et al., 2001.) (b) *Ensatina eschscholtzii eschscholtzii*. (c) *Ensatina eschscholtzii klauberi*.

Thus, what would normally be a temporal sequence of events, that we can only presume to have happened, becomes frozen in space for us to observe in the here and now. Salamanders of the *Ensatina eschscholtzii* species-complex in California provide a good example (Irwin et al., 2001). There are generally held to be seven subspecies (Figure 2.14a), following an initial invasion from the north, spreading southwards along the mountain ranges separated by the Central Valley of California. This has led, broadly, to two genetic lineages: *E. eschscholtzii oregonensis* (indicating subspecies *oregonensis*), *E. e. picta*, *E. e. xanthoptica*, *E. e. eschscholtzii*, and northern populations of *E. e. platensis*, on the one hand, and southern populations of *E. e. platensis*,

E. e. croceater, and *E. e. klauberi* on the other. For the most part, adjacent/overlapping subspecies hybridise readily: *E. e. oregonensis* hybridizes with *E. e. picta* in north-west California and further east with *E. e. platensis*, and so on. But in southern California, the subspecies *E. e. eschscholtzii* and *E. e. klauberi* coexist as separate species, hybridizing only rarely, and indeed looking strikingly different (Figure 2.14b, c), despite being linked to one another by the ring of subspecies between them.

Sympatric speciation?

As noted above (Figure 2.11), in principle, speciation may also occur sympatrically, without any initial geographic separation of populations. Convincing examples, however, remain scarce. One good example, though, is provided by two species of cichlid fish in Nicaragua: the Midas cichlid *Amphilophus citrinellus* and the Arrow cichlid *A. zaliosus* (Barluenga et al., 2006). The two species coexist in the isolated volcanic crater lake, Lake Apoyo (Figure 2.15a), which is small (c5 km in diameter), relatively homogeneous in terms of habitat, and of recent origin (less than 23,000 years). *A. zaliosus* is endemic to

Lake Apoyo (found nowhere else); *A. citrinellus* is found in a wide variety of water bodies in the region, including the largest. In the first place, genetic evidence from mitochondrial DNA (passed by mothers to their offspring in their eggs—see Chapter 8) indicates that the cichlids of Lake Apoyo, of both species, had a single common ancestor arising from the much more widespread stock of *A. citrinellus* (Figure 2.15b).

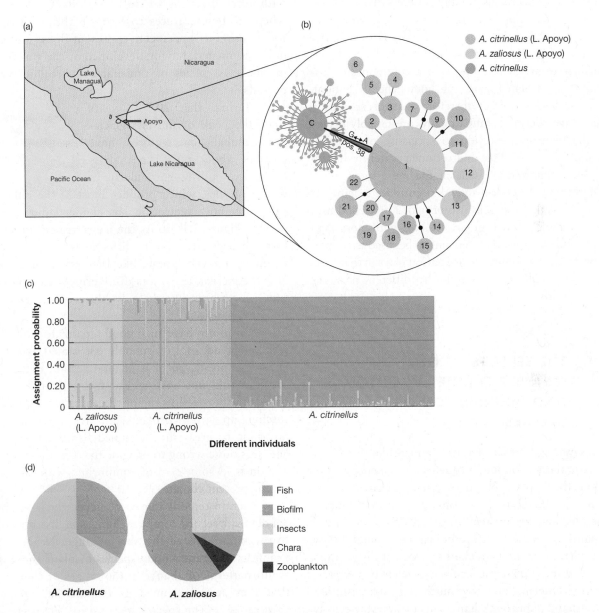

FIGURE 2.15 Sympatric speciation in cichlid fish, *Amphilophus citrinellus* and *A. zaliosus*. (a) The location of the study site, Lake Apoyo, near the Pacific coast of Nicaragua. (b) The relatedness network of 637 'haplotypes' of mitochondrial DNA sequences from individual fish, with *A. zaliosus* green, *A. citrinellus* from Lake Apoyo orange, and *A. citrinellus* from other lakes blue. The gray circles amongst the blue ones correspond to the Lake Apoyo haplotypes, enlarged toward the upper right. The size of a circle reflects the frequency with which a particular haplotype was found. The most common haplotype in Lake Apoyo ('1') is distinguished from the most common *A. citrinellus* haplotype ('C') by a single mutation (base substitution G –> A at position 38), but all Lake Apoyo haplotypes, of both species, share this mutation, indicating their origin from a single common ancestor. (c) The assignment of individuals to populations, color-coded as in (b), based on variation at ten microsatellite DNA loci. The clear separation of colours is indicative of partial or total reproductive isolation between them. (d) Stomach content analyses of the two species in Lake Apoyo (*Chara* are algae). (After Barluenga et al., 2006.)

Then, a variety of genetic evidence, including that from microsatellite DNA (for a definition, see Chapter 8) indicates that the two species in Lake Apoyo are reproductively isolated from one another and indeed from *A. citrinellus* in other lakes (Figure 2.15c). This reproductive isolation was further supported by mate choice experiments with the two Lake Apoyo species. And finally, *A. citrinellus* and *A. zaliosus* in Lake Apoyo are morphologically distinct from one another and have substantially different diets, indicative of *A. citrinellus* feeding more from the bottom of the lake and *A. zaliosus* feeding mostly in open water (Figure 2.15d). Hence, these are two distinct species, which have arisen recently from a common ancestor in a small, isolated patch of homogeneous habitat. There seems little doubt, therefore, that this speciation must have occurred sympatrically, presumably driven by the divergent selection to specialise on bottom-feeding in the one case and on open water-feeding in the other. Interestingly, one of the most staggeringly rich examples of **endemism**, the occurrence together of many endemic species, is also provided by cichlid fish: those of the East African Great Lakes, with more than 1,500 endemic species. It remains to be discovered how important a role sympatric speciation plays in that case, and whether divergent selection to different niches is the main driving force.

2.4 THE EFFECTS OF CLIMATIC CHANGE ON THE EVOLUTION AND DISTRIBUTION OF SPECIES

It is tempting to translate our confidence in the power of natural selection into a belief that each species represents the best possible adaptation to the environment in which it finds itself. We should resist that temptation. Techniques for analyzing and dating biological remains (particularly buried pollen) increasingly allow us to detect just how much of the present distribution of organisms is a precise locally evolved match to present environments, and how much is a fingerprint left by the hand of history. Changes in climate, particularly during the ice ages of the Pleistocene (the past 2–3 million years), bear much of the responsibility for the present distributions of plants and animals. As climates have changed, species populations have advanced and retreated, been fragmented into isolated patches, and may then have rejoined. Much of what we see in the present distribution of species represents a phase in the recovery from past climatic change.

For most of the past 2–3 million years, the Earth has been very cold. Evidence from the distribution of oxygen isotopes in cores taken from the deep ocean floor shows that there may have been as many as 16 glacial cycles in the Pleistocene, each lasting for up to 125,000 years (Figure 2.16a). Each cold (glacial) phase may have lasted for as long as 50,000–100,000 years, with brief intervals of only 10,000–20,000 years when the temperatures rose to, or above, those of today. From this perspective, present floras and faunas are unusual, having developed at the warm end of one of a series of unusual catastrophic warm periods.

cycles of glaciation have occurred repeatedly

During the 20,000 years since the peak of the last glaciation, global temperatures have risen by about 8°C. The analysis of buried pollen—particularly of woody species, which produce most of the pollen— can show how vegetation has changed during this period (Figure 2.16b). As the ice retreated, different forest species advanced in different ways and at different speeds. For some, like the spruce of eastern North America, there was displacement to new latitudes; for others, like the oaks, the picture was more one of expansion.

the distribution of trees has changed gradually since the last glaciation

We do not have such good records for the post-glacial spread of the animals associated with the changing forests, but it is at least certain that many species could not have spread faster than the trees on which they feed. Some of the animals may still be catching up with their plants, and tree species are still returning to areas they occupied before the last ice age! It is quite wrong to imagine that our present vegetation is in some sort of equilibrium with (adapted to) the present climate.

Even in regions that were never glaciated, pollen deposits record complex changes in distributions. In the mountains of the Sheep Range, Nevada, for example, different woody species of plant show different patterns of change in the ranges of elevations that they have occupied as climate has changed (Figure 2.17). The species composition of vegetation has continually been changing and is almost certainly still doing so.

The records of climatic change in the tropics are far less complete than those for temperate regions. It has been suggested, though, that during cooler, drier glacial periods, the tropical forests retreated to smaller patches, surrounded by a sea

species-rich forest refuges in the tropics?

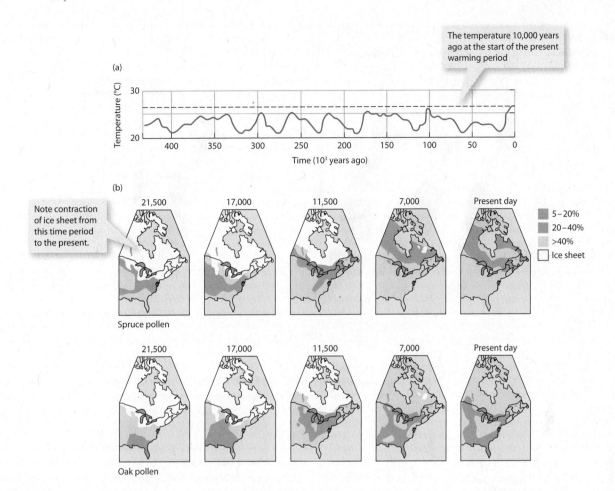

FIGURE 2.16   (a) An estimate of the temperature variations with time during glacial cycles over the past 400,000 years. The estimates were obtained by comparing oxygen isotope ratios in fossils taken from ocean cores in the Caribbean. Periods as warm as the present have been rare events, and the climate during most of the past 400,000 years has been glacial. (After Emiliani, 1966; Davis, 1976.) (b) Ranges in eastern North America, as indicated by pollen percentages in sediments, of spruce species (above) and oak species (below) from 21,500 years ago to the present. Note how the ice sheet contracted during this period. (After Davis & Shaw, 2001.)

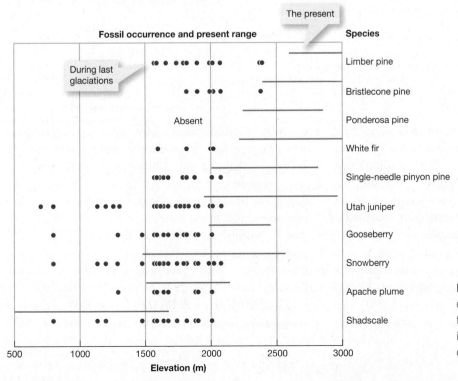

FIGURE 2.17   The elevation ranges of 10 species of woody plant from the mountains of the Sheep Range, Nevada. (Red dots indicate fossil occurrences; the blue line the current range) (After Davis & Shaw, 2001.)

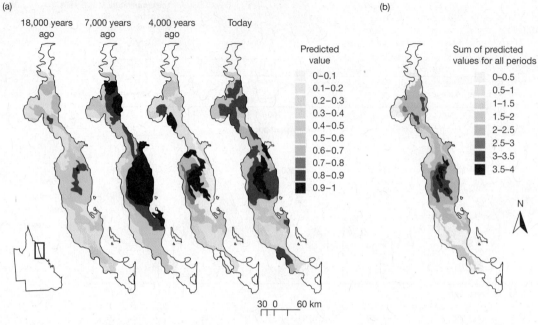

(a) 18,000 years ago | 7,000 years ago | 4,000 years ago | Today

Predicted value
- 0–0.1
- 0.1–0.2
- 0.2–0.3
- 0.3–0.4
- 0.4–0.5
- 0.5–0.6
- 0.6–0.7
- 0.7–0.8
- 0.8–0.9
- 0.9–1

30 0 60 km

(b)

Sum of predicted values for all periods
- 0–0.5
- 0.5–1
- 1–1.5
- 1.5–2
- 2–2.5
- 2.5–3
- 3–3.5
- 3.5–4

N

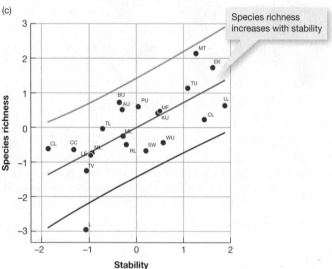

(c) Species richness increases with stability

FIGURE 2.18 (a) For the area of Australian Wet Forest in Queensland, northeast Australia (see inset) climatic conditions at 2,000 random points in the current forested region were used to predict the overall distribution of forest across the region in, from left to right, cool-dry (18,000 years ago), cool-wet (7,000 years ago), warm-wet (4,000 years ago), and current climatic conditions. The values between 0 and 1 indicate the probability, according to the model, of forest being found at a given point. (b) Forest stability, which is calculated simply as the sum of the values in the four figures in (a). (c) Species richness of a forest subregion (of mammals, birds, reptiles, and frogs) increases with stability. The axes are scaled around zero because both stability and species richness were standardized to take account of the fact that both increase with subregion area. Point labels refer to specific subregions, the names and locations of which may be found in the original article. (After Graham et al., 2006.)

of savanna, within which speciation was intense, giving rise to present-day 'hot-spots' of endemism. Support for this in, for example, the Amazonian rain forest now seems less certain than it once did, but there is support for the idea in other regions. In the Australian Wet Tropics of Queensland, northeastern Australia, it has been possible to use present-day distributions of forest to predict distributions in the cool-dry climate of the last Glacial Maximum when forest-contraction was greatest (about 18,000 years ago), the cool-wet period around 7,000 years ago when a massive expansion was likely, and the warm-wet period around 4,000 years ago when there was likely to have been another contraction (Figure 2.18a; Graham et al., 2006). Putting the distributions together, then, allows each subregion of the forest to be assigned a stability (Figure 2.18b)—the most stable being the one in which forest has been most constantly present—and these stabilities can in turn be compared to the species richness (of mammals, birds, reptiles, and frogs) of each subregion. Richness tended clearly to be greatest where stability was highest (Figure 2.18c), that is, where the forest refuges had been in the past. On this interpretation, the present distributions of species may again be seen as largely accidents of history (where the refuges were) rather than precise matches between species and their differing environments.

Evidence of changes in vegetation that followed the last retreat of the ice hint at the likely consequences of the global warming (0.7°C between 1970 and 2010, and a predicted additional 3°C or so in the next hundred years, if society does not take dramatic action soon) due to continuing increases in 'greenhouse' gases in the atmosphere (see Chapter 12). But the scales are quite different. Postglacial warming of about 8°C occurred over around 20,000 years—0.04°C per century, compared to the current rate of global warming of 1.75°C per century, and changes in the vegetation failed to keep pace even with this. But current projections for the 21st century require range shifts for trees at rates of 300–500 km per century compared to typical rates in the past of 20–40 km per century (and exceptional rates of 100–150 km). It is striking that the only precisely dated extinction of a tree species in the Quaternary Period, that of *Picea critchfeldii*, occurred around 15,000 years ago at a time of especially rapid postglacial warming (Jackson & Weng, 1999). Clearly, even more rapid change in the future could result in extinctions of many additional species (Davis & Shaw, 2001).

> human-caused global warming by the 'greenhouse effect' is nearly a hundred times faster than postglacial warming

2.5 CONTINENTAL DRIFT, PARALLEL AND CONVERGENT EVOLUTION

The patterns of species formation that occur on islands appear on an even larger scale in the distributions of species and much broader taxonomic categories across continents. Biologists, especially Wegener (1915), met outraged scorn from geologists and geographers when a century ago they argued that it must have been the continents that had moved rather than the organisms that had dispersed. Eventually, however, measurements of the directions of the Earth's magnetic fields required the same, apparently wildly improbable, explanation and the critics capitulated. The discovery that the tectonic plates of the Earth's crust move and

> land masses have moved

carry the migrating continents with them reconciles geologist and biologist (Figure 2.19). While major evolutionary developments were occurring in the plant and animal kingdoms, their populations were being split and separated, and land areas were moving across climatic zones. This was happening while changes in temperature were occurring on a vastly greater scale than the glacial cycles of the Pleistocene.

The established drift of large land masses over the face of the Earth explains many patterns in the distribution of species that would otherwise be difficult to understand. One classic example is provided by the placental and marsupial mammals. Marsupials arrived on what would become the Australian continent in the Cretaceous period (around 90 million years ago; see Figure 2.19), when the only other mammals present were the curious egg-laying monotremes (now represented only by the spiny anteaters and the duck-billed platypus). The splitting away of Australia, however, prevented the invasion of placental mammals once they had evolved on other, now quite separate continents. An evolutionary process of radiation then occurred among the Australian marsupials that in many ways paralleled what was occurring among the placental mammals on other continents (Figure 2.20). It is hard to escape the view that the geographically separate environments of the placentals and marsupials contained similar ecological pigeonholes (niches) into which the evolutionary process has neatly 'fitted' ecological equivalents. Because they started to diversify from a common ancestral line, and both inherited a common set of potentials and constraints, we refer to this as **parallel evolution**. Once again, though, these species are found where they are not simply because they are the best fitted to those particular environments but also because of accidents in (geological) history.

> and divided populations that have then evolved independently— parallel evolution

There are also, however, many examples of organisms that have evolved in isolation from each other and then converged on remarkably similar forms or behavior or function. Such similarity is particularly striking when similar roles are played by structures that have quite different evolutionary origins—that is, when the structures are *analogous* (similar in superficial form or function) but not *homologous* (derived from an equivalent structure in a common ancestry). When this occurs, it is termed **convergent evolution**. Bird and bat wings are a classic example (Figure 2.21).

> convergent evolution

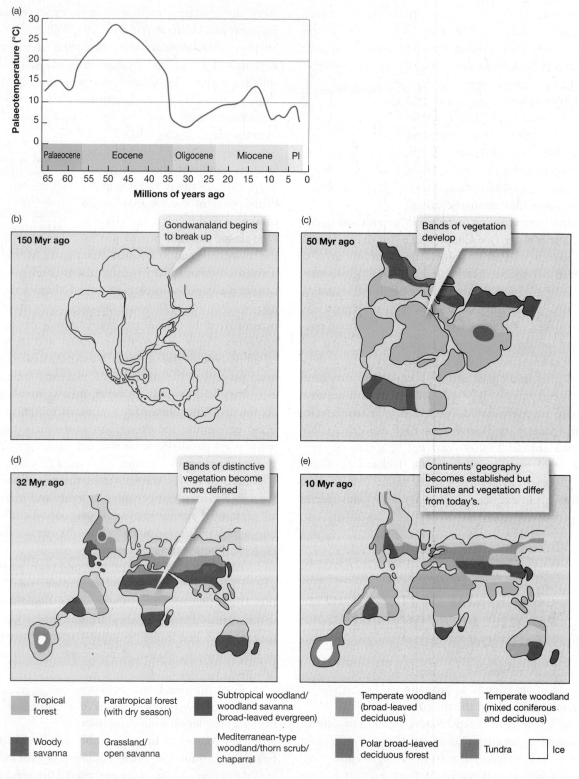

Tropical forest

Paratropical forest (with dry season)

Subtropical woodland/ woodland savanna (broad-leaved evergreen)

Temperate woodland (broad-leaved deciduous)

Temperate woodland (mixed coniferous and deciduous)

Woody savanna

Grassland/ open savanna

Mediterranean-type woodland/thorn scrub/ chaparral

Polar broad-leaved deciduous forest

Tundra

Ice

FIGURE 2.19 (a) Changes in temperature in the North Sea over the past 60 million years. During this period there were large changes in sea level that allowed dispersal of both plants and animals between landmasses. Pl = Pleistocene. (b–e) Continental drift. (b) The ancient supercontinent of Gondwanaland began to break up about 150 million years ago. (c) About 50 million years ago (early Middle Eocene) recognizable bands of distinctive vegetation had developed and (d) by 32 million years ago (early Oligocene) these had become more sharply defined. (e) By 10 million years ago (early Miocene) much of the present geography of the continents had become established but with dramatically different climates and vegetation from today: the position of the Antarctic ice cap is highly schematic. (Adapted from Janis, 1993; Norton & Sclater, 1979; and other sources.)

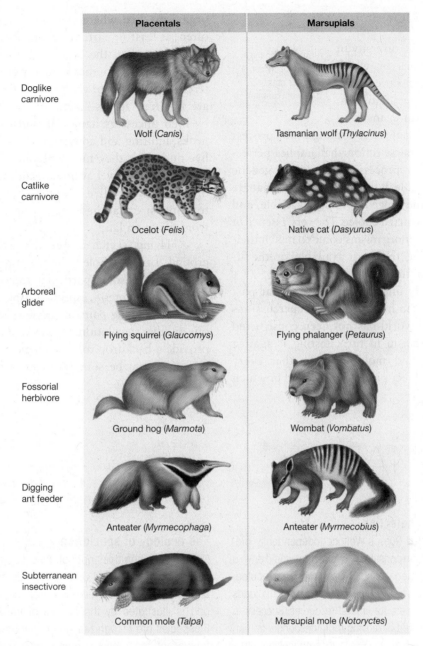

Placentals	Marsupials

Doglike carnivore — Wolf (*Canis*) / Tasmanian wolf (*Thylacinus*)

Catlike carnivore — Ocelot (*Felis*) / Native cat (*Dasyurus*)

Arboreal glider — Flying squirrel (*Glaucomys*) / Flying phalanger (*Petaurus*)

Fossorial herbivore — Ground hog (*Marmota*) / Wombat (*Vombatus*)

Digging ant feeder — Anteater (*Myrmecophaga*) / Anteater (*Myrmecobius*)

Subterranean insectivore — Common mole (*Talpa*) / Marsupial mole (*Notoryctes*)

FIGURE 2.20   Parallel evolution of marsupial and placental mammals. The pairs of species are similar in both appearance and habit and usually (but not always) in lifestyle.

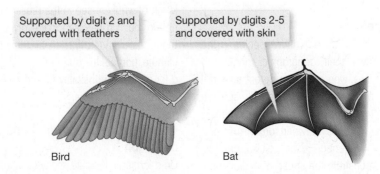

Supported by digit 2 and covered with feathers

Supported by digits 2–5 and covered with skin

Bird   Bat

FIGURE 2.21   Convergent evolution: the wings of bats and birds are analogous (not homologous). They are structurally different, as the labels show. (After Ridley, 1993.)

2.6 CONCLUSION

When we marvel at the diversity of complex specializations by which organisms match their varied environments, there is a temptation to regard each case as an example of evolved perfection. But there is nothing in the process of evolution by natural selection that implies perfection. The evolutionary process works on the genetic variation that is available. It favors those forms that are fittest from among the range of variety available, and this may be a very restricted choice. The very essence of natural selection is that organisms come to match their environments by being 'the fittest available' or 'the fittest yet': they are not 'the best imaginable.'

interpreting the match between organisms and their environment

It is particularly important to realize that past events on the Earth can have profound repercussions on the present. Our world has not been constructed by taking each organism in turn, testing it against each environment, and molding it so that every organism finds its perfect place. It is a world in which organisms live where they do for reasons that are often, at least in part, accidents of history. Moreover, the ancestors of the organisms that we see around us lived in environments that were profoundly different from those of the present. Evolving organisms are not free agents—some of the features acquired by their ancestors hang like millstones around their necks, limiting and constraining where they can now live and what they might become. It is very easy to wonder and marvel at how beautifully the properties of fish fit them to live in water—but just as important to emphasize that these same properties prevent them from living on land.

Having sketched out the evolutionary background for the whole of ecology in this chapter, we will return to some particular topics in evolutionary ecology in Chapter 8, especially aspects of coevolution, where interacting pairs of species play central roles in one another's evolution. However, since evolution does provide a backdrop to all ecological acts, its influence can of course be seen throughout the remainder of this book.

SUMMARY

Evolution by natural selection

Charles Darwin and Alfred Russell Wallace independently proposed that natural selection constituted a force that would drive a process of evolution. The theory of natural selection is an ecological theory. The reproductive potential of living organisms leads them inescapably to compete for limited resources. Success in this competition is measured by leaving more descendants than others to subsequent generations. When these ancestors differ in properties that are heritable, the character of populations will necessarily change over time and evolution will happen. Natural selection is a result of events in the past—it has no intentions and no aim.

Evolution within species

We can see natural selection in action within species in the variation within species over their geographic range and even over very short distances, where we can detect powerful selective forces in action and recognize ecologically specialized races. Transplanting plants and animals between habitats reveals tightly specialized matches between organisms and their environments. The evolutionary responses of animals and plants to pollution demonstrate the speed of evolutionary change, as do experiments on the effects of predators on the evolution of their prey.

The ecology of speciation

Natural selection does not normally lead to the origin of species unless it is coupled with the reproductive isolation of populations from each other—as occurs for example on islands and is illustrated by the finches of the Galapagos Islands. Biospecies are recognized when they have diverged enough to prevent them from forming fertile hybrids if and when they meet.

The effects of climatic change on the evolution and distribution of species

Much of what we see in the present distribution of organisms is not so much a precise local evolved match to present environments as a fingerprint left by the hand of history. Changes in climate, particularly during the ice ages of the Pleistocene, bear a lot of the responsibility for the present patterns of distribution of plants and animals.

Continental drift, parallel and convergent evolution

On a longer time scale, many distributions make sense only once we realize that while major evolutionary developments were occurring, populations were being split and separated, and land masses were moving across climatic zones. Evidence of the power of ecological forces to shape the direction of evolution comes from parallel evolution (in which populations long isolated from common ancestors have followed similar patterns of diversification) and from convergent evolution (in which populations evolving from very different ancestors have converged on very similar form and behavior).

REVIEW QUESTIONS

1 What was the contribution of Malthus to Darwin's and Wallace's ideas about evolution?

2 Why is 'the survival of the fittest' an unsatisfactory description of natural selection?

3 What is the essential difference between natural selection and the selection practiced by plant and animal breeders?

4 What are *reciprocal transplants*? Why are they so useful in ecological studies?

5 Is sexual selection, as exhibited by guppies, different from or just part of natural selection?

6 What is it about the Galapagos finches that has made them such ideal material for the study of evolution?

7 What is the difference between convergent and parallel evolution?

Challenge Questions

1 What do you consider to be the essential distinction between *natural selection* and evolution?

2 Review the utility and applicability of the biospecies concept to a range of groups, including a common species of plant, a rare animal species of conservation interest, and bacteria living in the soil.

3 The process of evolution can be interpreted as optimizing the fit between organisms and their environment or as narrowing and constraining what they can do. Discuss whether there is a conflict between these interpretations.

Part 2

Conditions and Resources

Anne Sorbas/Getty Images

Chapter 3

Physical conditions and the availability of resources

CHAPTER CONTENTS

KEY CONCEPTS

After reading this chapter you will be able to:

- describe how organisms respond to the range of conditions like temperature

- explain how the response of photosynthetic organisms to resources such as solar radiation, water, nutrients, and carbon dioxide are intertwined

- describe how contrasting body compositions affect the consumption of plants by animals,

and of overcoming defenses in the consumption of animals by other animals

- describe the effects of intraspecific competition for resources

- explain how responses to conditions and resources interact to determine ecological niches

The most fundamental prerequisites for life in any environment are that the organisms can tolerate the local conditions and that their essential resources are being provided. We cannot go very far in understanding the ecology of any species without understanding its interactions with conditions and resources.

3.1 ENVIRONMENTAL CONDITIONS

Conditions and resources are two quite distinct properties of environments that determine where organisms can live. **Conditions** are physicochemical features of the environment such as its temperature, humidity, pH, or, in aquatic environments, salinity. An organism often alters the conditions in its immediate environment—sometimes on a very large scale (a tree, for example, maintains a zone of higher humidity on the ground beneath its canopy) and sometimes only on a microscopic scale (an algal cell in a pond alters the pH in the microzone of water that surrounds it). But conditions are not consumed nor used up by the activities of organisms.

Environmental **resources**, by contrast, *are* consumed by organisms in the course of their growth and reproduction. Plants use the energy of sunlight to produce biomass through photosynthesis, consuming not only the quanta of solar radiation adsorbed by the chloroplast but also carbon dioxide, water, and nutrients such as phosphorus and nitrogen that become the organic matter of the plant. The plant's resources are thus carbon dioxide, water, the nutrients, and solar radiation. Correspondingly, rabbits are a resource for eagles, and a rabbit eaten by one eagle is not available to another eagle. This has an important consequence: organisms may *compete* with each other to capture a share of a limited resource.

In this chapter we consider, first, examples of the ways in which environmental conditions limit the function and distribution of organisms. We draw many of our examples from the effects of temperature, which serve to illustrate some general effects of environmental conditions. We consider next the resources used by photosynthetic organisms, and then we go on to examine the ways in which organisms

> resources, unlike conditions, are consumed

Penguins do not find the Antarctic in the least bit 'extreme.'

David Tipling/Getty Images

that are themselves resources have to be captured, grazed, or even inhabited before they are consumed. Finally we consider the ways in which organisms of the same species may compete with each other for limited resources.

What do we mean by 'harsh,' 'benign,' and 'extreme'?

It seems quite natural to describe environmental conditions as 'extreme,' 'harsh,' 'benign,' or 'stressful.' But these describe how we, human beings, feel about them. It may seem obvious when conditions are extreme: the midday heat of a desert, the cold of an Antarctic winter, the salt concentration of the Great Salt Lake. What this means, however, is only that these conditions are extreme *for us*, given our particular physiological characteristics and tolerances. But to a cactus there is nothing extreme about the desert conditions in which cacti have evolved; nor are the icy fastnesses of Antarctica an extreme environment for penguins. But a tropical rain forest *would* be a harsh environment for a penguin, though it is benign for a tropical bird such as a macaw; and a lake is a harsh environment for a cactus, though it is benign for a water hyacinth. There is, then, a relativity in the ways organisms respond to conditions; it is too easy and dangerous for the ecologist to assume that all other organisms sense the environment in the way we do. Emotive words like *harsh* and *benign*, even relativities such as *hot* and *cold*, should be used by ecologists only with care.

Effects of conditions

Temperature, relative humidity, and other physicochemical conditions induce a range of physiological responses in organisms, which determine whether the physical environment is habitable for them or not. There are three basic types of **response curve** (Figure 3.1). In the first, here illustrated by the response to temperature (Figure 3.1a), extreme conditions (either too hot or too cold) are lethal, but between the two extremes is a continuum of more favorable conditions. Different organisms will have different response curves, with for instance the optimum temperature for a penguin being considerably less than that for a macaw. For each species, an organism may be able to survive over a wide range of temperatures, but might grow actively only over a more restricted range and might reproduce only within an even narrower band. We can draw similar response curves for other conditions, such as pH, soil moisture, or water salinity. Note that although Figure 3.1a shows a symmetrical, bell-shaped distribution in response, this is by no means always the case.

In a second type of response curve, here illustrated by the response to a poison such as arsenic (Figure 3.1b), the condition may be lethal only at high intensities. When the poison is absent or present at low concentrations, the organism may be unaffected, but there is a threshold above which performance may decrease rapidly: first reproduction, then growth, and finally survival.

The third type (Figure 3.1c), shown here as a response to the amount of dissolved salts in water, applies to conditions that are required by organisms at low concentrations but become toxic at high concentrations. A freshwater mussel may require a little bit of salt and minerals in water in order to survive, but higher salinities again decrease performance and can be lethal if high enough.

Let's take a closer look at the response of organisms to temperature. For each 10°C rise in temperature, the rate of many biological processes such as respiration often roughly doubles, up to a certain point, and thus appears as an exponential curve on a plot of rate against temperature (Figure 3.2a). The increase is the result of the general tendency for chemical reactions to be faster at higher temperatures. However, at excessively high temperatures, biological processes will stop increasing in rate, and eventually will even fall or stop, due, for instance, to denaturing of proteins (as in cooking). Of greater importance from an ecological standpoint, effects on individual chemical reactions are often less important than effects on rates of growth or development or on final body size, since these tend to drive the core ecological activities of survival, reproduction, and movement (see Chapter 5). And when we plot rates of growth and development of whole organisms against the organism's temperature, we commonly observe an extended near-linear range (Figure 3.2b, c), again up to some level at which temperature becomes excessive.

> effectively linear effects of temperature on rates of growth and development

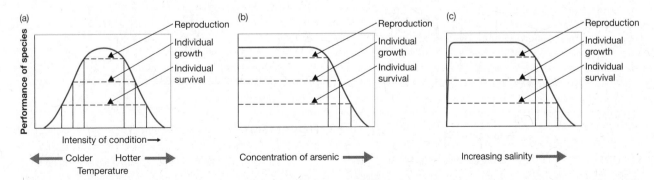

FIGURE 3.1 Response curves illustrating the effects of a range of environmental conditions on individual survival, growth, and reproduction. (a) Using temperate as an example condition, extreme conditions are lethal, less extreme conditions prevent growth, and only optimal conditions allow reproduction. (b) Response curve for a condition (arsenic poison, in this example) that has no effect at low doses but is lethal at high intensities; the reproduction–growth–survival sequence still applies. (c) Similar to (b), but the condition (salinity, or the amount of dissolved salts in water for this example) is required by organisms, as a resource, at low concentrations.

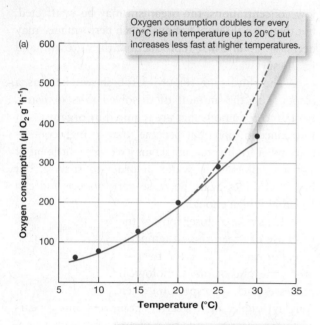

Oxygen consumption doubles for every 10°C rise in temperature up to 20°C but increases less fast at higher temperatures.

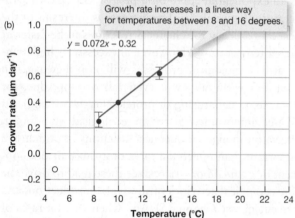

Growth rate increases in a linear way for temperatures between 8 and 16 degrees.

$y = 0.072x - 0.32$

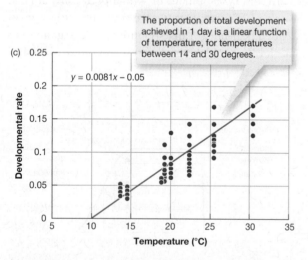

The proportion of total development achieved in 1 day is a linear function of temperature, for temperatures between 14 and 30 degrees.

$y = 0.0081x - 0.05$

FIGURE 3.2 (a) The rate of oxygen consumption of the Colorado beetle (*Leptinotarsa decemlineata*), which increases nonlinearly with temperature. (After Marzusch, 1952.) (b) Growth of the protist *Strombidinopsis multiauris*. The linear regression shown is highly significant. (After Montagnes et al., 2003.) (c) Egg-to-adult development in the mite *Amblyseius californicus*. The linear relationship is highly significant. (After Hart et al., 2002.)

At lower body temperatures (though 'lower' varies from species to species, as explained earlier), performance is likely to be impaired simply as a result of metabolic inactivity.

Together, rates of growth and development determine the final size of an organism. For instance, for a given rate of growth, a faster rate of development will lead to smaller final size. Hence, if the responses of growth and development to variations in temperature are not the same, temperature will also affect final size. In fact, development usually increases more rapidly with temperature than does growth, such that final size tends to decrease with rearing temperature for a very wide range of organisms (Figure 3.3).

These effects of temperature on growth, development, and size can be of practical importance in addition to academic interest. Increasingly, ecologists are called upon to predict. We may wish to know what the consequences would be, say, of a 2°C rise in global temperatures. We cannot afford to assume exponential relationships with temperature if they are actually linear, or to ignore the effects of changes in organism size on their role in ecological communities.

At extremely high temperatures, enzymes and other proteins become unstable and break down, and the organism dies. But difficulties may set in before these extremes are reached. At high temperatures, terrestrial organisms may be cooled by the evaporation of

temperature and final size

high and low temperatures

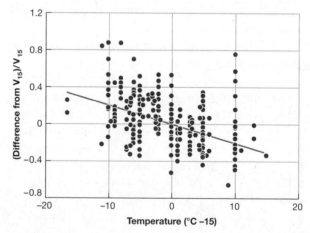

FIGURE 3.3 Final organism size decreases with increasing temperature, as illustrated in protists, single-celled organisms. Because the 72 data sets combined here were derived from studies carried out at a range of temperatures, both scales are 'standardized.' The horizontal scale measures temperature as a deviation from 15°C. The vertical scale measures size (cell volume, V) relative to the size at 15°C. The slope of the regression line is −0.025 (SE, 0.004; $P < 0.01$): cell volume decreased by 2.5% for every 1°C rise in rearing temperature. (After Atkinson et al., 2003.)

water (from open stomata on the surfaces of leaves, or through sweating), but this may lead to serious, perhaps lethal, problems of dehydration. Even where loss of water is not a problem, for example, among aquatic organisms, death is usually inevitable if temperatures are maintained for long above 60°C. The exceptions, **thermophiles**, are mostly specialized fungi, bacteria, and archaea. One of these, *Pyrodictium occultum*, can live at 105°C—something that is only possible because, under the pressure of the deep ocean where it lives, water does not boil at that temperature.

At body temperatures a few degrees above zero Celsius, organisms may be forced into extended periods of inactivity, and the cell membranes of sensitive species may begin to break down. This is known as **chilling injury**, which affects many tropical fruits. On the other hand, many species of both plants and animals can tolerate temperatures well below zero provided that ice does not form. If it is not disturbed, water can supercool to temperatures as low as −40°C without forming ice; but a sudden shock allows ice to form quite suddenly within plant cells, and this, rather than the low temperature itself, is then lethal, since ice formed within a cell is likely simply to disrupt and destroy it. If, however, temperatures fall slowly, ice can form between cells and draw water from within them. With dehydrated cells, the effects on plants are then very much like those of high temperature drought. Many animal species have adaptations to withstand body temperatures below zero: for instance, mussels of the species *Mytilis edulis*, when living in an intertidal area at low temperatures, produce antifreeze to prevent freezing when they are out in the air at low tide.

The absolute temperature that an organism experiences is important. But the timing and duration of temperature extremes may be equally important. For example, unusually hot days in early spring may interfere with fish spawning or kill the fry but otherwise leave the adults unaffected. Similarly, a late spring frost might kill seedlings but leave saplings and larger trees unaffected. The duration and frequency of extreme conditions are also often critical. In many cases, a periodic drought or tropical storm may have a greater effect on a species' distribution than the average level of a condition. To take just one example: the saguaro cactus is liable to be killed when temperatures remain below freezing for 36 hours, but if there is a daily thaw, it is under no threat. In Arizona, the northern and eastern edges of the cactus's distribution correspond to a line joining places where on occasional days it fails to thaw. Thus, the saguaro is absent where there are

| the timing of extremes |

The Saguaro cactus can only survive short periods at freezing temperatures.

James Randklev/Getty Images

occasionally lethal conditions—an individual need only be killed once.

Conditions as stimuli

Environmental conditions act primarily to modulate the rates of physiological processes. In addition, though, many conditions are important stimuli for growth and development and prepare an organism for conditions that are to come.

The idea that animals and plants in nature can anticipate, and be used by us to predict, future conditions ('a big crop of berries means a harsh winter to come') is the stuff of folklore. But there are important advantages to an organism that can predict and prepare for repeated events such as the seasons. For this, the organism needs an internal clock that can be used to check against an external signal. The most widely used external signal is the length of the period of daylight within the daily cycle—the **photoperiod**. On the approach of winter—as the photoperiod shortens—bears, cats, and many other mammals develop a thickened fur coat, birds such as ptarmigan put on winter plumage, and very many insects enter a dormant phase in which development is suspended and metabolic activity massively reduced, called **diapause**, within the normal activity of their life cycle. Temperate-zone insects may even slow down their development as daylength decreases in the fall (as

| photoperiod is commonly used to time dormancy, flowering, or migration |

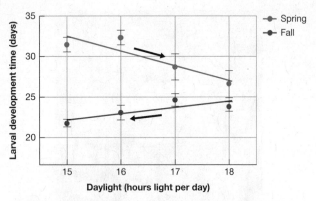

FIGURE 3.4 The effect of daylength on larval development time in the butterfly *Lasiommata maera* in the fall (third larval stage, before diapause) and spring. The arrows indicate the normal passage of time: The bars are standard errors. (After Gotthard et al., 1999.)

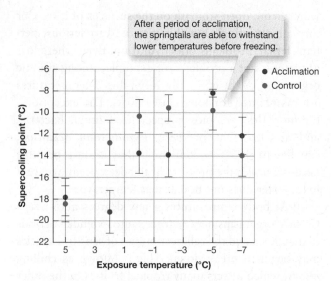

FIGURE 3.5 Acclimation to low temperatures. Samples of the Antarctic springtail *Cryptopygus antarcticus* were taken from field sites in the summer (ca. 5°C) on a number of days and their supercooling point (at which they froze) determined either immediately (controls, blue circles) or after a period of acclimation (red circles) at the temperatures shown. The supercooling points of the controls themselves varied because of temperature variations from day to day, but acclimation at temperatures in the range +2°C to −2°C (indicative of winter) led to a drop in the supercooling point, whereas no such drop was observed at higher temperatures (indicative of summer) or lower temperatures (too low for a physiological acclimation response). Bars are standard errors. (After Worland & Convey, 2001.)

harsh winter conditions approach), but then speed up development again in the spring as daylength increases, once the pressure is on to have reached the adult stage by the start of the breeding season (Figure 3.4). Other events with their timing controlled by the photoperiod are the seasonal onset of reproductive activity in animals, the onset of flowering, and seasonal migration in birds.

Many seeds need an experience of chilling before they will break dormancy. This prevents them from germinating during the moist, warm weather immediately after ripening and then being killed by the winter cold. As an example, temperature and photoperiod interact to control the seed germination of birch trees (*Betula pubescens*). Seeds that have not been chilled need an increasing photoperiod (indicative of spring) before they will germinate; but if the seed has been chilled, it starts growth without the light stimulus. Either way, growth should be stimulated only once winter has passed. The seeds of lodgepole pine, on the other hand, remain protected in their cones until they are heated by

forest fire. This stimulus is an indicator that the ground has been cleared and that new seedlings have a chance of becoming established.

Conditions may themselves trigger an altered response to the   acclimatization
same or even more extreme conditions: for instance, exposure to relatively low temperatures may lead to an increased rate of metabolism at such temperatures and/or to an increased tolerance of even lower temperatures. This is the process of **acclimatization** (called *acclimation* when induced in the laboratory). Antarctic springtails (tiny arthropods), for instance, when taken from 'summer' temperatures in the field (around 5°C in the Antarctic) and subjected to a range of acclimation temperatures, responded to temperatures in the range +2°C to −2°C (indicative of winter) by showing a marked drop in the temperature at which they froze (Figure 3.5); but at lower acclimation temperatures still (−5°C, −7°C), they showed no such drop because the temperatures were themselves too low for the physiological processes required to make the acclimation response. One way in which

NPS Photo by Jim Peaco

Seedlings of lodgepole pine growing up after a fire.

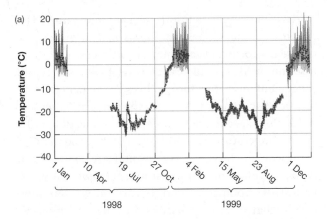

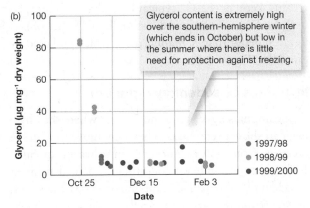

Glycerol content is extremely high over the southern-hemisphere winter (which ends in October) but low in the summer where there is little need for protection against freezing.

● 1997/98
● 1998/99
● 1999/2000

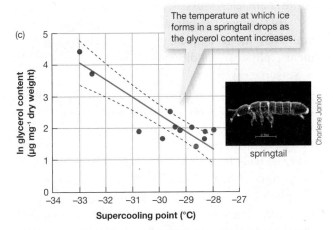

The temperature at which ice forms in a springtail drops as the glycerol content increases.

springtail

Charlene Janion

FIGURE 3.6   (a) Daily mean (points), maximum and minimum (tops and bottoms of lines, respectively) temperatures at Cape Bird, Ross Island, Antarctica. (b) Changes in the glycerol content of the springtail, *Gomphiocephalus hodgsoni*, from Cape Bird, which protect it from freezing. (c) Confirmation that the supercooling point (at which ice forms) drops in the springtail as glycerol concentration increases. (After Sinclair and Sjursen, 2001.)

such increased tolerance is achieved is by forming chemicals that act as antifreeze compounds: they prevent ice from forming within the cells and protect their membranes if ice does form (Figure 3.6). Acclimatization in some deciduous trees (frost hardening) can increase their tolerance of low temperatures by as much as 100°C.

The effects of conditions on interactions between organisms

Although organisms respond to each condition in their environment, the effects of conditions may be determined largely by the responses of other community members. Temperature, for example, does not act on one species alone: it also acts on its competitors, prey, parasites, and so on. Most especially, an organism will suffer if its food is another species that cannot tolerate an environmental condition. This result is illustrated by the distribution of the rush moth (*Coleophora alticolella*) in England. The moth lays its eggs on the flowers of the rush (*Juncus squarrosus*) and the caterpillars feed on the developing seeds. Above 600 m, the moths and caterpillars are little affected by the direct effects of low temperatures, but the rush, although it grows, fails to ripen its seeds. This, in turn, limits the distribution of the moth, because caterpillars that hatch in the colder elevations will starve as a result of insufficient food (Randall, 1982).

> conditions may affect the availability of a resource, . . .

The effects of conditions on disease may also be important. Conditions may favor the spread of infection (e.g., winds carrying fungal spores), or favor the growth of the parasite, or weaken or strengthen the defenses of the host. For example, fungal pathogens of grasshopper, *Camnula pellucida*, in the United States develop faster at warmer temperatures, but they fail to develop at all at temperatures around 38°C and higher (Figure 3.7a). Grasshoppers that regularly experience such temperatures effectively escape serious infection (Figure 3.7b), which they do by 'basking,' allowing solar radiation to raise their body temperatures by as much as 10–15°C above the air temperature around them (Figure 3.7c).

> . . . the development of disease . . .

Competition between species can also be profoundly influenced

> . . . or competition

The grasshopper, *Camnula pellucida*.

Joel Ledford

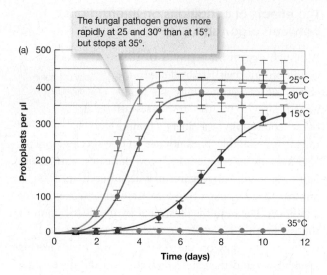

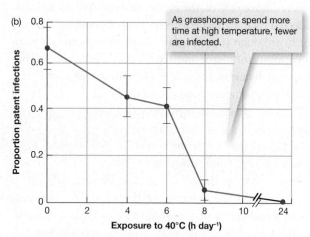

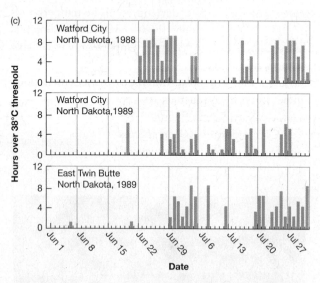

FIGURE 3.7 The effect of temperature on the interaction between the fungal pathogen, *Entomophaga grylli*, and the grasshopper, *Camnula pellucida*. (a) Growth curves over time of the pathogen (expressed as protoplasts per μl) at a range of temperatures. (b) The proportion of grasshoppers with patent (i.e., observable) infection with the pathogen drops sharply as grasshoppers spend more of their time at such higher temperatures. (c) Grasshoppers at two sites over 2 years did frequently raise their body temperatures to such high levels by basking. (After Carruthers et al., 1992.)

by environmental conditions such as temperature. Two stream salmonid fishes, *Salvelinus malma* and *S. leucomaenis*, coexist at intermediate altitudes (and therefore intermediate temperatures) on Hokkaido Island, Japan, but only the former lives at higher altitudes (lower temperatures) and only the latter at lower altitudes. Both species are quite capable, alone, of living at either temperature, and both species prefer cooler temperatures. However, in experimental streams at 6°C (a typical high-altitude temperature), *S. malma* proved to be a better competitor and its survival was far superior to that of *S. leucomaenis*. In other experimental streams kept at 12°C (typical low-altitude temperature), both species survived less well, but *S. leucomaenis* fared better than *S. malma*, all of which had died by 3 months (Figure 3.8).

Responses by sedentary organisms

Motile animals have some choice over where they live: they can show preferences. They may move into shade to escape from heat or into the sun to warm up. Such choice of environmental conditions is denied to fixed or sedentary organisms. Plants are obvious examples, but so are many aquatic invertebrates such as sponges, corals, barnacles, mussels, and oysters.

In temperate, boreal, arctic, and even many tropical environments, physical conditions follow a seasonal cycle, and organisms cope with this seasonality through a variety of mechanisms. Indeed, there has long been a fascination with organisms' responses to the seasons (Box 3.1). Morphological and physiological characteristics can never be ideal for all phases in the cycle, and the jack-of-all-trades is master of none. One solution is for the morphological and physiological characteristics to keep changing with the seasons (or even anticipating them, as in acclimatization). But change may be costly: a deciduous tree may have leaves ideal for life in summer but faces the cost of making new ones every year. An alternative is to economize by having long-lasting leaves like those of pines, heathers, and the perennial shrubs of deserts. Here, though, there is a cost to be paid in the form of more sluggish physiological processes. Different species have evolved different compromise solutions.

form and behavior may change with the seasons

Animal responses to environmental temperature

Like plants, algae, bacteria, and archaea, most species of animals are **ectotherms**: they rely on external sources

ectotherms and endotherms

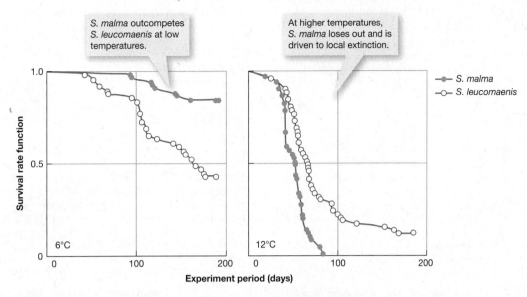

FIGURE 3.8   Changing temperature reverses the outcome of competition. At low temperature (6°C) on the left, the salmonid fish *Salvelinus malma* out-survives cohabiting *S. leucomaenis*, whereas at 12°C, right, *S. leucomaenis* drives *S. malma* to extinction. Both species are quite capable, alone, of living at either temperature. (After Taniguchi & Nakano, 2000.)

3.1   Historical Landmarks

Recording seasonal changes

Phenology, or the recording of the changing behavior of organisms through the season, was essential before agricultural activities could be intelligently timed. The earliest phenological records may have been the Wu Hou observations made in the Chou and Ch'in (1027–206 B.C.) dynasties. The date of the first flowering of cherry trees has been recorded at Kyoto, Japan, since A.D. 812.

A particularly long and detailed record was started in 1736 by Robert Marsham at his estate near the city of Norwich, England. He called these records 'Indications of the spring.' Recording was continued by his descendants until 1947. Marsham recorded 27 phenological events every year: the first flowering of snowdrop, wood anemone, hawthorn, and turnip; the first leaf emergence of 13 species of tree; and various animal events such as the first appearance of migrants (swallow, cuckoo, nightingale), the first nest building by rooks, croaking of frogs and toads, and the appearance of the brimstone butterfly.

Long series of measurements of environmental temperature are not available for comparison with the whole period of Marsham's records, but they are available from 1771 for Greenwich, about 160 km away. There is surprisingly close agreement between many of the flowering and leaf emergence events at Marsham and the mean January–May temperature at Greenwich (Figure 3.9). However, not surprisingly, events such as the time of arrival of migrant birds bears little relationship to temperature.

Analysis of the Marsham data for the emergence of leaves on six species of tree indicates that the mean date of leafing is advanced by 7.2 days for every 1°C increase in the mean temperature from February to May (Figure 3.10). Similarly, for the eastern United States, Hopkins's bioclimatic law states that the indicators of spring such as leafing and flowering occur 4 days later for every 1° latitude northward, 5° longitude westward or 120 m of altitude gain.

Collecting phenological records has now been transformed from the pursuit of gifted amateurs to sophisticated programs of data collection and analysis. At least 1,500 phenological observation posts are now maintained in Japan alone. The vast accumulations of data have suddenly become exciting and relevant as we try to estimate the changes in floras and faunas that will be caused by global warming.

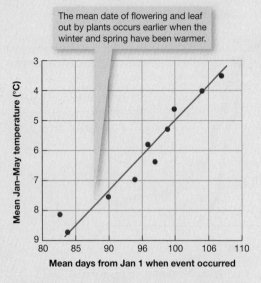

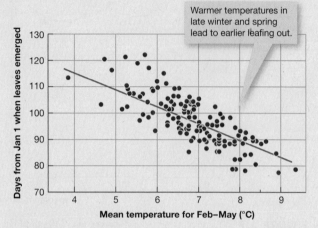

FIGURE 3.9 The relationships between mean January–May temperatures and the annual mean dates of 10 flowering and leafing events from the classic Marsham records started in 1736. (From Redrawn Figures of Margary, in Ford, 1982.)

FIGURE 3.10 The relationship between the mean temperature in the 4-month period, February–May, and the average date of six leafing events. The correlation coefficient is –0.81. (From Redrawn Figures of Kington, in Ford, 1982.)

of heat to determine the pace of their metabolism. This includes the invertebrates, amphibians, reptiles, and most fish. Others, mainly birds and mammals, are **endotherms**: they regulate their body temperature by producing heat within their body.

The distinction between ectotherms and endotherms is not absolute. Some insects, for example, can control body temperature through muscle activities (e.g., shivering flight muscles). Some reptiles can generate heat for limited periods of time; some fish such as tuna use countercurrent blood circulation to minimize their heat loss through their gills and maintain temperatures well above that of the ambient water; and even some plants can use metabolic activity to raise the temperature of their flowers. Some typical endotherms, on the other hand, such as dormice, hedgehogs, and bats, allow their body temperature to fall and become scarcely different from that of their surroundings when they are hibernating (Figure 3.11).

Despite these overlaps, endothermy is inherently a different strategy from ectothermy. Over a certain narrow temperature range, an endotherm consumes energy at a basal rate. But at environmental temperatures further above or below that zone, endotherms expend increasingly more energy to maintain their constant body temperature. This makes them

relatively independent of environmental conditions and allows them to stay longer at or close to peak performance. It makes them more effective in both searching for food and escaping from predators. However, there is a cost—a high requirement for food to fuel this strategy.

The idea that organisms are harmed (and limited in their distributions) not 'directly' by environmental conditions, but because of the energetic costs required to tolerate those conditions, is illustrated by a study examining the effect of a different condition: salinity. Two freshwater shrimp species, *Palaemonetes pugio* and *P. vulgaris*, co-occur in estuaries on the eastern coast of the United States at a wide range of salinities, but the former is more tolerant of lower salinities than the latter, occupying some habitats from which the latter is absent. Figure 3.12 shows the mechanism likely to be underlying this. Over the low salinity range (though not at the effectively lethal lowest salinity), metabolic expenditure was significantly lower in *P. pugio*. *P. vulgaris* requires far more energy simply to maintain itself at this lower salinity, putting it at a severe disadvantage in competition with *P. pugio*.

Endotherms have morphological modifications that reduce their energetic costs. In cold climates most

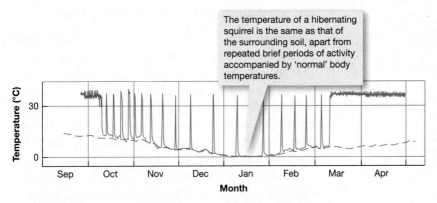

The temperature of a hibernating squirrel is the same as that of the surrounding soil, apart from repeated brief periods of activity accompanied by 'normal' body temperatures.

FIGURE 3.11 Changes in the body temperature over the 1996/97 winter of the European ground squirrel, *Spermophilus citellus* (solid, line) compared to ambient soil temperature (dotted line) at the same depth at which it was hibernating. (After Hut et al., 2002.)

Reinhard Hölzl/Imageb/Agefotostock

European ground squirrel.

fur (polar bears, mink, foxes) or feathers and extra layers of fat. In contrast, desert endotherms often have thin fur, and long ears and limbs, which help dissipate heat.

Variability of conditions can be as great a biological challenge as are extremes. Seasonal cycles, for example, can expose an animal to summer heat close to its thermal maximum, and winter chill close to its thermal minimum. Responses to these changing conditions include the laying down of different coats in the fall (thick and underlain by a thick fat layer) and in the spring (a thinner coat and loss of the dense fat layer) (Figure 3.13). Some animals also take advantage of each other's body heat as a means to cope with cold weather by huddling together. Hibernation—relaxing temperature control—allows some vertebrates to survive periods of winter cold and food shortage (see Figure 3.11) by avoiding

> temperatures that vary seasonally pose special problems

have low surface area to volume ratios (short ears and limbs), which reduces heat loss through surfaces. Typically, endotherms that live in polar environments are insulated from the cold with extremely dense

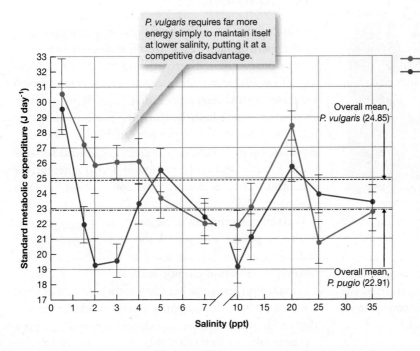

P. vulgaris requires far more energy simply to maintain itself at lower salinity, putting it at a competitive disadvantage.

Overall mean, *P. vulgaris* (24.85)

Overall mean, *P. pugio* (22.91)

FIGURE 3.12 Standard metabolic expenditure (estimated through minimum oxygen consumption) in two species of shrimp, *Palaemonetes pugio* and *P. vulgaris*, at a range of salinities. There was significant mortality of both species over the experimental period at 0.5 ppt (parts per thousand), especially *P. vulgaris* (75% compared to 25%). (After Rowe, 2002.)

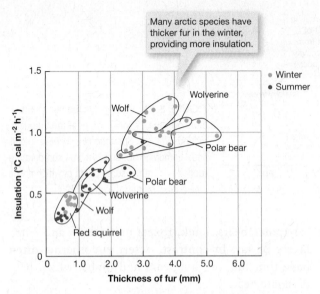

Many arctic species have thicker fur in the winter, providing more insulation.

FIGURE 3.13 Seasonal changes in the thickness of the insulating fur coats of some Arctic and northern temperate mammals.

W K Fletcher/Getty Images, Inc.

Corinna Stoeffl, Stoeffl Photography/ Getty Images, Inc.

The thick, white winter coat and the thinner, browner summer coat of the Arctic fox.

the difficulties of finding sufficient fuel over these periods. Migration is another avoidance strategy: the Arctic tern, to take an extreme example, travels from the Arctic to the Antarctic and back each year, never experiencing polar winters.

Microorganisms in extreme environments

Microorganisms survive and grow in all the environments that are lived in or tolerated by plants and animals, plus some that are too extreme for plants and animals, and they use many of the same strategies—avoid, tolerate, or specialize. Many microorganisms produce resting spores that survive drought, high temperature, or cold. Temperatures maintained higher than 45°C are lethal to almost all plants and animals and many bacteria and archaea, but **thermophilic** or temperature-loving microbes grow at much higher temperatures. Although similar in many ways to heat-intolerant microbes, the enzymes of these thermophiles are stabilized by especially strong ionic bonds.

Microbial communities that not only tolerate but grow at low temperatures are also known; these include diatoms and other photosynthetic algae as well as bacteria that have been found on Antarctic sea ice. Another form of microbial specialists are **acidophiles**, which thrive in environments that are highly acidic. One of them, *Thiobacillus ferroxidans*, is found in the waste from industrial metal-leaching processes and tolerates the extreme acidity of pH 1.0. At the other end of the pH spectrum, some cyanobacteria in highly alkaline soda lakes can grow at pH 13. Such life forms warn us against being too narrow-minded when we consider the kind of organism we might look for on other planets.

3.2 RESOURCES FOR PHOTOSYNTHETIC ORGANISMS

Resources may be either biotic or abiotic components of the environment: they are whatever an organism uses or consumes in its growth and maintenance, leaving less available for other organisms. When a photosynthesizing leaf intercepts radiation, it deprives some of the leaves or plants beneath it. When a caterpillar eats a leaf, less leaf material is available for other caterpillars. By their nature, resources are critical for survival, growth, and reproduction and also inherently a potential source of conflict and competition between organisms.

If an organism can move about, it has the potential to search for its food. Organisms that are fixed and 'rooted' in position cannot search. | resource requirements of nonmotile organisms

They must rely on growing toward their resources (like a shoot or root) or catching resources that move to them. Plants are an obvious example, depending upon: (i) energy that radiates to them; (ii) atmospheric carbon dioxide that diffuses to them; (iii) mineral cations that they obtain from soil colloids in exchange for hydrogen ions; and (iv) water and dissolved anions that the roots absorb from the soil. In the following sections, we concentrate on plants. But remember that many nonmobile animals, like corals, sponges, and bivalve mollusks, depend on resources that are suspended in the watery environment and are captured by filtering the water around them.

Solar radiation

Solar radiation is an essential resource for plants and other photosynthetic organisms, such as algae and cyanobacteria. We often refer to it loosely as 'light,' but plants actually use only about 44% of that narrow part of the spectrum of solar radiation that is visible to us between infrared and ultraviolet. In fact, leaves reflect rather than absorb much of the green wavelengths of light, which is why we see them as green! The rate of photosynthesis increases with the intensity of the radiation that a leaf receives, but with diminishing returns; and this relationship itself varies greatly between species (Figure 3.14), especially between those that usually live in shaded habitats (which reach saturation at low radiation intensities) and those that normally experience full sunlight and can take advantage of it. At high intensities, **photoinhibition** of photosynthesis may occur, such that the rate of fixation of carbon decreases with increasing radiation intensity.

sun and shade species

High intensities of radiation may also lead to dangerous overheating of plants. Radiation is an essential resource for plants, algae, and cyanobacteria, but they can have too much as well as too little.

The solar radiation that reaches a plant or algal cell is forever changing. Its angle and intensity change in a regular and systematic way annually, diurnally and with depth within the canopy or in a water body (Figure 3.15). In aquatic ecosystems, the reduction of light with depth varies greatly depending upon such things as the amount of algae and suspended clay in the water. There are also irregular, unsystematic variations due to changes in cloud cover or shadowing by the leaves of neighboring plants. As light flecks pass over leaves lower in the canopy, they receive seconds or minutes of direct bright light and then plunge back into shade. As algae and cyanobacteria are mixed up and down through the surface waters of a lake or ocean, they constantly see a different amount of light radiation. The daily photosynthesis of a leaf or algal cell integrates these various experiences. A whole plant integrates the diverse exposure of its various leaves.

There is enormous variation in the shapes and sizes of leaves. Most of the heritable variation in shape has probably evolved under selection not primarily for high photosynthesis, but rather for factors such as optimal efficiency of water use (photosynthesis achieved per unit of water transpired), minimization of the damage done by foraging animals, and reducing the likelihood of overheating. Not all the variations in leaf shape are heritable, though: many are responses by the individual to its immediate environment. Many trees, especially, produce different types of leaf in positions exposed to full sunlight ('sun leaves') and in places lower in the canopy where they are shaded ('shade leaves'). Sun leaves are thicker, with more densely packed chloroplasts (which process the incoming radiation) within cells and more cell layers. The more flimsy shade leaves intercept diffused and filtered radiation low in the canopy but may nonetheless supplement the main photosynthetic activity of the sun leaves high in the canopy.

sun and shade leaves

Among herbaceous plants and shrubs, specialist 'sun' or 'shade' *species* are much more common. Leaves of sun plants are commonly exposed at acute angles to the midday sun and are typically superimposed into a multilayered canopy, where even the lower leaves may have a positive rate of net photosynthesis. The leaves of shade plants are typically arranged in a single-layered canopy and angled horizontally, maximizing their ability fully to capture the available radiation.

sun and shade plants

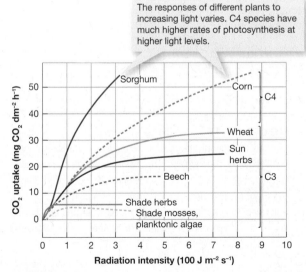

The responses of different plants to increasing light varies. C4 species have much higher rates of photosynthesis at higher light levels.

FIGURE 3.14 The response of photosynthesis by the leaves of various types of green plant (measured as carbon dioxide uptake) to the intensity of solar radiation at optimal temperatures and with a natural supply of carbon dioxide. (The different physiologies of C3 and C4 plants are explained later in this Section) (After Larcher, 1980, and Other Sources.)

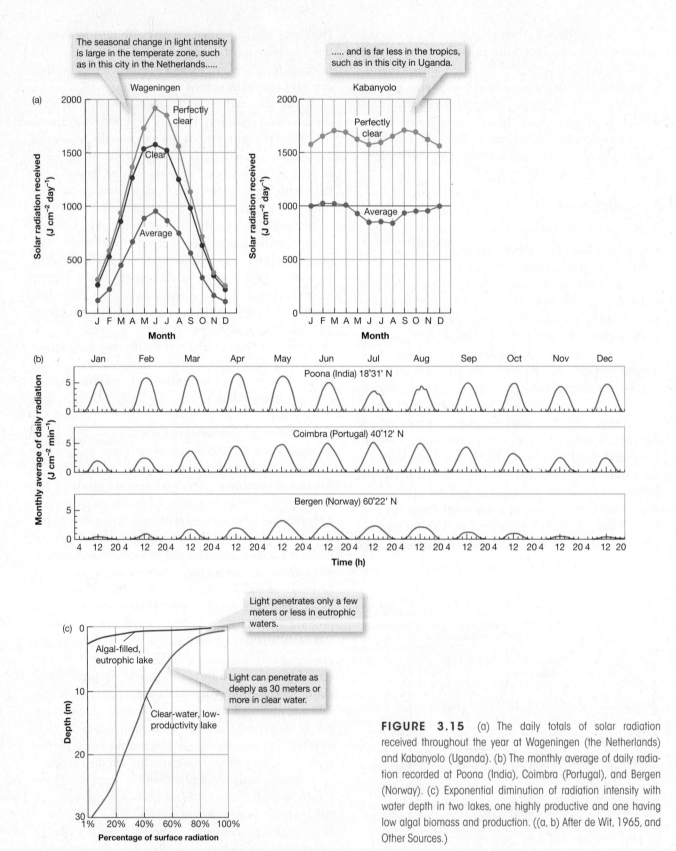

FIGURE 3.15 (a) The daily totals of solar radiation received throughout the year at Wageningen (the Netherlands) and Kabanyolo (Uganda). (b) The monthly average of daily radiation recorded at Poona (India), Coimbra (Portugal), and Bergen (Norway). (c) Exponential diminution of radiation intensity with water depth in two lakes, one highly productive and one having low algal biomass and production. ((a, b) After de Wit, 1965, and Other Sources.)

Other species develop as sun or shade plants, depending on where they grow. One such is the ever-green shrub, *Heteromeles arbutifolia*, which grows both in chaparral habitats in California, where shoots in the upper crown are regularly exposed to full sun-light and high temperatures, and also in shaded wood-land habitats, where it receives around one-seventh as much radiation (Figure 3.16). The leaves of sun plants are thicker and have a greater photosynthetic capacity (more chlorophyll and nitrogen) per unit leaf area than those of shade plants (Figure 3.16b). Sun-plant leaves are inclined at a much steeper angle to the horizontal, and therefore absorb the direct rays of the overhead summer sun over a wider leaf area than the more horizontal shade-plant leaves. The more angled leaves of sun plants, though, are also less likely than shade-plant leaves to shade other leaves

of the same plant from the overhead rays of the summer sun (Figure 3.16c). But in winter, when the sun is much lower in the sky, it is the shade plants that are much less subject to this self-shading. As a result, the proportion of incident radiation intercepted per unit area of leaf is higher in shade than in sun plants year round, in summer because of the more horizontal leaves and in winter because of the relative absence of self-shading.

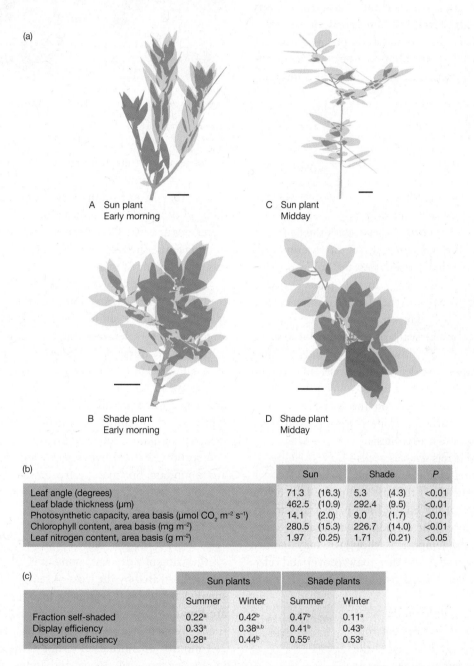

(a)

A Sun plant
   Early morning

C Sun plant
   Midday

B Shade plant
   Early morning

D Shade plant
   Midday

(b)

	Sun		Shade		P
Leaf angle (degrees)	71.3	(16.3)	5.3	(4.3)	<0.01
Leaf blade thickness (µm)	462.5	(10.9)	292.4	(9.5)	<0.01
Photosynthetic capacity, area basis (µmol CO$_2$ m^{-2} s^{-1})	14.1	(2.0)	9.0	(1.7)	<0.01
Chlorophyll content, area basis (mg m^{-2})	280.5	(15.3)	226.7	(14.0)	<0.01
Leaf nitrogen content, area basis (g m^{-2})	1.97	(0.25)	1.71	(0.21)	<0.05

(c)

	Sun plants		Shade plants	
	Summer	Winter	Summer	Winter
Fraction self-shaded	0.22[a]	0.42[b]	0.47[b]	0.11[a]
Display efficiency	0.33[a]	0.38[a,b]	0.41[b]	0.43[b]
Absorption efficiency	0.28[a]	0.44[b]	0.55[c]	0.53[c]

FIGURE 3.16   (a) Computer reconstructions of stems of typical sun (A, C) and shade (B, D) plants of the evergreen shrub *Heteromeles arbutifolia*, viewed along the path of the sun's rays in the early morning (A, B) and at midday (C, D). Darker tones represent parts of leaves shaded by other leaves of the same plant. Bars = 4 cm. (b) Observed differences in the leaves of sun and shade plants. Standard deviations are given in parentheses; the significance of differences are given following analysis of variance. (c) Consequent whole-plant properties of sun and shade plants. Letter codes indicate groups that differed significantly in analyses of variance ($P < 0.05$). (After Valladares & Pearcy, 1998.)

The properties of whole plants of *H. arbutifolia*, then, reflect both plant architecture and the morphologies and physiologies of individual leaves. The efficiency of light absorption per unit of biomass is massively greater for shade than for sun plants (Figure 3.16c). Despite receiving only one-seventh of the radiation of sun plants, shade plants reduce the differential in their daily rate of carbon gain from photosynthesis to only a half. They successfully counterbalance their reduced photosynthetic capacity at the leaf level with enhanced light-harvesting ability at the whole-plant level. The sun plants, on the other hand, can be seen as striking a compromise between maximizing whole-plant photosynthesis while avoiding photoinhibition and overheating of individual leaves.

A wilted plant.

Dr. Jeremy Burgess/Science Source

Water

Most plant parts are largely composed of water. In some soft leaves and fruits, as much as 98% of the volume may be water. Some water is also consumed in photosynthesis, with the hydrogen incorporated into organic matter. Yet this is a tiny fraction of the water that passes from the soil through a plant to the atmosphere during plant growth. Photosynthesis depends on the plant absorbing carbon dioxide. This can only happen across surfaces that are wet—most notably the walls of the photosynthesizing cells in leaves. If a leaf allows carbon dioxide to enter, some water vapor also leaves. Likewise, any mechanism that slows down the rate of water loss, such as closing the stomata (pores) on the leaf surface, will reduce the rate of carbon dioxide absorption and hence reduce the rate of photosynthesis.

water is lost from plants that photosynthesize

Terrestrial plants serve as wicks, conducting water from the soil and releasing it to the air. If the rate of uptake falls below the rate of release, the body of the plant starts to dry out. The cells lose their turgidity and the plant wilts. This may just be temporary (though it may happen every day in summer), and they may recover and rehydrate at night. But if the deficit accumulates, the plant may die.

wilting

Plant species differ in the ways in which they survive in dry environments. One strategy is to avoid the problems. **Avoiders** such as desert annuals, annual weeds, and most crop plants have a short lifespan: their photosynthetic activity is concentrated during periods when water is relatively available. For the remainder of the year, they remain dormant as

plant life in water deficit: avoiders and tolerators

seeds, a stage that requires neither photosynthesis nor transpiration. Some perennial plants, including many trees in seasonally dry tropical forests, shed their photosynthetic tissues during periods of low water availability. Other species may have a strategy to replace the shed leaves with new leaf forms that are less extravagant of water.

Other plants, **tolerators**, have evolved a different compromise, producing long-lived leaves that transpire slowly (for example, by having few and sunken stomata). They tolerate drought and low-water conditions, but of course their highest rate of photosynthesis is slower. These plants have sacrificed their ability to achieve rapid photosynthesis when water is abundant but gained the insurance of being able to photosynthesize throughout the seasons. This is not only a property of plants from arid areas but also of the pines and spruces that survive where water may be seasonally abundant but may be frozen and therefore inaccessible in winter, or low during the hottest summer periods.

The range of alternative strategies to solve the problem of photosynthesizing in a dry environment is nicely illustrated by the trees of seasonally dry tropical regions, which are found widely on many continents. **Deciduous** tree species, which lose all leaves for at least 1 and usually 2–4 months or more each year, are the most common in the savannas of Africa and India. The Llanos of South America are dominated by evergreens (a full canopy all year). And the savannas of Australia have roughly equal numbers of deciduous and evergreen species (Figure 3.17). There, the deciduous species

coexisting alternative strategies in Australian savannas

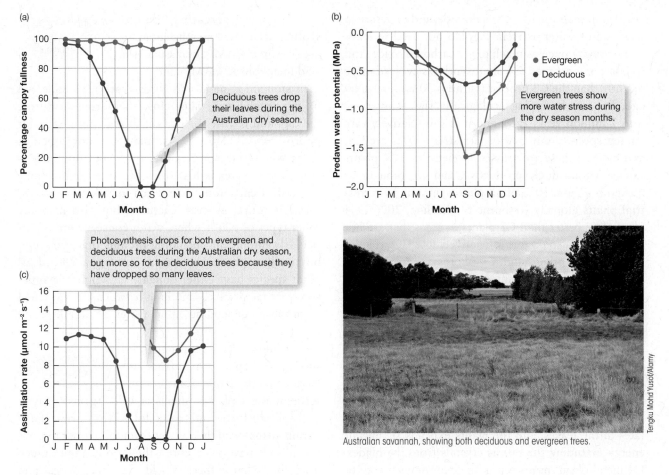

FIGURE 3.17   (a) Percentage canopy fullness for deciduous and evergreen trees in Australian savannas through the year. (b) Susceptibility to drought as measured by increasingly negative values of water potential just before dawn for deciduous and evergreen trees. (c) Net photosynthesis as measured by the carbon assimilation rate for deciduous and evergreen trees.

avoid drought in the dry season (April–November in Australia) by shedding their leaves to essentially stop water loss through transpiration (Figure 3.17a, b), whereas the evergreens tolerate the threat of drought in the dry season and continue to photosynthesize, albeit at a lower rate (Figure 3.17c).

The evaporation of water lowers the temperature of the body with which it is in contact. For this reason, if plants are prevented from transpiring, they may overheat. This, rather than water loss itself, may be lethal. The desert honeysweet (*Tidestromia oblongifolia*) grows vigorously in Death Valley, California, despite the fact that its leaves are killed if they reach 50°C, a temperature that is commonly reached in the surrounding air. Transpiration cools the surface of the leaf to a tolerable 40–45°C. Most desert plants bear hairs, spines, and waxes on the leaf surface. These reflect a high proportion of incident radiation and help to prevent overheating. Other more general modifications in desert plants include the characteristic 'chunky'

water and overheating

shape of succulents with few branches, giving a low surface area to volume ratio over which radiant heat is absorbed.

Specialized biochemical processes may increase the amount of photosynthesis that can be achieved per unit of water lost. The majority of plants on Earth, and all algae and photosynthetic bacteria, photosynthesize using what is termed the **C3 pathway**: their first product of photosynthesis is a sugar containing three carbon atoms. Although these plants are highly productive photosynthesizers, they are relatively wasteful of water, reach their maximum rates of photosynthesis at relatively low intensities of radiation, and are less successful in arid areas. Of course, this is no problem for photosynthetic organisms in lakes and oceans, where water availability is never an issue!

increasing the efficiency of water use: the C4 approach

One alternative approach to photosynthesis, the **C4 pathway**, produces a four-carbon sugar as the initial product. Plants with the C4 pathway are far more efficient in their use of carbon dioxide, and therefore in their use of water as well: they do not need to leave their stomata open as much to get carbon dioxide, and so less water escapes. C4 photosynthesis evolved more recently than the C3 process (perhaps 30 million years ago, but not becoming ecologically significant until 6 to 7 million years ago), and it is still relatively rare among species: only 3% of terrestrial plants are C4, and most of these are grasses. Nonetheless, C4 plants can be astonishingly productive, and are thought to make up almost 30% of all photosynthesis by terrestrial plants globally (Osborne & Beerling, 2006). C4 plants fare better than C3 plants in areas that are hot and dry, and also do fine at relatively low CO_2 (Figure 3.18). On the other hand, C4 photosynthesis is inefficient at low radiation intensities, making C4 plants poor shade plants. Several globally important agricultural crops that do well in hot and dry conditions, such as corn, sugar cane, and sweet sorghum, are C4 plants. It is interesting to note that the dominant grass in North America salt marshes, *Spartina alterniflora*, is a C4 grass. Even though the roots of these grasses are always submerged, they are submerged in seawater. During transpiration, the salt of the seawater accumulates in the leaves, and the plants expends significant energy extruding the salt as crystals from the blades. C4 metabolism makes sense for *Spartina* as a way to conserve the water that is transpired.

Another approach to photosynthesis is even more efficient in its use of water: **CAM**, or **crassulacean acid metabolism**. CAM plants open

> when water is really scarce: the CAM approach

their stomata at night and absorb carbon dioxide and fix it as malic acid. They close their stomata during the day and release the carbon dioxide internally for photosynthesis. Since the stomata are only open at night when temperatures are lower and humidity is higher, less water is lost from the leaf for the same amount of carbon dioxide taken in. CAM photosynthesis is found in most cactuses, but also in many other types of plants where water conservation is particularly important, including orchids that grow epiphytically without roots on trees. Some 7% of all plant species have either obligate CAM photosynthesis or are facultative, switching between C3 and CAM metabolism depending upon water availability and life stage.

Almost all falling water (rain, snow, etc.) bypasses the plants and passes to the soil. Some drains

> obtaining water from the soil

through the soil, but much is held against gravity by capillary forces and as colloids. Terrestrial plants obtain virtually all their water from this stored reserve. Sandy soils have wide pores: these do not hold much water but what is there is held with weak forces and plants can withdraw it easily. Clay soils have very fine pores. They retain more water against the force of gravity, but surface tension in the fine pores makes it more difficult for the plants to withdraw it. The

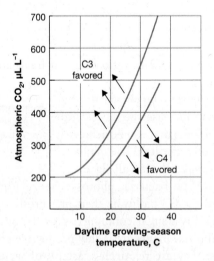

FIGURE 3.18 C4 plants are favored under conditions of high temperatures (and associated dryness) and low carbon dioxide. C3 plants, on the other hand, are favored by lower temperatures (and greater moisture) and high concentrations of carbon dioxide, as they are less efficient at using carbon dioxide and lose more water from their stomata for the same rate of photosynthesis. In the intermediate temperature and carbon dioxide ranges, both C3 and C4 plants do well. (Adapted from Ehleringer et al., 2002.)

Spartina marsh.

Paul Edmondson/Getty Images

primary water-absorbing zone on roots is covered with root hairs that make intimate contact with soil particles. As water is withdrawn from the soil, the first to be released is from the wider pores, where capillary forces retain it only weakly. Subsequent water is withdrawn from narrower paths in which the water is more tightly held. Consequently, the more the soil around the roots is depleted of water, the more resistance there is to water flow. As a result of water withdrawal, roots create water depletion zones (or, more generally, resource depletion zones or RDZs) around themselves. The faster the roots draw water from the soil, the more sharply defined the RDZs, and the more slowly water will move into that zone. In a soil that contains abundant water, rapidly transpiring plants may still wilt because water does not flow fast enough to replenish the RDZs around their root systems (or because the roots cannot explore new soil volumes fast enough).

The shapes of root systems are much less tightly programmed than those of shoots. The root architecture that a plant establishes early in its life can determine its responsiveness to later events. Plants that develop under waterlogged conditions usually set down only a superficial root system. If drought develops later in the season, these same plants can suffer because their root system did not tap the deeper soil layers. A deep tap root, however, will be of little use to a plant in which most water is received from occasional showers on a dry substrate. Figure 3.19 illustrates some characteristic differences between the root systems of plants from damp temperate and dry desert habitats.

Nutrients

Roots extract water from the soil, but they also extract key nutrients such as nitrogen (N), phosphorus (P), sulfur (S), potassium (K), calcium (Ca), magnesium (Mg), and iron (Fe), together with traces of manganese (Mn), zinc (Zn), copper (Cu), and boron (B). As discussed in Chapters 4 and 11, nitrogen and phosphorus are particularly important in regulating rates of photosynthesis and in structuring ecological communities (and human activity has greatly altered the availability of these nutrients across the globe; even more, indeed, than we have increased carbon dioxide in the atmosphere; see Chapter 12). All nutrients are obtained from the soil (or directly from the water in the case of free-living aquatic plants, algae, and cyanobacteria). Soils are patchy and heterogeneous, and as roots grow through them, they may meet regions that vary in nutrient and water content. They tend to branch profusely in the richer patches.

Root architecture is important in this process, because different nutrients behave differently and are held in the soil by different forces. Nitrate ions, the major form of nitrogen for many plants, diffuse readily in soil water. However, other key nutrients such as phosphate (the major form of phosphorus for plant uptake) and ammonium (another form of nitrogen often used by plants) are tightly bound in the soil, often through adsorption to mineral surfaces or in clay lattice structures. The phosphate RDZs of two roots 0.2 mm apart scarcely overlap, and the parts of a finely branched root system scarcely compete with each other. Consequently, if phosphate is in short supply, a highly branched surface root will greatly improve phosphate absorption. A more widely spaced extensive root system, in contrast, will tend to maximize access to nitrate.

the architecture of roots determines their foraging efficiency

Algal cells have no roots, and they assimilate nitrogen and phosphorus right across their cell surfaces. In both lake and ocean water, nitrate, ammonium, and phosphate all diffuse to these cell surfaces at similar rates, since there is no soil to bind the nutrients. Architecture is quite important, though. In ocean waters where nutrient concentrations are very low, very small algal species dominate: smaller cells have a greater surface area compared to their volume, and so the small size maximizes the rate of nutrient uptake across the surface area of the cell per volume of cell. In aquatic ecosystems where nutrient concentrations are higher, species with larger cells dominate. These have more chlorophyll per volume of cell, and can therefore photosynthesize and grow more rapidly, and less surface area per volume is necessary for sufficient nutrient uptake.

architecture matters to algae too

Carbon dioxide

Terrestrial plants take in carbon dioxide through the stomatal pores on leaf surfaces and, using the energy of sunlight, capture the carbon atoms during photosynthesis and release oxygen (submerged aquatic plants and algae take the carbon dioxide up by diffusion through their cell surfaces). Carbon dioxide varies in its concentration at a variety of scales. Globally, the concentration in the atmosphere has been increasing over time. In 1750, atmospheric carbon dioxide concentrations were approximately 280 μl l^{-1} (parts per

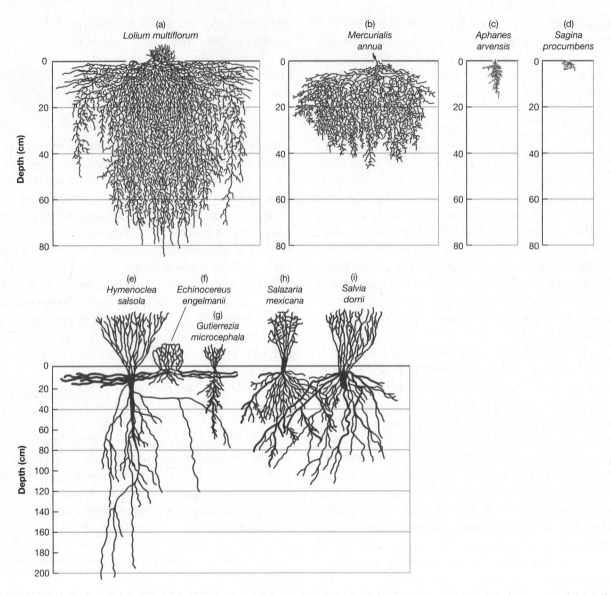

FIGURE 3.19 Profiles of root systems of plants from contrasting environments. (a–d) Northern temperate species of open ground: (a) *Lolium multiflorum*, an annual grass; (b) *Mercurialis annua*, an annual weed; and (c) *Aphanes arvensis* and (d) *Sagina procumbens*, both ephemeral weeds. (e–i) Desert shrub and semishrub species, Mid Hills, eastern Mojave Desert, California. ((a–d) From Fitter, 1991; (e–i) Redrawn from a variety of sources.)

million, or ppm). Since the advent of the Industrial Revolution, the concentration has grown exponentially, driven both by the carbon lost to the atmosphere as forests are cleared and even more so by the burning of fossil fuels. By early 2014, the figure had reached $400 \ \mu l \ l^{-1}$ and is increasing by 0.5% per year (Box 3.2). While the carbon dioxide concentrations were much higher yet further back in the geological past, the current levels are the highest seen in many millions of years. In fact, C4 plants probably evolved in response to the lowering carbon dioxide concentrations brought about by the storage in past geological eras of organic material in what eventually became fossil fuels.

Concentrations can also vary in space and over short time scales. At the global scale, the carbon dioxide concentrations change with the seasons, with a change as small as a few percent at the Mauna Loa observatory, high on a mountain in Hawaii (Figure 3.21), but much more over the major land masses. The change is driven by a seasonal uptake by photosynthesis and storage of carbon by terrestrial ecosystems in the Northern Hemisphere during summer, followed by release as dead plant leaves and other materials are respired away in the winter (the terrestrial ecosystems of the Southern Hemisphere have less influence simply because most of the land mass on Earth is in the Northern Hemisphere).

3.2   ECOncerns

Global warming: can we risk it?

Carbon dioxide is one of several global 'greenhouse gases,' the increasing concentrations of which are leading to rises in global mean temperatures, to a growth in the number of 'extreme' and 'record' weather events, and to the threat of the major biomes of Earth substantially changing their distribution.

The Intergovernmental Panel on Climate Change (IPCC) was established in 1988 by the World Meteorological Organization (WMO) and the United Nations Environment Programme (UNEP) and issues periodic reports on the current scientific understanding of the global climate system, providing estimates of future change and highlighting areas of uncertainty. The most recent synthesis report from the IPCC (2013) reflects the consensus work of over 3,000 expert scientists from over 130 countries. The report concludes that the Earth has indeed been warming over the past several decades (> 99% probability), that this warming is very likely a result of the increase of carbon dioxide and other greenhouse gases in the atmosphere (> 90% probability), and that continued release of greenhouse gases is very likely to result in further warming well beyond that already observed (> 90% probability). On average, the planet is now 0.7°C warmer than in the early 20th century, and the current rate of warming is probably the greatest the Earth has ever experienced. If current trends continue, the average temperature is expected to rise to 1.5°C above the early 20th-century baseline by 2030, and to 2°C above by 2045, with continued increases beyond then (Figure 3.20). At this level of global warming, many models predict severe consequences, including more extremes and much greater variability in weather, possible decreases in the global capacity to grow food, and the possibility of passing one or more tipping points in the climate system that could result in even more rapid warming (with less ability for humans to control it).

Policy-makers and law-makers are faced with different groups of scientific 'experts' offering different projections into the future, and with many interest groups, including a number of industries, resisting attempts to force them to change their behavior in order to reduce emissions of greenhouse gases. Even though a large majority of scientists believe the problem to be a very real one, the truth is that predictions of the future can never be made with absolute certainty.

Put yourself in the position of a politician. Would it be reasonable of you to demand major changes of significant sectors of the national economy, in order to avert a disaster that may never happen in any case? Or, since the consequences of the 'worst case' and even some of the 'middle of the road' scenarios are so profound, is the only responsible course of action to minimize risk – to behave as if disaster is certain if we do not change our collective behavior, even though it is not? One alternative might be to wait for better data. But suppose that by the time better data are available it is too late . . .

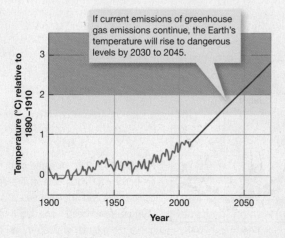

If current emissions of greenhouse gas emissions continue, the Earth's temperature will rise to dangerous levels by 2030 to 2045.

FIGURE 3.20 Average temperature of the Earth's atmosphere from 1900 to 2011, with projected future warming if greenhouse gas emissions are not reduced. By 2011, the Earth was 0.7°C warmer than in the early 20th century. Warming to 1.5°C above the early 20th century baseline is expected by 2030, and to 2.0°C above by 2045. The yellow bar shows this range of warming, which corresponds with an increased risk of reaching a tipping point in the planet's climate system, potentially leading to further runaway global warming. (Modified from Shindell et al., 2012.)

variations beneath a canopy At a smaller scale, carbon dioxide concentrations within a temperate-zone forest canopy are greatest very close to the ground in summer, where it is released rapidly from decomposing organic matter in the soil (Figure 3.22). Diffusion alone guarantees that the concentration quickly declines with increasing height, but in the daytime photosynthesizing plants also actively remove carbon dioxide from the air, whereas at night concentrations increase as plants respire and there is no photosynthesis. During late fall, when low temperatures mean that rates of photosynthesis, respiration, and decomposition are all slow, concentrations remain virtually constant through day and night at all heights. Thus, plants growing in

different parts of a forest will experience quite different carbon dioxide environments: the lower leaves on a forest shrub will usually experience higher carbon dioxide concentrations than its upper leaves, and seedlings will live in environments richer in carbon dioxide than mature trees.

Do higher concentrations of carbon dioxide encourage faster plant growth? When other resources are present at adequate levels, addi- **what will be the consequences of current rises on photosynthesis?** tional carbon dioxide scarcely influences the rate of photosynthesis of C4 plants but can dramatically increase the rate of C3 plants. Indeed, artificially increasing the carbon dioxide concentration in greenhouses is a commercial technique to increase crop yields of

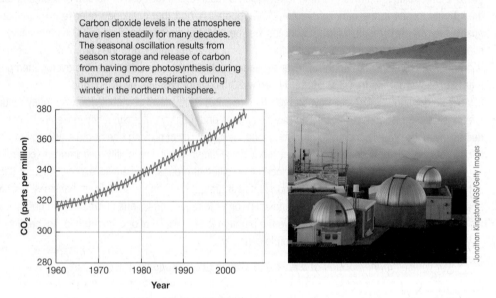

Carbon dioxide levels in the atmosphere have risen steadily for many decades. The seasonal oscillation results from season storage and release of carbon from having more photosynthesis during summer and more respiration during winter in the northern hemisphere.

Jonathan Kingston/NGS/Getty Images

FIGURE 3.21 Concentration of carbon dioxide in the air at the Mauna Loa observatory in Hawaii from 1958 to 2005. Note the seasonal oscillation, which is driven by photosynthesis and carbon storage in the terrestrial ecosystems of the northern hemisphere during the summer, and release the following winter.

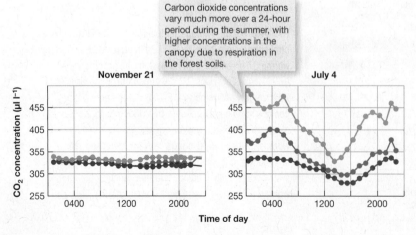

Carbon dioxide concentrations vary much more over a 24-hour period during the summer, with higher concentrations in the canopy due to respiration in the forest soils.

FIGURE 3.22 Average carbon dioxide concentrations for each hour of the day in a mixed deciduous forest (Harvard Forest, Massachusetts, USA) on November 21 and July 4 at three heights above the ground: ●, 0.05 m; ●, 1 m; ●, 12 m. (After Bazzaz & Williams, 1991.)

Chris Hildreth/Duke Photography

The FACE study at Duke Forest.

C3 plants. We might reasonably predict dramatic increases in the productivity of individual plants, and of whole crops and natural ecosystems, as atmospheric concentrations of carbon dioxide continue to increase, and in fact some scientists have made such projections. However, the responses of natural ecosystems are more complicated than that of individual plants grown under ideal conditions in a greenhouse, in part because other resources are often not present at adequate levels. For instance, a low availability of nitrogen may keep a forest from responding to higher levels of carbon dioxide.

Ecologists began a series of experiments in the 1990s to evaluate the consequences of elevated carbon dioxide on the dynamics of the community and ecosystem—not just the response of individual plants—by artificially increasing the concentration of carbon dioxide in the local environment over a scale of a few hundred square meters. By now, we have observations from a large number of such whole-ecosystem experiments in a variety of different types of systems, including forests, grasslands, and agricultural fields. It turns out that the variation in response across ecosystems is high, yet some generalities emerge. Any increase in production of agricultural cereal crops (maize or corn, wheat) is much less than would be predicted from studies of individual plants. On the other hand, forest plots responded much more to carbon dioxide than did crops, and the responses were more in line with predictions from smaller-scale experiments (Ainsworth & Long, 2005). It is essential that we continue to try to determine the long-term consequences

of increased carbon dioxide on photosynthesis at the community and ecosystem scale. To do so, we need to consider not only the direct physiological response to the carbon dioxide, but the interactions with water and nutrient availability, and also the possible influences from changes in herbivores (as the character of leaf tissues change), the influences of floods and of droughts, and factors such as changes in ozone toxicity to plants that may increase as temperatures rise.

Green plants, algae, cyanobacteria, and other photosynthetic bacteria are **autotrophs**: their major
$\quad$ *autotrophs and heterotrophs*

resources are photosynthetic radiation, water, nutrients, and carbon dioxide. These autotrophs (also known as **primary producers**) assemble carbon dioxide, water, and nutrients into complex molecules (carbohydrates, fats, proteins, nucleotides, etc.) and then package them into cells, tissues, organs, and whole organisms. It is these packages that form the food resources for virtually all other organisms, the **heterotrophs**. Heterotrophs unpack the organic packages, metabolize and excrete some of the contents (mostly as inorganic nutrients, carbon dioxide, and water), and reassemble the remainder into their own bodies. The heterotrophs in turn may be consumed, unpacked, and reconstituted in a chain of events in which each consumer becomes, in turn, a resource for some other consumer.

Another fascinating type of organisms are the autotrophs which get their energy by oxidizing the reduced energy found in inorganic compounds such as hydrogen sulfides. These chemo-autotrophic bacteria use this energy (generally using oxygen to oxidize the reduced chemical compound) to fix carbon dioxide into organic matter using the same basic enzymes as in C3 photosynthesis. The energy source is different—reduced chemical substances rather than light—but the rest of the autotrophy is similar, resulting in new biomass of living bacteria. This process can be locally important as a source of new organic production in some ecosystems, including many wetlands and some lakes and coastal marine ecosystems. But globally, the process is dwarfed by photosynthesis.

3.3 HETEROTROPHS AND THEIR RESOURCES

Heterotrophs include all animals, fungi, and archea as well as all bacteria except for the photosynthetic and chemo-autotrophic forms. Heterotrophs can generally be grouped as follows:

1 **Decomposers**, which feed on already dead plants and animals.

2 **Parasites**, which feed on one or a very few host plants or animals while they are alive but do not (usually) kill their hosts, at least not immediately.

3 **Predators**, which, during their life, eat many prey organisms, typically (and in many cases always) killing them.

4 **Grazers**, which, during their life, consume parts of many prey organisms, but do not (usually) kill their prey, at least not immediately.

Killer whales are an example of a predator, but the relationship encompasses a much wider array of consumer–resource interactions. For example, a squirrel is a predator when it eats an acorn (it kills the acorn embryo), and a planktonic crustacean is a predator when it eats an algal cell (even though many ecologists might informally call this consumption of algae "grazing"). In both cases, the predator kills its food resource as it consumes it.

Of course, nature can always confound simple distinctions. While lions might seem the premier example of a predator, much of their food comes from eating animals killed by others: they are both predators and decomposers. As an example of even greater complexity, JoAnn Burkholder and colleagues discovered a whole new genus of organisms in 1988: *Pfiesteria*. These fascinating creatures (two species have been identified) live in coastal marine ecosystems and have an amazingly complex life cycle (Figure 3.23). At various stages, cells may consume bacteria, or parasitize fish, or decompose already dead fish material, or carry

out photosynthesis. In the rest of this chapter, we concentrate on animal consumers (and take the subject further still in Chapter 7).

Some animals have very specialized diets, while others are generalists. The generalists feed on a wide variety of foods, although they very often have clear preferences and a rank order of what they will choose when there are alternatives available. Some specialists may focus on particular parts of their prey but range over a number of species. This is most common among herbivores because, as we shall see, different parts of plants are quite different in their composition. Thus, many birds specialize on eating seeds, though they are seldom restricted to a particular species. Other consumers may specialize on a single species or a narrow range of closely related species. Examples are caterpillars of the cinnabar moth (which eat the leaves, flower buds, and very young stems of a species of ragwort, *Senecio*) and many species of host-specific parasites.

Nutritional needs and provisions

The various parts of a plant have very different compositions (Figure 3.24) and so offer quite different resources. Bark, for example, is largely composed of dead cells with corky and lignified walls, packed with defensive phenolics, and is quite useless as a food for most herbivores (even species of 'bark beetle' specialize on the nutritious cambium layer just beneath the bark, rather than on the bark itself). The richest concentrations of plant proteins (and hence of nitrogen) are in the meristem in the buds at shoot tips and in leaf axils. Not surprisingly, these are usually heavily protected with bud scales and defended from herbivores by prickles and spines. Seeds are usually dried, packaged reserves rich in starch or oils as well as specialized storage proteins. And the very sugary and fleshy fruits are resources provided by the plant as 'payment' to animals that disperse the seeds. Very little of the plants' nitrogen is spent on these rewards.

plants as (a variety of) foods

The diversity of different food resources offered by plants is matched by the diversity of specialized mouthparts and digestive tracts that have evolved to consume them. This diversity is especially developed in the beaks of birds and the mouthparts of insects (Figure 3.25).

For a consumer, the body of a plant is a quite different package of resources from the body of an animal. First, plant cells are bounded by walls of cellulose, lignin, and other structural carbohydrates

from plants into animals

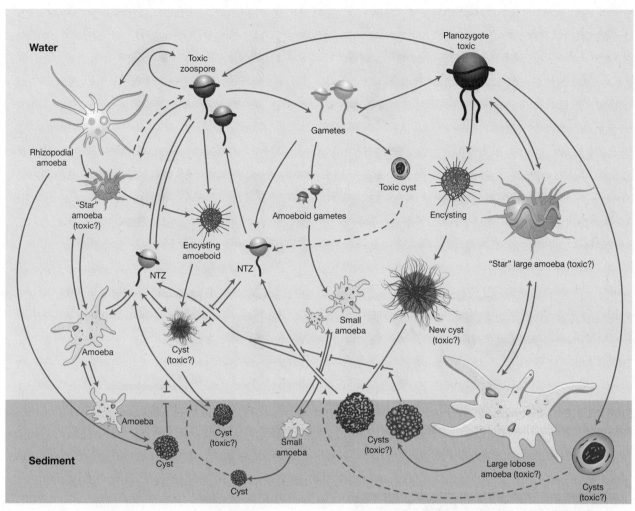

FIGURE 3.23   A depiction of the life cycle stages of *Pfiesteria*, a microorganism that lives in estuaries. At various times, this organism will be a consumer of bacteria, a parasite of fish, a decomposer of dead fish material, or a photosynthetic autotroph.

that give plants their high-fiber content and contribute to their high ratio of carbon to other elements. These large amounts of fixed carbon mean that plants are potentially rich sources of energy, although they are protein-poor. Yet the overwhelming majority of animal species lack cellulolytic and other enzymes that can digest these compounds: they are quite useless as a direct energy resource for most herbivores. Moreover, the cell wall material of plants hinders the access of digestive enzymes to the plant cell contents. The acts of chewing by the grazing mammal, cooking by humans, and grinding in the gizzard of birds are necessary precursors to digestion of plant food because they allow digestive enzymes access to the cell contents. The carnivore, by contrast, can more safely gulp its food.

Many herbivores have made up for their own lack of cellulolytic enzymes by entering into a **mutualistic** association – one that is beneficial to both parties—with cellulolytic bacteria and protozoa in their guts that do

have the appropriate enzymes. The rumen (or sometimes the cecum) of many herbivorous mammals is a temperature-regulated culture chamber for these microbes into which already partially fragmented plant tissues flow continually. The microbes receive a home and a supply of food. The herbivorous 'host' benefits by absorbing many of the major by-products of this microbial fermentation, especially fatty acids.

Unlike plants, animal tissues contain no structural carbohydrate or fiber component but are rich in fat and protein. The C : N ratio of plant tissues commonly exceeds 40 : 1, and is often far higher. In contrast, the ratios rarely exceed 8 : 1 in bacteria, fungi, and animals. Thus, herbivores on plants, which undertake the first stage of making animal bodies out of the plants, are involved in a massive burning off of carbon as the C : N ratio is lowered. The main waste products of herbivores are therefore carbon-rich compounds (carbon dioxide and fiber). Carnivores, on the other

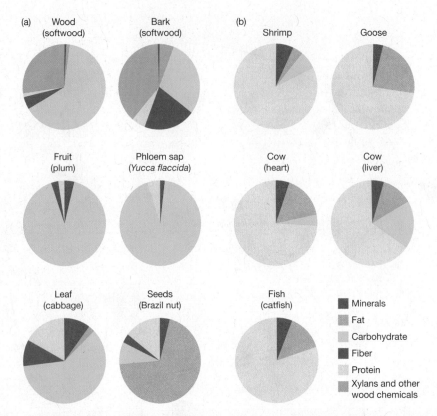

FIGURE 3.24 The composition of various plants (a) and animals (b) that may serve as food resources for herbivores or carnivores. Note that the different parts of a plant have very different compositions, whereas different species of animal (and their parts) are remarkably similar.

FIGURE 3.25 Examples of the variety of specialized mouthparts in herbivorous insects. (a) Honeybee with a long 'tongue' (glossia) for sucking. (b) Hawkmoth with an even longer sucking proboscis. (c) Leichhardt's grasshopper with large, plate-like chewing mandibles. (d) Acorn weevil with chewing mouthparts at the very end of its long rostrum. (e) Rose aphid with a piercing stylet.

hand, get most of their energy from the protein and fats of their prey, and their main excretory products are consequently nitrogenous.

The composition of algae is quite different from that of plants, making a very different food resource. Algae have very little structural material, as they are supported by the water in which they live. They are composed largely of chlorophyll, enzymes, and other proteins, held within a cell wall, which explains their low C : N ratio. Herbivores feeding on algae have a much more protein-rich diet than those feeding on plants. Their main waste products will be lower in carbon and richer in nitrogen.

Even if the structural component of plants is excluded from consideration, the C : N ratio is high in plants compared with other organisms. Aphids, which gain direct access to cell contents by driving their stylets into the phloem transport system, gain a resource that is rich in soluble sugars (see Figure 3.24a). In their search for valuable nitrogen, they use only a fraction of this energy resource and excrete the rest in sugary rich honeydew that may drip as a rain from an aphid-infested tree. For most herbivores and decomposers, the body of a plant is a superabundant source of energy and carbon; it is other components of the diet, especially nitrogen, that are more usually limiting.

The bodies of different species of animal have remarkably similar composition (see Figure 3.24b). In terms of protein, carbohydrate, fat, water, and minerals per gram, there is very little to choose between a diet of caterpillars or cod, or of earthworms, shrimps, or venison. The packages may be differently assembled (and the taste may be different), but the contents are essentially the same. Moreover, the different parts of an animal have very similar nutritional content. Unlike herbivores on plants, carnivores are not faced with difficult problems of digestion (and they vary very little in their digestive apparatus) but rather with difficulties in finding, catching, and overcoming the defenses of their prey.

| animals as food |

Defense

The value of a resource to a consumer is determined not only by what it contains but by how well its contents are defended. Not surprisingly, organisms have evolved physical, chemical, morphological, and behavioral defenses against being attacked. These serve to reduce the chance of an encounter with a consumer and/or increase the chance of survival in such encounters. The spiny leaves of holly are not eaten by larvae of the oak eggar moth, but if the spines are removed,

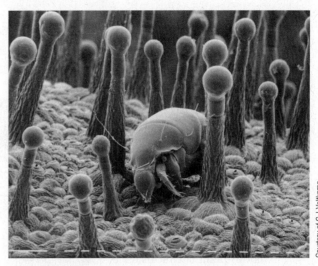

FIGURE 3.26   A mite trapped in the protective trichomes (hairs) on the surface of a *Primula* leaf. The trichomes themselves support capsules of irritant volatile oils at their tip. Each white bar towards the foot of the image represents 10 µm.

Courtesy of C.J. Veltkamp

the leaves are eaten quite readily. No doubt similar results would be achieved in equivalent experiments with foxes as predators and de-spined hedgehogs or porcupines as prey. On a smaller scale, many plant surfaces are clothed in epidermal hairs (trichomes) that may keep the smaller predators (such as thrips and mites) away from the leaf surface (Figure 3.26).

Any feature of an organism that increases the energy spent by a consumer in discovering or handling it is a defense if, as a consequence, the consumer eats less of it. The thick shell of a nut increases the time that an animal spends extracting a unit of effective food, and this may reduce the number of nuts that are eaten. We have already seen that most green plants are relatively overprovided with energy resources in the form of cellulose and lignin. It may therefore be cheap to build husks and shells around seeds (and woody spines on stems) if these defense tissues contain rather little protein, and if what is protected is far more valuable and protein rich.

| some resources are protected . . . |

Plants, animals, and even many algae and bacteria have a battery of chemical defenses. The plant kingdom, in particular, is very rich in 'secondary' chemicals that apparently play no role in normal plant biochemical pathways but do deter herbivory. Populations of white clover, for example, commonly contain some individuals that release hydrogen cyanide when their tissues are attacked (*cyanogenic* forms) and others that do not; those that cannot are eaten by slugs and snails. The cyanogenic forms, however, are nibbled but then rejected (Table 3.1).

| . . . or defended |

TABLE 3.1 Slugs (*Agriolimax reticulatus*) graze on the leaves of clover (*Trifolium repens*). There are forms of clover that release hydrogen cyanide when the cells are damaged. Slugs nibble clover leaves and reject cyanogenic forms but continue to consume the leaves of noncyanogenic forms. Two plants, one of each form, were grown together in plastic containers and slugs were allowed to graze for seven successive nights. The table shows the numbers of leaves in different conditions after slug grazing. +/− indicate deviation from random expectation; the difference from random expectation is significant at $P < 0.001$.

	Conditions of Leaves after Grazing			
	Not Damaged	Nibbled	Up to 50% of Leaf Removed	More than 50% of Leaf Removed
Cyanogenic plants	160 (+)	22 (+)	38 (−)	9 (−)
Noncyanogenic plants	87 (−)	7 (−)	30 (+)	65 (+)

Noxious plant chemicals have been classified into two broad types. The first are *quantitative* chemicals (so-called because they are most effective at relatively high concentrations), which make the tissues that contain them, such as mature oak leaves, relatively indigestible. They are also often called **constitutive** chemicals, since they tend to be produced even in the absence of herbivore attack. The second type are toxic or *qualitative* chemicals, which are poisonous even in small quantities but can be produced relatively rapidly and are therefore commonly **inducible**: produced only in response to damage itself, and hence with lower fixed costs to the plants.

> optimal defense theory: constitutive and inducible defenses

Plants differ in their chemical defenses from species to species and also from tissue to tissue within an individual plant. Broadly, relatively short-lived, ephemeral plants gain a measure of protection from consumers because of the unpredictability of their appearance in space and time. They therefore need to invest less in defense than predictable, long-lived species like forest trees. The latter, precisely because they are apparent for long periods to a large number of herbivores, tend to invest in constitutive chemicals that, while costly, afford them broad protection, whereas ephemeral plants tend to produce inducible toxins as required. Moreover, it may be predicted that, within an individual plant, the more important plant parts should be protected by costly, constitutive chemicals, whereas less important parts should rely on inducible toxins (McKey, 1979; Strauss et al., 2004). This is confirmed, for example, by a study of wild radish, in which plants were either attacked by caterpillars of the butterfly, *Pierisrapae*, or left as unmanipulated controls (Figure 3.27). Flower petals are known to be highly important to fitness in this insect-pollinated plant, and concentrations of toxic glucosinolates were twice as high in petals as in undamaged leaves: levels were maintained constitutively, irrespective of whether the petals were damaged by the caterpillars. Leaves, on the other hand,

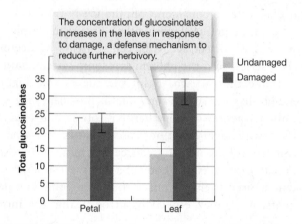

FIGURE 3.27 Concentrations of glucosinolates ($\mu g\ mg^{-1}$ dry mass) in the petals and leaves of wild radish, *Raphinus sativus*, either undamaged or damaged by caterpillars of *Pieris rapae*. Bars are standard errors. (After Strauss et al., 2004.)

have a much less direct influence on fitness: high levels of leaf damage can be sustained without any measurable effect on reproductive output. Constitutive levels of glucosinolates, as already noted, were low; but if leaves were damaged, the (induced) concentrations were even higher than in the petals.

Animals have more options than plants when it comes to defending themselves, but some still make use of chemicals. For example, defensive secretions of sulfuric acid of pH 1 or 2 occur in some marine gastropod groups, including the cowries. Other animals, which can tolerate the chemical defenses of their plant food, actually store the plant toxins and use them in their own defense. A classic example is the monarch butterfly, whose caterpillars feed on milkweeds, which contain cardiac glycosides, which are poisonous to mammals and birds. These caterpillars can store the poison, and it is still present in the adult. Thus, a blue jay will vomit violently after eating one, and, once it recovers, will reject all other monarch caterpillars on sight for some time. In contrast, monarch caterpillars reared on cabbage are edible.

> chemical defense in animals

Chemical defenses are not equally effective against all consumers. Indeed, what is unacceptable to some animals may be the chosen, even unique, diet of others. Many herbivores, particularly insects, specialize on one or a few plant species whose particular defense they have overcome. For example, females of the cabbage root fly, with eggs to lay, home in on a brassica crop from distances as far as 15 m downwind of the plants. It is probably hydrolyzed glucosinolates (toxic to many other species) that provide the attractive odor.

An animal may be less obvious to a predator if it matches its background, or possesses a pattern that disrupts its outline, or resembles an inedible feature of its environment. A straightforward example of such **crypsis** is the green coloration of many grasshoppers and caterpillars (Figure 3.28). Cryptic animals may be highly palatable, but their morphological traits and color (and their choice of the appropriate background) reduce the likelihood that they will be seen and used as a resource. In contrast, noxious or dangerous animals often advertise the fact by bright, conspicuous colors and patterns (Figure 3.28b). The monarch butterfly (see earlier), for example, has very conspicuous coloring. One attempt by a bird to eat an

| crypsis, aposematism, and mimicry |

adult monarch is so memorable that others are subsequently avoided for some time. The adoption of memorable body patterns by distasteful prey, moreover, immediately opens the door for deceit by other species—there will be a clear advantage to a palatable prey if it **mimics** an unpalatable species. Thus, the palatable viceroy butterfly mimics the distasteful monarch, and a blue jay that has learned to avoid monarchs will also avoid viceroys (even though the viceroys are not toxic).

By living in holes, animals (millipedes, moles) may avoid stimulating the sensory receptors of predators, and by 'playing dead' (opossum, African ground squirrel) animals may fail to stimulate a killing response. Animals that withdraw to a prepared retreat (rabbits and prairie dogs to their burrows, snails to their shells) or roll up and protect their vulnerable parts by a tough exterior (armadillos, hedgehogs) reduce their chance of capture. In the ocean, many zooplankton species will migrate vertically in the water column over the day, going into deeper and darker waters during the day, so they are less easily eaten by visually searching fish, and coming up to shallow surface waters at night to feed, as the algae they feed upon are much more common there. Other animals seem to try to bluff themselves out of trouble by

| behavior |

FIGURE 3.28 Lepidopterous caterpillars illustrate a range of defense strategies. (a) The irritating hairs of the gypsy moth. (b) Aposematism (advertizing distastefulness) in the black swallowtail. (c) A cryptic (camouflaged) noctuid, looking like bark. (d) Another swallowtail with mimicking eye spots (not eyes) possibly startling a potential predator.

threatening or startling displays (Figure 3.28d). Moths and butterflies that suddenly expose eye spots on their wings are one example. No doubt the most common behavioral response of an animal in danger of becoming a used resource is to run away.

3.4 EFFECTS OF INTRASPECIFIC COMPETITION FOR RESOURCES

Resources are consumed. The consequence is that there may not be enough of a resource to satisfy the needs of a whole population of individuals. Individuals may then compete with each other for the limited resource. **Intraspecific competition** is competition between individuals of the same species.

In many cases, competing individuals do not interact with one another directly. Rather, they deplete the resources that are available to | exploitation: competitors depleting each other's resources

each other. Grasshoppers may compete for food, but a grasshopper is not directly affected by other grasshoppers so much as by the level to which they have reduced the food supply. Two grass plants may compete, and each may be adversely affected by the presence of close neighbors, but this is most likely to be because their resource depletion zones overlap—each may shade its neighbors from the incoming flow of radiation, and water or nutrients may be less accessible than they would otherwise be around the plants' roots. The data in Figure 3.29, for example, show the dynamics of the

interaction between a single-celled algal species, a diatom, and one of the resources it requires, silicate. As diatom density increases over time, silicate concentration decreases: there is then less available for the many than there had been previously for the few. This type of competition—in which competitors interact only indirectly, through their shared resources—is termed **exploitation**.

On the other hand, competing individual vultures may fight | direct interference

one another over access to a newly found carcass. Individuals of other species may fight for ownership of a 'territory' and access to the resources that a territory brings with it. A barnacle that settles on a rock denies the space to another barnacle. This is called **interference competition**.

Whether competition occurs through exploitation, interference or a combination of the two, its | competition and vital rates

ultimate effect is on the **vital rates** of the competitors – their survival, growth, and reproduction – compared with what they would have been if resources had been more abundant. Competition typically leads to decreased rates of resource intake per individual, and thus to decreased rates of individual growth or development, perhaps to decreases in amounts of stored reserves or to increased risks of predation. Figure 3.30a shows how the mortality rate of steelhead trout increases, as the number of competing fish rises, at a range of food levels; Figure 3.30b shows how the birth rate of the sand dune grass *Vulpia* declines as individuals become increasingly crowded.

In practice, intraspecific competition is often a very one-sided affair: a strong early seedling will shade and suppress a stunted, late one; a large vulture is likely to fight off a smaller one. Some of the competitive strength of individuals is related to timing (the early seedling) or to random events (one seed may germinate in a depression where it obtains more water than its neighbors). Sometimes the winner and loser may be genetically different and then competition will be playing a role in natural selection.

The effects of intraspecific competition on any individual are typically greater the more crowded | density dependence

the individual is by its neighbors—the more the resource depletion zones of other individuals overlap its own. Hence, the effects of intraspecific competition are often said to be **density-dependent**. But it is doubtful that any organism has a way of detecting the density of its population. Rather, it responds to the effects of being crowded.

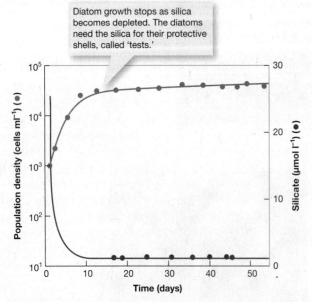

Diatom growth stops as silica becomes depleted. The diatoms need the silica for their protective shells, called 'tests.'

FIGURE 3.29 A population of the freshwater diatom *Asterionella formosa* was grown in flasks of culture medium. The diatom consumes silicate during growth and the population of diatoms stabilizes when the silicate has been reduced to a very low concentration. (After Tilman et al., 1981.)

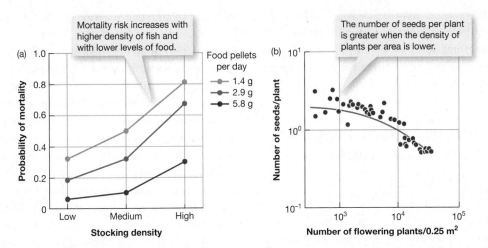

FIGURE 3.30   (a) The rate of mortality among steelhead trout (*Oncorynchus mykiss*) reared at a range of densities (32, 63, and 127 per m²) and at a range of food levels (1.4, 2.9, and 5.8 g of food pellets per day: orange, blue, and red lines, respectively). (After Keeley, 2001.) (b) The average number of seeds produced per plant of the dune grass *Vulpia fasciculata* growing at a range of densities. (After Watkinson & Harper, 1978.)

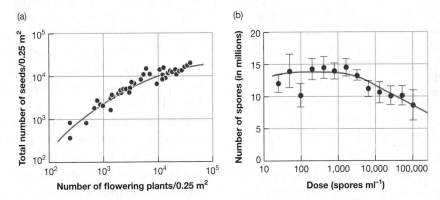

FIGURE 3.31   (a) An undercompensating effect on fecundity: the total number of seeds produced by *Vulpia fasciculata* continues to rise as density increases. (After Watkinson & Harper, 1978.) (b) When the planktonic crustacean *Daphnia magna* was infected with varying numbers of spores of the bacterium *Pasteuria ramosa*, the total number of spores produced per host in the next generation was independent of density (exactly compensating) at the lower densities, but declined with increasing density (overcompensating) at the higher densities. Standard errors are shown. (After Ebert et al., 2000.)

On the other hand, at low densities in the case of *Vulpia* (Figure 3.30b), the per capita birth rate or fecundity was *independent* of density (where per capita means literally 'per head' or 'per individual'). That is, the fecundity was effectively the same at a density of 1000 plants/0.25 m² as it was at a density of 500/0.25 m². Thus, there is no evidence at these densities that individuals are affected by the presence of other individuals, and hence no evidence of intraspecific competition. But as density increases further, the per capita birth rate progressively decreases. These effects are now density-dependent, and this may be taken as an indication that at these densities, individuals are suffering as a result of intraspecific competition.

The patterns in Figure 3.30 make the point that as crowding (or density) increases, the fecundity per individual is likely to decline and the

> competition and the total number of survivors

mortality per individual likely to increase (which would mean that the survival rate per individual would decrease). But what can we expect to happen to the *total* number of seeds or eggs produced by populations at different densities—or to the total number of survivors? In some cases, although the rate per individual declines with increasing density, the total fecundity or total number of survivors in the whole population continues to increase. We can see in Figure 3.31a that this was the case for the plant populations in Figure 3.30b—at least over the range of densities examined. In other cases, the rate per individual declines so rapidly with increasing density that the total fecundity or total number of survivors in the population actually gets smaller the greater the number of contributing individuals. We can see this in Figure 3.30b for the highest densities of a bacterial parasite of the planktonic crustacean *Daphnia magna*. In yet other cases, birth or mortality rates may be density-independent (no competition) throughout the range examined, in which case the total number of births or survivors will simply continue to rise in direct proportion to the original density.

3.5 CONDITIONS, RESOURCES, AND THE ECOLOGICAL NICHE

Finally, many of the ideas in this chapter can be brought together in the concept of the ecological **niche**. The term *niche*, though, is frequently misunderstood and used loosely to describe the sort of place in which an organism lives, as in the sentence 'Woodlands are the niche of woodpeckers.' However, strictly, where an organism lives is not its niche but its **habitat**. A niche is not a place but an idea: a summary of the organism's tolerances and requirements. The habitat of a gut microorganism would be an animal's alimentary canal; the habitat of an aphid might be a garden; and the habitat of a fish could be a whole lake. Each habitat, however, provides many different niches: many other organisms also live in the gut, the garden, or the lake—and with quite different lifestyles. The niche of an organism describes how, rather than just where, an organism lives.

The modern concept of the niche was proposed by Hutchinson in 1957 to address the ways in which tolerances and requirements interact to define the conditions and resources needed by an individual or a species

> the niche of an organism is defined by its needs and tolerances

in order to practice its way of life. Temperature, for instance, is a condition that limits the growth and reproduction of all organisms, but different organisms tolerate different ranges of temperature. This range is one dimension of an organism's ecological niche: Figure 3.32a shows how different species of plants vary in the temperature dimension of their niche. But there are many such dimensions for the niche of a species: its tolerance of various other conditions (relative humidity, pH, wind speed, waterflow, and so on), and its need for various resources (nutrients, water, food, and so on). Clearly the real niche of a species must be multidimensional.

The early stages of building such a multidimensional niche are relatively easy to visualize: Figure 3.32b illustrates the way in which two niche dimensions (temperature and salinity) together define a two-dimensional area that is part of the niche of a sand shrimp. Three dimensions, such as temperature, pH, and the availability of a particular food, may define a three-dimensional niche volume (Figure 3.32c). It is hard to imagine (and impossible to draw) a diagram of a more realistic, multidimensional niche. (Technically, we now consider a niche to be an *n*-**dimensional hypervolume**, where *n* is the number of dimensions that make up the niche.) But the simplified three-dimensional version nonetheless captures the idea of the ecological niche of a species. It is defined by the boundaries that limit where it can live, grow, and reproduce, and it is very clearly a concept rather than a place. The concept has become a cornerstone of ecological thought, as we shall see in later chapters.

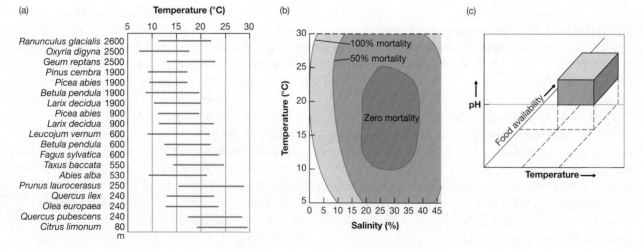

FIGURE 3.32 (a) A niche in one dimension. The range of temperatures at which a variety of plant species from the European Alps can achieve net photosynthesis at low intensities of radiation (70 W m⁻²). (After Pisek et al., 1973.) (b) A niche in two dimensions for the sand shrimp (*Crangon septemspinosa*) showing the fate of egg-bearing females in aerated water at a range of temperatures and salinities. (After Haefner, 1970.) (c) A diagrammatic niche in three dimensions for an aquatic organism showing a volume defined by the temperature, pH, and availability of food.

SUMMARY

Environmental conditions

Conditions are physicochemical features of the environment such as its temperature and humidity. They may be altered but are not consumed. Environmental resources such as nutrients and food are consumed by living organisms in the course of their growth and reproduction.

There are three basic types of response curve to conditions. Extreme conditions may be lethal with, between the two extremes, a continuum of more favorable conditions; or a condition may be lethal only at high intensities; or a condition may be required by organisms at low concentrations but become toxic at high concentrations. Many conditions are important stimuli for growth and development and prepare an organism for conditions that are to come.

Resources for photosynthetic organisms

Solar radiation, water, nutrients, and carbon dioxide are all critical resources for green plants, algae, and cyanobacteria. The shape of the curve that relates the rate of photosynthesis to the intensity of radiation varies greatly among species, and tends to be optimized for the environmental in which the organism lives.

Heterotrophs and their resources

The C : N ratio of plant tissues greatly exceeds those of bacteria, fungi, and animals. Thus, herbivores feeding on plants typically have a superabundant source of energy and carbon, but nitrogen is often limiting. Algae have a C : N ratio that is similar to that of animals, and herbivores feeding on these autotrophs gain proportionately more nitrogen and less carbon than do those feeding on plants.

The bodies of different species of animal have remarkably similar compositions. Carnivores are not faced with problems of digestion, but rather with difficulties in finding, catching, and overcoming the defenses of their prey.

Effects of intraspecific competition for resources

Individuals may compete indirectly, via a shared resource, through exploitation, or directly, through interference. The ultimate effect of competition is on survival, growth, and reproduction of individuals. Typically, the greater the density of a population of competitors, the greater is the effect of competition (density dependence). As a result, though, the total number of survivors, or of offspring, may increase, decrease, or stay the same as initial densities increase.

Conditions, resources, and the ecological niche

Where an organism lives is its habitat. A niche is a summary of an organism's tolerances and requirements. The modern concept, proposed by Hutchinson in 1957, is an *n*-dimensional hypervolume.

REVIEW QUESTIONS

1   Discuss whether you think the following statement is correct: 'A layperson might describe Antarctica as an extreme environment, but an ecologist should never do so.'

2   In what ways do ectotherms and endotherms differ, and in what ways are they similar?

3   Describe how plants' requirements to increase the rate of photosynthesis and to decrease the rate of water loss interact. Describe, too, the strategies used by different types of plants to balance these requirements.

4   Explain why different species of algae, with different surface to volume ratios, will dominate in different aquatic ecosystems having different nutrient levels.

5   Account for the fact that the tissues of plants, algae, and animals have such contrasting C : N ratios. What are the consequences of these differences?

6   Describe the various ways in which animals use color to defend themselves against attacks by predators.

7   Explain, with examples, what exploitation and interference intraspecific competition have in common and how they differ.

8   What is meant when an ecological niche is described as an *n*-dimensional hypervolume?

Challenge Questions

1   Explain, referring to a variety of specific organisms, how the amount of water in different organisms' habitats may define either the conditions for those organisms, or their resource level, or both.

2   Drawing examples from a variety of both animals and plants, contrast the responses of tolerators and avoiders to seasonal variations in environmental conditions and resources.

3   Describe and account for the differences in both root and shoot architecture exhibited by different plants.

Chapter 4

Climate and the world's biomes

CHAPTER CONTENTS

KEY CONCEPTS

After reading this chapter, you will be able to:

- explain how the differential heating of the Earth's surface by sunlight leads to large-scale movements of air in the atmosphere and water in the oceans, helping to define the planet's climate

- describe the terrestrial biomes of the Earth (such as tropical rain forest, desert, and tundra) and their recognizable patterns of life that align along gradients of temperature and precipitation

- identify the factors that control the ecological functioning of streams, rivers, ponds, lakes, and wetlands and how these ecosystems are linked to the land around them

- explain how the world's oceans consist of distinct biome types whose available light and nutrients are controlled by ocean currents and mixing patterns

The Earth's climate results from differential heating by sunlight, which drives winds and ocean currents. The climate and currents, in turn, lead to readily recognizable patterns of life—biomes—over large spatial scales on both the continents and in the oceans. On land, the biomes are structured along gradients of temperature and precipitation. In the oceans, biomes are structured by gradients in the availability of light and of nutrients.

4.1 THE WORLD'S CLIMATE

Having examined in Chapter 3 the way individual organisms are affected by conditions and resources, we now turn to the larger scale and examine how the interplay of conditions and resources influences whole communities and ecosystems, and how the communities and ecosystems of the world are patterned across the globe.

Not surprisingly, because of its influence on both conditions and resources, climate plays a dominant role in determining the large-scale distribution of different types of communities and ecosystems across the terrestrial face of the Earth. However, local factors such as soil type are responsible for variation and patchiness in communities and ecosystems on much smaller scales. It is worth remembering, then, that the Earth's surface would consist of a mosaic of different environments even if climate were identical everywhere. As we shall see repeatedly below, variations critical to the ecological community occur at a whole range of scales, from the global to the very local. For life in the ocean, the interplay of wind and currents is as important as the climate in determining the distribution of types of ecosystems, in part because water is very effective at absorbing heat and buffering temperatures, so the variation in temperature in the oceans is far less than on land. Of course, the moisture gradient of climate that is so important to terrestrial life has little influence on life in the oceans.

At the largest scale, the geography of life on the Earth's land masses is mainly a consequence of climate. Climate, in turn, is controlled primarily by the differential heating of the planet by sunlight across the planet, and secondarily by the interaction of the atmosphere with the oceans and with mountain ranges. At the time of the equinoxes in September and March, solar radiation hits the Equator at exactly 90°, while sunlight further north and south hits the Earth's surface at increasingly oblique angles. Consequently, the amount of solar radiation hitting the

> Climate controls life, but what controls climate?

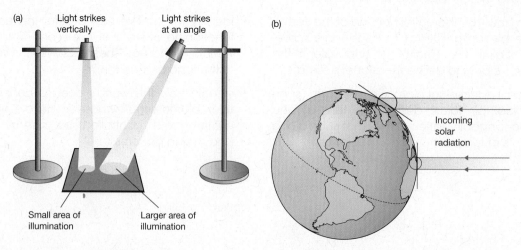

(a) Light strikes vertically Light strikes at an angle

Small area of illumination Larger area of illumination

(b) Incoming solar radiation

FIGURE 4.1   When light hits a surface at an oblique angle, the intensity of the light per surface area is less (left). This is true of sunlight hitting the surface of the Earth (right). One consequence is that heating by sunlight is greater at the Equator than at higher and lower latitudes.

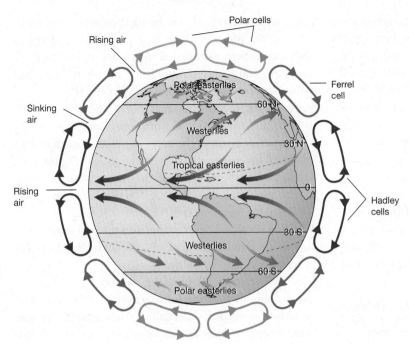

FIGURE 4.2 Heat is redistributed across the planet through the Hadley, Ferrel, and Polar cells. The Hadley cells move heat away from the Equator, and the Polar cells move heat towards the poles. The Ferrel cells are created by the interaction of the Hadley and Polar cells. Because of the Coriolis effect, the surface winds of the Hadley, Ferrel, and Polar cells are diverted to the right in the Northern Hemisphere and to the left in the Southern Hemisphere. As a result, the prevailing surface winds blow from the east in the tropics and in polar regions and from the west in temperate latitudes.

surface per unit area is much higher in the tropics, leading to greater heating there (Figure 4.1). This greater heating leads to warmer temperatures at the Equator, but the heat does not just accumulate there. Rather, it is redistributed across the planet to temperate and polar regions in movements either of air masses in the atmosphere (about 60% of the redistribution) or of water masses, currents, in the oceans (the remaining 40%).

Redistribution of heat through atmospheric movement

The movement of heat through the atmosphere is particularly important for the pattern of climate we observe across the Earth's surface. The heated air at the Equator rises in the atmosphere, cooling as it does so, and at a height of 10 to 15 km above the Earth's surface begins to move north and south. The air that has risen from the surface is displaced by other air from further north and south that flows towards the Equator along the ground (creating winds and weather on the surface). As the air at 10 to 15 km height reaches latitudes approximately 30° north and south, it descends to the surface, replacing the air that has moved along the surface towards the Equator and completing an atmospheric cycle referred to as the Hadley Cells (Figure 4.2).

Closer to the poles, similar atmospheric circulations – the Polar Cells – dominate, driven by rising, warmer air masses at latitudes 60° north and south and sinking, colder air at the poles. Like the Hadley Cells, air masses tend to move southward on the surface and northward at height in the northern

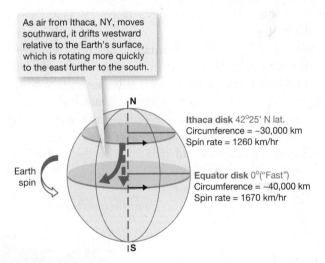

As air from Ithaca, NY, moves southward, it drifts westward relative to the Earth's surface, which is rotating more quickly to the east further to the south.

Ithaca disk 42°25' N lat.
Circumference = ~30,000 km
Spin rate = 1260 km/hr

Equator disk 0°("Fast")
Circumference = ~40,000 km
Spin rate = 1670 km/hr

Earth spin

FIGURE 4.3 The Coriolis effect affecting air moving south from Ithaca, New York, United States, towards the Equator.

hemisphere's Polar Cell, and vice a versa in the southern hemisphere. Between the Hadley Cells and the Polar Cells lies a more complicated zone of mixing, with varied air masses tending in general to move along the surface from the 30th to the 60th degrees of latitude. This circulation, referred to as Ferrel Cells, is driven by the interaction of the Hadley and Polar Cells.

The spinning of the Earth has an influence on the motion of air masses at the planet's surface. As demonstrated in Figure 4.3, the closer to the Equator an object (including an air mass) is on the surface of the Earth, the faster it is moving compared to an object further north or south, simply

> The Coriolis Effect diverts winds to the east and west

because the circumference of the Earth is larger at the Equator. For example, an air mass at Ithaca, New York (42.6° north), is spinning at only 75% of the rate of an object on the Equator, so as it moves south toward the Equator, the underlying surface is moving faster to the east than is the air mass, which has its eastern velocity set by its starting point in Ithaca. As a result, the air mass moves somewhat to the west, relative to the Earth's surface, as well as moving south. This tendency is termed the Coriolis Effect. It interacts with the Hadley, Ferrel, and Polar Cells to control the average direction of surface winds on the planet: from east to west between the Equator and 30° north and south, from west to east between 30° and 60° north and south, and again from east to west between 60° and the poles (Figure 4.2).

The rising and falling of these various air masses dramatically affects precipitation patterns on the Earth's surface. The pressure of air decreases rapidly with increasing height as it rises in the atmosphere. From chemistry's Universal Gas Laws, the temperature of a gas is proportional to its pressure, if volume is held constant. Therefore, temperature falls as pressure falls in the rising air masses of the Hadley and Polar Cells. Colder air cannot hold as much moisture, so as warm and moist air rises – particularly at the Equator because the starting air is so moist, but also near 60° north and south – the moisture condenses and falls back to the surface as

Rising and falling air and precipitation patterns

precipitation. On the other hand, the descending air at 30° and the poles – which had already been stripped of much of its moisture as it rose – becomes even drier as its pressure and temperature increases. The end result is very high rates of rainfall near the Equator, quite dry regions in the areas north and south of 30° in both the southern and northern hemispheres, high precipitation again in the areas north and south of 60°, and very little precipitation at the poles (Figure 4.4). Thus, for example, most of the major deserts of the world, including the Sahara, Kalahari, Mojave, and Sonoran, are found near 30° north or south. Of course, this is an idealized description, and the real patterns of climate on Earth have greater complexity, for a variety of reasons discussed below.

One complication is the seasonality caused by the change in tilt of the Earth's axis relative to the sun over the course of its annual rotation around the sun. During the Northern Hemisphere's summer, the Northern Hemisphere is tilted more towards the sun, and on the solstice in June, the direct right angle of sunlight hits the Earth further north than the Equator. The Northern Hemisphere gets more heating from sunlight than at the time of the equinoxes, while the Southern Hemisphere gets less. Of course, this whole pattern is reversed around the time of the December solstice. This seasonal oscillation in the distribution of heating shifts the Hadley Cells north and south from their position at the time of the equinoxes. And in turn, the regions of greatest and least rainfall shift as well.

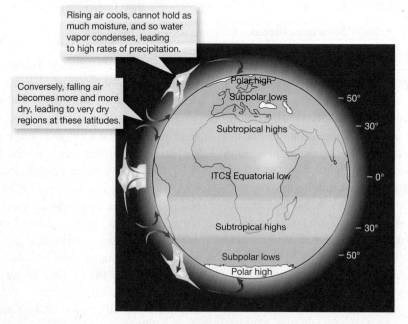

FIGURE 4.4 Air cools as it rises in the atmosphere and warms as it falls. Since cold air cannot hold as much moisture as warm air, water vapor condenses in areas of rising air, leading to high rates of precipitation. Conversely, falling air becomes more and more dry, leading to very dry regions on the surface of the Earth.

Another issue is the influence of mountain ranges. On the side of mountains hit by prevailing winds (that is the eastern side between the Equator and 30° and between 60° and the poles, and the western side between 30° and 60°), air masses rise to go over the mountains, and cool and drop their moisture as they do so. On the downwind side of the mountains, descending air will be very dry, creating a so-called **rain shadow** (Figure 4.5). Some of the most severe deserts on Earth such as Death Valley in California are created by this effect, while incredible rain and snowfall create rain forests on the upwind side of many mountain ranges, such as the Cascades in the northwestern United States.

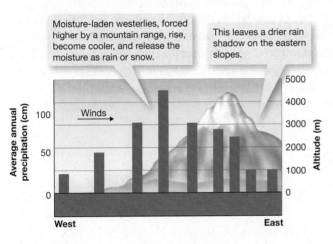

FIGURE 4.5 The typical influence of topography on rainfall (histogram bars) between 30° and 60°. (After Audesirk & Audesirk, 1996.)

Ocean currents and the redistribution of heat

Ocean currents have further powerful effects on climatic patterns, and through their redistribution of heat they can affect the climate not only over the oceans themselves but also over large areas of the continents adjacent to or downwind of ocean masses. In the southern hemisphere, currents carry cold Antarctic waters up along the western coasts of continents and distribute warmer waters from the tropics along their eastern coasts (Figure 4.6). In the North Atlantic Ocean, the Gulf Stream, which flows northward along the eastern side of North America, moderates not only the climate of that North American coast but also carries with it warm and moist air far into the Atlantic Ocean, affecting even the climate of Western Europe.

The surface currents of the oceans are driven in part by prevailing winds and, like the winds, are affected by the Coriolis effect. Global differences in density across the world's oceans also play a major role in controlling the pattern and strength of currents, including surface currents, and some evidence suggests that global warming may change this density-driven global circulation of the world's oceans (Box 4.1).

4.2 TERRESTRIAL BIOMES

The variety of influences on climate produces a mosaic of dry, wet, cool, and warm climates over the surface of the globe, with huge variation in the annual extent

FIGURE 4.6 The movements of the major ocean currents. Because of the Coriolis effect, the general circulation in the Northern Hemisphere is clockwise, in the Southern Hemisphere counterclockwise, with consequences for continental climate patterns.

4.1 ECOncerns

The global conveyor belt and climate warming

The world's oceans represent a continuous fluid, and as with any fluid, denser water in the oceans will tend to sink and less dense water rise throughout the world. The waters in the northern parts of the North Atlantic Ocean are unusually dense, both because they are cooled in the northern latitudes and because they are relatively salty: the result of greater evaporation than input of freshwaters in precipitation and rivers. Consequently, these waters sink, and other waters from further south move forward on the surface to replace them (Figure 4.7). The waters moving north in the Atlantic originate in the tropical waters of the Indian and Pacific Oceans, and flow through the South Atlantic Ocean on their way to the North Atlantic. This flow, bringing warm waters into the North Atlantic Ocean, is a main driver behind the Gulf Stream and helps to maintain a temperate climate far into the north of Europe. The deep waters of the North Atlantic Ocean flow southward at depth, through the South Atlantic Ocean and into the Indian and Pacific Oceans. There, they rise to the surface to replace the waters flowing on the surface toward the North Atlantic Ocean. This great conveyor belt of ocean circulation is critical to redistributing heat across the Earth. It is also very important in helping to mitigate human-driven global change: the cold North Atlantic waters absorb very large quantities of carbon dioxide from the atmosphere before they sink, and once these waters have sunk, that carbon dioxide stays trapped in the deep ocean water masses for centuries before eventually coming back to the surface and being released (see Chapter 12).

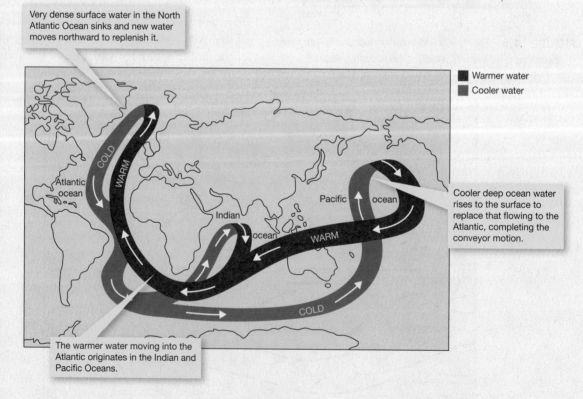

FIGURE 4.7 The great ocean conveyor belt moves water throughout the world's oceans, redistributing heat as it does so. The red stripe represents the movement of surface waters, and the blue stripe bottom water. This movement is driven by density gradients.

Climate scientists fear that the great conveyor belt may slow in response to global warming, and indeed there is some evidence that this may already have started: in fact, a paper in *Nature* published in 2005 by scientists from the UK National Oceanography Centre in Southampton (Bryden et al., 2005) reported a 30% slowing between 1957 and 2004, based on comparing ocean current measurements over time. The concern is that melting of Arctic sea ice and of glaciers in Greenland combined with greater river flows from Siberia into the Arctic Ocean will, as some of the Arctic Ocean waters reach the North Atlantic, make the North Atlantic Ocean less salty, and therefore less dense. The intensity of the great conveyor belt is driven by the density differences between the North Atlantic and the surface waters of the tropical Indian and Pacific Oceans, so this freshening would slow the flow globally. The consequences on the distribution of climate globally would be dramatic, and the mitigation of global warming by oceanic uptake of carbon dioxide would be greatly diminished.

One criticism that has been leveled at the 2005 paper published in Nature *is that the measured slowing of the conveyor belt across the decades is barely statistically significant (see Box 1.3). This should not be surprising, given the difficulty of measuring currents at the scale of the ocean. How would you respond to a relative or friend who says 'I am not going to worry about climate change and some abstract idea of change in the oceans until the science becomes stronger and more convincing'?*

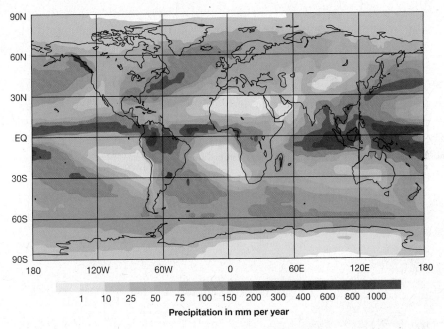

FIGURE 4.8   Average annual patterns of precipitation across both the land and ocean surfaces of the Earth.

http://www.physicalgeography.net/fundamentals/8g.html

of precipitation, as shown in Figure 4.8. Across this mosaic, distinctive terrestrial associations of vegetation and animals exist, reflecting particular combinations of temperature and precipitation. A world traveler repeatedly recognizes characteristic regions with particular types of vegetation, such as desert, savanna, and rain forest, which ecologists call **biomes**. These look similar and have similar adaptations to the environment, but the genetic and evolutionary history leading to the vegetation in, say, the deserts of the American southwest is very different from those for the deserts of Asia, Africa, or Australia. Remember, too, that neat pigeonholes and sharp categories are a convenience, not a reality of nature. Biomes are not homogeneous within their hypothetical boundaries; every biome has gradients of physicochemical conditions related to local topography and geology. Within the global mosaic are further mosaics at smaller and smaller scales. The communities of plants and animals that occur in different parts of this heterogeneous patchwork may be quite distinct.

Figure 4.9a recognizes a set of biomes and shows their distribution as a global map. We can also plot biome types as a function of the average temperature and precipitation (Figure 4.9b). In areas where the climate is warm, tropical forests dominate at higher levels of precipitation, deserts when the precipitation is low, and savannas and tropical seasonal forests in areas of intermediate rainfall. In somewhat colder regions, temperate forests dominate when the precipitation is greater, temperate grasslands at intermediate levels of precipitation, Mediterranean biomes at still lower levels of precipitation, and deserts when the precipitation is truly low. Moving into colder regions at higher latitudes, boreal forests dominate when the precipitation is sufficiently high, with deserts in regions of

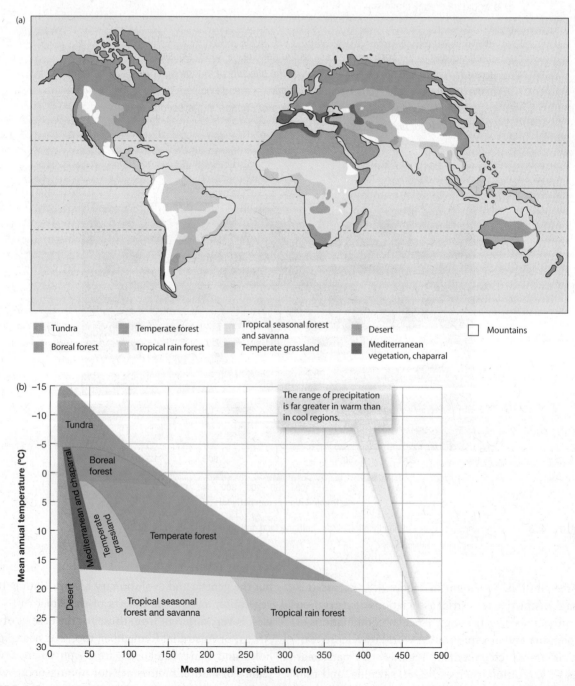

FIGURE 4.9 (a) World distribution of the Earth's biomes. (b) The terrestrial biomes of the planet fall along gradients of temperature and precipitation. Modified from R. Whittaker, 1975, Communities and Ecosystems.

extremely low precipitation. Tundra dominates in the coldest of the planet's regions. Note that the highest rates of rainfall occur in the warmest tropical regions, as this is where the air rising in the atmosphere—leading to condensation and precipitation—starts out with the greatest moisture content.

Local elevation can change the global patterns we see in Figure 4.9a, both because of the influence of rising and falling air on local patterns of precipitation and rain shadows (Figure 4.5), and because of the direct effect of elevation on temperature. As we rise higher in the atmosphere, by climbing up a mountain for instance, the temperature drops on average 6.4°C for every kilometer gain in elevation. As a result, the higher elevations in the Alps or the Rocky Mountains support ecosystems that have more in common with the tundra biome than with the temperate forests that occur there at lower elevations.

Biomes and convergent evolution

We pointed out in Chapter 2 the crucial importance of geographic isolation in allowing populations to diverge under selection. The geographic distributions of species, genera, families, and even higher taxonomic categories of plants and animals often reflect this geographic divergence. All species of lemurs, for example, are found on the island of Madagascar and nowhere else. Similarly, 230 species in the genus *Eucalyptus* (gum tree) occur naturally in Australia (and two or three in Indonesia and Malaysia). The lemurs and the gum trees occur where they do because they evolved there—not because these are the only places where they could survive and prosper. Indeed, many *Eucalyptus* species grow with great success and spread rapidly where they have been introduced to California or to Kenya. Another theme of Chapter 2 concerned the way species with quite different evolutionary origins have been selected to converge in their form and behavior. There were also examples of distinct taxonomic groups that have radiated into a range of species with strikingly similar form and behavior: parallel evolution, as in the marsupial and placental mammals.

A map of biomes, then, is not usually a map of the distribution of species. Instead, it shows where we find areas of land dominated by plants with characteristic shapes, forms, and physiological processes. These are the types of vegetation that can be recognized from an aircraft passing over them or from the windows of a fast car or train (Figure 4.10). It does not require a botanist to identify them. The scrubby chaparral vegetation characteristic of

describing and classifying vegetation

California provides a striking example. The spectrum of plant forms that gives this vegetation its distinctive nature also occurs in similar environments around the Mediterranean Sea and in Australia—but the species and genera of plants are quite different. We recognize different biomes from the functional types, not the species identities, of organisms that live in them.

Tropical rain forest

We have chosen to discuss tropical rain forests in greater depth than the other biomes because it represents the global peak of evolved biological diversity on land; all the other terrestrial biomes suffer from a relative poverty of resources or more strongly constraining conditions. Tropical rain forests cover approximately 12% of the Earth's land area and contain more than half the entire biomass of all terrestrial ecosystems.

Tropical rain forest is also the most productive of the Earth's biomes, with a rate of **primary production**—that is, the total amount of photosynthesis per area for a defined length of time—that can exceed 800 g of carbon fixed per square meter per year (see Chapter 11). Such exceptional productivity results from the coincidence of high solar radiation received throughout the year and regular, abundant, and reliable rainfall. The production is achieved, overwhelmingly, high in the dense forest canopy of evergreen foliage. It is dark at ground level except where fallen trees create gaps. A characteristic of this biome is that often many tree seedlings and saplings remain in a suppressed state from year to year, and only leap into action if a gap forms in the canopy above them.

Indeed, almost all the action in a rain forest (not just photosynthesis but also flowering, fruiting, predation, and herbivory) happens high in the canopy. Apart from the trees, the vegetation is largely composed of plant forms that reach up into the canopy vicariously, by climbing the trees (vines and lianas, including many species of fig) or growing as **epiphytes**, which are plants that grow on other plants, rooted on the damp upper branches. The epiphytes depend on the sparse resources of mineral nutrients they extract from crevices and pockets of humus on the tree branches. The rich floras and faunas of the canopy are not easy to study; even to gain access to the flowers in order to identify the species of tree is difficult without the erection of tree walks. It is a measure of the problems of doing research in a rain forest that botanists have trained monkeys to collect and throw down flowers, and a research team has used hot air balloons to move over the canopy and work in it.

Most species of animals and plants in tropical rain forest are active throughout the year, though the plants

FIGURE 4.10 Each biome is illustrated with two photographs, one focusing on the detail of the vegetation and the other providing a distant view and emphasizing the great structural variation to be found among the world's terrestrial communities. The animals found in each of these biomes also cannot be ignored; they are obvious in the savanna photo, but invertebrate and vertebrate animals are busy behind the scenes in all the biomes. (a) Desert biome, here illustrated for both the near and distant view by the Mohave Desert, Nevada, USA. (b) Temperate forest biome, with the distant view shown by the Great Smokey Mountains, North Carolina, USA in the summer and the closer view from the White Mountains, New Hampshire, USA in the fall. (c) Distant view of a boreal forest in British Columbia, Canada, and close up view of one in Quebec, Canada. (d) Distant and close views for a savanna biome from the Masai Mara National Park, Kenya, one of the world's great savannas. (e) Tropical forest biome, with a distant view from the Amazon in Eduador and the close up view from the Amazon in Henri Pittier National Park, Venezuela. (f) A tall grass prairie, one type of temperature grassland,

(e)

(f)

(g)

FIGURE 4.10   (continued)

in South Dakota, USA. (g) Tundra biome with a distant view in Denali National Park, Alaska, USA, and a close up view from the Arctic National Wildlife Refuge, Alaska, USA.

may flower and ripen fruit in sequence. In Trinidad, for example, the forest contains at least 18 trees in the genus *Miconia*, whose combined fruiting seasons extend throughout the year (Figure 4.11). This contrasts with the situation in temperate latitudes where there is a much narrower fruiting season that most species share.

Species richness, which describes the number of species, is dramatically high for tropical rain forest (see Chapter 10), and communities never become dominated by one or a few species – a very different situation from the low biodiversity of northern coniferous forests. This situation raises some fundamental

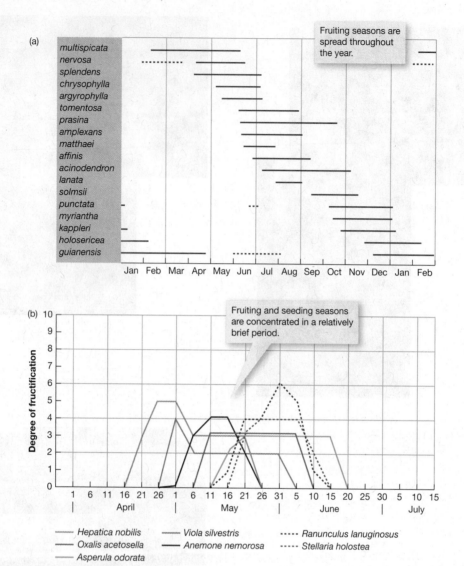

FIGURE 4.11 Contrasting patterns of fruit or seed production in tropical and temperate forests. (a) The fruiting seasons of 18 species of tree in the genus *Miconia* in the rain forest of Trinidad. (b) The seasonal production of fruit and seeds by herbs in a deciduous forest in Poland. (After Harper, 1977.)

questions that have proved very difficult to resolve. First, what is it about the evolutionary history of tropical rain forest that has allowed such diversity to evolve? Part of the answer relates to the comparative stability of patches of rain forest during the ice ages. As we saw in Section 2.5, it is thought that during these periods, drought forced tropical rain forests to contract into 'islands' (in a 'sea' of savanna), and these expanded and coalesced again as wetter periods returned. This would have promoted genetic isolation of populations, a phenomenon that is so important for speciation to occur (see Section 2.4). We may also ask why it is that among the diversity of species in tropical rain forests, a few have not dominated and suppressed the rest in a struggle for existence. We will see later (Chapter 10) that at least part of the answer is that populations of

specialized pathogens and herbivores develop near mature trees and attack new recruits of the same tree species nearby. Thus, the chance that a new seedling will survive can be expected to increase with its distance from a mature tree of the same species, reducing the likelihood of dominance by one or a few species in the forest.

The diversity of rain forest plants also provides for a corresponding diversity of resources for herbivores (Figure 4.12). A variety of fresh young leaves are available throughout the year, and a constant procession of seed and fruit production provides reliable food for specialists such as fruit-eating bats. Moreover, a diversity of flowers, such as epiphytic orchids with their specialized

tropical rain forest is also associated with high animal diversity . . .

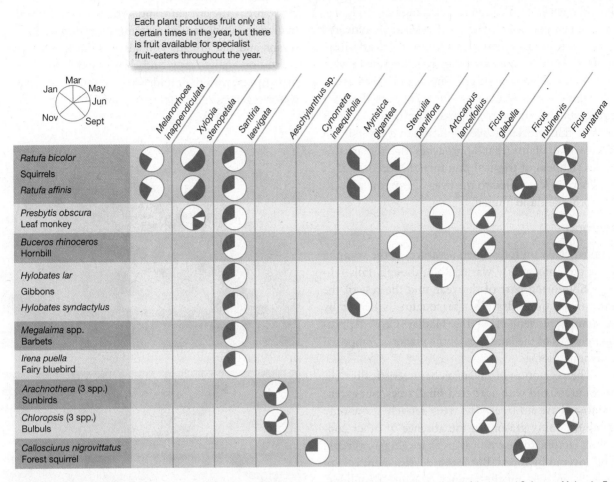

FIGURE 4.12   Animals (listed vertically) that feed on the fruit of trees (listed horizontally) at various times of the year at Selangor, Malaysia. Each circle is a calendar in which the feeding season is shown in blue. (After Harper, 1977.)

pollinating mechanisms, require a parallel specialized diversity of pollinating insects. Rain forests are the center of diversity for ants—43 species have been recorded on a single tree in a Peruvian rain forest. And there is even more diversity among the beetles (Figure 4.13); Erwin (1982) estimated that there are 18,000 species of beetle in 1 ha of Panamanian rain forest (compared with only 24,000 in the whole of the United States and Canada!).

There is intense biological activity in the soil of tropical rain forests. Leaf litter decomposes faster than in any other biome, and as a result the soil surface is often almost bare. The mineral nutrients in fallen leaves are rapidly released, and, as rainfall seeps down the soil profile, nutrients may be carried well below the levels at which roots can recover them. Almost all the mineral nutrients in a rain forest are held in the plants themselves, where they are safe from leaching. When such forests are cleared for agriculture, or the timber is felled or destroyed by fire, the nutrients are released and leached or washed away: on slopes the whole soil may go too.

. . . and intense soil activity

FIGURE 4.13   Evolution over the past 55 million years has led to approximately 200 species of neotropical leaf beetles in the genus *Cepaloleia*.

The full regeneration of soil and of a nutrient budget in a new forest may take centuries. Evidence of cultivated patches within rain forest can still be seen clearly from the air 40 years or more after they have been deserted.

All the other terrestrial biomes can be seen as the poor relations of tropical rain forest. They are all colder or drier and all are more seasonal. They have had prehistories that prevented the evolution of a diversity of animals and plants that approaches the remarkable species richness of tropical rain forest. Moreover, they are generally less suited to the lives of extreme specialists, both plant and animal.

Savanna

As in tropical rain forests, temperatures in the savanna biome are consistently warm. Rain, though, falls reliably only during part of the year, and the rest of the year can be quite dry. This seasonality is driven by the annual movement of the Hadley Cells. Approximately 9% of the land area of the planet is composed of savanna biome.

The vegetation of savanna characteristically consists of grassland with scattered small trees, but extensive areas have no trees. The trees actively grow only during the rainy season. In the absence of other controlling factors, these tropical areas would be expected to be covered by forest. But forest development is kept in check, in part by the limited moisture that limits their growth to the rainy season, but also by grazing and fire. In some savannas, herds of grazing herbivores (like zebra *Equus burchelli* and wildebeest *Connochaetes taurinus* in Africa; Figure 4.14) have a profound influence on the vegetation, favoring grasses (which protect their embryonic, actively dividing tissues in buds at or just below ground level) and hindering the regeneration of trees (because these same tissues are exposed to browsing animals). In other cases, fire is the critical thing. Fire, whether natural or human-induced, can be a common hazard in the dry season and, like grazing animals, tips the balance in the vegetation against trees and shrubs and favors perennial grasses, with their underground rhizomes and protected regenerating surfaces (Figure 4.15).

FIGURE 4.14 Great herds of grazing animals such as wildebeest are a dominant feature of the savannas of Africa.

(a)

(b)

(c)

FIGURE 4.15 Frequent fires such as this one in Kruger National Park, South Africa (a, top), help to maintain grasses, the primary food of many savanna animal populations. The middle picture (b) shows savanna in this park that has been purposely maintained with annual fires over many decades. When fires were suppressed over 35 years as part of an ecological experiment to study the effects of fire, brush encroaches and begins to outcompete the grasses (c, bottom).

Seasonal rainfall places the most severe restrictions on the diversity of plants and animals in savanna. Plant growth is limited for part of the year by drought, and there is a seasonal glut of food, alternating with shortage. As a consequence, the larger grazing animals suffer extreme famine (and mortality) in drier years. The strong seasonality of savanna ecology is well illustrated by its bird populations. An abundance of seeds and insects supports large populations of birds that migrate in from higher latitudes, but only a few species can find sufficiently reliable resources to be resident year round.

> seasonal glut and food shortage are characteristic of savanna

FIGURE 4.16 A horde of grasshoppers. Such invertebrates can be as important as vertebrate animals as grazers in temperate grasslands.

Temperate grasslands

Temperate grassland is the natural vegetation over large areas in every continent except Antarctica. These include the tall grass *prairie* of North America and *pampas* of South America, where rainfall is moderate and soils are rich, and the short grass *steppes* of Russia, typical of drier, more semiarid conditions. Temperate grasslands now only cover some 5% of the Earth's land surface, but another 9% used to be temperate grasslands but is now dominated by agriculture instead.

Temperate grasslands experience seasonal drought, but grazing animals also have a powerful impact. Populations of invertebrates, such as grasshoppers (Figure 4.16), are often very large and their biomass may exceed that of grazing vertebrates. The latter include bison (*Bison bison*), pronghorn antelope (*Antilocapra americana*), and gophers (*Thomomys bottae*) in North America, and saiga antelope (*Saiga tatarica*) and marmots (*Marmota bobac*) in Russia.

Many of these natural grasslands have been cultivated and replaced by arable annual grasslands of wheat, oats, barley, rye, and corn. Such annual grasses, together with rice in the tropics, provide the staple food of human populations worldwide. In fact, the vast increase in the size of the human population in historical times (see Chapter 14) has depended on the domestication of grasses for human food and feed for domestic animals, with a large amount of global crop production from these grains occurring on former temperate grasslands. At the drier margins of the biome, where cultivation is not economical, many of the grasslands are managed for meat or milk production, sometimes requiring a nomadic human lifestyle. The natural populations of grazing animals, especially buffalo in North America, have been driven back in favor of cattle, sheep, and goats. Of all the biomes, temperate

> of all the biomes, temperate grassland has been most transformed by humans

grassland is one of the most coveted, used, and transformed by humans.

Desert

In their most extreme form, the deserts are too arid to bear any vegetation, including both some hot deserts such as Death Valley in California and some of the cold deserts of Antarctica. Where there is sufficient rainfall to allow plants to grow in arid deserts, its timing is unpredictable. Approximately 10% of the surface area of Earth is covered by deserts.

Desert vegetation falls into two sharply contrasted patterns of behavior. Many species have an opportunistic lifestyle, stimulated into germination by the unpredictable rains (physiological 'internal' clocks are useless in this environment). They grow fast and complete their life history by starting to set new seed after a few weeks. These are the species that can occasionally make a desert bloom; the ecophysiologist Fritz Went called them "belly plants" because only someone lying on the ground can appreciate their individual charm.

> contrasting patterns of behavior of desert plants

A different pattern of behavior of arid desert plants is to be long-lived with sluggish physiological processes. Cacti and other succulents, and small shrubby species with small, thick, and often hairy leaves, can close their stomata and tolerate long periods of physiological inactivity and have the CAM specialized photosynthetic systems to maximally conserve water (see Chapter 3). In many deserts, freezing temperatures are common at night and tolerance of frost is almost as important as tolerance of drought.

The relative poverty of animal life in deserts reflects the low productivity of the vegetation and

> animal diversity is low in deserts

the indigestibility of much of it. Desert perennials including species of wormwood (*Artemisia*) and creosote plant (*Larrea mexicana*) in the southwestern United States, and mallee species of *Eucalyptus* in Australia, carry high concentrations of chemicals that are repellent to herbivores (Figure 4.17). Ants and small rodents rely on seeds as a relatively reliable source of both energy and water, whereas bird species are largely nomadic, driven by the need to find water. Among carnivores globally, only those of the desert can survive on the water they obtain from their food.

Temperate forest

Like all biomes, temperate forest includes, under one name, a variety of types of vegetation. At its low-latitude limits in Florida and New Zealand, winters are mild, frosts and droughts are rare, and the vegetation largely consists of broad-leaved evergreen trees. At its northern limits in the forests of Maine and the upper Midwest of the United States, the seasons are strongly marked, winter days are short, and there may be 6 months of freezing temperatures. Deciduous trees, which dominate in many, temperate forests, lose their leaves in the fall and become dormant after transferring much of their mineral content to the woody body of the tree. On the forest floor, diverse floras of perennial herbs often occur, particularly those that grow

FIGURE 4.17 The slow-growing creosote bush (*Larrea tridentate*) is well defended with chemicals against herbivory.

quickly in the spring, before the new tree foliage has developed. Temperate forests make up 8% of the terrestrial biosphere, with a slightly larger area covered by temperate deciduous forests than by temperate evergreen forests.

All forests are patchy because old trees die, providing open environments for new colonists. This patchiness is on an especially large scale after hurricanes fell the older and taller trees or after fire kills the more sensitive species. In temperate forests the canopies are often composed of a mixture of long-lived species, such as red oaks (*Quercus rubra*) in the Midwest of the United States, and colonizers of gaps, such as sugar maple (*Acer saccharum*).

Temperate forests provide food resources for animals that are usually very seasonal in their occurrence (compare Figure 4.11b with 4.11a), and only species with short life cycles, such as leaf-eating insects, can be dietary specialists. Many of the birds of temperate forests are migrants that return in spring but spend the remainder of the year in warmer biomes.

Soils are usually rich in organic matter that is continually added to, decomposed, and often churned by a rich community of other decomposers. Waterlogging and low pH, in some locations, inhibit the decomposition of organic matter and cause the accumulation of peat.

> temperate forest soils are rich in organic matter

Large swathes of deciduous forest in Europe and the United States have been cut down to provide land for agriculture, but these have sometimes been allowed to regenerate as farmers abandoned the land (a conspicuous feature in New England and New York state, where many farmers abandoned the land and moved further west to more productive lands between the 1890s and 1930s). It takes time for the forest to regrow from agricultural land, often going through several different types of species of trees in the process of **succession** (discussed further in Chapter 9). The first species to colonize are rapidly growing species, but these cannot compete with the more slowly growing species that eventually come to dominate and shade them out. Some legacy of the old farming activities continues in these recovered forests, as the farmers of the 1800s increased the amount of phosphorus and other nutrients in the soil beyond the natural levels.

Boreal forest (taiga)

Boreal forests (also known as taiga, particularly in Russia) are coniferous and occur in regions where the short growing season and the polar cold of winter limit the vegetation and its associated fauna. The summers tend to be warmer, though, with some influence from temperate air masses as the Ferrel cells and Polar cells move further north. Boreal forests make up approximately 8% of the land surface of the planet.

The tree flora is quite limited. Where winters are less severe, the forests may be dominated by evergreen pines (*Pinus* species) and firs (*Abies*) as well as deciduous trees such as larch (*Larix*), birch (*Betula*), or aspens (*Populus*), often as mixtures of species. Farther north, these species give way to monotonous single-species forests of spruce (*Picea*) over immense areas of North America, Europe, and Asia. This provides an extreme contrast to the biodiversity of tropical rain forests.

Much of the areas now occupied by boreal forests (as well as much of the northern temperate forests) were occupied by the ice sheet during the last ice age, which started to withdraw only 20,000 years ago. Temperatures are now as high as they have ever been since that time, but the vegetation has not yet caught up with the changing climate and the forests are still spreading north. The very low diversity of northern floras and faunas is in part a reflection of a slow recovery from the catastrophes of the ice ages.

Permafrost is a major environmental constraint in boreal spruce forests. The water in the soil remains frozen throughout the year, creating permanent drought except when the sun warms the very surface. The root system of spruce can develop in only the superficial soil layer, from which the trees derive all their water during the short growing season.

Low-diversity communities provide ideal conditions for the development of disease and epidemics of pests. For example, the spruce budworm (*Choristoneura fumiferana*) lives at low densities in immature northern forests of spruce. As the forests mature, the budworm populations explode in devastating epidemics. These wreck the old forest, which then regenerates with young trees. This cycle takes about 40 years to run its course.

> the low diversity of northern coniferous forest provides ideal conditions for pest outbreaks

Periodic and often catastrophic fires are another major disturbance in boreal forests (Figure 4.18). During major fires, even some of the organic matter in the soil burns, which can contribute significantly to the global flux of carbon dioxide to the atmosphere.

FIGURE 4.18   Major fires have long tended to occur every 75 to 100 years in many boreal forests, and are critically important to the functioning of this biome.

In many boreal forests, major fires occur every 75 to 100 years. Most of the mature trees die resetting the stage for ecological succession. The result is not only a forest of low species diversity but a very even age distribution of trees as well.

Tundra

To the north of the boreal forest, polar air masses and cold temperatures dominate even in the summer. Here, the vegetation changes to tundra, with its low shrubs, grasses, sedges, and small flowering plants, as well as mosses and lichens. Permafrost is a dominant feature, as in the boreal forest, but even less melting of surface soil occurs in the summer. In the very coldest areas, rooted plants such as grasses and sedges disappear because of the permafrost, leaving only mosses and lichens that grow on rather than in the soil. The tundra makes up some 5% of the terrestrial land surface of Earth.

The number of species of higher plants (excluding mosses and lichens) decreases from 600 species in the Low Arctic of North America to 100 species in the High Arctic (north of 83°) of Greenland and Ellesmere Island. In contrast, the flora of Antarctica contains only two native species of vascular plant and some lichens and mosses that support a few small invertebrates. The biological productivity and diversity of Antarctica are concentrated at the coast and depend almost entirely on resources derived from the sea.

The tundra fauna has long intrigued ecologists because populations of lemmings, mice, voles, and hares (herbivores), and the fur-bearing carnivores (such as lynx and ermine) that feed on them, pass through remarkable cycles of expansion and collapse (see Chapter 7). Lemmings (*Lemmus*) are famous for their population cycles and the role they play in the tundra. When the snow melts during a period when the lemming cycle is at a high point, the animals are exposed and they support large migratory populations of predatory birds (owls, skuas, gulls) and mammals such as weasels. Reindeer and caribou (they are the same species, *Rangifer tarandus*) occur in migrant herds capable of foraging on lichens of the tundra, which they can reach through the snow cover.

> dramatic animal population cycles are characteristic of northern biomes

Permafrost and generally cold temperatures severely limit decomposition in tundra ecosystems, so soils tend to be very organic rich. Until recently, fires were very rare in the tundra. However, over the past decade, global warming has led to many more and much larger fires in the Arctic tundra (Figure 4.19), for two reasons. First, somewhat warmer temperatures have led to higher rates of evapotranspiration, and therefore drier soils that can more easily burn. And second, convective thunderstorms—once almost unheard of in the Arctic—are becoming more common. The resulting lightning provides the spark to set off a fire. These tundra fires, as with fires in boreal forests, may become an increasingly important source

FIGURE 4.19 Historically, fires were very rare in the tundra biome, but they are now becoming more common because of global climate change. This shows a fire along the Anaktuvuk River in Alaska in 2007, a fire set by lightning. In the past, thunderstorms and associated lightning rarely occurred in tundra biomes, as the air was too cool to support the development of these storms. Lightning and associated fires will become more common as the tundra region warms further.

of carbon dioxide to the atmosphere. That is, global warming is providing a feedback in the tundra biome, where increasing fires further increase concentrations of atmospheric carbon dioxide.

The future distribution of terrestrial biomes

It is clear that the distribution of biomes has changed in the past in response to the ebb and flow of the ice ages. Nowadays, we are also acutely aware that their boundaries are probably on the move again. Predicted changes in global climate over the next few decades can be expected to result in dramatic changes to the distribution of biomes over the face of the Earth (Box 4.2). But the exact nature of these changes remains uncertain.

4.3 AQUATIC ECOSYSTEMS ON THE CONTINENTS

The aquatic ecosystems on the continents—lakes, ponds, rivers, streams, and wetlands—cover less than 1% of the land mass of the Earth and represent only 0.007% of the water on the planet. Most of the water on Earth is in the oceans, and most freshwater is frozen in glaciers and ice sheets. Nonetheless, these aquatic ecosystems are immensely important in the ecological functioning of the landscape around them, are critical resources of human society, and of course are ecologically interesting in their own right.

Streams and rivers

Streams and rivers throughout much of the world have been tapped, dammed, straightened, rerouted, dredged, and polluted since the beginning of civilization. Understanding the impacts and sustainability of some of these practices begins with understanding the basics of stream and river ecology.

Streams and rivers are characterized by their linear form, unidirectional flow, fluctuating discharge, and unstable beds. Streams and rivers form a continuum, with several small streams joining together in their downstream flow to form a river, and smaller rivers joining together to form larger and larger rivers. The water entering a stream comes from its **watershed**—the land area where all the water draining from it comes to the particular stream or river—through groundwater flows, surface runoff, or both. The ecological functioning of streams and rivers is intimately connected

had been on the *Alvin* were found to have suffered no discernible degradation, and a cheese sandwich—although soggy with seawater—was still edible enough to actually be eaten. Despite the slow rate of microbial decomposition, little organic matter persists in the deep ocean, and the organic content of sediments is remarkably low, in part because the inputs of organic matter to the deep ocean from the surface photic zone are relatively small.

The deep ocean displays extraordinary biological productivity, including worms, crustaceans, mollusks, and fish found nowhere else. Many of the invertebrate animals are tiny, have very low metabolic rates, and possess a lifespan that may last for decades.

Hot vents (see Box 2.2) provide an ecologically fascinating oasis of biological diversity along tectonic plate spreading centers in the deep ocean. Only discovered in the past 50 years, these vents pump out extremely hot water laden with reduced chemical substances, such as hydrogen sulfide. Special chemo-autotrophic bacteria use the chemical energy of these substances, oxidizing it to gain energy for producing new organic matter from carbon dioxide (see Chapter 3). At the deep-sea hot vents, many of these chemo-autotrophic bacteria are **symbionts** of animals, meaning they live inside them, providing them with nutrients, such as tubeworms that can grow to more than 2 meters tall. Large animal populations are commonly supported off this chemical energy at the hot vents. But these vents are a rarity in the deep oceans, and an exception to the general rule that life in the deep sea is driven by the slow use of a slow rain of dead organic matter material from the photic zone into very cold water.

Subtropical gyres

The planet has five subtropical ocean **gyres**, in the north and south Atlantic and Pacific Oceans as well as the Indian Ocean. These are large masses of semi-isolated surface water surrounded by a circular current of water moving clockwise for the two gyres in the northern hemisphere and counterclockwise for the three gyres in the southern hemisphere (Figure 4.6). Combined, they cover approximately half the surface area of the world's oceans, or 35% of the entire surface of the planet: more than the entire land area on all continents. The gyre in the North Atlantic Ocean is called the Sargasso Sea, named after the large mats of floating *Sargassum* algae commonly found there. It is the only gyre named a "sea," and the only named sea whose boundaries are defined by ocean currents rather than land masses. But the term "sea" is a fitting description for these gyres, since their water has only very modest exchanges across the boundary currents with neighboring ocean masses. Only the occasional eddy carries parcels of the gyre water out and parcels of neighboring water masses into the gyres.

Rates of primary productivity in the gyres are the lowest for any aquatic ecosystems on Earth, and rival the rates found in any but the most arid deserts on land (Figure 4.30 and Chapter 11). The water is clear and blue, in part because primary production

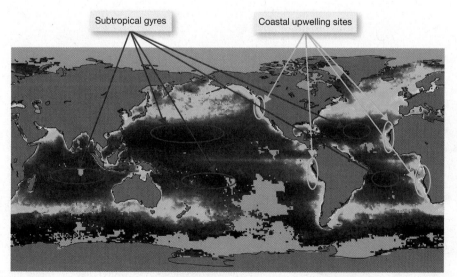

FIGURE 4.30 Rates of primary productivity in the oceans of the world. The general areas of the five subtropical gyres and many of the larger upwelling ecosystems are indicated.

http://geoserver.isciences.com/DataBlog/?p=25

Net Primary Productivity (grams Carbon per m² per year)

0   200   400   600   800

and phytoplankton biomass are low, and in part because of the isolation from the input of particles from erosion on land carried to the oceans in rivers. The clarity of the water allows light to penetrate as deeply as anywhere in the world's oceans, up to 200 meters (Figure 4.27). However, nutrients are in very short supply. In fact, in the surface waters, the concentrations of inorganic nitrogen and phosphorus are as low as in distilled water. Abundant nutrients are near, in the waters of the deep ocean just a few hundred meters below these surface-water gyres. But those nutrients are largely trapped in the deep ocean, with very little exchange by limited mixing of waters across the strong thermocline. The very low nutrient supply severely limits the rate of primary productivity.

Most of the phytoplankton species in these subtropical gyres are very small, an adaptation to the low nutrient availability: small cells have a high surface to volume ratio, which provides a lot of area for nutrient uptake relative to the volume of chlorophyll and other photosynthetic apparatus within the cell (see Section 3.2). These very small phytoplankton are consumed by very small zooplankton, which in turn are then consumed by somewhat larger animals, leading to long food chains. Since some energy is lost at each step in the food chain (see Chapter 11), the flow of energy to top predators has a rather low efficiency compared to most other aquatic or terrestrial biomes. Compared to other parts of the oceans, the production of fish is low even relative to the low rate of net primary productivity.

Biodiversity, however, is very high in these subtropical gyres, perhaps the highest seen in any plankton-dominated ecosystems on Earth. Ecologists remain intrigued by what has led to this high diversity. Candidate explanations include the relative lack of disturbance over geological time scales, the low disturbance on shorter times, and perhaps the specialization that results from the low-nutrient status. Patterns in biodiversity generally, and their potential causes, are discussed in Chapter 10.

Coastal upwelling systems

Upwelling systems provide a sharp contrast to the nutrient-limited but light-abundant subtropical gyres. In upwelling systems, nutrient-rich deep ocean waters move steadily upward into the photic zone. Primary productivity and phytoplankton biomass are very high, which limits the penetration of light into the water column. The rising water keeps the phytoplankton buoyed near the surface, though, so they continue to have plenty of light.

Coastal upwelling is caused by strong prevailing winds that blow parallel to coastlines, but due to the Coriolis effect, the water moves at a 90° angle to the prevailing wind. When this is away from land, as it is off the coast of Oregon and California in the United States and Baja California in Mexico, the offshore-flowing waters are replaced with new water rising (upwelling) from the depths (Figure 4.31). Other major coastal upwelling areas occur off the western coast of Chile and Peru in the South Pacific Ocean, the western coast of South Africa in the South Atlantic Ocean, and the western coast of Mauritania and Morocco in the North Atlantic Ocean (Figure 4.30). A very strong upwelling sets up seasonally to the east of Somalia in

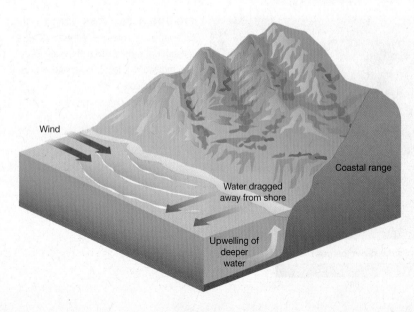

Wind

Coastal range

Water dragged away from shore

Upwelling of deeper water

FIGURE 4.31 Prevailing winds blowing from the north along western coast lines in the Northern Hemisphere move water away from shore. Newly upwelled water from below rises to replace this water. The upwelled water is both cold and extremely nutrient rich.

the Indian Ocean, driven by the southwestern monsoon winds. Cumulatively, these five upwelling areas produce 25% of the world's fishery catch, yet they cover only 5% of the surface area of the oceans.

The Southern Ocean—a large water mass that surrounds Antarctica—is another site of large-scale coastal upwelling. Strong prevailing winds blow from the west around the coast of Antarctica. Interacting with the Coriolis effect, surface waters move northward away from the continent, again being replaced by deep ocean water.

In contrast to the gyres, the phytoplankton in upwelled waters tend to be relatively large, with a low ratio of surface area to volume, and are grazed by relatively large species of zooplankton, which in turn can be grazed by plankton-eating fish and whales, leading to relatively short food chains that efficiently transfer energy. This efficiency, combined with the high rates of primary productivity, is what leads to the extraordinarily productive fisheries of upwelling areas. While productivity is high, however, biodiversity is low. This association of low diversity with nutrient-rich, productive ecosystems is something else taken up in Chapter 10.

Broad continental shelves

The major coastal upwelling systems all occur along coasts that have a narrow continental shelf. Deep ocean waters occur, for example, within 1 km along parts of the California coast. Along other coastlines, however, the continental shelf extends much further. Globally, the continental shelves stretch out on average 80 km from shore. The water depth on the average shelf is less than 150 m. Often, high nutrient inputs lead to high, and sometimes very high, rates of primary productivity. The water column is often stratified, with a relatively shallow photic zone that favors high production even though phytoplankton biomass (and resulting light absorption) can be high.

The nutrients come from several sources, as we discuss in detail in Chapter 11. One of these is rivers. A notable example is in the input of nutrients from the Mississippi River to the shelf in the northern Gulf of Mexico. The Mississippi carries huge quantities of nitrogen and phosphorus from the agricultural breadbasket of the American Midwest, and since the 1970s, this elevated nutrient load has over-stimulated primary production on the shelf. As the phytoplankton die in the bottom waters during summer when the water is stratified, oxygen is depleted, leading to a "dead zone" of 20,000 km^2 or more, an area where fish and other animals cannot survive.

Hans-Petter Fjeld

FIGURE 4.32   Large schools of fish are a dominant feature of the productive shoals of the continental shelves. The cod shown here once had tremendously high numbers on Georges Bank and elsewhere along the New England coast, but decades of overfishing has greatly diminished their populations.

The deep ocean also provides nutrients to the continental shelves, notably in areas such as the North Sea and Georges Bank and the Grand Banks off the northeastern coast of North America, some of the most productive fisheries in the world. Georges Bank, an area of some 30,000 km^2, is, even at its closest, 100 km off the coast of Massachusetts, yet in places it is only a few meters deep. This shallowness contributes to high rates of primary production, since phytoplankton are always near the surface where the light is most intense. The incredible fisheries of Georges Bank and the Grand Banks lured European fishers 1,000 years ago, and were a major attraction to the earlier European settlers of New England and Nova Scotia in the 16th and 17th centuries. But the once-abundant halibut of Georges Bank had already been overfished and depleted by 1850 (Bigelow & Shroeder, 1953), and overfishing of other fish species has continued since (see Chapter 14). The cod fish that once swarmed on Georges Bank, causing the surface water to "boil" (and that gave nearby Cape Cod its name), are now largely gone (Figure 4.32).

As with the coastal upwelling biome, most of the productive waters on the broader continental shelves are dominated by large phytoplankton and have efficient energy transfer in relatively short food webs. Biodiversity tends to be relatively low.

Nearshore coastal marine ecosystems

The final types of ocean ecosystems we consider in this chapter are the diverse types that lie right on the coastal fringe, including the intertidal zone, the near-shore submerged littoral zone, and estuaries. These ecosystems serve as nursery grounds for many species of fish and invertebrates and provide critical habitat for other species, including sea turtles, shore birds, and many species of marine mammals. They are often beautiful, and provide enjoyment to millions of people who live near the coast or travel to the coast on holidays. And since they are accessible, they are also the parts of the ocean most studied by ecologists. Here, we briefly describe some of these coastal marine ecosystems, but the list is in no way exhaustive.

The intertidal zone is the part of the shoreline that is submerged at high tide but exposed to air at low tide. The extent of this zone depends on the height of tides and the slope of the shore. Away from the shore, the tidal rise and fall are rarely greater than 1 m, but closer to shore the shape of the land mass can funnel the ebb and flow of the water to produce extraordinary tidal ranges of, for example, nearly 20 m in the Bay of Fundy (between Nova Scotia and New Brunswick, Canada). In contrast, the shores of the Mediterranean Sea experience scarcely any tidal range. Both algae and animals are profoundly affected by the physical force of wave action, and by the extremes of hot and cold that can occur during low tide (changes in air temperature and far less moderated than those of water). Low tide also presents a continued challenge of desiccation.

> The intertidal is underwater for only part of each day . . .

On rocky shores, anemones, barnacles, and mussels attach themselves securely and permanently to the substrate and filter planktonic plants and animals from the water when the tides cover them. Other animals, such as limpets and snails, move to graze micro-algae from the rocks, and crabs move with the tides and use rock crevices as refuges. In addition to micro-algae, the flora in a rocky intertidal zone may include larger algae, such as kelps that fix themselves to the rock with specialized 'holdfasts.'

Environments are quite different on shallow sloping shores on which the tides deposit and stir up sand and mud. Here the dominant animals are mollusks and polychaete worms, living buried in the substrate and feeding by filtering the water when they are covered by the tides. Where sandy or muddy intertidal areas are adequately protected from waves, wetlands often develop. In the temperate zone, these are commonly salt marshes, such as the extensive ones dominated by the flowering grass *Spartina* along the east coast of the United States. In the tropics, swamps dominated by mangrove trees are common.

Ralph A. Clevenger/Corbis

FIGURE 4.33   The giant kelp (*Macrocystis pyrifera*) grows rapidly to lengths of 50 m and more. This alga is not only extremely productive but provides structure to the kelp bed ecosystems that occur commonly along the temperate coasts of many regions in the world.

Below the low-tide mark begins the submerged littoral zone. A variety of types of communities occur in this zone, depending upon the characteristics of the bottom (rocky vs. sandy or muddy, etc.). Generally light penetrates to the bottom, so primary producer organisms that are attached to the bottom are very important, providing structured habitat in addition to often very high rates of primary production. The giant kelp (*Macrocystis pyrifera*) is the largest algal species in the world and grows to lengths of over 50 m along several coastlines including the west coasts of North and South America, South Africa, New Zealand, and southern Australia, where it can adequately attach its holdfast to rocks and where the water temperatures are below 20°C (Figure 4.33). Rates of primary production are among the highest found anywhere on Earth, commonly exceeding 1,000 g C per square meter per year.

> The nearshore submerged littoral is permanently underwater. . . .

In other regions of the world, smaller but still highly productive beds of other types of macro-algae—including other species of kelp, but also red and green algae—thrive on rocky bottoms. On sandy bottoms, rooted grass species such as eelgrass (*Zostera marina*) in the temperate zone and turtle grass (*Thalassia testudinum*) in the tropics can form vast meadows (Figure 4.34). The diversity of these seagrass meadows, particularly in the tropics, can be high. Coral reefs are also a common feature of the submerged littoral zone along many tropical coastlines, where water quality is sufficiently high to support their growth. Coral reefs boast some of the highest biodiversity for any type of ecosystem on Earth. Global change, however, poses huge risk

Robert W. Howarth, Ph.D

FIGURE 4.34   This tropical seagrass meadow in the Caribbean Sea in Berlize shows remarkable diversity of both rooted vascular plants and algae attached by holdfasts to the sand.

to corals, both because warmer temperatures can lead to coral "bleaching"—the loss of algal symbionts from the coral—and because ocean acidification from the increasing amount of carbon dioxide in the atmosphere makes it more difficult for the coral animals to secrete their carbonate reefs.

Estuaries are salty, semi-enclosed water bodies that exchange water with more open coastal waters. Often estuaries occur near the mouth of a river, where outflowing freshwater mixes with seawater. Other estuaries have rather small inputs of freshwater, as for example the waters behind barrier islands that exchange water tidally with coastal waters. Estuaries provide an intriguing mix of the conditions normally experienced in rivers, shallow lakes, and ponds, and the nearshore littoral zone.

> Estuaries are where rivers meet the sea . . .

In one common type of estuary, the salt-wedge estuary, dense salt water enters from the ocean and travels along the bottom, up the estuary, while freshwater (which is less dense than seawater) entering from a river flows downstream over the salt water. Some mixing of the fresh and salt water occurs over the distance of the salt wedge, and so the surface waters gradually become saltier down the estuary and the bottom waters gradually fresher up the estuary. The shape of the salt water wedge is largely determined by the size of the discharge of the river flowing into the estuary; high discharge tends to create a smaller wedge of salt water and less mixing. The strong gradients in salinity, in both space and time, are reflected in a specialized estuarine fauna. Some animals cope through particular physiological mechanisms. Others avoid the variable salt concentrations by burrowing, closing protective shells, or moving away when conditions do not favor them.

The salinity gradient over depth at any place in the salt-wedge estuary provides very strong stratification—year round—in addition to any temperature stratification that may develop in summer. This often makes estuaries prone to low-oxygen conditions in the bottom waters, and even anoxia (the complete absence of oxygen). Further, the dynamics of the salt wedge act to trap particles in the estuary. This can aggravate pollution from toxic substances, such as PCBs, which are adsorbed to particles. Estuaries also frequently receive very large inputs of nutrients in rivers and from the sewage of major cities – such as Venice, New York, Washington, San Francisco, and Hong Kong – that often lie on the estuary: the estuary is the harbor of the city. In the United States, two-thirds of estuaries are moderately to severely degraded from nutrient pollution. As with all biomes, human impacts add an additional dimension to the variations that make each environment unique, notwithstanding our broad-brush categorization.

SUMMARY

The world's climate

The intense heating near the Equator is redistributed across the planet through the movement of air in the atmosphere and ocean currents. In the atmosphere, heat is moved from the Equator toward the poles by sets of interacting cells: the Hadley, Ferrel, and Polar Cells. As air rises in these cells, it cools, leading to condensation of water vapor and high rates of precipitation. Where the air in cells falls back to the surface, it warms and becomes very dry. The resulting patterns of temperature and precipitation—interacting with ocean currents and mountain ranges—create the Earth's mosaic of climate.

Terrestrial biomes

A map of biomes shows where we find areas of land dominated by plants with characteristic life forms. Tropical rain forests, savannas, temperate grasslands, deserts, temperate and boreal forests, and tundra are each found where their characteristic combinations of temperature and precipitation patterns favor them and each displays a characteristic profile of productivity and biological diversity.

Aquatic ecosystems on the continents

Water makes up a very small area of the continents, but these aquatic ecosystems play a critical role in human society and in the ecological functioning of the landscape around them. Streams and rivers form a continuum that drains their watersheds and their ecological functioning is intimately connected to the watersheds. In ponds and small lakes, sunlight often reaches the bottom, and rooted plants can be important primary producers. In deeper lakes, no sunlight can reach the bottom sediments, and phytoplankton are the dominant primary producers. Wetlands ecosystems are intermediate between aquatic and terrestrial ecosystems. Some types, particularly marshes and swamps, are among the most productive ecosystems on Earth.

Ocean biomes

Ocean currents and stratification result in a series of types of water masses, generating ocean biomes structured along gradients in the availability of nutrients and of light. The deep ocean is dark and cold and life is dependent on a rain of dead organic matter from the ocean's surface waters. In the subtropical gyres, light is plentiful, and the low nutrient availability leads to very low rates of primary production, but biological diversity is very high. In coastal upwelling systems, cold nutrient-rich water from the deep oceans rises to the surface providing nutrients, where extremely high rates of primary production result, often leading to very productive fisheries. In the broad continental shelves, light in the shallow surface layer is high, and nutrient supply is also often high.

REVIEW QUESTIONS

1   From your understanding of the global climate system, why are deserts more likely to be found at around 30° latitude than at other latitudes?

2   How would you expect the climate to change as you crossed from west to east over the Rocky Mountains?

3   Why do we never see a tundra or boreal forest ecosystem that has rainfall as high as in tropical forests? Why don't we ever see cold-climate biomes with high rainfall?

4   Compare and contrast the role of fires in tropical forests, savannas, deserts, boreal forests, and tundra.

5   Describe how the logging of a forest may influence the community of organisms inhabiting a stream running through the affected area.

6   Why do we never see an ocean biome that has both deep penetration of light into the water column and high nutrient availability?

7   Compare and contrast the role of nutrients in structuring the diversity and food webs of subtropical gyres and coastal upwelling ecosystems.

Challenge Questions

1   Biomes are differentiated by gross differences in the nature of their communities, not by the species that happen to be present. Explain why this is so.

2   Why is diversity so high in the deep ocean?

3   Diversity is much lower in most lakes than in the subtropical gyres of the oceans. Why do you think this might be?

Part 3

Individuals and Populations

Andrea Gingerich/Getty Images

Chapter 5

Birth, death, and movement

CHAPTER CONTENTS

KEY CONCEPTS

After reading this chapter you will be able to:

- explain how counting individuals is essential in helping us understand the distribution and abundance of organisms and populations

- describe the range of life cycles and patterns of birth and death exhibited by different organisms

- explain the role of life tables and fecundity schedules

- describe how dispersal and migration can affect the dynamics of populations

- explain the impact of intraspecific competition on birth, death, and movement, and hence on whole populations

- identify the benefits and limitations of identifying life history patterns linking types of organism to types of habitat

M any questions in ecology, including some of the scientifically most fundamental and those most crucial to immediate human needs and aspirations, amount to attempts to understand the distributions and abundances of organisms, and the processes—birth, death, and movement – that determine distribution and abundance. In this chapter, these processes, methods of monitoring them, and their consequences are introduced.

5.1 POPULATIONS, INDIVIDUALS, BIRTHS AND DEATHS

As ecologists, we try to describe and understand the distribution and abundance of organisms. We may

> what is a population?

do so because we wish to control a pest or conserve an endangered species, or simply because we are fascinated by the world around us. A major part of our task, therefore, involves studying changes in the size of populations. We use the term **population** to describe a group

(a)

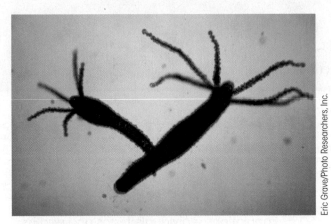

Nigel Cattlin/Photo Researchers, Inc.

Eric Grave/Photo Researchers, Inc.

(b)

Jane Legate/Getty Images, Inc.

© Sanamyan/Alamy Limited

FIGURE 5.1 Modular plants (on the left) and animals (on the right), showing the underlying parallels in the various ways they may be constructed. (a) Modular organisms that fall to pieces as they grow: duckweed (*Lemna* sp.) and *Hydra* Sp. (b) Freely branching organisms in which the modules are displayed as individuals on 'stalks': a vegetative shoot of a higher plant (*Lonicera japonica*) with leaves (feeding modules) and a flowering shoot, and a hydrozoa (*Extopleura larynx*) bearing both feeding and reproductive modules.

FIGURE 5.1 (continued)
(c) Stoloniferous organisms in which colonies spread laterally and remain joined by 'stolons' or rhizomes: strawberry plants (*Fragaria*) reproducing by means of runners and a colony of the hydroid *Tubularia crocea*. (d) Tightly packed colonies of modules: a tussock of yellow marsh saxifrage (*Saxifraga hirculus*) and a segment of the sea fan, *Acanthogorgia*. (e) Modules accumulated on a long, persistent, largely dead support: an oak tree (*Quercus robur*) in which the support is mainly the dead woody tissues derived from previous modules and a gorgonian coral in which the support is mainly heavily calcifed tissues from earlier modules.

of individuals of one species. What actually constitutes a population, though, varies from species to species and from study to study. In some cases, the boundaries of a population are obvious: the sticklebacks occupying a small lake are 'the stickleback population of the lake.' In other cases, boundaries are determined more by an investigator's purpose or convenience. Thus, we may study the population of lime aphids inhabiting one leaf, one tree, one stand of trees, or a whole woodland. In all cases, though, the most fundamental characteristic of a population is the number of individuals that compose it: populations grow or decline by changes in those numbers.

The processes that change the size of populations are birth, death, and movement (either into or out of that population). Trying to understand changes in population size is important because ecology is concerned not just with comprehending nature but often also with predicting or controlling it. We might, for example, wish to reduce the size of a population of rabbits that can do serious harm to crops. We might do this by increasing the death rate by introducing the myxomatosis virus to the population, or by decreasing the birth rate by offering them food that contains a contraceptive. We might encourage their emigration by bringing in dogs, or prevent their immigration by fencing.

birth, death, and movement change the size of populations

Similarly, a nature conservationist may wish to increase the population of a rare endangered species. In the 1970s, a rapid decline became apparent in the numbers of bald eagles and other birds of prey in the United States. This might have occurred because their birth rate had fallen, or their death rate had risen, or patterns or rates of movement had changed. Eventually the decline was traced to reduced birth rates. The insecticide DDT was widely used at the time (it is now banned in the United States, although it is still widely used in many developing countries) and had been absorbed by many species on which the birds preyed. As a result, it accumulated in the bodies of the birds themselves, leading the shells of their eggs to become so thin that the chicks often died before hatching (discussed further in Chapter 14). Conservationists charged with helping populations of the bald eagle to recover had to find a way to increase the birds' birth rate. The banning of DDT did this.

What is an individual?

A population is characterized by the number of individuals it contains, but for some kinds of organism it is not always clear what we mean by an individual. Often there is no problem, especially for *unitary* organisms, such as birds, insects, reptiles, and mammals. The whole form of **unitary** organisms, and their program of development, is predictable and 'determinate.' An individual spider has eight legs. A spider that lived a long life would not grow more legs.

unitary and modular organisms

But none of this is so simple for *modular* organisms such as trees, shrubs, and herbs, chain-forming bacteria and algae, corals, sponges, and very many other marine invertebrates. **Modular** organisms grow by the repeated production of 'modules' (leaves, individual cells, coral polyps, etc.). Most are rooted or fixed, not motile (Figure 5.1), and both their structure and their precise program of development are not predictable but *in*determinate. After several years' growth, depending on circumstances, the same germinating tree seed could either give rise to a stunted sapling with a handful of leaves or a thriving young tree with many branches and thousands of leaves. Similarly, we could count the individual trees in a forest, but unless we also noted the numbers of branches and leaves on each, or at least whether they were stunted or thriving, we would have only a very limited idea of the 'size' of the tree population.

In modular organisms, then, we need to distinguish between the *genet* and the module. **Genet** is short for genetic individual, starting life as a single-celled zygote and considered dead only when all its component modules have died. An individual **module**, by contrast, starts life as a multicellular outgrowth from another module and proceeds through its own life cycle to maturity and death, even though the form and development of the whole genet are indeterminate. When we write or talk about populations, we usually think of unitary organisms, perhaps because we ourselves are unitary. There are certainly many more unitary than modular species. But modular organisms are not rare exceptions or oddities. Most of the living matter (biomass) on Earth and a large part of that in the sea is composed of modular organisms: the forests, grasslands, coral reefs, algal mats and colonies, and peat-forming mosses.

modular organisms are themselves populations of modules

Counting individuals, births, and deaths

Even with unitary organisms, we face enormous problems when we try to count what is happening to populations in nature, and a great many ecological questions remain unanswered because of these problems. If we want to know how many fish there are in a pond, we might obtain an accurate count by putting in poison and counting the dead bodies. But apart from the questionable morality of doing this, we usually want to continue studying a population after we have counted it. It is not too difficult to count the numbers of large mammals such as deer on an isolated island. And occasionally it may be possible to trap alive all the individuals in a population, count them, and then release them. With birds, for example, it may be possible to mark all

the difficulties of counting

Quantitative Aspects

Mark–recapture methods for estimating population size

If we capture a representative sample of individuals from a population, mark them in some way (paint spots, leg rings) and then release them, then later, if another representative sample is captured, the proportion that is marked gives an estimate of the size of the whole population (see Figure 5.2). For example, we might capture and mark 100 individuals from a population of sparrows and release them back into the population. If we later sample a further 100 individuals from the population and find half are marked, we could argue in the following way. Half the sample are marked; the sample is representative of the whole population; therefore, half the population are marked; 100 individuals were given a mark; therefore, the whole population is composed of about 200 individuals. In practice, however, there are many pitfalls in the sampling process and in interpretation of the data. If, for example, many of the individuals we marked died between our first and second visits, then modifications of the method would be needed to take account of this. Often, though, it is the only technique that we have to estimate the size of a population.

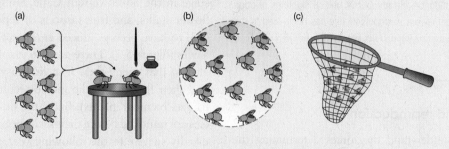

FIGURE 5.2 The mark and recapture technique for estimating the size of a population of mobile organisms (in simplified form). (a) On a first visit to a population of unknown total size N, a representative sample is caught (r individuals) and given a harmless mark. (b) These are released back into the population, where they remix with the unknown number of unmarked individuals. (c) On a second visit, a further representative sample is caught. Because it is representative, the proportion of marks in the sample (m out of a total sample of n) should, on average, be the same as that in the whole population (r out of a total of N). Hence N can be estimated.

nestlings in a small woodland with leg rings, and subsequently recognize every individual (except immigrants) in the population. But it is very much more difficult to count the numbers of lemmings in a patch of tundra, or even trap them, because they spend a large part of the year (and may reproduce) under thick snow cover. And most other species are so small, or cryptic, or hidden, or fast moving that they are even more difficult to count.

Ecologists, therefore, are almost always forced to estimate rather than count. We may estimate the numbers of aphids on a crop, for example, by counting the number on a representative sample of leaves, then estimating the number of leaves per square meter of ground, and from this estimating the number of aphids per square meter. Sometimes more

estimates from representative samples

complex methods are used (Box 5.1); and at other times we may rely on indirect 'indices' of abundance. These can provide information on the relative size of a population, but they usually give little indication of absolute size. Figure 5.3 shows an example in which the relative abundance of grizzly bears (*Ursos arctos*) in the 'Greater Yellowstone Ecosystem' in the northwestern United States has been estimated each year by the numbers of 'unique' sightings of adult females with cubs-of-the-year (i.e., if the same female was observed more than once, these additional sightings were ignored). Despite their shortcomings, even indices of abundance can provide valuable information: in this case, it seems clear that the population has been increasing since the mid-1980s.

Moreover, as we have already noted, for modular organisms it is often not even clear what it is we should be counting.

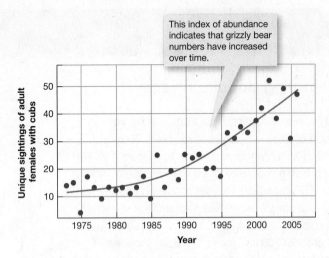

FIGURE 5.3 The abundance of grizzly bear (*Ursos arctos*) in the Greater Yellowstone Ecosystem of the northwestern United States, including the Yellowstone and Grand Teton National Parks, appears to have increased significantly since the mid-1980s, as evidenced by an index of overall abundance, the number of unique sightings of adult females with cubs-of-the-year. A 'smoothed' line has been fitted to the points to capture the overall trend. (After Eberhardt & Breiwick, 2010.)

5.2 LIFE CYCLES

Life cycles and reproduction

If we wish to understand the forces determining the abundance of a population, we need to know the important phases of the lives of the organisms concerned when these forces act most significantly. For this, we need to understand the sequences of events that occur in those organisms' life cycles.

There is a point in the life of any individual when, if it survives that long, it should be in a position to start reproducing and leaving progeny. A highly simplified, generalized life history (Figure 5.4) comprises birth, followed by a pre-reproductive period, a period of reproduction, a post-reproductive period, and then death as a result of senescence (though of course other forms of mortality may intervene at any time). The life histories of all unitary organisms can be seen as variations around this simple pattern, though a post-reproductive period (as seen in humans) is probably rather unusual.

Some organisms fit several or many generations within a single year, some have just one generation each year (annuals), and others (ourselves, perennial plants, etc.) have a life cycle extended over several or many years. For all organisms, though, a period of growth occurs before there is any reproduction, and growth usually slows down (and in some cases stops altogether) when reproduction starts. Growth and reproduction both require resources and there is clearly some conflict between them. Thus, as the perennial plant *Sparaxis grandiflora* enters its reproductive stage in the Southwestern Cape, South Africa, flowers, flower stalks, and fruit (aspects of reproduction) can be seen to have been produced *at the expense* of roots and leaves (Figure 5.5). There are also many plants (e.g., foxgloves) that spend their first year in vegetative growth, and then flower and die in the second or a later year (called 'biennial' plants). But if the flowers of these species are removed before their seeds begin to set, the plants usually survive to the following year, when they flower again and set seed even more vigorously. It seems to be the cost of provisioning the offspring (seeds) rather than flowering itself that is lethal. Similarly, pregnant women are advised to increase their caloric intake by as much as half their normal consumption: when nutrition is inadequate, pregnancy can harm the health of the mother.

the conflict between growth and reproduction

Among both annuals and perennials, there are some—**iteroparous species**—that breed repeatedly, devoting some of their resources

iteroparous and semelparous species

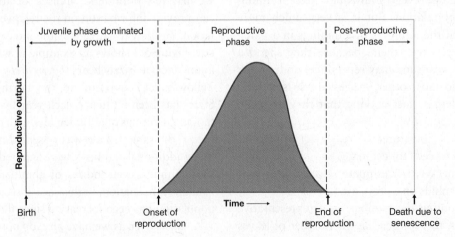

FIGURE 5.4 An outline life history for a unitary organism. Time passes along the horizontal axis, which is divided into different phases. Reproductive output is plotted on the vertical axis.

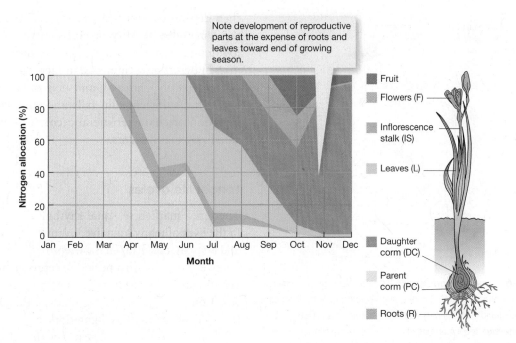

FIGURE 5.5   Percentage allocation of the crucial resource nitrogen to different structures throughout the annual cycle of the perennial plant *Sparaxis grandiflora* in South Africa, where it sets fruit in the Southern Hemisphere spring (September–December). The plant grows each year from a corm, which it replaces over the growing season. The plant parts themselves are illustrated here for a plant in early spring. (After Ruiters & Mckenzie, 1994.)

during a breeding episode not to breeding itself, but to survival to further breeding episodes (if they manage to live that long). We ourselves are iteroparous. There are others, though, **semelparous species**, like the biennial plants already described and some species of salmon, in which there is a single reproductive episode, with no resources set aside for future survival, so that reproduction is inevitably followed quickly by death.

Annual life cycles

In strongly seasonal, temperate latitudes, most annuals germinate or hatch as temperatures start to rise in spring, grow rapidly, reproduce, and then die before the end of summer. The European common field grasshopper *Chorthippus brunneus* is an example of an annual species that is iteroparous. It emerges from its egg in late spring and passes through four juvenile stages of nymph before becoming adult in midsummer and dying by mid-November. During their adult life, the females reproduce repeatedly, each time laying egg pods containing about 11 eggs, and recovering and actively maintaining their bodies between the bursts of reproduction.

Many annual plants, by contrast, are semelparous: they have a sudden burst of flowering and seed set, and then they die. This is commonly the case among the weeds of arable crops. Others, such as groundsel, are iteroparous: they continue to grow and produce new flowers and seeds through the season until they are killed by the first lethal frost of winter. They die with their buds on.

Most annuals spend part of the year dormant as seeds, spores, cysts, or eggs. In many cases these dormant stages may remain viable for many years; there are reliable records of seeds of the annual weeds *Chenopodium album* and *Spergula arvensis* remaining viable in soil for 1600 years. Similarly, the dried eggs of brine shrimps remain viable for many years in storage. This means that if we measure the length of life from the time of formation of the zygote, many so-called 'annual' animals and plants live very much longer than a single year. Large populations of dormant seeds form a **seed bank** buried in the soil: as many as 86,000 viable seeds per square meter have been found in cultivated soils. The species composition of the seed bank may be very different from that of the mature vegetation above it (Figure 5.6). Species of annuals that seem to have become locally extinct may suddenly reappear after the soil has been disturbed and these seeds germinate.

Dormant seeds, spores, or cysts are also necessary to the many ephemeral plants and animals of sand dunes and deserts that complete most of their life cycle in less than 8 weeks. They

the ephemeral 'annuals' of deserts

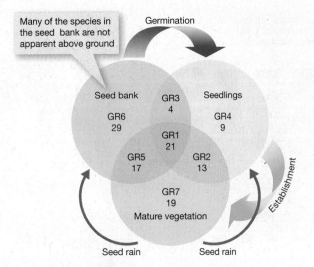

FIGURE 5.6   Species recovered from the seed bank, from seedlings and from the mature vegetation in a coastal grassland site on the western coast of Finland. Species may germinate from the buried seed bank into seedlings, and seedlings may establish themselves in the mature vegetation. Mature plants may contribute seeds (in the 'seed rain') that germinate into seedlings immediately or enter the buried seed bank. Seven species groups (GR1-GR7) are defined on the basis of whether they were found in only one, two, or all three life stages. The marked difference in composition, especially between the seed bank and the mature vegetation, is readily apparent, in that 32 species in the mature vegetation (19 + 13) were not represented in the seed bank, while 33 species in the seed bank (29 + 4) were not found in the mature vegetation, and 29 of these were not found as seedlings either. (After Jutila, 2003.)

then depend on the dormant stage to persist through the remainder of the year and survive the hazards of low temperatures in winter and the droughts of summer. In desert environments, in fact, the rare rains are not necessarily seasonal, and it is only in occasional years that sufficient rain falls and stimulates the germination of characteristic and colorful floras of very small ephemeral plants.

Longer life cycles

There is a marked seasonal rhythm in the lives of many long lived plants and animals, especially in their reproductive activity, with a period of reproduction once per year (Figure 5.7a). Mating (or the flowering of plants) is commonly triggered by the length of the **photoperiod**: the light phase in daily light–dark cycle, which varies continuously through the year. The time that young are born, eggs hatch, or seeds are ripened tends thus to be synchronized with the period when seasonal resources are likely to be abundant.

repeated, seasonal breeders

In populations of perennial species, the generations overlap and individuals of a range of ages breed side by side. The population is maintained in part by survival of adults and in part by new births. A study of wood thrushes, *Hylocichla mustelina*, for example, in

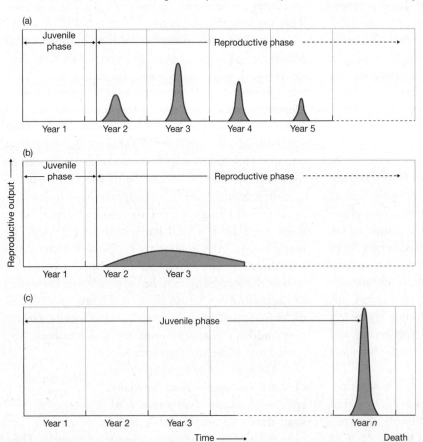

FIGURE 5.7   Simplified life histories for organisms living more than one year. (a) An iteroparous species breeding seasonally once per year. Deaths tend not to occur predictably at a particular time, though a decline toward senescence is often observed. (b) An iteroparous species breeding continuously throughout the year. The pattern of death and decline is similar to that in (a). (c) A semelparous species passing several or many years in a pre-reproductive juvenile phase, followed by a burst of reproduction, followed in turn by inevitable death.

Delaware, showed that for every 1000 eggs laid by a breeding population in one season, only 462 hatchlings survived to become fully fledged on average, and only around 150 of these survived to adulthood the following year. These one-year-old birds were joined in that second year, though, by a very similar number of birds aged between 2 and 8 years – the survivors from previous years (Figure 5.8).

In wet equatorial regions, on the other hand, where there is very little seasonal variation in temperature and rainfall and scarcely any variation in photoperiod, we find species of plant that are in flower and fruit throughout the year – and continuously breeding species of animal that subsist on this resource (Figure 5.7b). There are several species of fig (*Ficus*), for instance, that bear fruit continuously and form a reliable year-round food supply for birds and primates. In more seasonal climates, humans of course breed continuously throughout the year, but numbers of other species, cockroaches for example, also do so in the stable environments that humans have created.

| continuous breeders |

Other plants and animals (Figure 5.7c) may spend almost all their lives in a long nonreproductive (juvenile) phase and then have one lethal burst of reproductive activity. We saw such semelparity earlier in biennial plants, but it is also characteristic of some species that live much longer than 2 years. The Pacific salmon is a familiar example. Salmon are spawned in rivers. They spend the first phase of their juvenile life in fresh water and then migrate to the sea, often travelling thousands of miles. At maturity they return to the stream in which they were hatched. Some mature and return to reproduce after only 2 years at sea; others mature more slowly and return after 3, 4, or 5 years. At the time of reproduction the population of salmon is composed

| semelparous species like salmon and bamboo |

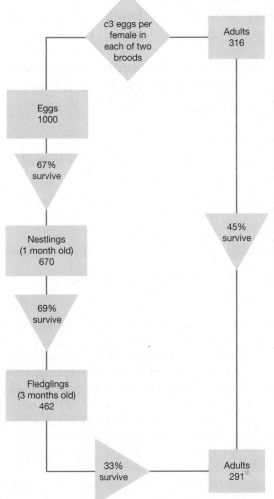

FIGURE 5.8 A diagrammatic life history for a population of wood thrush, *Hylocichla mustelina*, in Delaware. Individuals typically live for several years. Hence, the population in any one year is a combination of survivors from previous years and newborn individuals. Population sizes for different stages are in rectangles; the proportions surviving from one stage to the next are in triangles; the rate of egg production per female is shown in the diamond. The population was studied from 1974 to 2004 and the figures quoted are averages calculated to represent a 'typical' year. Note, though, that the population tended to decline year on year over the study period as a whole. The actual population size was in excess of 1000 but has been 'standardized' to generate a typical cohort of 1000 eggs per year. (From data in Brown & Roth, 2009.)

Bamboo: semelparous plants

of overlapping generations of individuals. But all are semelparous: they lay their eggs and then die; their bout of reproduction is terminal.

There are even more dramatic examples of species that have a long life but reproduce just once. Many species of bamboo form dense clones of shoots that remain vegetative, with no sexual reproduction, for many years: in some species, 100 years. The whole population of shoots then flowers simultaneously in a mass suicidal orgy. Even when shoots have become physically separated from each other, the parts still flower synchronously.

Organisms of long-lived species that are the same age, however, are not necessarily the same size – especially in modular organisms. Some individuals may be very old but have been suppressed in their growth and development by predators or by competition. Age, then, is often a particularly poor predictor of fecundity. An analysis that classifies the members of a population according to their size rather than their age (Figure 5.9) is often more useful in suggesting whether they will survive or reproduce.

size matters

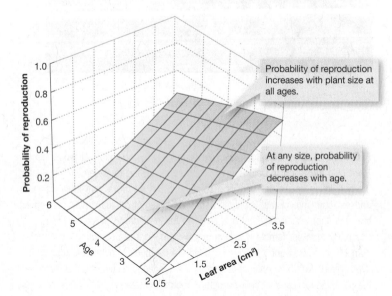

Probability of reproduction increases with plant size at all ages.

At any size, probability of reproduction decreases with age.

FIGURE 5.9 The effect of plant age (years) and plant size (as measured by leaf area) on the probability of *Rhododendron lapponicum* shoots entering their reproductive phase. The relationships have been 'smoothed' by a statistical technique called 'logistic regression.' Also, older shoots are overall more likely to enter their reproductive phase because they tend to be bigger. Note that age is a much poorer predictor of shoot-fate than size. (After Karlsson & Jacobson, 2001.)

5.3   MONITORING BIRTH AND DEATH: LIFE TABLES AND FECUNDITY SCHEDULES

The previous sections have outlined the different patterns of births and deaths in different species. But patterns are just a start. What are the *consequences* of these patterns in specific cases, in terms of their effects on how a population might grow to pest proportions, say, or shrink to the brink of extinction? To determine these consequences, we need to monitor the patterns in a quantitative way.

There are different ways of doing so. To monitor and quantify survival, we may follow the fate of individuals from the same **cohort** within a population: that is, all individuals born within a particular period. A **cohort life table** then records the survivorship of the members of the cohort over time (Box 5.2). A different approach is necessary when we cannot follow cohorts but we know the ages of all the individuals in a population. We can then, at one time, describe the numbers of survivors of different ages in what is called a **static life table** (Box 5.2).

Quantitative Aspects

The basis for cohort and static life tables

In Figure 5.10, a population is portrayed as a series of diagonal lines, each line representing the life 'track' of an individual. As time passes, each individual ages (moves from bottom-left to top-right along its track) and eventually dies (the dot at the end of the track). Here, individuals are classified by their age. In other cases it may be more appropriate to split the life of each individual into different developmental stages.

Time is divided into successive periods: t_0, t_1, etc. In the present case, three individuals were born (started their life track) prior to the time period t_0, four during t_0, and three during t_1. To construct a *cohort life table*, we direct our attention to a particular cohort and monitor what happens to them subsequently. Here we focus on those born during t_0. The life table is constructed by noting the number surviving to the start of each time period. So, four were there at the beginning of t_0, two of the four survived to the beginning of t_1; only one of these was alive at the beginning of t_2; and none survived to the start of t_3. The first data column of a cohort life table for these individuals would thus comprise the series of declining numbers in the cohort: 4, 2, 1, 0.

A different approach is necessary when we cannot follow cohorts but we know the ages of all the individuals in a population (perhaps from some clue such as the condition of the teeth in a species of deer). We can then, as the figure shows, direct our attention to the whole population during a single period (in this case, t_1) and note the numbers of survivors of different ages in the population. These may be thought of as entries in a life table *if* we assume that rates of birth and death are, and have previously been, constant—a very big assumption. What results is called a *static life table*. Here, of the seven individuals alive during t_1, three were actually born during t_1 and are hence in the youngest age group, two were born in the previous time interval, two in the interval before that, and none in the interval before that. The first data column of the static life table thus comprises the series 3, 2, 2, 0. This amounts to saying that over these time intervals, a typical cohort will have started with three and declined over successive time intervals to two, then two again, then zero.

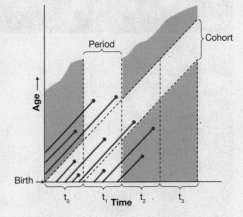

FIGURE 5.10   See text for details.

The fecundity of individuals also changes with their age, and to understand properly what is going on in a population we need to know how much individuals of different ages contribute to births in the population as a whole: these can be described in **age-specific fecundity schedules**.

Cohort life tables

The most straightforward life table to construct is a cohort life table for annuals, because with nonoverlapping generations it is indeed often possible to follow a single cohort from the first birth to the death of the last survivor. Two very simple life tables, for inland and coastal subspecies of the annual plant *Gilia capitata*, growing in California, are shown in Table 5.1. Initial cohorts of around 750 seeds were followed from seed germination to the death of the last adult.

> an annual life table for a plant

Even when generations overlap, if individuals can be marked early in their life so that they can be recognized subsequently, it can be possible to follow the fate of each year's cohort separately. It may then be possible to merge the cohorts

> a cohort life table for marmots

from the different years of a study so as to derive a single, 'typical' cohort life table. We saw one example, in diagrammatic form, for the wood thrushes in Figure 5.8. Another is shown in Table 5.2: females from a population of the yellow-bellied marmot, *Marmota flaviventris*, which were live-trapped and marked individually from 1962 through to 1993 in the East River Valley of Colorado.

The first column in each life table is a list of the stages or age classes of the organism's life: for *Gilia* these are the simple stages 'seed,' 'emerged plants,' and 'flowering plants,' and for the marmots, years. The second column is then the raw data from each study, collected in the field. It reports the number of individuals surviving to the beginning of each stage or age class (see Box 5.2).

Ecologists are typically interested not just in examining populations in isolation but in comparing the dynamics of two or more perhaps rather different populations. This was precisely the case for the plant populations in Table 5.1. Hence, it is necessary to standardize the raw data so that comparisons can be made. This is done in the third column of the table, which is said to contain l_x values, where l_x is defined as the proportion of the original cohort surviving to the start of age class.

TABLE 5.1 Two simplified cohort life tables for the annual plant *Gilia capitata*. One is for the 'inland' subspecies, *G. capitata capitata*, and one for the 'coastal' subspecies, *G. capitata chamissonis*, growing at an inland site in Napa County, California, and being easily distinguishable morphologically, despite being cross-fertile. Cohorts of seeds were planted at the beginning of the season in 1993 and the life cycle divided simply into seeds, plants that emerged from those seeds, and emerged plants that went on to flower. Other column entries are explained in the text. (After Nagy & Rice, 1997.)

Stage (x)	Number alive at the start of each age class a_x	Proportion of original cohort surviving to the start of each age class l_x	Number of female young produced by each age class F_x	Number of female young produced per surviving individual in each age class m_x	Number of female young produced per original individual in each age class $l_x m_x$
Inland subspecies:					
Seed (0)	746	1.00	0	0	0
Emergence (1)	254	0.34	0	0	0
Flowering (2)	66	0.09	28,552	432.61	38.29
Coastal subspecies:					
Seed (0)	754	1.00	0	0	0
Emergence (1)	204	0.27	0	0	0
Flowering (2)	19	0.03	8645	455.00	11.47

TABLE 5.2   A simplified cohort life table for female yellow-bellied marmots, *Marmota flaviventris* in Colorado. The columns are explained in the text. (After Schwartz et al., 1998.)

Age Class (Years) x	Number alive at the start of each age class a_x	Proportion of original cohort surviving to the start of each age class l_x	Number of female young produced by each age class F_x	Number of female young produced per surviving individual in each age class m_x	Number of female young produced per original individual in each age class $l_x m_x$
0	773	1.000	0	0.000	0.000
1	420	0.543	0	0.000	0.000
2	208	0.269	95	0.457	0.123
3	139	0.180	102	0.734	0.132
4	106	0.137	106	1.000	0.137
5	67	0.087	75	1.122	0.098
6	44	0.057	45	1.020	0.058
7	31	0.040	34	1.093	0.044
8	22	0.029	37	1.680	0.049
9	12	0.016	16	1.336	0.021
10	7	0.009	9	1.286	0.012
11	3	0.004	0	0.000	0.000
12	2	0.003	0	0.000	0.000
13	2	0.003	0	0.000	0.000
14	2	0.003	0	0.000	0.000
15	1	0.001	0	0.000	0.000
Total			519		0.670

$$R_0 = \sum l_x m_x = \frac{\sum F_x}{a_0} = 0.67.$$

The first value in this column, l_0 (spoken: *L*-zero), is therefore the proportion surviving to the beginning of this original age class. Obviously, in Tables 5.1 and 5.2, and in every life table, l_0 is 1.00 (the whole cohort is there at the start). Thereafter, in the marmots for example, there were 773 females observed in this youngest age class. The l_x values for subsequent age classes are therefore expressed as proportions of this number. Only 420 individuals survived to reach their second year (age class 1: between 1 and 2 years of age). Thus, in Table 5.2, the second value in the third column, l_1, is the proportion 420/773 = 0.543 (that is, only 0.543 or 54.3% of the original cohort survived this first step). In the next row, $l_2 = 208/773 = 0.269$, and so on. For *Gilia* (Table 5.1), $l_1 = 254/746 = 0.340$ for the inland subspecies and 204/754 = 0.271 for the coastal subspecies. That is, 34% and 27.1% survived the first step to become

established plants in the two cases: a slightly higher survival rate at this inland site for the inland than for the coastal subspecies.

In a full life table, subsequent columns would then use these same data to calculate the proportion of the original cohort that died at each stage and also the mortality rate for each stage, but for brevity these columns have been omitted here.

Tables 5.1 and 5.2 also include fecundity schedules for *Gilia* and for the marmots (columns 4 and 5). Column 4 in each case shows F_x, the total number of the youngest age class produced by each subsequent age class. This youngest class is seeds for *Gilia*, produced only by the flowering plants. For the marmots, these are independent juveniles, fending for themselves outside of their burrows, produced when adults were between 2 and 10 years old. The fifth column is then said to contain m_x values, *fecundity*: the mean number of the youngest age class produced per surviving individual of each subsequent class. For the marmots, fecundity was highest for eight-year old females —1.68—that is, 37 young produced by 22 surviving females.

In the final column of a life table, the l_x and m_x columns are brought together to express the overall extent to which a population increases or decreases over time – reflecting the dependence of this on both the survival of individuals (the l_x column) and the reproduction of those survivors (the m_x column). That is, an age class contributes most to the next generation when a large proportion of individuals have survived and they are highly fecund. The sum of all the $l_x m_x$ values, $\Sigma l_x m_x$, where the symbol Σ means 'the sum of,' is therefore a measure of the overall extent by which this population has increased or decreased in a generation. We call this the **basic reproductive rate** and denote it by R_0 ('R-nought').

> combined to give the basic reproductive rate

For *Gilia* (Table 5.1), R_0 is calculated very simply (no summation required) since only the flowering class produces seed. Its value is 38.29 for the inland subspecies and 11.47 for the coastal subspecies: a clear indication that the inland subspecies thrived, comparatively, at this inland site. (Though the annual rate of reproduction would not have been this high, since, no doubt, a proportion of these would have died before the start of the 1994 cohort. In other words, another class of individuals, 'winter seeds,' was ignored in this study.)

For the marmots, $R_0 = 0.67$: the population was declining, each generation, to around two-thirds its former size. However, whereas for *Gilia* the length of a generation is obvious, since there is one generation each year, for the marmots the generation length must itself be calculated. The details of that calculation are beyond our scope here, but its value, 4.5 years, matches what we can observe ourselves in the life table: that a 'typical' period from an individual's birth to giving birth itself (i.e., a generation) is around four and a half years. Thus, Table 5.2 indicates that each generation, every four and a half years, this particular marmot population was declining to around two-thirds its former size.

It is also possible to study the detailed pattern of decline in a cohort. Figure 5.11a, for example, shows the numbers of marmots surviving relative to the original population—the l_x values—plotted against the age of the cohort. However, this can be misleading. If the original population is 1000 individuals, and it decreases by half to 500 in one time interval, then this decrease looks more dramatic on a graph like Figure 5.11a than a decrease from 50 to 25 individuals later in the season. Yet the risk of death to individuals is the same on both occasions. If, however, l_x values are replaced by $\log(l_x)$ values, that is, the logarithms of the values, as in Figure 5.11b (or, effectively the same thing, if l_x values are plotted on a log scale), then it is a characteristic of logs that the reduction of a population to half its original size will always look the same. **Survivorship curves**

> logarithmic survivorship curves

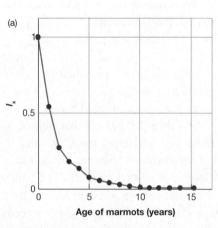

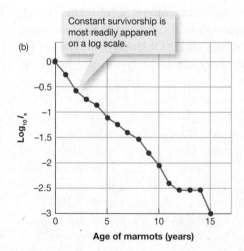

> Constant survivorship is most readily apparent on a log scale.

FIGURE 5.11 Following the survival of a cohort of the yellow-bellied marmot (Table 5.2). (a) When l_x is plotted against cohort age, it is clear that most individuals are lost relatively early in the life of the cohort, but there is no clear impression of the risk of mortality at different ages. (b) By contrast, a survivorship curve plotting $\log(l_x)$ against age shows a virtually constant mortality risk until around age 8, followed by a brief period of slightly higher risk, and then another brief period of low risk after which the remaining survivors died.

are, therefore, conventionally plots of log(l_x) values against cohort age. Figure 5.11b shows that for the marmots, there was a steady, more or less constant rate of decline until around the eighth year of life, then three further years at a slightly higher rate (until breeding ceased), followed by a brief period with effectively no mortality, after which the few remaining survivors died.

It is possible to see, therefore, even from these two examples, how life tables can be useful in characterizing the 'health' of a population—the extent to which it is growing or declining—and identifying where in the life cycle, and whether it is survival or birth, that is apparently most instrumental in determining that rate of increase or decline. Either or both of these may be vital in determining how best to conserve an endangered species or control a pest.

Life tables for populations with overlapping generations

Many of the species for which we have important questions, and for which life tables may provide an answer, have repeated breeding seasons like the marmots, or continuous breeding as in the case of humans, but constructing life tables here is complicated, largely because these populations have individuals of many different ages living together. Building a cohort life table is sometimes possible, as we have seen, but this is relatively uncommon. Apart from the mixing of cohorts in the population, it can be difficult simply because of the longevity of many species.

Another approach is to construct a static life table (Box 5.2). The data look like a cohort life table – a series of different numbers of individuals in different age classes – but these come simply from the **age structure** of the population captured at one point in time. Hence, great care is required: they can only be treated and interpreted in the same way as a cohort life table if patterns of birth and survival in the population have remained much the same since the birth of the oldest individuals—and this will happen only rarely. Nonetheless, there is often no alternative and useful insights can still be gained. This is illustrated for a population of small dinosaurs, *Psittacosaurus lujiatunensis*, recovered as fossils from the Lower Cretaceous Yixian Formation in China, where the alternative of following a cohort really isn't available (Figure 5.12). They appear to have perished simultaneously in a volcanic mudflow, which might therefore have captured a representative snapshot of the population at the time, around 125 million years ago.

> a static life table— useful if used with caution

It appears that mortality rates were high amongst the dinosaurs until around the age of three, after which there was another period of around five years during which mortality rates were low even though the animals continued to grow rapidly (Figure 5.12b). Mortality rates then seem to have increased again, just as the animals were attaining their maximum size, and broadly coinciding with the appearance in the fossils of characteristic associated with sexual maturity (e.g., enlarged, flaring 'jugal' horns). As we shall see in Section 5.6, many organisms suffer

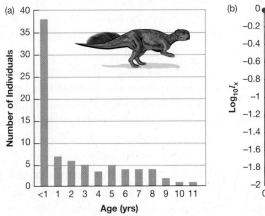
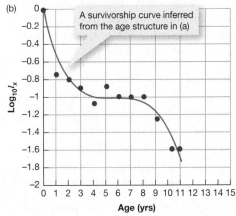

FIGURE 5.12 (a) The age structure (and hence the static life table) of a population of dinosaurs, *Psittacosaurus lujiatunensis*, recovered as fossils from the Lower Cretaceous Yixian Formation in China. Age was estimated from the length of the femur, which had been shown in a subsample of specimens to correlate very strongly with the number of 'growth lines' (one per year) in the bone. (b) A survivorship curve [log(l_x) plotted against age, similar to that in Figure 5.11b] derived from the life table. The animals continued to grow rapidly in size until around the age of nine or ten. (After Erickson et al., 2009.)

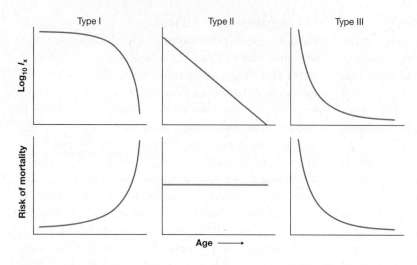

FIGURE 5.13 A classification of survivorship curves plotting log(l_x) against age, above, with corresponding plots of the changing risk of mortality with age, below. The three types are discussed in the text. (After Pearl, 1928; Deevey, 1947.)

a cost of reproduction in terms of reductions in growth and/or survival.

A classification of survivorship curves

Life tables provide a great deal of data on specific organisms. But ecologists search for generalities – patterns of life and death that we can see repeated in the lives of many species – conventionally dividing survivorship curves into three types in a scheme that goes back to 1928, generalizing what we know about the way in which the risks of death are distributed through the lives of different organisms (Figure 5.13).

- In a type I survivorship curve, mortality is concentrated toward the end of the maximum life span. It is perhaps most typical of humans in developed countries and their carefully tended zoo animals and pets.

- A type II survivorship curve is a straight line signifying a constant mortality rate from birth to maximum age. It describes, for instance, the survival of buried seeds in a seed bank.

- In a type III survivorship curve there is extensive early mortality, but a high rate of subsequent survival. This is typical of species that produce many offspring. Few survive initially, but once individuals reach a critical size, their risk of death remains low and more or less constant. This appears to be the most common survivorship curve among animals and plants in nature.

These types of survivorship curve are useful generalizations, but in practice, patterns of survival are usually more complex. We saw with the marmots, for example, that survivorship was broadly type II throughout much of their lives, but not at the end (Figure 5.11b); and with the dinosaurs, survivorship followed the typical type III pattern until they reached sexual maturity, but again failed to conform to such a simple classification thereafter.

5.4 DISPERSAL AND MIGRATION

Birth is only the beginning. If we were to stop there in our studies, many crucial ecological questions would remain unanswered. From their place of birth, all organisms move to locations where we eventually find them. Plants grow where their seeds fall, but seeds may have been moved by the wind, water, animals, or shifting soil. Animals move in search of food and safe havens, whether it is to move only one centimeter along a leaf from where their egg was deposited, or halfway around the globe. The effects of those movements are varied. In some cases they aggregate members of a population into clumps; in others they continually redistribute and shuffle them; and in still others they spread the individuals out. Three generalized spatial patterns that result from this movement—aggregated (clumped), random, and regular (evenly) spaced—are illustrated in Figure 5.14. Clearly, movement and spatial distribution (the latter sometimes, confusingly, called 'dispersion') are intimately related.

Technically, the term **dispersal** describes the way individuals spread away from each other, such as when seeds are carried away from a parent plant or young lions leave the pride in search of their own territory. **Migration** refers to the mass directional movement of large numbers of a species from one location to another. Migration therefore describes the movement of locust swarms but also includes the smaller-scale movements of intertidal organisms, back and forth twice a day as they follow their preferred level of immersion or exposure.

> patterns of distribution

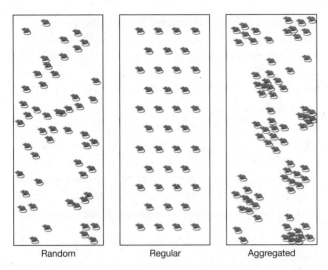

FIGURE 5.14 Three generalized spatial patterns that may be exhibited by organisms across their habitat.

Our view of dispersal and migration, and of the resulting distributions, is determined by the scale on which we are working. For | the perception of pattern depends on the spatial scale |

example, consider the distribution of an aphid living on a particular species of tree in a woodland. On a large scale, the aphids appear to be aggregated in the woodlands and nonexistent in the open fields. If the samples we took were smaller, and taken only in woodlands, the aphids would still appear to be aggregated, but now aggregated on their host trees rather

than on trees in general. However, if samples were collected at an even smaller scale—the size of a leaf within a canopy—the aphids might appear to be randomly distributed over the tree as a whole. And on the scale experienced by the aphid itself (1 cm^2), the distribution might appear regular as individuals on a leaf spread out to avoid one another (Figure 5.15).

This example also illustrates the difference between the 'average density' and the crowding experi- | density and crowding |

enced by individuals in a population. The **average density** is simply the total number of individuals divided by the total size of the habitat—but it depends very much on how we define the habitat. For the aphids, if it includes everything, woodland and non-woodland, then average density will be low. It will higher, but still quite low, if we include only woodland but every species of tree. It will be much higher, however, if we include only the aphids' host trees.

The average density of individuals in the United States is about 75 persons/km^2. Yet there are vast areas of the United States—rural and wilderness areas—within which the density is low, but also crowded cities and towns within which the density is much higher. And because the majority of people live in urban and suburban settings, the density actually experienced by people, on average, has been calculated at 3630 persons/km^2. There may be little impetus for dispersal, or migration, at the relatively low population pressure

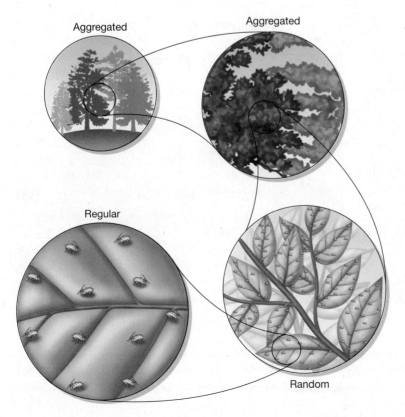

FIGURE 5.15 Are aphids distributed evenly, randomly, or in an aggregated fashion? It all depends on the spatial scale at which they are viewed.

of 75 individuals/km². At 3630 people/km², however, individuals are much more likely to find ways to escape from their neighbors. Real measures of crowding as experienced by individuals are likely to be more important forces driving dispersal and migration than some average value of population density.

Dispersal determining abundance

Compared to birth and death, relatively few studies have examined the role of dispersal in determining the abundance of populations. However, studies that *have* looked carefully at dispersal have tended to bear out its

importance. In a long-term and intensive investigation of a population of great tits, *Parus major*, near Oxford, England, it was observed

dispersal: important but frequently neglected

that 57% of breeding birds were immigrants rather than born in the population (Greenwood et al., 1978). And in many cases, including numbers of pests, the rapid spread of a species into new areas is a compelling testament to the power of dispersal in determining the abundance we observe (see, for example, Figure 5.16). Indeed, most populations are more affected by immigration and emigration than is commonly imagined. Within the United

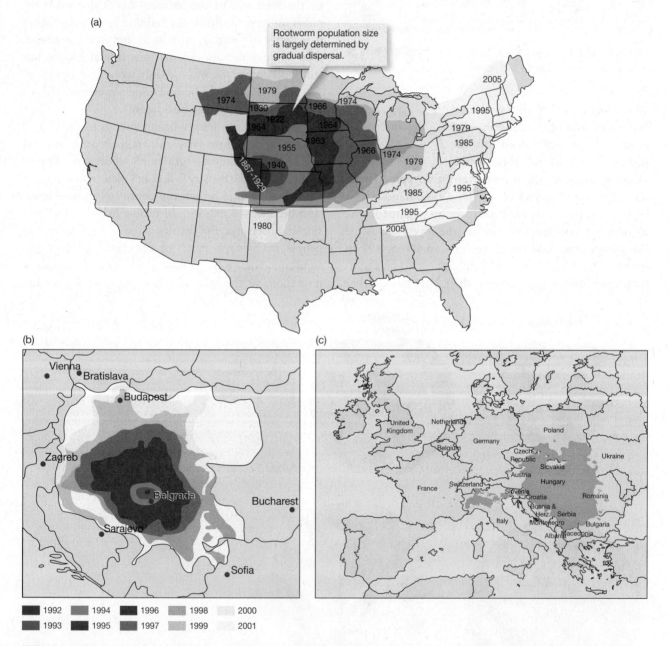

FIGURE 5.16 The spread of the western corn rootworm, *Diabrotica virgifera virgifera*, one of the most important pests of maize worldwide (a) in the United States following its breakout from a stronghold around eastern Nebraska in the 1940s; (b) in south eastern Europe, from 1992–2001, following its probable introduction from material carried by plane to Belgrade airport; and (c) to elsewhere in Europe by 2007. ((a) and (c) After Gray et al., 2009, (b) After European Environment Agency, 2002.)

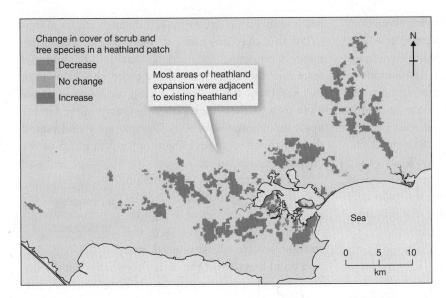

FIGURE 5.17 The invasion (i.e., increase in abundance) of most of the 116 patches of lowland heath in Dorset, England, by scrub and tree species between 1978 and 1987, which largely depended on the availability of dispersers in vegetation neighboring the patches. (After Bullock et al., 2002.)

States, for example, it is said that over 40% of U.S. residents, over 100 million people, can trace their roots to the 12 million immigrants who entered the United States through the Ellis Island port from 1870 to 1920.

In fact, often the most important role played by dispersal in a population is to get the organisms there in the first place, as we have seen for the western corn rootworms in Figure 5.16. Another example is the invasion by scrub and tree species of 116 patches of lowland heath vegetation in southern England, which was studied for the period from 1978 to 1987 (Figure 5.17). Was the extent to which patches were invaded determined by the nature of the heath patches themselves or by the abundance of the scrub and tree species in the vegetation bordering those patches? It was the latter that was consistently most important. Invasions, and thus the subsequent dynamics of patches, were being driven by initiating acts of dispersal, which were most common when there was a ready availability of dispersers.

One key force provoking dispersal is the more intense competition suffered by crowded individuals (Section 3.5) and the direct interference between such individuals even in the absence of a shortage of resources. We frequently observe, therefore, that the highest rates of dispersal are away from the most crowded patches (Figure 5.18): emigration dispersal is commonly density dependent. But this is by no means a general rule, and in some cases the converse pattern is observed—most dispersal at the lowest densities or *inverse* density dependence—a pattern often attributed to the avoidance of inbreeding between closely related individuals (and the lowered offspring

dispersal as invasion

density-dependent dispersal—and its converse

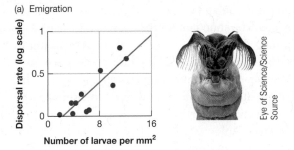

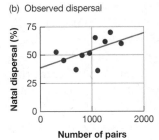

FIGURE 5.18 Density-dependent dispersal. (a) The dispersal rates of newly hatched black fly, *Simulium vittatum*, larvae increase with increasing density. (Data from Fonseca & Hart, 1996.) (b) The percentage of juvenile male barnacle geese, *Branta leucopsis*, dispersing from breeding colonies on islands in the Baltic Sea to nonnatal breeding locations increased as density increased. (Data from van der Jeugd, 1999.), (After Sutherland et al., 2002.)

fitness that would result), since on average, at low densities, a high proportion of those you grow up with are likely to be your close relatives.

Furthermore, immigrants and emigrants not only influence the numbers in a population – they can also affect its composition. Dispersers are often the

age- and sex-biased dispersal

young, and males frequently do more moving about than females. In mammal dispersal, for instance, age- and sex-biases, and the forces of inbreeding-avoidance and competition-avoidance may all be tied intimately together. Thus, in an experiment with gray-tailed voles, *Microtus canicaudus*, 87% of juvenile males and 34% of juvenile females dispersed within four weeks of initial capture at low densities, but only 16% and 12%, respectively, dispersed at high densities (Wolff et al., 1997). There was massive juvenile dispersal; this was particularly pronounced in males; and the especially high rates at low densities argue in favor of inbreeding-avoidance as a major force shaping the pattern.

The role of migration

The mass movements of populations that we call migration are almost always from regions where the food resource is declining to regions where it is abundant (or where it will be abundant for the progeny). By day, some planktonic algae live in the upper layers of the water in lakes where the light needed for photosynthesis is brightest. At night, they migrate to lower, nutrient-rich depths. Crabs migrate along the shore with the tides, following the movement of their food supply as it is washed up in the waves. At longer time scales, some shepherds still follow the ages-old practice of 'transhumance,' moving their flocks of sheep and goats up to mountain pastures in summer and down again in the fall to track the seasonal changes in climate and food supply.

The long-distance migrations of terrestrial birds in many cases involve movement between areas that supply abundant food, but only for a limited time. They are areas in which seasons of comparative glut and famine alternate, and that cannot support large all-year-round resident populations. For example, swallows (*Hirundo rustica*) migrate seasonally from northern Europe in the fall, when flying insects start to become rare, to South Africa when they are becoming common. In both areas the food supply that is reliable throughout the year can support only a small population of resident species. The seasonal glut supports the populations of invading migrants, which make a large contribution to the diversity of the local fauna.

5.5 THE IMPACT OF INTRASPECIFIC COMPETITION ON POPULATIONS

The concept of intraspecific competition was introduced in Section 3.5 because its intensity is typically dependent on resource availability. It reemerges here because its effects are expressed through the focal topics of this chapter—rates of birth, death, and movement. Competing individuals that fail to find the resources they need may grow more slowly or even die; survivors may reproduce later and less; or, as we have seen, if they are mobile, they may move farther apart or migrate elsewhere. We can rarely understand the dynamics of a species without a firm grasp of the effects of competition.

The intensity of competition for limiting resources is often related to the density of a population, though, as we have seen, the straightforward density need not be a good measure of the extent to which its individuals are crowded. Modular sessile organisms are particularly sensitive to competition from their immediate neighbors: they cannot withdraw from each other and space themselves more evenly or escape by dispersal or migration. Thus, when silver birch trees (*Betula pendula*) were grown in small groups, there were more suppressed and dying branches on the sides where neighbors' branches shaded each other, whereas on the sides away from neighbors there was more vigorous growth (Figure 5.19).

crowding not density—especially in modular organisms

We saw in Section 3.5 that, over a sufficiently large density range, as density increases, competition between individuals generally reduces the per capita birth rate and increases the death rate, and that this effect is described as **density-dependent**. Thus, when birth and death rate curves are plotted against density on the same graph, and either or both are density-dependent, the curves must cross (Figure 5.20a–c). They do so at the density at which birth and death rates are equal, and because they are equal, there is no overall tendency at this density for the population either to increase or to decrease (ignoring, for convenience, both emigration and immigration). The density at the crossover point is called the **carrying capacity** and is denoted by the symbol K. At densities below K, births exceed deaths and the population increases. At densities above K, deaths exceed births and the population decreases. There is therefore an overall tendency for the density of a population under the influence of intraspecific competition to settle at K.

density-dependent birth and death and the carrying capacity

In fact, because of the natural variability within populations, the birth rate and death rate curves are best represented by broad lines, and K is best thought of not as a single

population regulation by competition – but not to a single carrying capacity

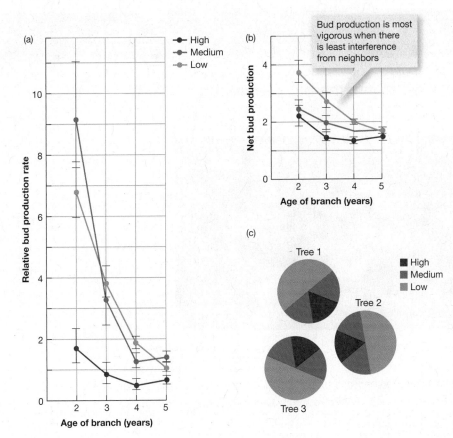

FIGURE 5.19 Mean relative bud production (new buds per existing bud) for silver birch trees (*Betula pendula*), expressed (a) as gross bud production and (b) as net bud production (birth minus death), in different interference zones (i.e., where they interfered to differing extents with their neighbors). (c) Plan of three trees, explaining these zones. •, high interference; •, medium; •, low. Bars represent standard errors. (After Jones & Harper, 1987.)

density, but as a range of densities (Figure 5.20d). Thus, intraspecific competition does not hold natural populations to a single, predictable, and unchanging level (*K*), but it may act upon a very wide range of starting densities and bring them to a much narrower range of final densities. It therefore tends to keep density within certain limits and may thus be said to play a part in *regulating* the size of populations.

Of course, graphs like those in Figure 5.20 are generalizations on a grand scale. Many organisms, for example, have seasonal life cycles. For part of the year births vastly outnumber deaths, but later, after the period of peak births, there is likely to be a period of high juvenile mortality. Most plants, for example, die as seedlings soon after germination. Thus, although births may balance deaths over the year, a population that is stable from year to year will often change dramatically over the seasons.

Patterns of population growth

When populations are sparse and uncrowded they may grow rapidly (and this can cause real problems—even with species that were previously endangered, such as sea otters, which have recovered rapidly from the brink of extinction: Box 5.3)—it is only as crowding increases that density-dependent changes in birth and death rates start to take effect. In essence,

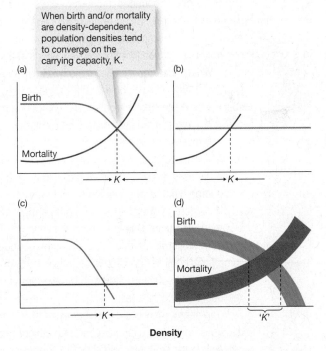

FIGURE 5.20 Density-dependent birth and mortality rates lead to the regulation of population size. When both are density-dependent (a), or when either of them is (b, c), their two curves cross. The density at which they do so is called the *carrying capacity* (*K*). However, the real situation is closer to that shown by thick lines in (d), where mortality rate broadly increases, and birth rate broadly decreases, with density. It is possible, therefore, for the two rates to balance not at just one density, but over a broad range of densities, and it is toward this broad range ('*K*') that other densities tend to move.

ECOncerns

On sea otters, we need to see the big picture

Greg Vaughn/Alamy Limited

The competing priorities of aiding otter recovery and support for ailing coastal fishing communities appear on a collision course. But sea otters are more than just our cute competition.

February 21, 2012 | *By James A. Estes*

Sea otters once abounded in coastal waters around the North Pacific Coast from Russia to Mexico. Over decades, they were hunted to near-extinction as a popular target of the maritime fur trade. Yet their dramatic brush with extinction, along with their intrinsically appealing characteristics, has provided a powerful emotional foundation for otter recovery efforts in the United States.

While these conservation initiatives have shown impressive results, they have also created conflicts with fishermen over valuable shellfish resources such as clams, mussels and abalone, traditional food sources for these small predators. The competing priorities of aiding otter recovery and support for ailing coastal fishing communities appear on a collision course.

The seemingly conflicting values of preserving an important element of nature's biodiversity versus our reluctance to incur the associated costs can be at least partly resolved through a better understanding of the greater ecological services provided by many of these animals. Otters, now at the center of some rather large national policy debates, offer an example.

Leaders in Washington are considering several important but highly contentious and contrasting policy decisions. One is a proposal by the U.S. Fish and Wildlife Service to eliminate the 24-year-old Southern California 'no-otter zone' designation under which any otters are captured and relocated. At the same time, a bill in Congress would allow resumption of trade in fur from sea otters that are trapped for subsistence in southeast Alaska, in an effort to reduce the competition between humans and otters over key local shellfish fisheries.

In nearly every case in which recovery of a predator is occurring or has been proposed, the policy debate is framed as a contrast between ethical and monetary values: a choice between our hearts and our wallets. Yet the many scientifically known and suspected positive influences of these animals on the health of regional ecosystems are all too often overlooked. Indeed, research has shown that these ecological services provide very real dollars-and-cents benefits for society that policymakers should take into consideration.

For instance, kelp forests are the foundation of many North Pacific marine ecosystems. These wildernesses of the deep support regional fish populations, serving both as a vital habitat and a source of nourishment for many species. Perhaps even more important, research indicates that these forests also provide humans with a number of crucial natural services. Their role in helping to buffer shorelines from wave exposure, reducing rates of coastal erosion and shoreline recession, is just one example. And sea otters are essential to keeping these coastal kelp forests healthy.

Much like the way wolves cull deer and elk populations on land, sea otters eat urchins. But when otter populations are allowed to plummet, urchins are left unchecked. Without any natural predators, urchins can become so numerous that they overgraze the lush kelp forests that otherwise abound along the West Coast. When this happens, the lost ecological benefits—both to society and the environment—are dramatic.

An ever-growing body of research shows that the ecological and economic influences of predators in nature, from sea otters to wolves, extend well beyond the things they eat. Although debates over managing these species are likely to always be controversial, invoking deep passions on both sides, science provides the best basis for policy decisions, as it affords a common currency in valuing the perceived costs and benefits for predator conservation.

Shellfisheries are of considerable importance to commercial, recreational, and tribal fishers. How would you weigh up the competing demands of conservation and fishing? Should the sea otters remain absolutely protected or is there a case for culling or some other form of control?

Many of the arguments for conservation are based not simply on sentiment but on genuine ecological concerns. Should the public (and politicians) be better educated so as to appreciate those concerns? How might this be done? Or should the public (and politicians) simply take the advice of ecologists?

James A. Estes is a professor with the Department of Ecology and Evolutionary Biology at UC Santa Cruz, and a Pew marine fellow.

populations at these low densities grow by simple multiplication over successive intervals of time. This is **exponential growth** (Figure 5.21; see also Box 5.4) and the rate of increase is the population's **intrinsic rate of natural increase** (denoted by r). Of course, any population that behaved in this way would soon run out of resources, but as we have seen, the rate of increase tends to become reduced by competition as the population grows, and it falls to zero when the population reaches its carrying capacity (since birth rate then equals death rate). A steady reduction in the rate of increase as densities move toward the carrying capacity gives rise to population growth that is not exponential but S-shaped (Figure 5.21). The pattern is also often called *logistic* growth after the so-called logistic equation (Box 5.4).

The S-shaped curve can best be seen in action in laboratory studies of microorganisms or animals with very short life cycles (Figure 5.23a). In these kinds of experiment it is easy to have experimental control

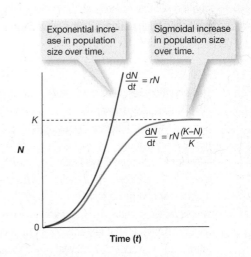

FIGURE 5.21 Exponential and S-shaped or *sigmoidal* increase in the size of a population over time. These patterns describe the growth to be expected in general in populations in the absence (exponential) and under the influence (sigmoidal) of intraspecific competition but are also generated, specifically, by the exponential and logistic equations shown (see also Box 5.4).

Quantitative Aspects

The exponential and logistic equations of population growth

In this box, simple mathematical models are derived for populations first in the absence of, and then under the influence of, intraspecific competition. These and other mathematical models play an important part in ecology (see Chapter 1). They help us to follow through the consequences of assumptions we may wish to make, and to explore the behavior of ecological systems that we may find it hard to observe in nature or construct in the laboratory. The particular models in this box themselves form the basis for more complex models of *interspecific* competition (competition between individuals of different species, Chapter 6) and predation: they are important building blocks. It is essential to appreciate, however, that a pattern generated by such a model—for example, the S-shaped pattern of population growth under the influence of intra-specific competition—does not derive its interest, or its importance, from having been generated by the model. There are many other models that could generate very similar (indistinguishable) patterns. Rather, the point about the pattern is that it reflects important, underlying ecological processes—and the model is useful in that it appears to capture the essence of those processes.

We start with a model of a population in which there is no intraspecific competition and then incorporate that competition later. Our models are in the form of differential equations, describing the net rate of increase of a population, which will be denoted by *dN/dt* (spoken: *dN by dt*). This represents the speed at which a population increases in size, N, as time, t, progresses.

The increase in size of the whole population is the sum of the contributions of the various individuals within it. Thus, the average rate of increase per individual, or the per capita rate of increase (*per capita* means per head) is given by $dN/dt \cdot (1/N)$: the population rate of increase divided by the number in the population. In the absence of intraspecific competition (or any other force that increases the death rate or reduces the birth rate), this rate of increase is a constant and as high as it can be for the species concerned. It is called the *intrinsic rate of natural increase* and denoted by r. Thus:

$$dN/dt(1/N) = r$$

and the net rate of increase for the whole population is therefore given by the per capita rate of increase multiplied by the number in the population:

$$dN/dt = rN$$

This equation describes a population growing *exponentially* (Figure 5.21).

Intraspecific competition can now be added. In doing this, we will be deriving the logistic equation, using the method set out in Figure 5.22. The net rate of increase per individual is unaffected by competition when N is very close to zero, because there is no crowding nor a shortage of resources. It is still therefore given by r (point A). When N rises to K (the carrying capacity) the net rate of increase per individual is, by definition, zero (point B). For simplicity, we assume a straight line between A and B; that is, we assume a linear reduction in the per capita rate of increase, as a result of intensifying intraspecific competition, between N = 0 and N = K.

Thus, on the basis that the equation for any straight line takes the form $y = \text{intercept} + \text{slope } x$, where x and y are the variates on the horizontal and vertical axes, here we have

$$dN/dt(1/N) = r - (r/K)N$$

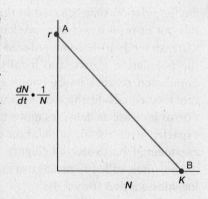

FIGURE 5.22 An ideal linear decline in the net rate of increase per individual with increasing population (N).

or, rearranging,

$$dN/dt = r N [1- (N/K)]$$

This is the *logistic equation,* and a population increasing in size under its influence is shown in Figure 5.21. It describes a 'sigmoidal' or S-shaped growth curve approaching a stable carrying capacity, but it is only one of many reasonable equations that do this. Its major advantage is its simplicity. Nevertheless, it has played a central role in the development of ecology.

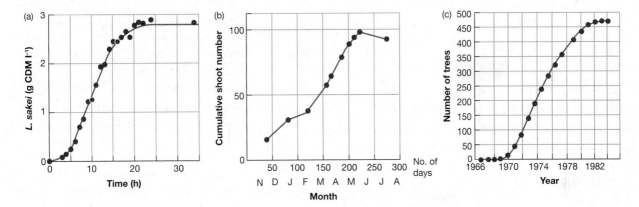

FIGURE 5.23   Real examples of S-shaped population increase. (a) The bacterium *Lactobacillus sakei* (measured as grams of 'cell dry mass' per liter) grown in nutrient broth. (After Leroy & de Vuyst, 2001.) (b) The population of shoots (i.e., modules – see Section 5.1.1) of the annual plant *Juncus gerardi* in a salt marsh habitat on the west coast of France. (After Bouzille et al., 1997.) (c) The population of the willow tree (*Salix cinerea*) in an area of land after myxomatosis had effectively prevented rabbit grazing. (After Alliende & Harper, 1989.)

of environmental conditions and resources. In the real world, outside the laboratory and the mind of the mathematician, the world is less simple. The complex life cycles of organisms, seasonal changes in conditions and resources, and the patchiness of habitats all introduce complications. In nature, populations often follow a very bumpy ride along the path of perfect logistic growth (Figure 5.23b), though not always (Figure 5.23c).

Another way to summarize the ways in which intraspecific competition affects populations is to look at **net recruitment**—the number of births minus the number of deaths in a population over a period of time. When densities are low, net recruitment will be low because there are few individuals available either to give birth or to die. Net recruitment will also be low at much higher densities as the carrying capacity is approached. Net recruitment will be at its peak, then, at some intermediate density. The result

is a 'humped' or dome-shaped curve (Figure 5.24). Again, of course, as with the ideal logistic curve, real data from nature never fall on a single line: indeed, sometimes the lines drawn through the data seem somewhat imaginative! But the dome-shaped curve reflects the essence of net recruitment patterns when density-dependent birth and death are the result of intraspecific competition.

5.6   LIFE HISTORY PATTERNS

One of the ways in which we can try to make sense of the world around us is to search for repeated patterns. In doing so, we are not pretending that the world is simple or that all categories are watertight, but we do need a description that is more than just a catalog of unique special cases. This final section of the chapter describes some simple, useful, though by no means

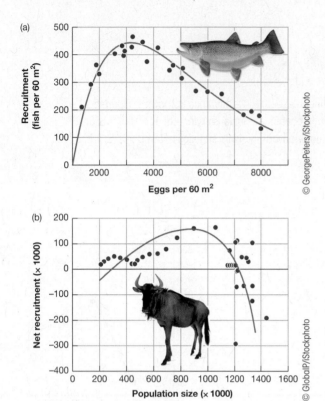

FIGURE 5.24 Dome-shaped net-recruitment curves. (a) Six-month-old brown trout, *Salmo trutta*, in Black Brows Beck, England, between 1967 and 1989. (After Myers, 2001; following Elliott, 1994.) (b) Wildebeest, *Connochaetes taurinus*, in the Serengeti, from 1959–1995. (Data from Mduma et al., 1999.)

perfect patterns linking different types of life history and different types of habitat.

First, though, we return to a point made earlier: that in any life history there is a limited total amount of energy (or some other resource) available to an organism for growth and reproduction. Some trade-off may therefore be necessary: either grow more and reproduce less, or reproduce more and grow less. Specifically, there may be an observable **cost of reproduction** in that when reproduction starts, or increases, growth may slow or stop completely, as resources are diverted. We can, of course, look at this trade-off the other way around: an organism that makes vigorous growth, and so thrives in competition with its neighbors, may have to pay the price by reducing reproductive activity. In many forest trees, for example, growth rings in the trunk may be narrower in 'mast' years, when very heavy crops of seeds are produced (Figure 5.25a). Furthermore, the diversion of resources to present reproduction may jeopardize subsequent survival (Figure 5.25b, and as also seen in the salmon and foxgloves described earlier), or simply reduce the capacity for future reproduction (Figure 5.25c).

> the 'cost' of reproduction—a life history trade-off

Yet it would be quite wrong to think that such trade-offs abound in nature, only waiting to be observed. The examples in Figure 5.25 tell a different story. Only the youngest and oldest squirrels paid a survival cost of reproduction (Figure 5.25b); and only mountain goats in the poorest condition had their chances of future reproduction reduced when they produced offspring (Figure 5.25c). Note especially that if there is variation between individuals in the amount of resource they have at their disposal, then there is likely to be a positive, not a negative correlation between two apparently alternative processes—some individuals will be good at everything, others consistently awful. For instance, in Figure 5.26, the snakes in the best condition produced larger litters but also recovered from breeding more rapidly, ready to breed again.

We can now turn to the life history patterns themselves. The potential of a species to multiply rapidly is advantageous in environments that are short-lived, allowing the organisms to colonize new habitats quickly and exploit new resources. This rapid multiplication is a characteristic of the life cycles of terrestrial organisms that invade disturbed land (for example, many annual weeds), or colonize newly opened habitats such as forest clearings, and of the aquatic inhabitants of temporary puddles and ponds. These are species whose populations are usually found expanding after the last disaster or exploiting the new opportunity. They have the life cycle properties that are favored by natural selection in such conditions: the production of large numbers of progeny, early in the life cycle, rather than investing heavily in either growth or survival. They have been called **r species**, because they spend most of their life in the near-exponential, *r*-dominated phase of population growth (Box 5.4), and the habitats in which they are likely to be favored have been called *r*-selecting. They are also commonly called opportunistic species, as they take advantage of the opportunities presented by newly created environments.

> *r* and *K* species

Organisms with quite different life histories survive in habitats where there is often intense competition for limited resources. The individuals that are successful in leaving descendants are those that have captured, and often held on to, the larger share of resources. Their populations are usually crowded and those that win in a struggle for existence do so because they have grown faster and/or larger (rather than reproducing) or have spent more of their resources in aggression or some other activity that has favored their survival under crowded conditions. They are called **K species** because their populations spend most of their lives in the *K*-dominated phase of population growth

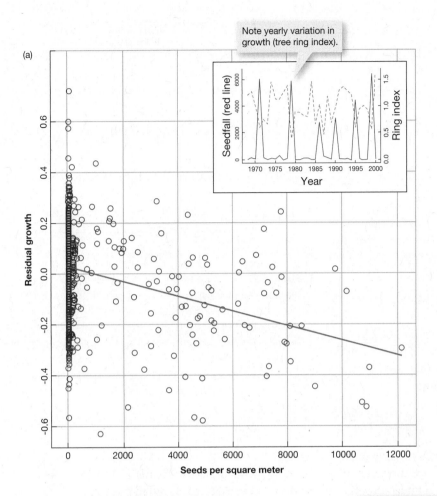

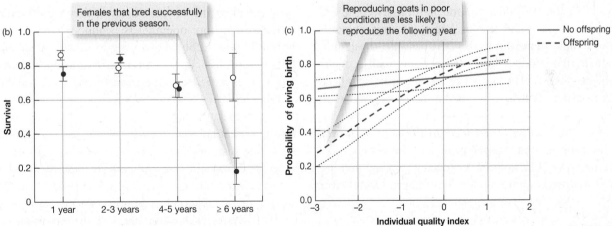

FIGURE 5.25   Costs of reproduction. (a) The tree *Nothophagus truncata*, growing in Orongorongo, New Zealand, exhibits 'mast seeding' (inset): years of massively increased seed production (i.e., reproduction) exhibited by whole groups of trees growing together. Much of this growth variation was attributable to prevailing weather conditions, but once this had been accounted for, the remaining 'residual growth' was significantly negatively correlated with seed output ($P < 0.0001$): trees that reproduced most grew least; trees that grew most reproduced least. (After Monks & Kelly, 2006.) (b) Among North American red squirrels, *Tamiasciurus hudsonicus*, from the southern Yukon, Canada, survival was lower in females that bred successfully in the previous season (means ± standard errors) than in those that did not, but only among the youngest and oldest: those in their first year or more than five years old. (After Descamps et al., 2009.) (c) In mountain goats, *Oreamnos americanus*, from Alberta, Canada, the probability of giving birth one year was reduced among females that had given birth the previous year, but only for females in relatively poor condition, measured on a scale according to which females were either relatively light and subordinate (negative values) or heavy and dominant (positive values). The lines are not the original data but the output of best-fit statistical models, with associated standard errors. (After Hamel et al., 2009.)

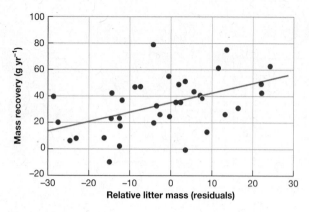

FIGURE 5.26 Female aspic vipers (*Vipera aspis*) that produced larger litters ('relative' litter mass because total female mass was taken into account) also recovered more rapidly from reproduction (not 'relative' because mass recovery was not affected by size) ($r = 0.43$; $P = 0.01$). (After Bonnet et al., 2002.)

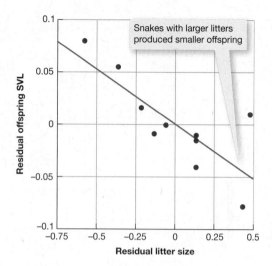

FIGURE 5.27 Evidence for a trade-off between the number of offspring produced in a clutch by a parent and the individual fitness of those offspring: a negative correlation between the size of offspring (as measured by their snout-vent length, SVL) and the number of them in a litter in the Australian highland copperhead snake, *Austrelaps ramsayi* ($r^2 = 0.63$, $P = 0.006$). 'Residual' offspring and litter sizes have been used: these are the values arrived at after variations in maternal size have been allowed for, since both increase with maternal size. (After Rohr, 2001.)

(Box 5.4), close to the limits of environmental resources, and the habitats in which they are likely to be favored have been called *K*-selecting.

A further common distinction between *r* and *K* species is whether they produce many small progeny (characteristic of *r* species) or few large progeny (characteristic of *K* species). This is another example of a life history trade-off: an organism has limited resources available for reproduction, and natural selection will influence how these are packaged. In environments where rapid population growth is possible, those individuals that produce large numbers of small progeny will be favored. The size of progeny can be sacrificed because they will usually not be in competition with others. However, in environments in which the individuals are crowded and there is competition for resources, those progeny that are well provided with resources by the parent will be favored. Producing progeny that are well endowed requires the trade-off of producing fewer of them (see, for example, Figure 5. 27).

The *r/K* concept has been useful as an organizing principle in helping ecologists think about life histories, and there have certainly been studies that have provided broad support for the scheme. For instance, in a study of the common dandelion, *Taraxacum officinale*, plants were either left for five years in crowded, multi-species communities (*K*-selecting), or were 'weeded out' twice per year, early in their adult lives, such that they had to regrow again from root stumps or reestablish from newly deposited seed (*r*-selecting). They were then grown side by side, in a common garden, either from seeds or as cuttings

r, K, and progeny size and number

evidence for the *r/K* scheme?

from established plants (Figure 5.28). Plants from the *r*-selecting environment invested more in reproduction, as opposed to growth and survival: they were smaller (Figure 5.28a) but nonetheless produced more flower heads (Figure 5.28b) having begun to reproduce earlier (Figure 5.28c). As a result, they produced more seeds (Figure 5.28d), but those seeds were smaller (Figure 5.28e).

On the other hand, there are also many other examples that fail to fit the *r/K* scheme, and recent studies of life histories that mention it are likely to do so in order to explain how the patterns described depart from the expectations it generates. For example, earlier studies of the Trinidadian guppies we met in Section 2.3.1, *Poecilia reticulata*, emphasized the importance of the predation pressures acting on the fish (Figure 5.29). Two contrasting habitats were again compared: high predation habitats where predation pressure was intense and directed at all age classes, including adults, and low predation habitats where predation pressure was only moderate and was directed more at juveniles. In the former, fish tended to mature at a smaller size, they allocated more to reproduction and reproduced more often, and they produced smaller embryos. Increased and early investment in reproduction in high predation habitats (Figure 5.29a, b, c) is understandable as a response to the premium there must be on reproducing before

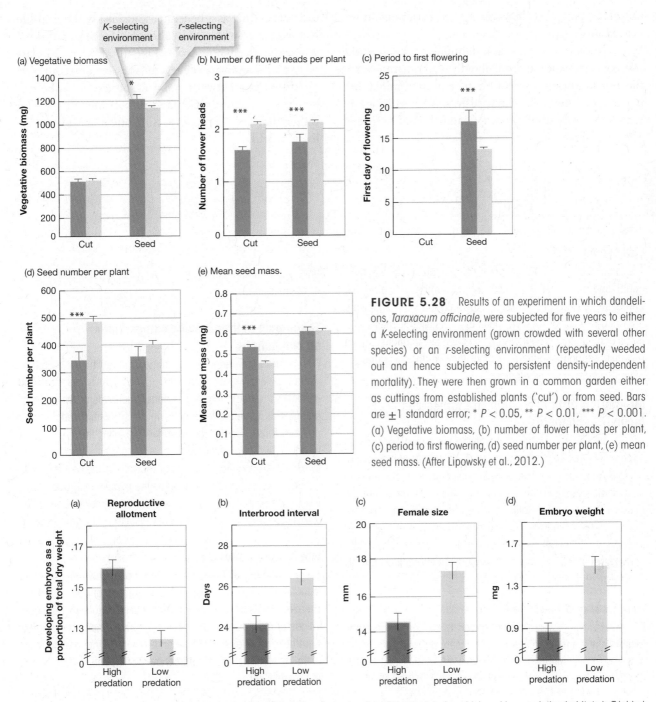

FIGURE 5.28   Results of an experiment in which dandelions, *Taraxacum officinale*, were subjected for five years to either a *K*-selecting environment (grown crowded with several other species) or an *r*-selecting environment (repeatedly weeded out and hence subjected to persistent density-independent mortality). They were then grown in a common garden either as cuttings from established plants ('cut') or from seed. Bars are ±1 standard error; * $P < 0.05$, ** $P < 0.01$, *** $P < 0.001$. (a) Vegetative biomass, (b) number of flower heads per plant, (c) period to first flowering, (d) seed number per plant, (e) mean seed mass. (After Lipowsky et al., 2012.)

FIGURE 5.29   A comparison of the life histories characteristics of guppies, *Poecilia reticulata*, from high and low predation habitats in Trinidad: (a) reproductive allotment: developing embryos as a proportion of total dry weight; (b) mean interbrood interval; (c) mean minimum size of reproducing females; (d) mean embryo dry weight. (After Reznick & Endler, 1982.)

succumbing to those intense predation pressures, but neither predation nor the age-profile of mortality figure in the *r/K* classification.

However, subsequent studies of the system have indicated that contrasting densities and resource availabilities in the two habitats—central to the *r/K* concept—seem also to play a key role in the evolution of guppy life histories. As a result of the intense predation there, the high predation environments also

have lower guppy densities and greater availability of resources, and so in this sense they are relatively *r*-selecting. The smaller embryos in this habitat (Figure 5.29d) conform to the predictions of the scheme, while the greater investment in reproduction (Figure 5.29a, b), and the higher intrinsic growth rate of juveniles, bringing them more rapidly to maturity (Arendt &Reznick, 2005), are consistent both with the *r/K* scheme and as a response to the unavoidable predation pressures. There is clearly no single, simple explanation for the variations in life history observed.

We can regard this as a damning criticism of the r/K concept, since it undoubtedly demonstrates that the explanatory powers of the scheme are limited. But it is equally possible to regard it as very satisfactory that a relatively simple concept can help make sense of a large proportion of the multiplicity of life histories, even when only as a benchmark against which real-life histories are set. Nobody, though, can regard the r/K scheme as the whole story. Like all attempts to classify species and their characteristics into pigeonholes, the distinction between r and K species has to be recognized as a convenient (and useful) human creation rather than an all-encompassing statement about the living world.

SUMMARY

Life cycles

The processes that change the size of populations are birth, death, and movement. A population is a number of individuals, but for some kinds of organisms, especially modular organisms, it is not always clear what we mean by an individual. The life histories of all unitary organisms can be seen as variations around a simple, sequential pattern, whether they be annuals or perennials, iteroparous or semelparous.

Monitoring birth and death: life tables and fecundity schedules

Life tables can be useful in identifying what in a life cycle is apparently most instrumental in determining rates of increase or decline. A cohort life table records the survivorship of members of a single cohort. When we cannot follow cohorts, it may be possible to construct a static life table, but great care is required. The fecundity of individuals also changes with age, described in age-specific fecundity schedules. A useful set of survivorship curves (types 1–3) has been developed, but in practice patterns of survival are usually more complex.

Dispersal and migration

Dispersal is the way individuals spread away from each other. Migration is the mass directional movement of large numbers of a species from one location to another. Movement and spatial distribution are intimately related. Dispersal and migration can have a profound effect on the dynamics of a population and on its composition.

The impact of intraspecific competition on populations

Over a sufficiently large density range, competition between individuals generally reduces the birth rate as density increases and increases the death rate (i.e. is density-dependent). Intraspecific competition therefore tends to keep density within certain limits and may thus be said to play a part in regulating the size of populations. When populations are sparse and uncrowded they tend to exhibit exponential growth, but the rate of increase tends to become reduced by competition as the population grows, giving rise to population growth that is logistic. Intraspecific competition also affects net recruitment, typically resulting in a humped curve.

Life history patterns

There is typically a limited total amount of energy or some other resource available to an organism for growth and reproduction. There may thus be an observable cost of reproduction. But populations of individuals that reproduce early in their life can grow extremely fast. The r/K concept can be useful in interpreting many of the differences in form and behavior of organisms, but it is not the whole story.

REVIEW QUESTIONS

1 Contrast the meaning of the word *individual* for unitary and modular organisms.

2 In a mark–recapture exercise during which a population of butterflies remained constant in size, an initial sample provided 70 individuals, each of which was marked and then released back into the population. Two days later, a second sample was taken, totaling 123 individuals of which 47 bore a mark from the first sample. Estimate the size of the population. State any assumptions that you have had to make in arriving at your estimate.

3 Contrast the derivation of *cohort* and *static life tables* and discuss the problems of constructing and/or interpreting each.

4 Describe what are meant by *aggregated*, *random*, and *regular* distributions of organisms in space, and outline, with actual examples where possible, some of the

behavioral processes that might lead to each type of distribution.

Stage (x)	Numbers at start of stage (a_x)	Proportion of original cohort alive at start of stage (l_x)	Mean no. of eggs produced per individual in stage (m_x)
Eggs	173	?	0
Nestlings	107	?	0
Fledglings	64	?	0
1-year-olds	31	?	2.5
2-year-olds	23	?	3.7
3-year-olds	8	?	3.1
4-year-olds	2	?	3.5
		$R = ?$	

5   What is meant by the *carrying capacity* of a population? Describe where it appears, and why, in (a) S-shaped population growth, (b) the logistic equation, and (c) dome-shaped net-recruitment curves.

6   Explain why an understanding of life-history trade-offs is central to an understanding of life-history evolution. Explain the contrasting trade-offs expected to be exhibited by *r*-selected and *K*-selected species.

7   The following is an outline life table and fecundity schedule for a cohort of a population of sparrows. Fill in the missing values (wherever there is a question mark).

Challenge Questions

1   Define *annual*, *perennial*, *semelparous*, and *iteroparous*. Try to give an example of both an animal and a plant for each of the four possible combinations of these terms. In which cases is it difficult (or impossible) to come up with an example and why?

2   Why is the average density of people in the United States lower than the density *experienced* by people, on average, in the United States? Is a similar contrast likely to apply to most species? Why? Under what conditions might it not apply?

3   Compare unitary and modular organisms in terms of the effects of intraspecific competition both on individuals and on populations.

Andrea Gingerich/Getty Images

Chapter 6

Interspecific competition

CHAPTER CONTENTS

KEY CONCEPTS

After reading this chapter you will be able to:

- elaborate the differences between fundamental and realized niches and the importance and limitations of the Competitive Exclusion Principle

- explain the potential role of the evolutionary effects of competition in species coexistence but also the difficulty of proving that role

- explain the nature and importance of niche complementarity

- elaborate the difficulties in determining the prevalence of current competition in nature, and in distinguishing between the effects of competition and mere chance

nterspecific competition is one of the most fundamental phenomena in ecology, affecting not only the current distribution and success of species but also their evolution. Yet the existence and effects of interspecific competition are often remarkably difficult to establish and demand an armory of observational, experimental and modeling techniques.

6.1 ECOLOGICAL EFFECTS OF INTERSPECIFIC COMPETITION

Having been introduced to *intra*specific competition in previous chapters, we can readily understand what **inter**specific **competition** is. Individuals of one species suffer a reduction in fecundity, survivorship, or growth as a result of either exploitation of resources or interference by individuals from another species. These individual effects are likely to influence the population dynamics and distributions of the competing species, which go to determine the compositions of the communities of which they are part. But those individual effects will also influence those species' evolution, which, in turn, can influence the species' distributions and dynamics.

This chapter, then, is about both the ecological and the evolutionary effects of interspecific competition on individuals, on populations, and on communities. But

> two separate questions—the possible and actual consequences of competition

it also addresses a more general issue in ecology and indeed in science—that there is a difference between what a process can do and what it *does* do: a difference between what, in this case, interspecific competition is capable of doing and what it actually does in practice. These are two separate questions, and we must be careful to keep them separate.

The way these different questions can be asked and answered will be different, too. To find out what interspecific competition is capable of doing is relatively easy. Species can be forced to compete in experiments, or they can be examined in nature in pairs or groups chosen precisely because they seem most likely to compete. But it is much more difficult to discover how important interspecific competition actually is. It will be necessary to ask how realistic our experiments are, and whether the pairs and groups of species we

choose to study are really typical of pairs and groups more generally.

We begin, though, with some examples of what interspecific competition can do.

Competition amongst phytoplankton for phosphorus

Competition was investigated in the laboratory between five single-celled freshwater phytoplankton species, competed together in pairs: *Chlorella vulgaris, Selenastrum caprocornutum, Monoraphidium grofithii, Monodus subterraneus,* and *Synechocystis,* all of which require phosphorus as an essential resource for their growth. The population densities of the different species were monitored over time as resources were continuously being added to the liquid medium, but their impact on their limiting resource (phosphorus) was also recorded. When any of the species was grown alone, it established a steady population density, reducing the phosphorus to a constant low concentration (Figure 6.1a). However, there was a clear rank order in the steady-state concentration to which phosphorus was reduced: *Synechocystis* (0.030 μmol/L) < *Chlorella* (0.059 μmol/L) < *Monoraphidium* (0.117 μmol/L) < *Selenastrum* (0.160 μmol/L) < *Monodus* (0.182). Then, when any two species were grown together, there was only one survivor – and that survivor was whichever species had previously reduced phosphorus to the lower level (Figure 6.1b).

Thus, although all species were capable of living alone in the laboratory habitat, when they competed, one species always excluded

> more efficient exploiters exclude less efficient ones

the other, because it was the more effective exploiter of their shared, limiting resource, reducing it to a level too low for the other species to survive. A similar result has been obtained for the nocturnal, insectivorous gecko *Hemidactylus frenatus,* an invader of urban habitats

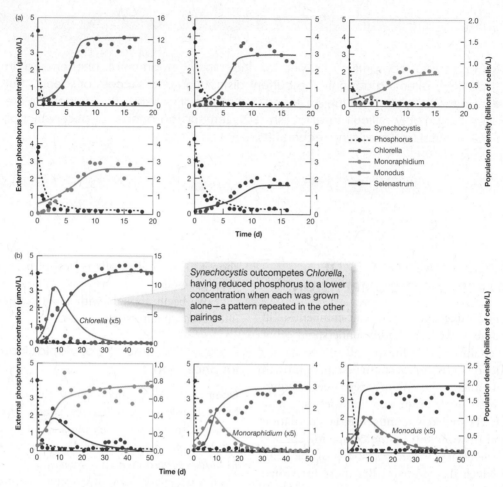

FIGURE 6.1 Competition between phytoplankton species. (a) Each of five species (names and symbols shown in each panel), when grown alone in a laboratory chemostat, establishes a stable population and maintains a resource, phosphorus (red dots), at a constant low level: *Synechocystis* (0.030 µmol/L) < *Chlorella* (0.059 µmol/L) < *Monoraphidium* (0.117 µmol/L) < *Selenastrum* (0.160 µmol/L) < *Monodus* (0.182 µmol/L). (b) With symbols retained from (a), it is apparent that when species are grown together in pairs, the only one to survive is the one that had previously reduced phosphorus to the lower level. (Densities have been multiplied by 5 in the cases indicated, so they can be visualized in the same figure. The lines are drawn simply to indicate general trends.) (After Passarge et al., 2006.)

across the Pacific basin, where it is responsible for population declines of the native gecko *Lepidodactylus lugubris* (Petren & Case, 1996). The diets of the two geckos overlap substantially and insects are a limiting resource for both. The invader is capable of depleting insect resources in experimental enclosures to lower levels than the native gecko, and the native suffers reductions in body condition, fecundity, and survivorship as a result.

Coexistence and exclusion of competing salmonid fishes

The Dolly Varden charr (*Salvelinus malma*) and the white-spotted charr (*S. leucomaenis*) are two morphologically similar and closely related species of salmonid fish (see Section 3.2) found in many streams on Hokkaido Island, Japan. The Dolly Varden are distributed further

The gecko, *Lepidodactylus lugubris*

upstream than the white-spotted charr. Water temperature, which has profound consequences for fish ecology, increases downstream. In streams where they

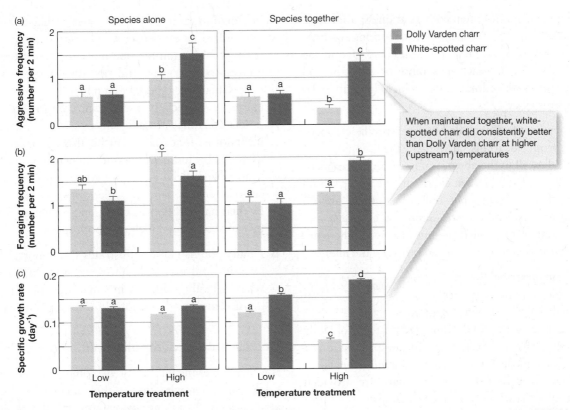

FIGURE 6.2 (a) Frequency of aggressive encounters initiated by individuals of each of two fish species (Dolly Varden charr, Light blue histograms, and white-spotted charr, dark blue histograms) during a 72-day experiment in artificial stream channels with two replicates each. Left, either 50 Dolly Varden or 50 white-spotted charr alone (allopatry); right: 25 of each species together (sympatry); (b) foraging frequency; and (c) specific growth rate in length. Different letters indicate means are significantly different from each other. (After Taniguchi & Nakano, 2000.)

occur together, there is a zone of overlap at intermediate altitudes, but in streams where one species is absent, the other expands its range beyond this intermediate zone.

In laboratory streams, higher temperatures (12°C as compared to 6°C) led to increased aggression in both species when they were tested alone, but these temperatures led to the Dolly Varden being *less* aggressive when white-spotted charr were also present (Figure 6.2a). Consequently, Dolly Varden charr failed to obtain favorable foraging positions when white-spotted charr were present at the higher temperature and so foraged far less effectively (Figure 6.2b). Also, when alone, neither species' growth rates were influenced by temperature. But when both species were present, growth of Dolly Varden charr decreased with increasing temperature, whereas that of white-spotted charr increased (Figure 6.2c). Thus, the growth rate of the Dolly Varden was much lower than that of white-spotted charr at the higher temperature.

These results all support the idea that the lower altitudinal boundary of Dolly Varden charr in the Japanese streams was due to temperature-mediated competition favoring white-spotted charr, which were more aggressive, foraged more effectively, and grew far faster. But the Dolly Varden did not outcompete the

white-spotted charr in any of the experiments, even at the lower temperatures. Thus, the results provide no evidence for the upper boundary of the white-spotted charr also being due to temperature-mediated competitive effects. Further work will be needed to determine why Dolly Varden exclude white-spotted charr upstream.

Some general observations

These three examples illustrate several points of general importance.

1 Competing species often coexist at one spatial scale but are found to have distinct distributions at a finer scale of resolution. Here, the fishes coexisted in the same stream, but each was more or less confined to its own altitudinal zone (which had different temperatures).

2 Species are often excluded by interspecific competition from locations at which they could exist perfectly well in the absence of interspecific competition. Here, Dolly Varden charr can live in the white-spotted charr zone—but only when there are no white-spotted charr there. Similarly, *Chlorella* can live in laboratory cultures—but only when there are no *Synechocystis* there (and so on down the line).

3 We can describe this by saying that the conditions and resources provided by the white-spotted charr zone are part of the fundamental niche of Dolly Varden charr (see Section 3.6 for an explanation of ecological niches) in that the basic requirements for the existence of Dolly Varden charr are provided there. But the white-spotted charr zone does not provide a realized niche for Dolly Varden when white-spotted charr are present. Likewise, the laboratory cultures provided the requirements of the fundamental niches of both *Chlorella* and *Synechocystis* (and the other three phytoplankton species), but when *Chlorella* and *Synechocystis* were both present, provided a realized niche for only *Synechocystis*.

> fundamental and realized niches

4 Thus, a species' **fundamental niche** is the combination of conditions and resources that allow that species to exist, grow, and reproduce when considered in isolation from any other species that might be harmful to its existence; whereas its **realized niche** is the combination of conditions and resources that allow it to exist, grow, and reproduce in the presence of specified other species that might be harmful to its existence—especially interspecific competitors. Even in locations that provide a species with the requirements of its fundamental niche, that species may be excluded by another, superior competitor that denies it a realized niche there. But competing species can coexist when both are provided with a realized niche by their habitat. For the charr, the stream as a whole provided a realized niche for both species.

5 Finally, the fish study illustrates the importance of experimental manipulation if we wish to discover what is really going on in a natural population—'nature' may need to be prodded to reveal its secrets.

Coexistence of competing diatoms

Another experimental study of competition amongst single-celled phytoplankton, in this case diatoms, looked at species coexisting on not one but two shared, limiting resources. The two species were *Asterionella formosa* and *Cyclotella meneghiniana*, and the resources, which were both capable of limiting the growth of both diatoms, were silicate and phosphate. *Cyclotella* was the more effective exploiter of silicate (reducing its concentration to a lower level), but *Asterionella* was the more effective exploiter of phosphate. Hence, in cultures where there were especially low supplies of silicate, *Cyclotella* excluded *Asterionella* (Figure 6.3): such cultures failed to provide a realized niche for *Asterionella*, the inferior

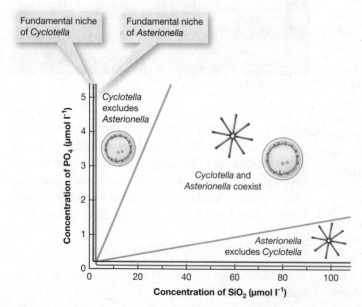

FIGURE 6.3 *Asterionella formosa* and *Cyclotella meneghiniana* coexist when there are roughly balanced supplies of silicate (SiO_2) and phosphate (PO_4), but *Asterionella* excludes *Cyclotella* when there are especially low supplies of phosphate, whereas *Cyclotella* excludes *Asterionella* when there are especially low supplies of silicate. (After Tilman, 1982.)

FIGURE 6.4   Four species of warbler that coexist on the islands of Corsica and Sardinia: top left Marmora's warbler (*Sylvia sarda*), top right the Dartford warbler (*S. undata*), bottom left the Sardinian warbler (*S. melanocephala*), and bottom right the Subalpine warbler (*S. cantillans*).

competitor there. Conversely, in cultures where there were especially low supplies of phosphate, *Asterionella* excluded *Cyclotella*. However, in cultures with relatively balanced supplies of silicate and phosphate, the two diatoms coexisted (Figure 6.3): when both species were provided with sufficient supplies of a resource on which they were inferior, there was a realized niche for both.

Coexistence of competing birds

It is not always so easy to identify the **niche differentiation** (i.e., differentiation of their realized niches) or 'differential resource utilization' that allows competitors to coexist. Ornithologists, for example, are well aware that closely related species of birds often coexist in the same habitat. For example, four *Sylvia* species (warblers) occur together on the Mediterranean islands of Corsica and Sardinia (Figure 6.4): Marmora's warbler (*Sylvia sarda*), the Dartford warbler (*S. undata*), the Sardinian warbler (*S. melanocephala*), and the Subalpine warbler (*S. cantillans*). All look similar and all are insectivorous, with just the occasional use of berries or soft fruit; and all live in open bush-dominated landscapes, using the bushes there for nesting. Yet, the closer we look at the details of the ecology of such coexisting species, the more likely we are to find ecological differences. In Corsica and Sardinia, all four species are found in habitats with bushes of medium height; but where the bushes are lower, Marmora's warblers are most often observed with Subalpine warblers largely absent, whereas the reverse is true where bushes are higher. And even within the particular habitat types, the species differ

significantly in the different plant species and plant heights from which they prefer to forage (Martin & Thibault, 1996). Hence, it is tempting to conclude that such species compete but coexist by utilizing slightly different resources in slightly different ways: **differential resource utilization**. But in complex natural environments, such conclusions, while plausible, are difficult to prove.

Indeed, it is often not easy to prove even that the species compete. To do so, it is usually necessary to remove one or more of the species and monitor the responses of those that remain. This was done, for example, in a study of two very similar bird species, again warblers: the orange-crowned warbler (*Vermivora celata*) and the Virginia's warbler (*V. virginiae*), whose breeding territories overlap in central Arizona. On plots where one of the two species had been removed, the remaining species fledged between 78% and 129% more young per nest (Figure 6.5a). The enhanced performance was largely due to improved access to preferred nest sites and consequent decreases in the loss of nestlings to predators (Figure 6.5b). The effects of interspecific competition, normally, are revealed by experimentally eliminating that competition.

> coexistence through niche differentiation—and even competition—may be difficult to prove

Competition between unrelated species

The examples described so far have all involved pairs or groups of closely related species – phytoplankton, salmonid fish, or birds. But competition may also occur

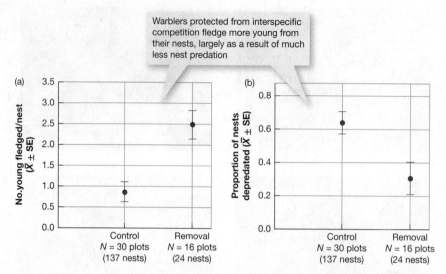

FIGURE 6.5 Competition between orange-crowned warblers and Virginia's warblers, examined through a comparison of control plots (both species present) and plots from which one or other of the species was removed (males removed before pairing such that females neither settle nor nest). (a) The number of young fledged from nests (means ± SEs) i.e. leaving to be self-supporting, was significantly higher ($P = 0.0004$) when one species was removed (data from both species combined). (b) The proportion of nests in which all offspring were taken by predators (means ± SEs) was significantly lower ($P = 0.002$) when one species was removed (data from both species combined). (After Martin & Martin, 2001.)

between completely unrelated species, as shown for two examples in Figure 6.6. In the first (Figure 6.6a), removing sea urchins from experimental plots in the coastal waters of St. Thomas, U.S. Virgin Islands, increased the abundance of both parrotfish and surgeonfish, with which the sea urchins compete for their shared seaweed food; though one year later the urchins had returned, and levels of food and fish abundance had returned to control levels. In the second example (Figure 6.6b), allowing insects to compete at natural densities with tadpoles of the Pine Barrens tree frog demonstrated that the competition from this very distantly related group was comparable in intensity to that provided by the much more closely related tadpoles of Fowler's toads. Close relatives are most similar, and hence, all other things being equal, more likely to compete. But clearly, interspecific competition can also occur across huge taxonomic divides.

The competitive exclusion principle

The patterns that are apparent in these examples have also been uncovered in many others, and have been elevated to the status of a principle: the **Competitive Exclusion Principle** or Gause's principle (named after an eminent Russian ecologist – see History Box 6.1). It can be stated as follows:

- If two competing species coexist in a stable environment, then they do so as a result of niche differentiation.

- If, however, there is no such differentiation, or if it is precluded by the habitat, then one competing species will eliminate or exclude the other.

The principle has emerged here from examining patterns evident in real sets of data. However, the establishment of the principle had its roots in a simple mathematical model of interspecific competition, usually known by the names of its two (independent) originators: Lotka and Volterra (Quantitative Box 6.2).

There is no question that there is some truth in the principle that competitor species can coexist as a result of niche differentiation, and that one competitor species may exclude another by denying it a realized niche. But it is crucial also to be aware of what the Competitive Exclusion Principle does *not* say.

It does *not* say that whenever we see coexisting species with different niches, it is reasonable to jump to the conclusion that this is the principle in operation. All species, on close inspection, have their own unique niches. Niche differentiation does not prove that there are coexisting competitors. The species may not be competing at all and may never have done so in their evolutionary history. We require proof of interspecific competition. In the examples above, this was provided by experimental manipulation – for example, remove one species (or one group of species) and the other species increases its abundance or its survival. But most cases for competitors coexisting as a result of niche differentiation have not been subjected to experimental proof. So just how important is the Competitive Exclusion Principle in practice? We return to this question in Section 6.5.

> The Competitive Exclusion Principle—what it does and does not say

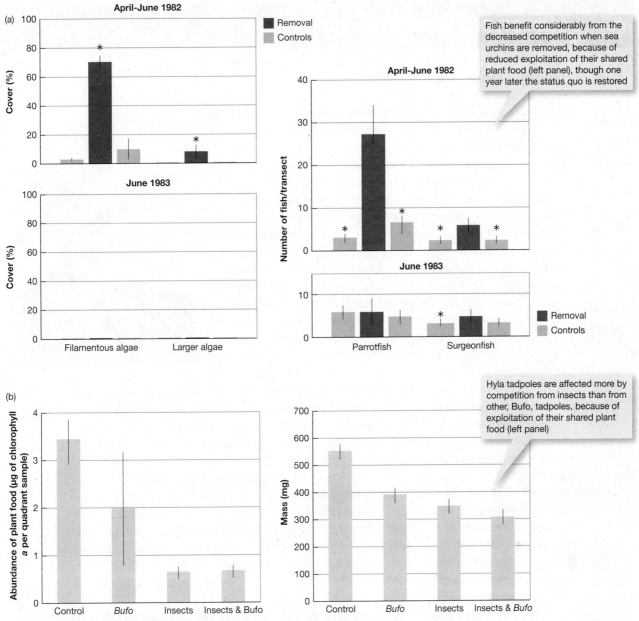

Fish benefit considerably from the decreased competition when sea urchins are removed, because of reduced exploitation of their shared plant food (left panel), though one year later the status quo is restored

Hyla tadpoles are affected more by competition from insects than from other, Bufo, tadpoles, because of exploitation of their shared plant food (left panel)

FIGURE 6.6   (a) Competition between sea urchins and fish. When sea urchins, *Diadema antillarum*, were removed from experimental plots in St. Thomas, U.S. Virgin Islands, in February 1982, two months later amounts of their seaweed food had significantly increased (left panel) as had numbers of parrotfish and surgeonfish, which share the food resource with the urchins (right panel). But one year later, urchins had re-invaded and reached 70% of their original abundance, and food and numbers of fish had reverted to control levels. Bars are 2 SEs. (After Hay & Taylor, 1985.) (b) Competition between insects and tadpoles. Tadpoles of the Pine Barrens tree frog, *Hyla andersonii*, were allowed to develop and metamorphose either alone in covered experimental tanks ('Control'), or in covered tanks to which tadpoles of Fowler's toad, *Bufo woodhousei fowleri*, had been added ('*Bufo*'), or in uncovered tanks a natural assemblage of insects could therefore colonize ('Insects'), or in uncovered tanks to which *Bufo* had been added ('Insects plus *Bufo*'). The presence of *Bufo* reduced the abundance of the *Hyla* tadpoles' plant food (left panel) but the insects (and the insects and *Bufo* together) did so even more. Consequently (right panel) the *Hyla* tadpoles suffered more from competition with insects than they did from competition from *Bufo*, in terms of their mean mass at metamorphosis to froglets, and even more from the two combined. Bars are SEs. (After Hay & Taylor, 1985 and Morin et al., 1988.)

Part of the problem is that although species may not be competing now, their ancestors may have competed in the past, so that the mark of inter-specific competition is left imprinted by evolution on either their niches, or their behavior, or their morphology. This particular question is taken up in Section 6.3.

Finally, the Competitive Exclusion Principle, as stated above, includes the word 'stable.' That is, in the habitats envisaged in the principle, conditions and the

Georgyi Frantsevich Gause

G. F. Gause was, by any measure, a quite remarkable scientist. Born in Moscow in 1910 into an academic family, he entered university in 1927 and graduated in 1931, but by 1928 was already undertaking field research, subsequently published, on the distribution of Russian grasshoppers. His tutor at university, Vladimir Alpatov, had around this time been visiting Raymond Pearl's laboratory in the United States (Baltimore), where they studied ecology in laboratory microcosms, and Gause seems himself to have grasped, more or less immediately, the possibilities this opens up for teasing apart and elaborating the various forces acting on the abundance of populations, away from the complexities of the natural world. Alpatov had brought back fruit flies, *Drosophila*, as 'model organisms' from Pearl's laboratory, but he, and especially Gause, soon turned to microorganisms, yeast and protozoa, for their experimental material.

Gause Institute of New Antibiotics

These were early days for ecological modelling. The logistic equation (Chapter 5) and the Lotka–Volterra equations of interspecific competition (Box 6.2) and of predator–prey interactions (Chapter 7) were all hot topics. Part of Gause's genius, and of his legacy to Ecology, was to see so clearly, and so early, how relatively simple, controlled, laboratory systems can form a bridge between mathematical models like these and the complexities of nature. Sometimes his experiments confirmed a model's ability to capture the essence of an ecological pattern; but more often they revealed not so much the inadequacy of a model, but what precisely was required in a biological system for the model patterns to be apparent. His experiments were quickly published in two books: *The Struggle for Existence* in 1934, and (in French) *Experimental Verifications of the Mathematical Theory of the Struggle for Existence* in 1935.

His 1934 book also focused the attention of ecologists onto what we now know as the Competitive Exclusion Principle, but what for many years usually had Gause's name attached to it: Gause's principle, Gause's postulate, and so on. Charles Darwin had recognized it in his *Origin of Species* and others had recognized a basic truth in it, but nobody previously had seriously understood that it could act as an organizing principle in the construction of ecological communities. Indeed, it was only after another ten years or so that the principle moved toward the center of the ecological stage, where it has, in one way or another, remained ever since.

Yet, having established his legacy within Ecology so early in his career, Gause did not remain an ecologist. Perhaps because of a perceived need in the Soviet Union for their best scientists to be working on behalf of the State and the people, or perhaps because Gause simply felt it was time to move on, from the late 1930s his work had turned to studies of chemicals produced by microbes in their own struggle for existence. In 1942, with his wife, he discovered gramicidin S, the first original Soviet antibiotic, for which he was awarded the Stalin Prize, and the rest of his life was devoted to the discovery and development of antibiotics. He was Director of the Institute of New Antibiotics in Moscow when he died in 1986. Kingsland (1985) provides a much fuller account of Gause's contribution to the development of the science of Ecology in her book, *Modeling Nature*.

supply of resources remain more or less constant – if species compete, then that competition runs its course, either until one of the species is eliminated, or until the species settle into a pattern of coexistence within their realized niches. Sometimes this is a realistic view of a habitat, especially in laboratory or other controlled environments where the experimenter consciously *holds* conditions and the supply of resources constant. However, most environments are not stable for long periods of time. How does the outcome of competition change when environmental heterogeneity in space and time are taken into consideration? This is the subject of the next section.

Quantitative Aspects

The Lotka–Volterra model of interspecific competition

The most widely used model of interspecific competition is the Lotka–Volterra model (Volterra, 1926; Lotka, 1932). It is an extension of the logistic equation described in Box 5.4. Its virtues are (like the logistic) its simplicity, and its capacity to shed light on the factors that determine the outcome of a competitive interaction.

In the logistic equation,

$$\frac{dN}{dt} = rN\frac{(K-N)}{K}$$

intraspecific competition is incorporated by the use of the term $(K-N)/K$. Within this, the greater the value of N (the bigger the population), the greater the strength of intraspecific competition. The basis of the Lotka–Volterra model is the replacement of this term by one that incorporates both intra- and interspecific competition. We call the population size of the first species N_1, and that of a second species N_2. Their carrying capacities and intrinsic rates of increase are K_1, K_2, r_1, and r_2.

By analogy with the logistic, we expect the *total* competitive effect on species 1, for example, (intra- *and* interspecific) to be greater the larger the values of N_1 and N_2. But we cannot just add the two Ns together, since the strengths of the competitive effects of the two species on species 1 are unlikely to be the same. Suppose, though, that 10 individuals of species 2 have, between them, only the same competitive effect on species 1 as does a single individual of species 1. The total competitive effect on species 1 will then be equivalent to the effect of $(N_1 + N_2 * 1/10)$ species 1 individuals. The constant (1/10 in the present case) is called a **competition coefficient** and is denoted by α_{12} (alpha one two). Multiplying N_2 by α_{12} converts it to a number of N_1 equivalents, and adding N_1 and $\alpha_{12}N_2$ together gives us the total competitive effect on species 1. (Note that $\alpha_{12} < 1$ means that individuals of species 2 have less inhibitory effect on individuals of species 1 than individuals of species 1 have on others of their own species, and the reverse for $\alpha_{12} > 1$.)

The equation for species 1 can now be written:

$$\frac{dN_1}{dt} = r_1N_1\frac{(K_1-[N_1+\alpha_{12}N_2])}{K_1}$$

and for species 2 (with its own competition coefficient, converting species 1 individuals into species 2 equivalents):

$$\frac{dN_2}{dt} = r_2N_2\frac{(K_2-[N_2+\alpha_{21}N_1])}{K_2}$$

These two equations constitute the Lotka–Volterra model.

The best way to appreciate its properties is to ask the question, 'Under what circumstances does each species increase or decrease in abundance?' In order to answer, it is necessary to construct diagrams in which all possible combinations of N_1 and N_2 can be displayed. This has been done in Figure 6.7. Certain combinations (certain regions in Figure 6.7) give rise to increases in species 1 and/or species 2, whereas other combinations give rise to decreases. There are also therefore dividing lines in the figures for each species, with combinations leading to increase on one side and those leading to decrease on the other, but along which there is neither increase nor decrease. We call these lines **zero isoclines**.

We can map out the regions of increase and decrease in Figure 6.7a for species 1 if we can draw its zero isocline, and we can do this by using the fact that *on* the zero isocline $dN_1/dt = 0$ (the rate of change of species 1 abundance is zero, by definition). Rearranging the equation, this gives us, as the zero isocline for species 1:

$$N_1 = K_1 - \alpha_{12} N_2$$

Below and to the left of this, species 1 increases in abundance (arrows in the figure, representing this increase, point from left to right, since N_1 is on the horizontal axis). It increases because numbers of both species are relatively low, and species 1 is thus subjected to only weak competition. Above and to the right of the line, however, numbers are high, competition is strong, and species 1 decreases in abundance (arrows from right to left). Based on an equivalent derivation, Figure 6.7b shows the species 2 zero isocline. The arrows, like the N_2 axis, run vertically.

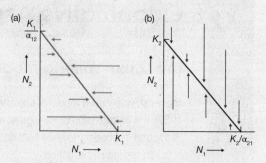

FIGURE 6.7 The zero isoclines generated by the Lotka–Volterra competition equations. (a) The N_1 zero isocline: species 1 increases below and to the left of it, and decreases above and to the right of it. (b) The equivalent N_2 isocline: species 2 increases below and to the left of it, and decreases above and to the right of it.

In order to determine the outcome of competition in this model, it is necessary to determine, at each point on a figure, the behavior of the joint species 1—species 2 population, as indicated by the pair of arrows. There are, in fact, four different ways in which the two zero isoclines can be arranged relative to one another, and these can be distinguished by the intercepts of the zero isoclines on the two axes of the graph (Figure 6.8). The outcome of competition will be different in each case.

Looking at the intercepts in Figure 6.8a, for instance,

$$\frac{K_1}{\alpha_{12}} > K_2 \quad \text{and} \quad K_1 > \frac{K_2}{\alpha_{21}}$$

Rearranging these slightly gives us:

$$K_1 > K_2 \alpha_{12} \quad \text{and} \quad K_1 \alpha_{21} > K_2$$

The first inequality ($K_1 > K_2 \alpha_{12}$) indicates that the inhibitory intraspecific effects that species 1 can exert on itself (denoted by K_1) are greater than the interspecific effects that species 2 can exert on species 1 ($K_2 \alpha_{12}$). This means that species 2 is a weak interspecific competitor. The second inequality, however, indicates that species 1 can exert more of an effect on species 2 than species 2 can on itself. Species 1 is thus a *strong* interspecific competitor. As the arrows in Figure 6.8a show, strong species 1 drives weak species 2 to extinction and attains its own carrying capacity. The situation is reversed in Figure 6.8b. Hence, Figure 6.8a and b describe cases of predictable competitive exclusion, in which the environment is such that one species invariably outcompetes the other, because the first is a strong interspecific competitor and the other weak.

In Figure 6.8c, by contrast:

$$K_1 > K_2 \alpha_{12} \quad \text{and} \quad K_2 > K_1 \alpha_{21}$$

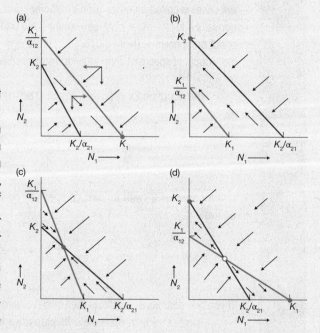

FIGURE 6.8 The outcomes of competition generated by the Lotka–Volterra competition equations for the four possible arrangements of the N_1 and N_2 zero isoclines. Black arrows refer to joint populations, and are derived as indicated in (a). The solid circles show stable equilibrium points. The open circle in (d) is an unstable equilibrium point. For further discussion, see box text.

In this case, both species have less competitive effect on the other species than those other species have on themselves. In this sense, both are weak interspecific competitors. This would happen, for example, if there were niche differentiation between the species – each competed mostly 'within' its own niche. The outcome, as Figure 6.8c shows, is that all arrows point towards a stable, equilibrium combination of the two species, which all joint populations therefore tend to approach. That is, the outcome of this type of competition is the stable coexistence of the competitors. Indeed, it is only this type of competition (both species having more effect on themselves than on the other species) that does lead to the stable coexistence of competitors.

Finally, in Figure 6.8d:

$$K_2\alpha_{12} > K_1 \quad \text{and} \quad K_1\alpha_{21} > K_2$$

Thus, individuals of both species have a greater competitive effect on individuals of the other species than those other species do on themselves. This will occur, for instance, when each species is more aggressive toward individuals of the other species than toward individuals of its own species. The directions of the arrows are rather more complicated in this case, but eventually they always lead to one or other of two alternative stable points. At the first, species 1 reaches its carrying capacity with species 2 extinct; at the second, species 2 reaches its carrying capacity with species 1 extinct. In other words, both species are capable of driving the other species to extinction, but which actually does so cannot be predicted with certainty. It depends on which species has the upper hand in terms of densities, either because they start with a higher density or because density fluctuations in some other way give them that advantage. Whichever species has this upper hand capitalizes on that and drives the other species to extinction.

Environmental heterogeneity

As explained in previous chapters, spatial and temporal variations in environments are the norm rather than the exception. Environments are usually a patchwork of favorable and unfavorable habitats; patches are often only available temporarily; and patches often appear at unpredictable times and in unpredictable places. Under such variable conditions, competition may only rarely 'run its course,' and the outcome cannot be predicted simply by application of the Competitive Exclusion Principle. A species that is a 'weak' competitor in a constant environment might, for example, be good at colonizing open gaps created in a habitat by fire, or a storm, or the hoofprint of a cow in the mud—or may be good at growing rapidly in such gaps immediately after they are colonized. It may then coexist with a strong competitor, as long as new gaps occur frequently enough. Thus, a realistic view of interspecific competition must acknowledge that it often proceeds not in isolation, but under the influence of, and within the constraints of, a patchy, cyclical, impermanent, or even unpredictable world.

competition may only rarely 'run its course'

The following examples illustrate just two of the many ways in which environmental heterogeneity ensures that the Competitive Exclusion Principle is very far from being the whole story when it comes to determining the outcome of an interaction between competing species.

The first concerns the coexistence of superior competitors with superior colonizers in a patchy environment. Four species of ant occupy acacia trees in Kenya: *Crematogaster sjostedti*, *C. mimosae*, *C. nigriceps*, and *Tetraponera penzigi*. The species are intolerant of one another. Less than 1% of trees are occupied by more than one species, and these cohabitations are only transient. Nonetheless, they coexist on a very fine spatial scale: all four are likely to be found within any 100 m² area. In battles amongst them for the occupation of trees, there is a clear competitive hierarchy: *C. sjostedti* > *C. mimosae* > *C. nigriceps* > *T. penzigi* (Palmer et al., 2000). However, trees naturally become available for recolonization, for example following fire or elephant damage or drought. Examination of their

a colonization-competition trade-off in ants

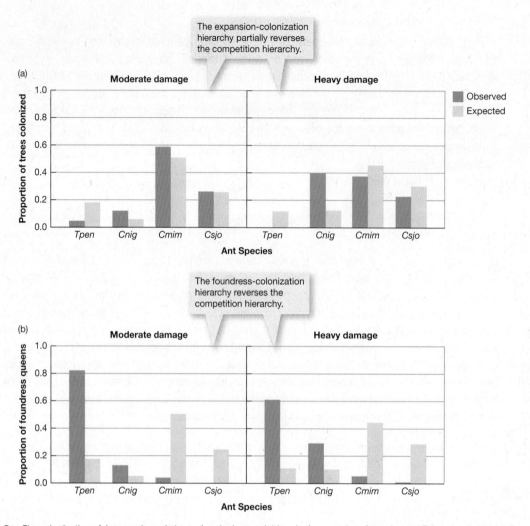

FIGURE 6.9   The colonization of damaged acacia trees, *Acacia drepanolobium*, by four species of acacia ants: *Tetraponera penzigi, Crematogaster nigriceps, C. mimosae*, and *C. sjostedti*, where damage was either moderate (ants absent, but plant growth intact) or heavy (ants absent and plants recovering from damage by regrowth). (a) Colonization by expansion of nearby colonies. Observed proportions (out of 51 and 32 for moderately and heavily damaged tees, respectively) are compared to proportions expected from occupancies of trees in the near locality. (b) Colonization by foundress queens. Observed proportions (out of 247 and 285 for moderately and heavily damaged tees, respectively) are compared to proportions expected from occupancies of trees in the near locality. (After Stanton et al., 2002.)

abilities to colonize damaged trees by the expansion of mature colonies nearby showed a pattern very different from the competitive hierarchy (Figure 6.9a), with *C. nigriceps* being the most effective, followed by *C. mimosae*. Moreover, the effectiveness of colonizing these trees from a greater distance, with new 'foundress' queen ants, showed a perfect reversal of the competitive hierarchy: *T. penzigi* > *C. nigriceps* > *C. mimosae* > *C. sjostedti* (Figure 6.9b).

Thus, if new trees never became available, either as a result of damage (Figure 6.9), or simply via recruitment from saplings, it seems clear that *C. sjostedti* would steadily outcompete and ultimately exclude the other species. But because the species that were best in direct competition were worst at colonization (a competition-colonization 'trade-off': when one improves, it tends to be at the expense of the other), and because trees were continuously becoming available for colonization, all four species were able to coexist.

A perhaps more widespread path to the coexistence of a superior and an inferior competitor is based simply on the idea that the two species may have independent, aggregated (i.e., clumped) distributions over the available habitat. This would mean that the powers of the superior competitor were mostly directed against members of its own species (in the high-density clumps), but that this aggregated superior competitor would be absent from many areas—within which

coexistence as a result of aggregated distributions

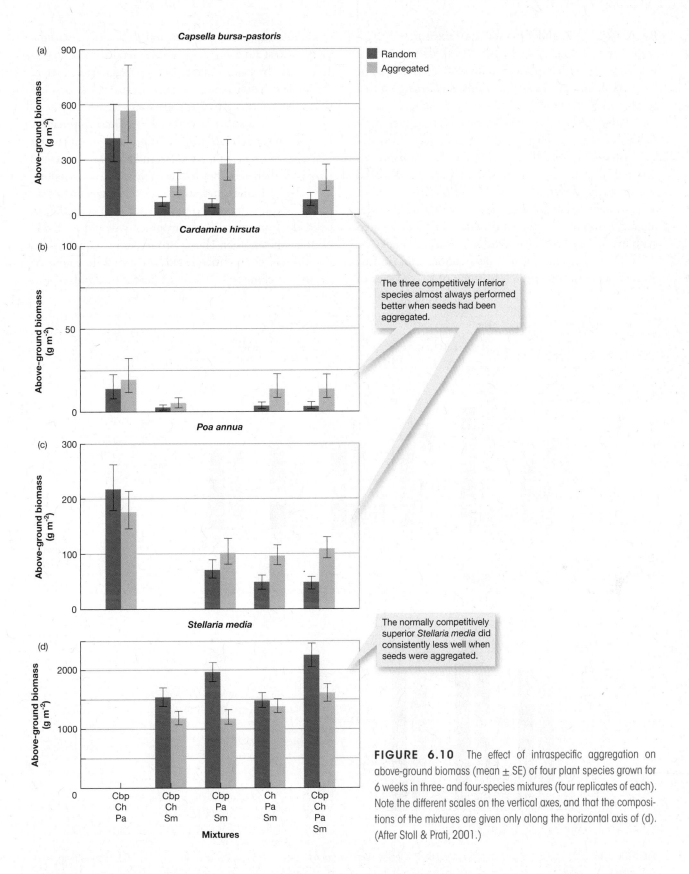

FIGURE 6.10 The effect of intraspecific aggregation on above-ground biomass (mean ± SE) of four plant species grown for 6 weeks in three- and four-species mixtures (four replicates of each). Note the different scales on the vertical axes, and that the compositions of the mixtures are given only along the horizontal axis of (d). (After Stoll & Prati, 2001.)

the inferior competitor could escape competition. An inferior competitor may then be able to coexist with a superior competitor that would rapidly exclude it from a continuous, homogeneous environment.

Experimental confirmation of this is provided by a study of artificial communities of four annual terrestrial plants—*Capsella bursa-pastoris*, *Cardamine hirsuta*, *Poa annua*, and *Stellaria media* (Figure 6.10).

Replicate three- and four-species mixtures of these were sown at high density, but the seeds were sown in two ways: either placed completely at random, or sown as single-species aggregations. *Stellaria* is known to be the superior competitor among these species, but the intraspecific aggregation harmed its performance in the mixtures, whereas aggregation improved the performance of the three inferior competitors in all but one case. Coexistence of competitors was favored not by niche differentiation but simply by a type of heterogeneity that is typical of the natural world: aggregation ensured that most individuals competed with members of their own and not of another species.

That such aggregated distributions are indeed a reality is illustrated by a field study of two species of sand-dune plant, *Aira praecox* and *Erodium cicutarium*, in northwest England. Both species tended to be aggregated, and the smaller plant, *Aira*, was aggregated even at the smallest spatial scales (Figure 6.11a). At these smallest scales, however, the two species were negatively associated with one another (Figure 6.11b). Thus, *Aira* tended to occur in small single-species clumps and was therefore much less liable to competition from *Erodium* than would have been the case if they had been distributed at random.

These studies, and others like them, go a long way toward explaining the co-occurrence of species that in constant, homogeneous environments would probably exclude one another. The environment is frequently too variable for competitive exclusion to run its course or for the outcome to be the same across the landscape.

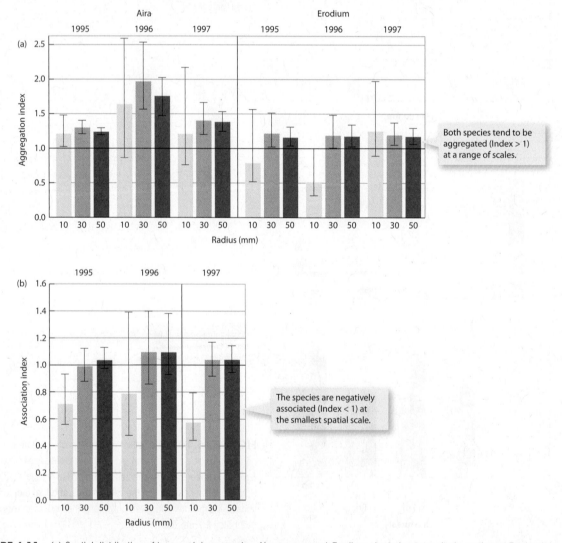

FIGURE 6.11 (a) Spatial distribution of two sand-dune species, *Aira praecox* and *Erodium cicutarium*, at a site in northwest England. An aggregation index of 1 indicates a random distribution. Indices greater than 1 indicate aggregation (clumping) within patches with the radius as specified; values less than 1 indicate a regular distribution. Bars represent 95% confidence intervals. (b) The association between *Aria* and *Erodium* in each of the 3 years. An association index greater than 1 would indicate that the two species tended to be found together more than would be expected by chance in patches with the radius as specified; values less than 1 indicate a tendency to find either one species or the other. Bars represent 95% confidence intervals. (After Coomes et al., 2002.)

6.2 EVOLUTIONARY EFFECTS OF INTERSPECIFIC COMPETITION

Putting to one side the fact that environmental heterogeneity ensures that the forces of interspecific competition are often much less profound than they would otherwise be, it is nonetheless the case that the potential of interspecific competition to adversely affect individuals is considerable. We have seen in Chapter 2 that natural selection in the past will have favored those individuals that have, by their behavior, physiology, or morphology, avoided adverse effects that act on other individuals in the same population. The adverse effects of extreme cold, for example, may have favored individuals with an enzyme capable of functioning effectively at low temperatures. Similarly, in the present context, the adverse effects of interspecific competition may have favored those individuals that managed to avoid those competitive effects. We can, therefore, expect species to have evolved characteristics that ensure that they compete less, or not at all, with members of other species.

How will this look to us at the present time? Coexisting species, with an apparent potential to compete, will exhibit differences in behavior, physiology, or morphology that ensure that they compete little or not at all. Connell has called this line of reasoning 'invoking the Ghost of Competition Past.' Yet the pattern it predicts is precisely the same as

> invoking the Ghost of Competition Past and the difficulty of distinguishing ecological and evolutionary effects

that supposed by the Competitive Exclusion Principle to be a prerequisite for the coexistence of species that *still* compete. Coexisting present-day competitors, and coexisting species that have evolved an avoidance of competition, can look the same.

The question of how important either past or present competition are, as forces structuring natural communities, will be addressed in the last section of this chapter (Section 6.5). For now, we examine some examples of what interspecific competition *can* do as an evolutionary force. Note, however, that by invoking something that is not easily observed directly (evolution), it may be impossible to 'prove' an evolutionary effect of interspecific competition in the way that we might get proof, say, from carefully controlled experiments in the laboratory. Nonetheless, we consider some examples where an evolutionary (rather than an ecological) effect of interspecific competition is the most reasonable explanation for what is observed.

Character displacement and ecological release in the Indian mongoose

In western parts of its range, the small Indian mongoose (*Herpestes javanicus*) coexists with one or two slightly larger species in the same genus (*H. edwardsii* and *H. smithii*), but these species are absent in the eastern part of its range (Figure 6.12a). The upper canine teeth are the mongoose's principal prey-killing organ, and these vary in size within and between species and between the

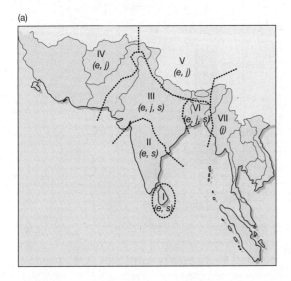

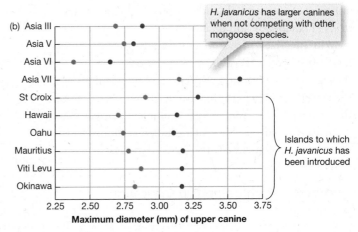

> H. javanicus has larger canines when not competing with other mongoose species.

> Islands to which H. javanicus has been introduced

FIGURE 6.12 (a) Native geographic ranges of *Herpestes javanicus* (*j*), *H. edwardsii* (*e*), and *H. smithii* (*s*). (b) Maximum diameter (mm) of the upper canine for *Herpestes javanicus* in its native range [data only for areas III, V, VI, and VII from (a)] and islands on which it has been introduced. Symbols in blue represent mean female size and in maroon mean male size. Compared to area VII (*H. javanicus* alone), animals in areas III, V, and VI, where they compete with the two larger species, are smaller. On the islands, they have increased in size since their introduction, but are still not as large as in area VII. (From Simberloff et al., 2000.)

The Java mongoose

sexes (female mongooses are smaller than males). In the east, where *H. javanicus* occurs alone (area VII in Figure 6.12a), both males and females have larger canines than in the western areas (III, V, VI) where it coexists with the larger species (Figure 6.12b). This is consistent with the view that where similar but larger mongoose species are present, the prey-catching apparatus of *H. javanicus* has been selected for reduced size (**character displacement**: a morphological response to competition from another species), reducing the strength of competition with other species in the genus because smaller predators tend to take smaller prey. Where *H. javanicus* occurs in isolation, since no character displacement has occurred, its canine teeth are much larger.

In fact, *H. javanicus* was introduced about a century ago to many islands outside its native range (often as part of a naive attempt to control introduced rodents). In these places, the larger competitor mongoose species were absent. Within 100–200 generations *H. javanicus* had increased in size (Figure 6.12b), so that the sizes of island individuals are now intermediate between those in the region of origin (where they coexisted with other species and were small) and those in the east where they occur alone. Their size on the islands is consistent with **ecological release** from competition with larger species (i.e., a response to the *absence* of ecological effects of other species).

Character displacement in Canadian sticklebacks

If character displacement is an evolutionary response to interspecific competition, then the effects of competition should decline with the degree of displacement. Brook sticklebacks, *Culaea inconstans*, coexist in some

Canadian lakes with nine-spine sticklebacks, *Pungitius pungitius* ('**sympatry**'—living in the same place), whereas in other lakes, brook sticklebacks live alone. In sympatry, the brook sticklebacks possess significantly shorter gill rakers (more suited for capturing food from open waters), longer jaws and deeper bodies than they do when they live alone. It is therefore reasonable to suppose that brook sticklebacks living alone possess pre-displacement characteristics, whereas those coexisting with nine-spine sticklebacks display character displacement. To test this, each type was placed separately in enclosures in the presence of nine-spine sticklebacks. The pre-displacement brook sticklebacks grew significantly less well than their sympatric post-displacement counterparts (Figure 6.13). This is clearly consistent with the hypothesis that the post-displacement phenotype has evolved to avoid competition, and hence enhance fitness, in the presence of nine-spine sticklebacks.

Evolution in action: selection on microorganisms

The most direct way of demonstrating the evolutionary effects of competition within a pair of competing species would be for the experimenter to induce these effects – impose the selection pressure (competition) and observe the outcome. In fact, there have been very few

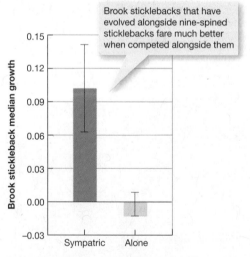

FIGURE 6.13 Means (with standard errors) of group median growth (natural log of the final mass of fish in each enclosure divided by the initial mass of the group) for sympatric brook sticklebacks, representing post-displacement phenotypes (dark blue bar), and brook sticklebacks living alone, representing pre-displacement phenotypes (pale blue bar), both reared in the presence of nine-spine sticklebacks. In competition with nine-spine sticklebacks, growth was significantly greater for post-displacement versus pre-displacement phenotypes ($P = 0.012$). (After Gray & Robinson, 2001.)

successful experiments of this type. In large part, this is due to the sheer practical difficulty of maintaining competing species together for long enough, and with sufficient replication, for any changes to be detected. It is no surprise, therefore, that most of these few examples are for bacteria (e.g., Figure 6.14a), but these have suffered the disadvantage of using competing strains of the *same* bacterial species, which, while they do not interbreed, may fail to satisfy whole-organism biologists seeking evidence from quite separate species.

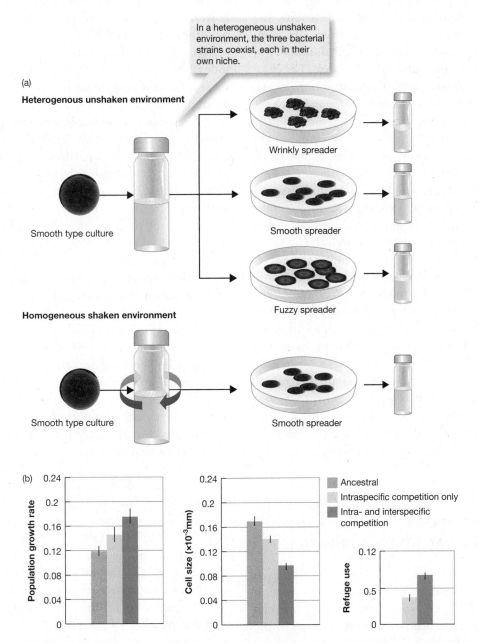

FIGURE 6.14   (a) Pure cultures of three types of the bacterium, *Pseudomonas fluorescens* (smooth, wrinkly spreader, and fuzzy spreader) concentrate their growth in different parts of a liquid culture vessel, and hence compete less with one another than they would otherwise do. In unshaken culture vessels (above), an initial smooth-type culture gives rise to wrinkly spreader, and fuzzy spreader mutants and all three coexist. But in shaken vessels (below), there is only one niche, and only the smooth type survives, excluding its competitors. (After Rainey & Trevisano, 1998 (Figure After Buckling et al., 2009.)) (b) When the pitcher-plant protozoan, *Colpoda*, was subjected either to a period of intraspecific competition or of competition against another protozoan, *Pseudocyrtolophis alpestris*, it evolved a higher population growth rate (left panel) and a smaller cell size (middle panel), especially when there had been interspecific competition, associated with an increased tendency to make greater use of a glass-bead refuge from the competitor (right panel). Bars are standard errors. (Measurements were made in the absence of the competitor, though the original study also made measurements in their presence.) (After Terhorst, 2011.)

In one example, though, a species of protozoan, *Colpoda* sp., that lives in the water-filled leaves of purple pitcher plants, *Sarracenia purpurea*, was competed for 20 days (60-120 generations) against another protozoan species, *Pseudocyrtolophis alpestris*, in replicate laboratory microcosms designed to mimic the pitcher plant environment. *Colpoda* was taken from natural pitcher plant populations and maintained either in monoculture (intraspecific competition only) or with *P. alpestris* added, initially at about twice the starting density of *Colpoda*. The characteristics of *Colpoda* from the ancestral population, and after selection both in monoculture and with the competitor, were then compared in the monoculture environment. Clearly, *Colpoda* made an evolutionary response to competition – population growth rate was greater and cell size was smaller (Figure 6.14b, upper and middle panels) – and while this response was apparent following exposure to intraspecific competition only, it was significantly greater when *P. alpestris* had been present throughout the selection period. What is less clear though, unlike the bacterial example in Figure 6.14a, is whether this response was associated with any niche differentiation. The competitor, *P. alpestris*, is mostly found in the water column rather than in a 'refuge' provided by glass beads in the laboratory tubes. Following its period of selection with *P. alpestris*, there was an increased tendency for *Colpoda* to be found in this refuge (Figure 6.14b, lower panel), but this effect was not statistically significant. Overall, therefore, demonstrating directly the evolutionary effects of interspecific competition remains a considerable challenge for ecologists.

6.3   INTERSPECIFIC COMPETITION AND COMMUNITY STRUCTURE

Interspecific competition, then, has the potential to either keep apart (Section 6.2) or drive apart (Section 6.3) the niches of coexisting competitors. How can these forces express themselves when it comes to the role of interspecific competition in molding the shape of whole ecological communities—who lives where and with whom?

Limiting resources and the regulation of diversity in phytoplankton communities

We begin by returning to the question of coexistence of competing phytoplankton species. Earlier (Section 6.2), we saw how two diatom species could coexist in the laboratory on two shared limiting resources—silicate and phosphate. In fact, theory predicts that the diversity of coexisting species should be proportional to the number of resources in a system that are at physiological limiting levels (Tilman, 1982): the more limiting resources, the more coexisting competitors. A direct test of this hypothesis examined three lakes in the Yellowstone region of Wyoming, using an index (Simpson's index) of the species diversity of phytoplankton there (diatoms and other species). If one species exists on its own, the index equals 1; in a group of species where biomass is strongly dominated by a single species, the index will be close to 1; when two species exist at equal biomass, the index is 2; and so on. According to the theory, therefore, this index should increase in direct proportion to the number of resources limiting growth. The spatial and temporal patterns in phytoplankton diversity in the three lakes for 1996 and 1997 are shown in Figure 6.15a.

The principal limiting resources for phytoplankton growth are nitrogen, phosphorus, silicon, and light. These parameters were measured at the same depths and times that the phytoplankton were sampled, and it was noted where and when any of the potential limiting factors actually occurred at levels below threshold limits for growth. Consistent with the theory, species diversity increased as the number of resources at physiologically limiting levels increased (Figure 6.15b). This suggests that even in the highly dynamic environments of lakes, where equilibrium conditions are rare, resource competition plays a role in continuously structuring the phytoplankton community. It is heartening that the results of experiments performed in the artificial world of the laboratory (Section 6.2) are echoed here in the much more complex natural environment. On the other hand, we should also remember that these environmental heterogeneities and absences of equilibria in natural systems mean that the number of coexisting competing species may exceed the number of different resources available to them.

Niche complementarity amongst anemone fish in Papua New Guinea

The region near Madang in Papua, New Guinea, has the highest reported species richness of both anemone fish (nine) and of sea anemones with which they

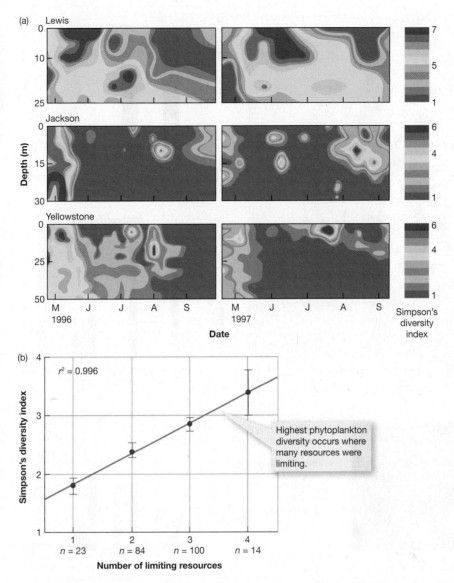

FIGURE 6.15   (a) Phytoplankton species diversity (Simpson's index) at a range of depths in 1996 and 1997 in three large lakes in the Yellowstone region. Purple denotes high species diversity, blue denotes low species diversity. (b) Phytoplankton diversity in samples from (a) (mean ± SE) plotted against the number of limiting resources shown to be limiting at the location of those samples. The number of samples (n) in each limiting resource class is shown. (From Interlandi & Kilham, 2001, with Permission.)

are associated (10). Anemones seem to be a limiting resource for the fish in that almost all anemones are occupied, and if some are transplanted to new sites, they are quickly colonized. However, each individual anemone tends to be occupied by individuals of just one species of anemone fish, because the residents aggressively exclude intruders, though aggressive interactions are less frequently observed between anemone fish of very different sizes (likely reflecting their differing food particle sizes). Surveys in four zones (nearshore, mid-lagoon, outer barrier reef, and offshore: Figure 6.16a) showed that each species of

anemone fish was primarily associated with a particular species of anemone. Each also showed a characteristic preference for a particular zone (Figure 6.16b). Crucially, though, anemone fish that lived with the same anemone were typically associated with different zones. For example, *Amphiprion percula* occupied the anemone *Heteractis magnifica* in nearshore zones, but the related *A. perideraion* occupied *H. magnifica* in offshore zones. Finally, small anemone fish species (*A. sandaracinos* and *A. leucokranos*) were able to cohabit the same anemone with larger species.

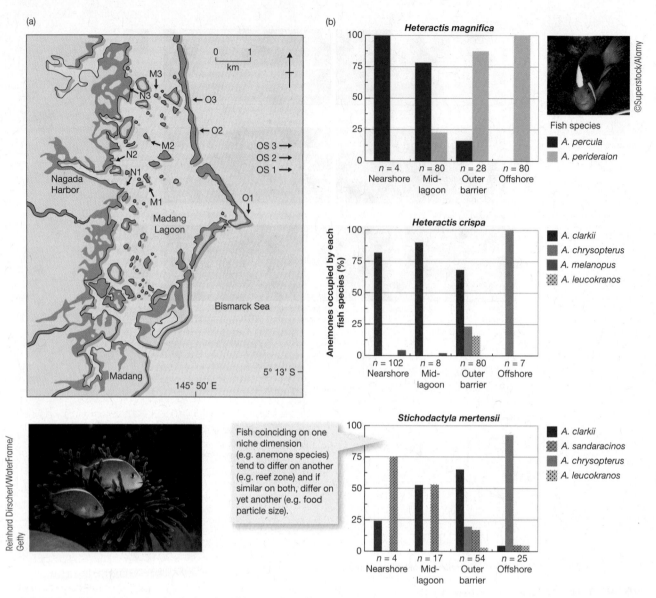

FIGURE 6.16 (a) Map showing the location of three replicate study sites in each of four zones within and outside Madang Lagoon, Papua, New Guinea (N1-3, nearshore; M1-3, mid-lagoon; O1-3, outer barrier reef; OS1-3, offshore reef). The blue areas indicate water, brown shading represents coral reef, and green represents land. (b) The percentage of three common species of anemone (*Heteractis magnifica*, *H. crispa*, and *Stichodactyla mertensii*) occupied by different anemone fish species (*Amphiprion* spp., in key below) in each of the four zones. The number of anemones censused in each zone is shown by *n*. Small species have stippled shading. (After Elliott & Mariscal, 2001.)

Two important points are illustrated here. First, the anemone fish demonstrate **niche complementarity**. That is, niche differentiation in a community of species involves several niche dimensions, with fish species occupying a similar position along one dimension tending to differ along another dimension, be it species of anemone, zone on the shore, or food particle size (reflected in the size of the fish). Second, the fish can be considered to be a **guild**: a group of species that exploit the same class of environmental resource in a similar way. Insofar as interspecific competition plays a role in structuring communities, it tends to do so, as here, not by affecting some random sample of the members of that community, nor by affecting every member, but by acting within guilds.

Species separated in space or in time

We have seen that there is commonly, but not necessarily, an association between niche differentiation and the coexistence of species within natural communities, and that niches may be differentiated through resource partitioning or differential resource utilization. In many cases, the resources used by ecologically similar species are separated spatially. Differential resource utilization will then express itself as either a microhabitat differentiation between the species (different species of fish, say, feeding at different depths) or even a difference in geographic distribution. Alternatively, the availability of the different resources may be separated in time. Different resources may become available at different times of the day or in different seasons, and differential resource utilization will then express itself as a temporal separation between the species.

Alternatively, niches may be differentiated on the basis of conditions. Two species may use precisely the same resources, but if their ability to do so is influenced by environmental conditions (as it is bound to be), and if they respond differently to those conditions, then each may be competitively superior in different environments. This too can express itself as either a microhabitat differentiation, or a difference in geographic distribution, or a temporal separation, depending on whether the appropriate conditions vary on a small spatial scale, a large spatial scale or over time.

Spatial separation in trees and tree-root fungi

A study of 11 tree species in the genus *Macaranga* in Borneo showed marked differentiation in light requirements, from extremely light-demanding species to the strongly shade-tolerant. The shade-tolerant species were also smaller, persisting lower down in the understory but rarely establishing in disturbed microsites (e.g., *M. kingii* in Figure 6.17), in contrast to some of the larger, high-light species that are pioneers of large forest gaps (e.g., *M. gigantea*). Other species were of intermediate height and were associated with intermediate light levels and can be considered small-gap specialists (e.g., *M. trachyphylla*). The species were also differentiated along a second niche gradient, with some species being more common on clay-rich soils and others on sand-rich soils (Figure 6.17). Hence, as with the anemone fish, there was evidence of niche complementarity: species with similar light requirements tended to differ in terms of preferred soil textures and vice versa. In addition, though, this niche partitioning expressed itself as a differentiation in space, partly horizontally (variation in soil types and in light levels from place to place) and partly vertically (height in the canopy, depth of the root mat).

Differential resource utilization in the vertical plane is also evident for fungi intimately associated with plant roots (ectomycorrhizal fungi; see Section 8.4)

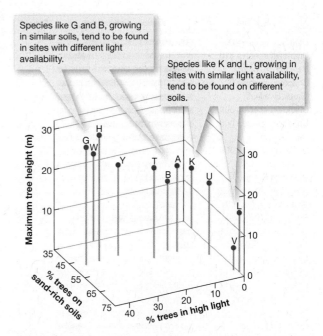

FIGURE 6.17 The three-dimensional distribution of 11 species of *Macaranga* tree in Borneo with respect to maximum height, the proportion of stems in high light levels, and proportion of stems in sand-rich soils. Each species of *Macaranga* is denoted by a single letter. G, *gigantean*; W, *winkleri*; H, *hosei*; Y, *hypoleuca*; B, *beccariana*; T, *triloba*; A, *trachyphylla*; V, *havilandii*; U, *hullettii*; L, *lamellate*; K, *kingii*. (After Davies et al., 1998.)

in the floor of a forest of pine, *Pinus resinosa* (Figure 6.18), where DNA analyses have recently made it possible to study the distribution of ectomycorrhizal species in their natural environment. The forest soil has a well-developed litter layer below which is a fermentation layer (F layer), then a thin humified layer (H layer), with mineral soil beneath (B horizon). Of the 26 species separated by the DNA analysis, some were very largely restricted to the litter layer (group I in Figure 6.18), others to the F layer (group IV), the H layer (group V), or the B horizon (group VI). The remaining species were more general in their distributions (groups II and III). The species are differentiated in (vertical) space, but that separation cannot simply be ascribed to one resource or condition: there are no doubt several that vary with the soil layers.

Temporal separation in mantids and tundra plants

Resources are commonly partitioned over time through a staggering of life cycles throughout the year. It is notable that two species of mantids, *Tenodera sinensis* and *Mantis religiosa*, which feature as predators in many parts of the world, commonly

staggered life cycles in mantids

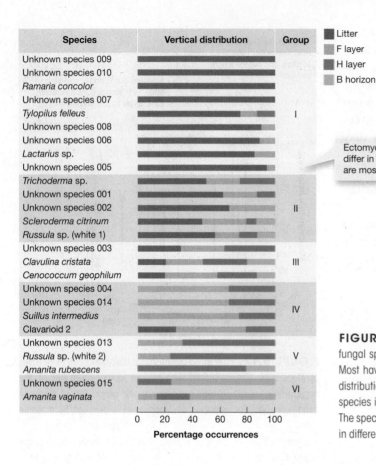

FIGURE 6.18 The vertical distribution of 26 ectomycorrhizal fungal species in the floor of a pine forest determined by DNA analysis. Most have not formally been named but are shown as a code. Vertical distribution histograms show the percentage of occurrences of each species in the litter, the F layer, the H layer, and the B horizon of the soil. The species have been grouped according to their tendencies to be found in different soil layers. (After Dickie et al., 2002.)

coexist both in Asia and North America. The two species have life cycles that are 2–3 weeks out of phase, and so to test the hypothesis that this asynchrony serves to reduce interspecific competition between them, the timing of their egg hatch was experimentally synchronized in replicated field enclosures (Hurd & Eisenberg, 1990). *T. sinensis*, which normally hatches earlier, was unaffected by *M. religiosa*. However, the survival and body size of *M. religiosa* declined when denied its normal time-differentiation and forced to compete with *T. sinensis*.

In plants too, resources may be partitioned in time. Thus, tundra plants growing in nitrogen-limited conditions in Alaska are differentiated in their timing of nitrogen uptake, as well as in the soil depth from which nitrogen is extracted and the chemical form of nitrogen used. To trace how tundra species differed in their uptake of different nitrogen sources, McKane et al. (2002) injected three chemical forms labeled with the rare isotope ^{15}N (ammonium, nitrate, and glycine) at two soil depths (3 and 8 cm) on two occasions (June 24 and August 7). Concentration of the ^{15}N tracer was measured in each of five common tundra plants 7 days after application. The five plants proved to be well differentiated in their use of nitrogen sources (Figure 6.19). Cottongrass (*Eriophorum vaginatum*)

and the cranberry bush (*Vaccinium vitis-idaea*) both relied on a combination of glycine and ammonium, but cranberry obtained more of these forms early in the growing season and at a shallower depth than cottongrass. The evergreen shrub *Ledum palustre* and the dwarf birch (*Betula nana*) used mainly ammonium, but *L. palustre* obtained more of this form early in the season while the birch exploited it later. Finally, the grass *Carex bigelowii* was the only species to use mainly nitrate. Here, niche complementarity can be seen along three niche dimensions: nitrogen form, depth, and time.

6.4 HOW SIGNIFICANT IS INTERSPECIFIC COMPETITION IN PRACTICE?

Competitors may exclude one another, or they may coexist if there is ecologically significant differentiation of their realized niches (Section 6.2). On the other hand, interspecific competition may exert neither of these effects if environmental heterogeneity prevents the process from running its course (also Section 6.2). Evolution may drive the niches

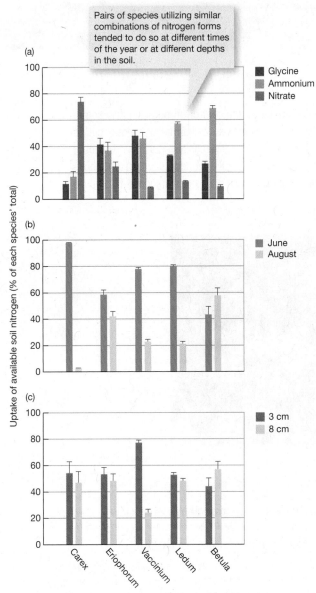

FIGURE 6.19 Mean uptake of available soil nitrogen (± *SE*) in terms of (a) chemical form, (b) timing of uptake, and (c) depth of uptake by the five most common species in tussock tundra in Alaska. Data are expressed as the percentage of each species' total uptake. (After Mckane et al., 2002.)

of competitors apart until they coexist but no longer compete (Section 6.3); and all these forces may express themselves at the level of the ecological community (Section 6.4). Interspecific competition, moreover, sometimes makes a high-profile appearance by having a direct impact on human activity (Box 6.3). In this sense, competition can certainly be of practical significance.

In a broader sense, however, the significance of interspecific competition rests not on a limited number of high-profile effects, but on an answer to the question 'How widespread are the ecological and evolutionary consequences of interspecific competition in practice?' We address this question in two ways. First, we ask 'How prevalent is current competition in natural communities?' Answering this, demonstrating current competition, requires experimental field manipulations, in which one species is removed from or added to the community and the responses of the other species are monitored. However, even if current competition is *not* prevalent, past competition, and therefore competition generally, may still have played a significant role in structuring communities. The second question, then, is to distinguish between interspecific competition (past *or* present) and 'mere chance': species differing not as a reflection of interspecific competition but simply because they *are* different species. Patterns observed in nature can be examined to see if they provide strong evidence for a role for competition, or are open to alternative interpretations.

The prevalence of current competition

There have been several surveys of field experiments on interspecific competition over the years. Two early and very influential ones, by Schoener and by Connell, were both published in 1983. Schoener (1983) examined the results of all the experiments he could find—164 in all. Studies of freshwater organisms, and amongst terrestrial studies those dealing with phytophagous (plant-eating) insects, were relatively underrepresented. Any conclusions were therefore limited, to a degree, by what ecologists had chosen to look at. Nevertheless, Schoener found that approximately 90% of the studies had demonstrated the existence of interspecific competition. Moreover, even if he looked at small groups of species (of which there were 390) rather than at whole studies, which may have dealt with several groups of species, he found that 76% showed effects of competition at least sometimes. Connell's (1983) review was less extensive than Schoener's: 72 studies, dealing with a total of 215 species and 527 different experiments. Interspecific competition was demonstrated in most of the studies, more than half of the species and approximately 40% of the experiments. A subsequent survey, using more sophisticated forms of analysis of multiple, disparate studies ('meta-analysis') reached essentially similar conclusions (Gurevitch et al., 1992).

> surveys of published studies of competition indicate that current competition is widespread . . .

ECOncerns

Competition in action

When exotic plant species are introduced to a new environment, by accident or on purpose, they sometimes prove to be exceedingly good competitors and many native species suffer harmful consequences as a result. Some of them have even more far-reaching consequences for native ecosystems. This newspaper article by Beth Daley, published in the *Contra Costa Times* on June 27, 2001, concerns grasses that have invaded the Mojave Desert in the southern United States. Not only are the invaders outcompeting native wild flowers, they have also dramatically changed the fire regime.

Nomad/SuperStock

Invader grasses endanger desert by spreading fire

The newcomers crowd out native plants and provide fuel for once-rare flames to damage the delicate ecosystem.

Charred creosote bushes dot a mesa in the Mojave Desert, the ruins of what was likely the first fire in the area in more than 1000 years.

Though deserts are hot and dry, they aren't normally much of a fire hazard because the vegetation is so sparse there isn't much to burn or any way for blazes to spread.

But, underneath these blackened creosote branches, the cause of the fire seven years ago has already grown back: flammable grasses fill the empty spaces between the native bushes, creating a fuse for the fire to spread again.

Tens of thousands of acres in the Mojave and other southwestern deserts have burned in the last decade, fueled by the red brome, cheat grass and Sahara mustard, tiny grasses and plants that grow back faster than any native species and shouldn't be there in the first place.

. . . The grasses brought to America from Eurasia more than a century ago have no natural enemies, and little can stop their spread across empty desert pavement. And, once an area is cleared of native vegetation by one or repeated fires, the grasses grow in even thicker, sometimes outcompeting native wildflowers and shrubs.

. . . 'These grasses could change the entire makeup of the Mojave Desert in short order', said William Schlesinger of Duke University, who has studied the Mojave Desert for more than 25 years. When he began his research in the 1970s, the grasses were in the Mojave, but there still were vast areas left untouched. Now, he said, the grasses are virtually everywhere and soon will be in concentrations large enough to fuel massive fires. 'This is not an easy problem to solve', he said.

. . . Despite the harsh conditions, a rainbow of wildflowers blooms regularly in the desert, sometimes carpeting the ground with blossoms after a rainstorm. Zebra-tailed lizards, rattlesnakes, desert tortoises and kangaroo rats are able to get by for long periods without water and bear up under the sun. But the innocuous-looking grasses threaten all these species by choking out wildflowers and killing off shelter and food that they rely on.

. . . Esque [of the US Geological Survey] has roped off 12 experimental sites, six of which he burned in 1999 to see how quickly invasive species re-establish themselves. But the result only showed the unpredictability of the desert: the first year, the invasive red brome took hold, but this year, native wildflowers came back in force.

∴ . . Esque said 'It's not black and white with what is going on. We don't know if we are looking at coexistence or competition.'

(All content © 2001 *Contra Costa Times* (Walnut Creek, CA) and may not be republished without permission.)

1 *Some people have suggested bringing sheep into the desert to graze the invading grasses. Do you think this is a sensible idea? What further information would help you make a decision?*

2 *The U.S. Geological Survey scientist found that red brome grass appeared to be outcompeting native flowers one year but not the next. Suggest some factors that may have changed the competitive outcome.*

Taken together, these reviews certainly seem to indicate that active, current interspecific competition is widespread. Connell also found, however, that in studies of just one pair of species, interspecific competition was almost always apparent, whereas with more species the prevalence dropped markedly (from more than 90% to less than 50%). In part, this is a result of many pairs of species being chosen for study because competition between them is suspected, whereas if none is found, this is simply not reported. Judging the prevalence of competition from such studies is rather like judging the prevalence of debauchery amongst clergymen from the 'gutter press.' The results of these surveys are likely to exaggerate, to an unknown extent, the frequency of competition.

> . . . though these surveys may exaggerate the true frequency of competition

Nonetheless, evidence for interspecific competition has been strong even in surveys of studies on groups often suspected of being little influenced by competition. Phytophagous (plant-eating) insects, in particular, have in the past been quoted as a group within which the forces of interspecific competition are weak. Yet, of 333 interactions from 145 experimental studies that satisfied criteria of rigor and experimental design for inclusion in a meta-analysis, 205 (62%) provided evidence of interspecific competition (Kaplan & Denno, 2007). Moreover, using those cases where it was possible to make a comparison, the effects of interspecific competition, while less, were not significantly less than those of intraspecific competition (Figure 6.20a); and competition between species from different guilds, while less, was not significantly less than that between species in the same guild (Figure 6.20b). It was also telling that competitive effects were strongest when there was neither spatial nor temporal separation between the species but also detectable when their niches were separated in either or both of these ways (Figure 6.20c). And finally, out of 231 competitive interactions where it was possible to decide, 30 were the result of direct interference between competing individuals, 47 were the result of exploitation of the plant food diminishing the quantity of resource available to the competing species, but 116 were the indirect result of exploitation diminishing the *quality* of food available for the other species.

Taken overall, therefore, current interspecific competition has been reported in studies on a wide

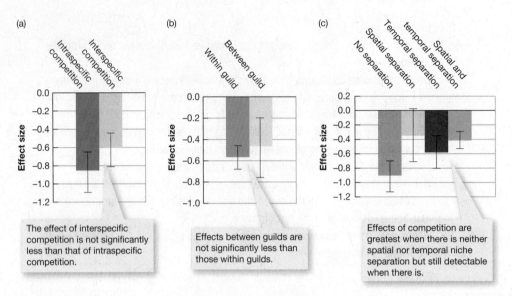

FIGURE 6.20   Results from a meta-analysis of field experimental studies of interspecific competition amongst phytophagous insects. Effect sizes quantify the negative effects of competition on a composite group of fitness traits (size, survival, fecundity, etc.) relative to a competition-free control. An effect size of 0.5 may be considered moderate, one of 0.8 strong. Bars are 95% confidence intervals. (a) Comparative effects of intra- and interspecific competition. (b) Effects of interspecific competition within and between guilds. (c) Effects of interspecific competition with and without spatial and/or temporal separation of niches. (After Kaplan & Denno, 2007.)

range of organisms, resulting from various types of interaction, and may be detected even across guilds and when niches are spatially and/or temporally differentiated.

Competition or mere chance?

There is a tendency to interpret differences between the niches of coexisting species as confirming the importance of interspecific competition. But the theory of interspecific competition predicts not simply that the niches of competing species differ, but that they differ more than would be expected from chance alone. A rigorous investigation of the role of interspecific competition must therefore address the question: 'Does the observed pattern, even if it appears to implicate competition, differ significantly from the sort of pattern that could arise in the community even in the absence of any interactions between species?' This question has been the driving force behind analyses that seek to compare real communities with so-called neutral or **null models**. The idea is that the data are rearranged into a form (the null model) representing what the data would look like in the absence of interspecific competition. Then, if the actual data show a significant statistical difference from the null model, the action of interspecific competition is strongly inferred. Hence, the null model analyses are attempts to follow a much more general approach to scientific investigation, namely, the construction and testing of 'null hypotheses.'

In fact, the approach has been applied to three different predictions of what a community structured by interspecific competition should look like: (i) potential competitors that coexist in a community should exhibit niche differentiation; (ii) this niche differentiation will often manifest itself as morphological differentiation; and (iii) within a community, potential competitors with little or no niche differentiation should not coexist, so each should tend to occur only where the other is absent ('negatively associated distributions'). The application of null models to community structure—that is, the reconstruction of natural communities with interspecific competition removed—has not been achieved to the satisfaction of all ecologists. But a brief examination of a very early study of niche differentiation in lizard communities shows the potential and rationale of the null model approach (Box 6.4). For these lizard communities, niches are more spaced out than would be expected by chance alone and interspecific competition therefore appears to play an important role in community structure.

> niche differentiation, morphological differentiation, and negatively associated distributions

If niche differentiation is reflected in morphological differentiation, then the spacing out of niches can be expected to have its counterpart in a corresponding spacing out of morphologies amongst species belonging to a single guild. One example concerns four species of fossil strophomenide brachiopod

6.4   **Quantitative Aspects**

Null models of lizard communities

Lawlor (1980) investigated differential resource utilization in 10 North American lizard communities, each consisting of between four and nine species. For each community, there were estimates of the amounts of each of 20 food categories consumed by each species. This pattern of resource use allowed the calculation, for each pair of species in a community, of an index of *resource use overlap*, which varied between 0 (no shared food categories) and 1 (perfect correspondence between a pair of species in the proportions of different foods used). Each community was then characterized by a single value: the mean resource overlap calculated over all species-pairs.

A number of 'null models' of these communities were then created. They were of four types. The first type, for example, retained the minimum amount of original community structure. Only the original number of species and the original number of resource categories were retained. Beyond that, species were allocated food preferences completely at random, such that there were far fewer species completely ignoring food in particular categories than in the real community. Hence, species tended to be categorized as eating a far broader range of food

Dan Suzio/Getty Images, Inc.

A desert lizard of the southwestern United States.

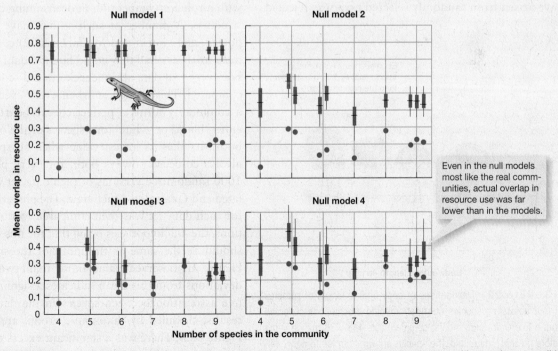

Even in the null models most like the real communities, actual overlap in resource use was far lower than in the models.

FIGURE 6.21   The mean indices of resource use overlap for each of 10 North American lizard communities are shown as solid circles. These can be compared, in each case, with the mean (horizontal line), standard deviation (vertical rectangle), and range (vertical line) of mean overlap values for a corresponding set of 100 randomly constructed communities. The random reconstructions ranged from those retaining least of the original community structure (Null Model 1) to those retaining most (Null Model 4), as explained in the box text. (After Lawlor, 1980.)

items than they do naturally, and pairs would therefore be expected to overlap more. At the other extreme, the fourth type retained most of the original community structure. If a species ignored food in a particular category, then that was left unaffected, but among those categories where food was eaten, preferences were reassigned at random. These null models were then compared with the real data in terms of their patterns of resource use overlap. If competition is a significant force in determining community structure, then resource use overlap in the real communities should be less – and statistically significantly less—than that in the null models.

The results (Figure 6.21) were that in all communities, and for all four null models, the model mean overlap was higher than that observed for the real community, and that in almost all cases this was statistically significant. For these lizard communities, therefore, the observed low overlaps in resource use suggest that niches are more segregated than would be expected by chance alone, and that interspecific competition plays an important role in community structure.

(so-called 'lamp shells' that resemble bivalve molluscs). These appear from the fossil record to have coexisted, though they are just four out of 74 strophomenide brachiopods in the complete fossil fauna. If these four are compared, they have a consistent length ratio (largest: next largest, etc.) of around 1.5 (Figure 6.22). However, when groups of four species were drawn at random from the complete fauna 100,000 times, and the size ratios between adjacent species calculated, these ratios tended to be much smaller. The null hypothesis that the observed ratios could have arisen from randomly selected taxa was rejected

$(P < 0.03$; Hermoyian et al., 2002). Morphologically, the species were more different from one another than would be expected by chance, supporting the idea that competition had played a key role in structuring this community.

The null model approach can also be applied to distributional differences, comparing the pattern of co-occurrence of two species at a suite of locations with what would be expected by chance. An excess of negatively associated distributions (either one species or the other but not both) would then be consistent with a role for competition in determining community structure. A number of such studies have been conducted, and Gotelli and McCabe (2002) carried out a meta-analysis of all they could find (96 data sets in all). For every real data set, a 'checkerboard score,' C, was computed. This is highest when every species-pair in a community forms a perfect checkerboard: sites are either 'black' or 'white' (one species or the other, never both). It takes its lowest value when all species-pairs always co-occur (always both, for every pair). Next, 1000 randomized versions of each data set were simulated and C calculated each time. The observed C-value for each data set was then compared to these simulations, the null hypothesis being that the observed value should be the same as the mean of the simulations. Hence, an observed value more than two standard deviations from this mean indicates a significant negative association between species in the data set. The results, classified by taxonomic group, are shown in Figure 6.23. There was a significant excess of negative associations for plants, birds, bats, other mammals, and for ants, but the excess was not significant for invertebrates (other than ants), fish, amphibians, and reptiles. Competition is often, but not always, implicated.

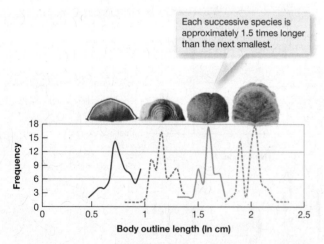

Each successive species is approximately 1.5 times longer than the next smallest.

FIGURE 6.22 Distributions of body outline length of samples of four coexisting species of strophomenide brachiopods collected from a late Ordovician marine sediment (ca. 448–438 million years before present) in Indiana. The species shown, from left to right, with their mean lengths in cm, are *Eochonetes clarksvillensis* (2.2 cm), *Leptaena richmondensis* (3.3 cm), *Strophomena planumbona* (4.9 cm), and *Rafinesquina alternata* (7.7 cm). (After Hermoyian et al., 2002.)

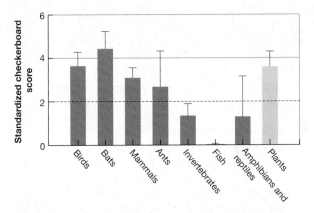

FIGURE 6.23 An analysis of data sets of species distributions across sites, classified by taxonomic group (mean ± SE), seeking evidence of an excess of negative associations, as measured by the standardized 'checkerboard score' (see text). A value of zero indicates no tendency for observed and null-model distributions to be different. A value greater than 2 indicates that observed distributions, for that taxonomic group, tend significantly to be more checkerboard-like than those generated by the null models. (After Gotelli & Mccabe, 2002.)

This kind of pattern—sometimes a role for competition is confirmed, sometimes not—has been the general conclusion from the null model approach. What then should be our verdict on it? Perhaps most fundamentally, the aim of the null model approach is undoubtedly worthy. It concentrates the minds of investigators, stopping them from jumping to conclusions too readily. It is important to guard against the temptation to see competition in a community simply because we are looking for it. On the other hand, the approach can never take the place of a detailed understanding of the field ecology of the species in question, or of manipulative experiments designed to reveal competition by increasing or reducing species abundances. It, like so many other approaches, can only be part of the community ecologist's armory.

SUMMARY

Ecological effects of interspecific competition

The essence of interspecific competition is that individuals of one species suffer a reduction in fecundity, survivorship, or growth as a result of exploitation of resources or interference by individuals of another species. A fundamental niche is the combination of conditions and resources that allow a species to exist when considered in isolation from any other species. Its realized niche is the combination of conditions and resources that allow it to exist in the presence of other species that might be harmful to its existence—especially interspecific competitors. The Competitive Exclusion Principle states that if two competing species coexist in a stable environment, then they do so as a result of differentiation of their realized niches. Environments are usually a patchwork of favorable and unfavorable habitats; patches are often only available temporarily; and patches often appear at unpredictable times and in unpredictable places. Under such variable conditions, competition may only rarely 'run its course.'

Evolutionary effects of interspecific competition

Although species may not be competing now, their ancestors may have competed in the past. We can expect species to have evolved characteristics that ensure that they compete less, or not at all, with members of other species. Coexisting present-day competitors, and coexisting species that have evolved an avoidance of competition, can look, at least superficially, the same. It is impossible to prove an evolutionary effect of interspecific competition. However, careful observational studies have sometimes revealed patterns that are difficult to explain in any other way.

Interspecific competition and community structure

Interspecific competition tends to structure communities by acting within guilds—groups of species exploiting the same class of resource in a similar fashion. Niche complementarity can be discerned in some communities, where coexisting species that occupy a similar position along one niche dimension tend to differ along another dimension. Niches can be differentiated through differential resource utilization or on the basis of conditions. Either can express itself as a microhabitat differentiation, or a difference in geographic distribution, or a temporal separation.

How significant is interspecific competition in practice?

Surveys of published studies of competition indicate that current competition is widespread but these exaggerate to an unknown extent the true frequency of competition. Theory predicts that the niches of competing species should be arranged regularly rather than randomly in niche space, that as a reflection of this they should be more distinct morphologically than expected by chance, and that competitors should be negatively associated in their distributions. Null models have been developed to determine what the community pattern would look like in the absence of interspecific competition. Real communities are sometimes structured in a way that makes an influence of competition difficult to deny.

REVIEW QUESTIONS

1 Some experiments concerning interspecific competition have monitored both the population densities of the species involved and their impact on resources. Why is it helpful to do both?

2 Interspecific competition may be a result of exploitation of resources or of direct interference. Give an example of each and compare their consequences for the species involved.

3 Define fundamental niche and realized niche. How do these concepts help us to understand the effects of competitors?

4 With the help of one plant and one animal example, explain how two species may coexist by holding different resources

at levels that are too low for effective exploitation by the other species.

5 Explain how environmental heterogeneity may permit an apparently 'weak' competitor to coexist with a species that might be expected to exclude it.

6 Provide one example each of niche differentiation involving physiological, morphological, and behavioral properties of coexisting species. How may these differences have arisen?

7 Define 'niche complementarity' and, with the help of an example, explain how it may help to account for the coexistence of many species in a community.

Challenge Questions

1 Define the Competitive Exclusion Principle. When we see coexisting species with different niches is it reasonable to conclude that this is the principle in action?

2 What is the 'Ghost of Competition Past'? Why is it impossible to prove an evolutionary effect of interspecific competition?

3 Discuss the pros and cons of the neutral model approach to evaluating the effects of competition on community composition.

Andrea Gingerich/Getty Images

Chapter 7

Predation, grazing, and disease

CHAPTER CONTENTS

KEY CONCEPTS

After reading this chapter you will be able to:

- differentiate between true predators, grazers, and parasites

- explain how, for each type of predator, individual victims and their populations may be adversely affected

- describe the subtleties of predation, including the capacity of prey to compensate for being preyed upon

- explain the value of the optimal foraging approach for analyzing predator choices

- describe the underlying tendency of populations of predators and prey to cycle in abundance and the 'damping' effect on cycles of crowding and patchy distributions

- identify some of the consequences of predation for community composition

Every living organism is either a consumer of other living organisms, is consumed by other living organisms, or—in the case of most animals—is both. We cannot hope to understand the structure and dynamics of ecological populations and communities until we understand the links between consumers and their prey.

7.1   WHAT DO WE MEAN BY PREDATION?

Ask most people to name a predator and they are almost certain to say something like lion, tiger or grizzly bear—something big, ferocious, instantly lethal. However, from an ecological point of view, a **predator** may be defined as any organism that consumes all or part of another living organism (its **prey** or host) thereby benefiting itself, but reducing the growth, fecundity, or survival of the prey. Thus, this definition extends beyond the likes of lions and tigers by including organisms that consume all *or part* of their prey and also those that merely *reduce* their prey's growth, fecundity, or survival. Predators are not all large, aggressive, or instantly lethal—they need not even be animals. Here we examine these consumers together and try to understand the part they play in determining the structure and dynamics of ecological systems.

> predator: a term extending beyond the obvious examples

Within the broad definition, we can distinguish three main types of predator.

1 True predators:

- invariably kill their prey and do so more or less immediately after attacking them;

- consume several or many prey items in the course of their life.

    True predators therefore include lions, tigers, and grizzly bears, but also spiders, baleen whales that filter plankton from the sea, zooplanktonic animals that consume phytoplankton, birds that eat seeds (each one an individual organism), and carnivorous plants.

2 Grazers:

- attack several or many prey items in the course of their life;

- consume only part of each prey item;

- do not usually kill their prey, especially in the short term.

    Grazers therefore include cattle, sheep, and locusts, but also, for example, blood-sucking leeches that take a small, relatively insignificant blood meal from several vertebrate prey over the course of their life.

3 Parasites:

- consume only part of each prey item, usually called their **host**;

- do not usually kill their prey, especially in the short term;

- attack one or very few prey items in the course of their life, with which they therefore often form a relatively intimate association.

    Parasites therefore include some obvious examples: animal parasites and pathogens such as tapeworms and the tuberculosis bacterium, plant pathogens like tobacco mosaic virus, parasitic plants like mistletoes, and the tiny wasps that form 'galls' on oak leaves. But aphids that extract sap from one or a very few plants with which they enter into an intimate association, and even caterpillars that spend their whole life on one host plant, are also, in effect, parasites.

    On the other hand, these distinctions between 'true' predators, grazers, and parasites, like most categorizations of the living world, have been drawn in large part for convenience – certainly not because every organism fits neatly into one and only one category. We could, for example, have included a fourth class, the parasitoids, which are little known to nonbiologists but are extensively studied by ecologists (and are immensely important in

> parasitoids—and the artificiality of boundaries

A parasitoid wasp, which uses its long ovipositor to insert its eggs into the larvae of other insects, where they develop by consuming their host.

the biological control of insect pests; see Chapter 14). **Parasitoids** are flies and wasps whose larvae consume their insect larva host from within, having been laid there as eggs by their mother. Having only one host individual, which they always kill, parasitoids therefore straddle the 'parasite' and 'true predator' categories, fitting neatly into neither and confirming the impossibility of constructing clear boundaries.

Despite these distinctions, however, for convenience throughout this chapter, we will use 'predator' as a shorthand term to encompass true predators, grazers, and parasites in general, and also to refer to predators in the more conventional sense.

7.2   PREY FITNESS AND ABUNDANCE

The fundamental similarity between true predators, grazers, and parasites is that each, in obtaining the resources it needs, reduces either the fecundity or the chances of survival of individual prey and may therefore reduce prey abundance. The effects of true predators on the survival of individual prey hardly need illustrating—the prey die. But the effects of grazers and parasites can be equally profound, if more subtle, as illustrated by the following two examples.

> predators reduce the fecundity and/or survival of individual prey

Lathyrus vernus is a long-lived perennial herb found in the forest understory throughout much of Europe. Molluscs, especially the slug *Arion fasciatus*, often graze on emerging shoots of the plant, below the litter layer, early in the spring. The effects of this grazing have been monitored in southeast Sweden both naturally, through a direct comparison of grazed and ungrazed plants (Figure 7.1a), and also through an experimental manipulation, in which plots that had all molluscs removed, followed by repeated molluscicide treatment, were compared with untreated control plots (Figure 7.1b). The natural and experimental data tell very similar stories. Grazing had some effects on the subsequent survival of plants, on their tendency to remain dormant (not appearing above ground after grazing, but reappearing in the following season), and on their growth. But the largest effects of this early grazing were exhibited much later in the life cycle: on flowering and on the subsequent production of seeds. On the other hand, although these effects were the largest, they were not the ones that had the greatest effect on the overall rate of plant population increase, because the different processes themselves make very different contributions to population increase. In this respect it was the grazing effects on survival and growth that were most profound (Figure 7.1). Clearly, the effects of early-season grazing played out in complex ways as the season progressed.

Philornis downsi is a fly parasite whose larvae inhabit birds' nests, chewing through the skin of nestlings and consuming blood and other fluids. They were seen first on the Galapagos Islands in 1997, where they attack a number of species including nine species of Darwin's finches. One of these, the medium ground finch, *Geospiza fortis*, has been the subject of field experiments to determine the effect of the parasite. Twenty-four nests constructed by the birds in tree cacti on Santa Cruz Island were fitted with a nylon liner to prevent the larvae reaching the nestlings from the nest material; a further 24 acted as unmanipulated (unlined) controls. The manipulation was only partially successful; there were around 22 *P. downsi* larvae per lined nest compared to around 37 in the unlined ones. Nonetheless, the experiment demonstrated that parasitism by the larvae was adversely affecting the finch nestlings. The nestlings grew faster in lined than in unlined nests (Figure 7.2a), and more lined nests fledged birds than unlined nests, while the total number of birds fledged from lined nests was also higher (Figure 7.2b).

It is not so straightforward, though, to demonstrate that reductions in the survival or fecundity of individual prey translate into reductions in prey abundance—we need to be able to compare prey populations in the presence and the absence of predators. As so often in ecology, we cannot rely simply on observation.

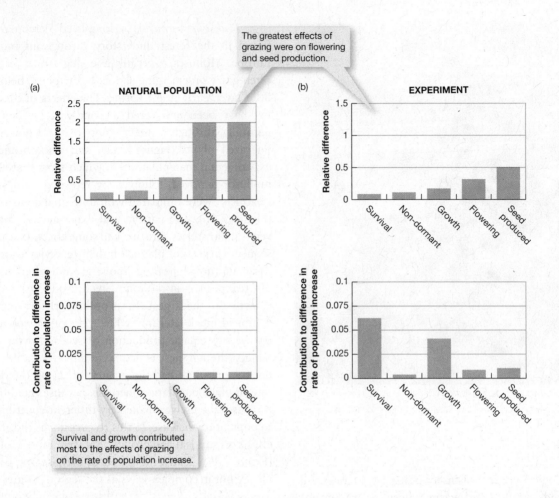

FIGURE 7.1 The effects of mollusc grazing on the perennial herb, *Lathyrus vernus*. (a) A comparison between grazed and ungrazed plants in a natural population. (b) A comparison between plots where molluscs were excluded and control plots. In both cases, the upper panel shows the relative difference (the difference between ungrazed and grazed divided by the mean) for a number of plant traits: survival of established individuals from one summer to the next, the proportion not remaining dormant for one year, individual growth [(size in one year—size in the previous year)/ size in the first year], the proportion flowering, and the number of seeds set per plant. The lower panels show the contributions of each of these traits to the overall effect of grazing on the rate of population increase. (After Ehrlen, 2003.)

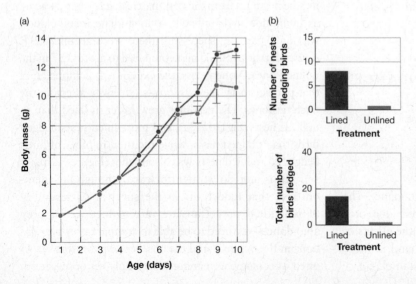

FIGURE 7.2 Effects of fly larva parasitism on the medium ground finch, *Geospiza fortis*. (a) Nestlings in lined nests (●), providing protection from parasitism, grew faster than those in unlined nests (●); bars show standard errors. (b) With equal numbers of nests lined and unlined, the number of lined nests fledging birds was higher (above; $P = 0.02$) as was the total number of bird fledged from lined nests (below; $P < 0.001$). (After Koop et al., 2011.)

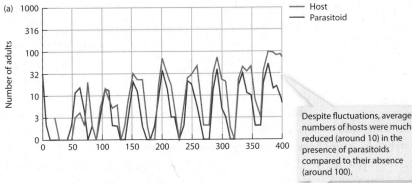

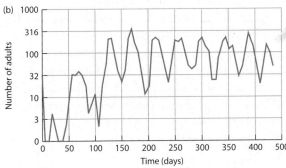

Despite fluctuations, average numbers of hosts were much reduced (around 10) in the presence of parasitoids compared to their absence (around 100).

FIGURE 7.3 Long-term population dynamics in laboratory population cages of a host (*Plodia interpunctella*), with and without its parasitoid (*Venturia canescens*). (a) Host and parasitoid, and (b) the host alone. Note the logarithmic vertical scale. (After Begon et al., 1995.)

Lake Moondarra, Australia, choked with weed before biological (beetle) control.

Peter Room, The University of Queensland, Bugwood.org

Lake Moondarra following control.

CSIRO

We need experiments—either ones we set up ourselves, or natural experiments set up for us by nature. For example, Figure 7.3 contrasts the dynamics of laboratory populations of an important pest, the Indian meal moth, with and without a parasitoid wasp, *Venturia canescens*. Ignoring the rather obvious regular fluctuations (cycles) in both moth and wasp, it is apparent that the wasp reduced moth abundance to less than one-tenth of what it would otherwise be (notice the logarithmic scale in the figure).

Turning to a natural experiment, we find a particularly graphic example of the impact grazers can have in the story of the invasion of Lake Moon Darra in North Queensland (Australia) by *Salvinia molesta*, a water fern that originated in Brazil (see photos). In 1978, the lake carried an infestation of 50,000 metric tons of the fern. In *Salvinia*'s native habitat in Brazil, the black long-snouted weevil (*Cyrtobagous* spp.) was known to

graze only on *Salvinia*. In June 1980, therefore, 1500 adults were released at an inlet to the lake, and a further release was made in January 1981. By April 1981, *Salvinia* was dying throughout the lake, supporting an estimated population of one billion beetles. By August 1981, less than 1 metric ton of *Salvinia* remained. This was a controlled experiment, in that other lakes in the region continued to bear large populations of *Salvinia*.

All sorts of predators can cause reductions in the abundance of their prey. We shall see as this chapter develops, however, that they do not *necessarily* do so.

7.3 THE SUBTLETIES OF PREDATION

We learn much from studying the similarities between different types of predators. On the other hand, we must avoid oversimplification, since there *are* important differences between true predators, grazers, and parasites. Nor should we give the impression that the scenario in all acts of predation is simply, "prey dies, predator takes one step closer to the production of its next offspring."

Interactions with other factors

Grazers and parasites, in particular, often exert their harm not by killing their prey immediately like true predators, but by making the prey more vulnerable to some other form of mortality. For example, grazers and parasites may have a more drastic effect than is initially apparent because they decrease the competitive abilities of their victims. We see this in a southern Californian salt marsh, where the parasitic plant, dodder (*Cuscuta salina*) attacks a number of plants including *Salicornia* (Figure 7.4). *Salicornia* tends to be the strongest competitor in the marsh, but it is also the preferred host of dodder. We can therefore understand the distribution

of plants in the marsh only as a result of the interaction between competition and parasitism (Figure 7.4).

Infection or grazing may also make hosts or prey more susceptible to predation. For example, snowshoe hares, *Lepus americanus*, in Canada (see Section 7.5, below) are parasitized by nematode worms in their guts, their lungs, and even in the tendons of their ankles. In a large field experiment, hares were live-trapped and fitted with tags that allowed them to be individually identified. Half were treated with a drug, ivermectin, to rid them of nematodes, while the other half acted as unmanipulated (parasitized) controls. A number of them were fitted with radio collars so they could be followed and their ultimate cause of death determined; 95% were eaten by predators, including coyotes, foxes, and several species of birds. The survival of parasitized and unparasitized hares in the face of this predation could then be compared.

In three of four runs of the experiment, there was no difference between the two groups (Figure 7.5a), but

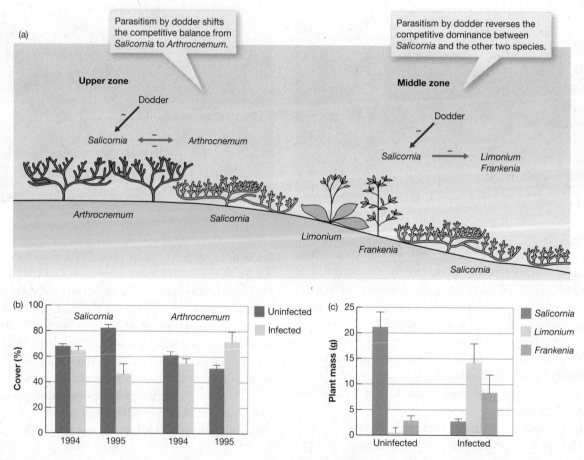

FIGURE 7.4 The effect of dodder, *Cuscuta salina*, on competition between *Salicornia* and other species in a southern Californian salt marsh. (a) A schematic representation of the main plants in the community in the upper and middle zones of the marsh and the interactions between them (red arrows: parasitism; blue arrows: competition). *Salicornia* (the relatively low growing plant in the figure) is most attacked by, and most affected by, dodder (which is not itself shown in the figure). When uninfected, *Salicornia* competes strongly and symmetrically with *Arthrocnemum* in the upper zone, and is a dominant competitor over *Limonium* and *Frankenia* in the middle zone. However, dodder significantly shifts the competitive balances. (b) Over time, in plots infected with dodder, *Salicornia* decreased and *Arthrocnemum* increased. (c) Large patches of dodder suppress *Salicornia* and favor *Limonium* and *Frankenia*. (After Pennings & Callaway, 2002.)

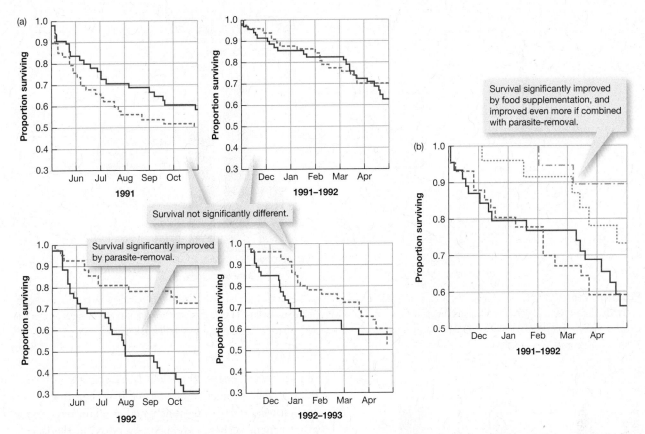

FIGURE 7.5 (a) When snowshoe hares, *Lepus americanus*, were treated for parasitic nematodes (blue dashed line) and compared with untreated (parasitized) controls (red solid line), their survival was significantly increased in summer 1992 but not over three other periods (95% of all mortality being due to predation). (b) When hares were provided with supplementary food in winter 1991–92, this significantly improved their survival (green dotted line) compared to controls (red solid line), but this was improved still further by the removal of nematodes as well (orange dashed-dotted line). (After Murray et al., 1997.)

in the fourth, the parasites greatly increased the hares' vulnerability to predation. An explanation for this apparent inconsistency was then offered by a further experiment in which half the experimental field plots were provided with supplementary food for the hares (Figure 7.5b). There again, removal of parasites alone did not alter the vulnerability to predators. But food supplementation improved it significantly, and parasite removal then improved it further still. This reminds us that the harm parasites do is very often dependent on various other aspects of the lives of their hosts.

Compensation and defense by individual prey

The effects of parasites and graz- | compensatory
ers, however, are not always more | plant responses
profound than they first seem: they
are often *less* profound. For example, individual plants can compensate in a variety of ways for the effects of herbivory (Strauss & Agrawal, 1999). The removal of leaves from a plant may decrease the shading of other leaves and thereby increase their rate of photosynthesis. Or, following herbivore attack, many plants compensate by utilizing stored reserves. Often, there

is compensatory regrowth of defoliated plants when buds that would otherwise remain dormant are stimulated to develop. There is also commonly a reduced subsequent death rate of surviving plant parts.

For example, when herbivory on the biennial plant field gentian (*Gentianella campestris*) was simulated by clipping to remove half its biomass (Figure 7.6a), compensation was such that subsequent production of fruits could actually be increased (Figure 7.6b). But this happened only if clipping took place at the time when damage by herbivores normally occurs, between July 12 and 20. If clipping occurred later than this, fruit production was less in clipped plants than in unclipped controls: no compensation was apparent.

Plants may also respond by ini- | defensive plant
tiating or increasing their production | responses
of defensive structures or chemicals.
For example, a few weeks of grazing by snails (*Littorina obtusata*) on the brown seaweed *Ascophyllum nodosum* induced substantially increased concentrations of phlorotannins (Figure 7.7a), which have a deterrent effect on snail grazing, as evidenced by the fact that they ate significantly less when offered grazed material than when offered ungrazed plants (Figure 7.7b). Such induced

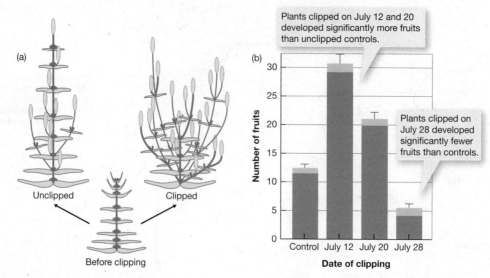

FIGURE 7.6 (a) Clipping of field gentians to simulate herbivory causes changes in the architecture and numbers of flowers produced. (b) Production of mature (dark green histograms) and immature fruits (light green histograms) of unclipped control plants and plants clipped on different occasions from July 12 to 28, 1992. Means and standard errors are shown and all means are significantly different from each other ($P < 0.05$). (After Lennartsson et al., 1998.)

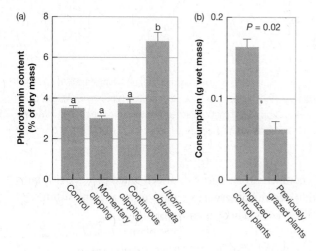

FIGURE 7.7 (a) Phlorotannin content of *Ascophyllum nodosum* plants after exposure to simulated herbivory (removing tissue with a hole punch) or grazing by the snail *Littorina obtusata*. Only the snail had the effect of inducing increased concentrations of the defensive chemical in the seaweed. Means and standard errors are shown. Different letters indicate that means are significantly different ($P < 0.05$). (b) In a subsequent experiment, the snails were presented with algal shoots from the control and the snail-grazed treatments in (a)—the snails ate significantly less of plants with high phlorotannin content. (After Pavia & Toth 2000.)

responses take time to develop, but they can still be effective in reducing damage, either because the snails stay and feed on the same plant for long time periods, as here, or because other grazers move in to replace those that induced the response. Interestingly, simple clipping of the seaweeds did not have the same effect.

The snails in Figure 7.7 suffer as a consequence of the seaweed's response (they eat less), and the plants benefit in that less of them is eaten. But in order to obtain their benefit, the plants pay the cost of producing their protective chemicals. It is therefore never straightforward to determine whether plants experience a *net* benefit (benefits greater than costs) in the longer term. One attempt to address this question looked at lifetime fitness of wild radish plants (*Raphanus sativus*) assigned to one of three treatments: (i) grazed by caterpillars, *Pieris rapae*, (ii) artificially damaged by having an equivalent amount of biomass removed using scissors, and (iii) undamaged controls. The response induced by the caterpillars protected the plants, relatively, from subsequent attack both by other chewing herbivores (Figure 7.8a) and by phloem-sucking aphids (Figure 7.8b). Artificial damage, by contrast, induced no such protective effect. Moreover, despite the costs of making this response, the lifetime fitness of the caterpillar-grazed plants was 60% greater than that of the control plants (Figure 7.8c), whereas artificially damaged plants (with no induced protection) had 38% *lower* fitness, emphasizing the negative effect of tissue loss in its own right. This fitness benefit to caterpillar-grazed plants, however, occurred only when the other herbivores were present. In their absence, the costs of producing the chemicals outweighed the benefits: plants suffered a reduction in fitness (Karban et al., 1999). Clearly, then, the benefits of inducible defences were *net* benefits: benefits outweighed costs.

From individual prey to prey populations

In spite of these various qualifications, the general rule is that predators are harmful to individual prey.

compensatory reactions among surviving prey

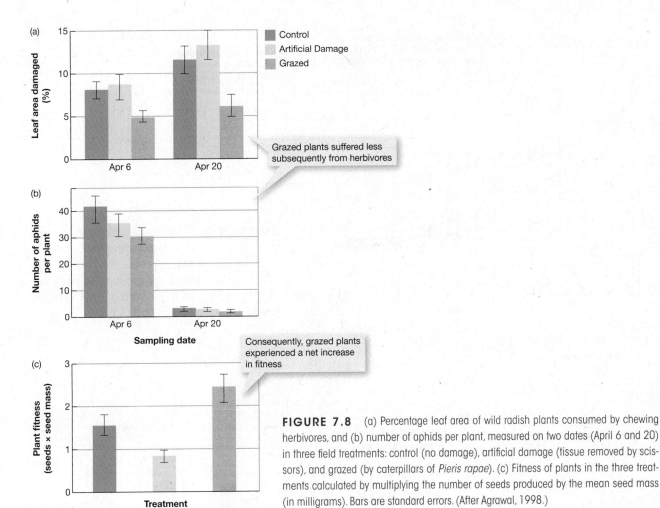

FIGURE 7.8 (a) Percentage leaf area of wild radish plants consumed by chewing herbivores, and (b) number of aphids per plant, measured on two dates (April 6 and 20) in three field treatments: control (no damage), artificial damage (tissue removed by scissors), and grazed (by caterpillars of *Pieris rapae*). (c) Fitness of plants in the three treatments calculated by multiplying the number of seeds produced by the mean seed mass (in milligrams). Bars are standard errors. (After Agrawal, 1998.)

But the effects of predation on a population of prey are not always so predictable. The most common reason is that the impact of predation is limited by compensatory reactions among the survivors as a result of reduced intraspecific competition. So, when food is short and competition intense, predation may relieve competitive pressures and allow individuals to survive that would not otherwise do so. The results of an experiment confirming this are shown in Figure 7.9. The survival of grasshoppers (*Ageneotettix deorum*) was monitored in caged plots with grass food that was either plentiful (fertilized plots) or limited (not fertilized), and in the presence or absence of predatory spiders. As predicted, with plentiful food, spider predation reduced the numbers surviving. But with limited food, spider predation and food limitation were compensatory; spiders ate grasshoppers, but this allowed other grasshoppers to survive, and so the same numbers of grasshoppers were recovered at the end of the 31-day experiment.

Predators may also have little impact on prey populations as a whole because of the particular individuals they attack. Many large carnivores, for example, concentrate their attacks on the old (and infirm), on the young (and naive) or on the sick. Thus, a study in the Serengeti found that cheetahs and wild dogs killed a disproportionate number from the younger age classes of Thomson's gazelles (Figure 7.10a) because: (i) these young animals were easier to catch (Figure 7.10b); (ii) they had lower stamina and running speeds; (iii) they were less good at outmaneuvering the predators (Figure 7.10c); and (iv) they may even have failed to recognize the predators. The effects of predation on the prey population will therefore have been less than would otherwise have been the case, because these young gazelles will have been making no present reproductive contribution to the population, and many would have died anyway from other causes before they were able to do so.

predators often attack the weakest and most vulnerable

Nonetheless, there is no doubt that predators can affect not only individual prey but also the abundance of whole prey populations. As so often in ecology, though, what is going on naturally may be apparent only if we intervene, in this case,

predators can reduce prey abundance

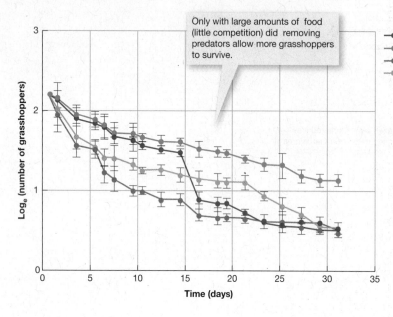

Only with large amounts of food (little competition) did removing predators allow more grasshoppers to survive.

- No spiders, no fertilizer
- No spiders, fertilizer
- Spiders, no fertilizer
- Spiders, fertilizer

FIGURE 7.9 Trajectories of numbers of grasshoppers surviving (mean ± SE) for treatments with greater or lesser amounts of food (with and without fertilizer), and with and without predatory spiders, in a field experiment involving caged plots in the Arapaho Prairie, Nebraska, USA. (After Oedekoven & Joern, 2000.)

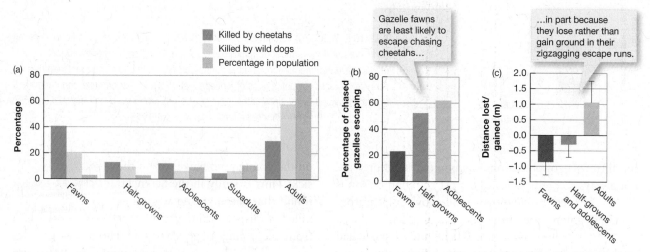

Killed by cheetahs
Killed by wild dogs
Percentage in population

Gazelle fawns are least likely to escape chasing cheetahs…

…in part because they lose rather than gain ground in their zigzagging escape runs.

FIGURE 7.10 (a) The proportions of different age classes of Thomson's gazelles in cheetah and wild dog kills is quite different from their proportions in the population as a whole. (b) Age influences the probability for Thomson's gazelles of escaping when chased by cheetahs. (c) When prey (Thomson's gazelles) zigzag to escape chasing cheetahs, prey age influences the mean distance lost by the cheetahs. (After Fitzgibbon & Fanshawe, 1989; Fitzgibbon, 1990.)

by removing the predators and observing the response of the prey population. For example, field plots of the perennial shrub, the bush lupine, were established both in grassland and duneland habitats in California. These were either protected from, or exposed to, voles that ate their leaves, mice that ate their seeds, and insects that attacked their roots. Then, when the plant populations were followed for more than five years, the effects of these various herbivores on plant abundance were clear (Figure 7.11a). With a single exception (the effects of rodents on adult plants in grasslands), both the insects and the rodents decreased the abundance of both adult

Thomson's gazelles

Fuse/Getty Images

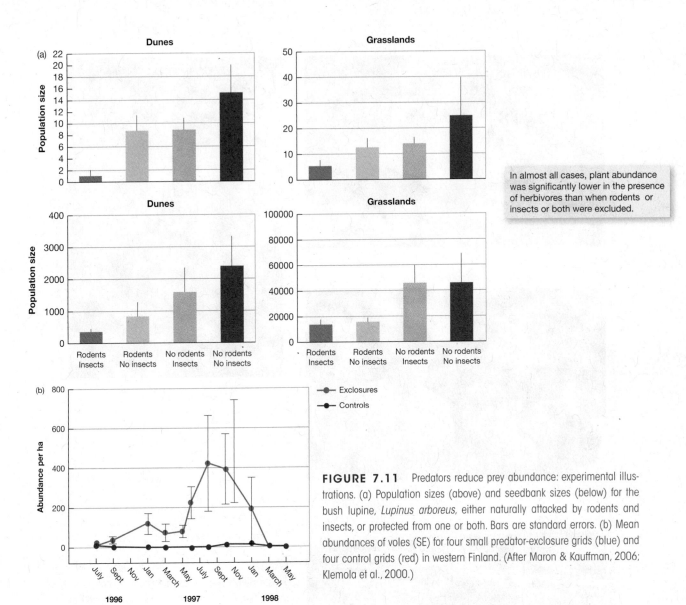

In almost all cases, plant abundance was significantly lower in the presence of herbivores than when rodents or insects or both were excluded.

FIGURE 7.11 Predators reduce prey abundance: experimental illustrations. (a) Population sizes (above) and seedbank sizes (below) for the bush lupine, *Lupinus arboreus*, either naturally attacked by rodents and insects, or protected from one or both. Bars are standard errors. (b) Mean abundances of voles (SE) for four small predator-exclosure grids (blue) and four control grids (red) in western Finland. (After Maron & Kauffman, 2006; Klemola et al., 2000.)

plants and the number of seeds in the soil, and rodents and insects together depressed abundance even more than either did alone. In another experiment, voles were protected for two years from all predators, in a series of four one-hectare enclosures in Finland that were fenced and had net roofs. Their abundance increased more than 20-fold compared to control grids (Figure 7.11b). On the other hand, at the end of this period, as a result of a shortage of food, the populations of voles collapsed to an abundance similar to that of the control.

It is apparent that predators can reduce the abundance of their prey. It is equally apparent, though, that as a result of various compensatory effects, the consumption of a prey or plant individual may not inevitably translate into a prey population one individual smaller than it would otherwise have been, either in the short or in the longer term. We return to take a closer look at the dynamics of predator and prey populations in Section 7.5.

7.4 PREDATOR BEHAVIOR: FORAGING AND TRANSMISSION

So far we have been looking at what happens *after* a predator finds its prey. Now, we take a step back and examine how contact is established in the first place. This is crucially important, because this pattern of contact is critical in determining the predator's consumption rate, which goes a long way to determining its own level of benefit and the harm it does to the prey, which determine, in turn, the impact on the dynamics of predator and prey populations.

True predators and grazers typically forage. Many move around within their habitat in search of their prey, and their pattern of contact is therefore itself determined by the predators' behavior—and sometimes by the evasive behavior of the prey (Figure 7.12a). This foraging behavior is discussed below. Other predators 'sit and wait' for their prey, often in a location they

(a)

Anup Shah/Corbis

(b)

Michael P. Gadomski/Photo Researchers/Getty Images

(c)

Peter Parks/AFP/Getty Images

(d)

Sanjay Kanojia/AFP/Getty Images

FIGURE 7.12 The different types of foraging and transmission. (a) An active predator seeking (possibly active) prey: a lion chasing a wildebeest. (b) A sit-and-wait predator (a web-spinning spider) waiting for active prey to come to it. (c) Direct parasite transmission—infectious and uninfected hosts 'bumping into each other' (a crowded population in Hong King at the time of the 2002–03 SARS outbreak). (d) Transmission between free-living stages of a parasite shed by a host and new, uninfected hosts (children in India playing in a river polluted with human excrement).

have selected, web-spinning spiders for instance, though often too, barnacles and corals for example, where their young stages just happen to have settled (Figure 7.12b).

With parasites and pathogens, on the other hand, we usually talk about transmission rather than foraging. This may be direct transmission between infectious and uninfected hosts when they come into contact with one another (Figure 7.12c), or free-living stages of the parasite may be released from infected hosts, so that it is the pattern of contact between these and uninfected

hosts that is important (Figure 7.12d). The simplest assumption we can make for directly transmitted parasites – and one that often is made when attempting to understand their dynamics (discussed in Section 7.5)—is that transmission depends on infectious and uninfected hosts 'bumping into one another.' In other words, the overall rate of parasite transmission depends both on the density of uninfected, susceptible hosts (since these represent the size of the 'target') and on the density of infectious hosts (since this represents the risk of the target being 'hit') (Figure 7.12c). With free living stages, then (Figure 7.12d), the focus is usually on the densities of the stages themselves and of susceptible hosts.

Foraging behavior

There are many questions we might ask about the behavior of a foraging predator. Where, within the habitat available to it, does it concentrate its foraging? How long does it tend to remain in one location before moving on to another? And so on. Ecologists address all such questions from two points of view. The first is from the viewpoint of the *consequences* of the behavior for the dynamics of predator and prey populations. We turn to this in Section 7.5.

The second is from the viewpoint of 'optimal foraging.' The aim is to seek to understand why particular patterns of foraging behavior have been favored by natural selection. We apply the same general approach as we would apply, for example, to the anatomy of the bird's wing, where we might seek to understand why a particular surface area, or a particular arrangement of feathers, has been favored by natural selection for the effectiveness they bring to the bird's powers of flight. Of course, nobody suggests that the birds have even a basic understanding of aerodynamics theory—only that those birds with the most effective wings have been favored in the past by natural selection and have passed their effectiveness on to their offspring. Likewise, in applying this approach to foraging behavior, there is no question of suggesting 'conscious decision-making' on the predator's part.

What, though, is the appropriate measure of 'effectiveness' in foraging behavior? For many aspects of foraging, though not all, the best measure is the *net rate of energy intake*—that is, the amount of energy obtained per unit time, *after* account has been taken of the energy expended by the predator in carrying out its foraging. Armed with this measure of effectiveness we can ask a number of questions, as follows.

How long does a predator tend to remain in one location – one patch, say, of a patchy environment – before moving on to another, especially given that moving between patches is, in itself, unproductive? Details aside, we expect foragers simply to leave once they have depleted the patch to the point where they have more to gain by leaving ('for pastures new') than by staying, and this will take longest to do in the most productive patches. Figure 7.13a shows evidence for such a pattern for bison foraging in patches of meadow (only 10% of the total habitat) in Price Albert National Park, Canada: they remained significantly longer in the more productive patches containing most of their preferred food, the sedge *Carex atherodes*, since it took them

longer there to deplete the patch to the point where it paid them to move.

Next, what are the effects of other competing predators foraging in the same habitat? The expected net energy intake from a location is now presumably a reflection of both its intrinsic productivity and the number of competing foragers. How can we expect predators to distribute themselves over various habitat patches differing in their intrinsic productivity? Figure 7.13b shows a commonly observed pattern of predator distribution (the so-called 'ideal free distribution'), though in an uncommon setting: street children (predators) foraging for cars (prey) to whose drivers they can sell water. The foraging children distributed themselves between two lanes in the road in the same ratio as their prey, the cars. This is an optimal strategy in the sense that any child departing from this ratio (behaving suboptimally) would have experienced more competition and hence a lower intake of money.

Next, does the location chosen by a predator reflect just the energy intake it can expect there? Or does there appear to be some balancing of this against the risk of being preyed upon by its own predators? Balancing of risk is often apparent as, for example, for the large African herbivores in Figure 7.13c. In the presence of lions, each showed an increased preference for foraging in grasslands, which might otherwise not have even been a preferred habitat, where the lions could more easily be seen and evasive action more easily taken.

And finally here, there are questions of diet width. No predator can be capable physically of consuming all types of prey. However, most animals consume a narrower range of food types than they are morphologically capable of consuming. What determines what is included and excluded from a predator's diet? Again putting details to one side, it should pay a predator always to include the most profitable prey type in its diet—profitable in terms of energy intake per unit time spent searching for and 'handling' its prey (pursuing, subduing, and consuming prey). But does it pay the predator then to add further items to its diet? The answer is that it should add the item as long as, having found it, its rate of energy intake handling it is greater than the average rate of intake from handling *and* searching for items already in the diet. This will serve to maximize its overall rate of energy intake. This optimal diet width model, then, leads to a number of predictions.

1 Predators with handling times that are typically short compared to their search times should be generalists (that is, have broad diets), because in the short time it takes them to handle a prey item that has already been found, they can barely begin to search for another

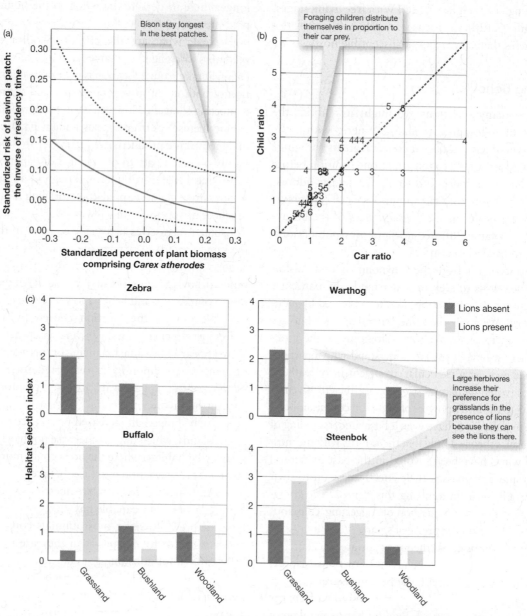

FIGURE 7.13 Optimal foraging. (a) The risk of bison, *Bison bison*, leaving a food patch declines as the quality of the food patch (the proportion of their preferred food) increases. The solid line is the output of a statistical model fitted to field data after other contributory factors have been taken into account; the dotted lines are 95% confidence intervals. (b) Street children in Istanbul, Turkey, thought of as 'predators,' adjust the ratio according to which they distribute themselves across two lanes in a highway when they 'prey' upon cars (try to sell the drivers water), such that their ratio matches that of the cars. (c) Habitat preferences of large herbivores in Hwange National Park, Zimbabwe, where a value of the selection of index exceeding 1 represents positive selection, and a value less than 1 represents avoidance. (After Courant & Fortin, 2011; Disma et al., 2011; Valeix et al., 2009.)

vario images GmbH & Co.KG/Alamy

prey item. This prediction seems to be supported by the broad diets of many insectivorous birds feeding in trees and shrubs. Searching is always moderately time-consuming, but handling the minute, stationary insects takes negligible time and is almost always successful. A bird, therefore, has something to gain and virtually nothing to lose by consuming an item once found, and overall profitability is maximized by a broad diet.

2 By contrast, predators with handling times that are long relative to their search times should be specialists:

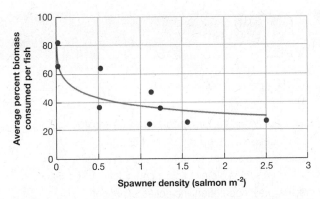

FIGURE 7.14   Bears become more specialized as their prey (salmon) abundance increases: as the spawning density of salmon increases, the average percentage of each salmon consumed by bears decreases. (After Gende et al., 2001.)

maximizing the rate of energy intake is achieved by including only the most profitable items in the diet. For instance, lions live much of the time in sight of their prey, so that search time is negligible; handling time, on the other hand, and particularly pursuit time, can be long (and very energy-consuming). Lions consequently specialize on those prey that can be pursued most profitably: the immature, the lame, and the old.

3   Other things being equal, a predator should have a broader diet in an unproductive environment (where prey items are relatively rare and search times relatively large) than in a productive environment (where search times are generally smaller). This prediction is supported by a study of brown and black bears (*Ursos arctos* and *U. americanus*) feeding on salmon in Bristol Bay, Alaska (Figure 7.14). When salmon availability was high, bears consumed less biomass per captured fish, targeting energy-rich fish (those that had not spawned) or energy-rich body parts (eggs in females, brain in males). That is, their diet became more specialized when prey were more abundant.

Overall, then, we can see how an evolutionary, optimal foraging approach can help us make sense of predators' foraging behavior. The approach makes predictions of what that behavior might be expected to be, and there is good evidence from real examples to support these predictions.

7.5   POPULATION DYNAMICS OF PREDATION

What roles do predators play in driving the dynamics of their prey, or prey play in driving the dynamics of their predators? Are there common patterns of dynamics that emerge? The preceding sections should have made it plain that there are no simple answers to these questions.

It depends on details of the behavior of individual predators and prey, on possible compensatory responses at individual and population levels, and so on. We can build an understanding of these complex dynamics, however, by starting simply and then adding additional features one by one to construct a more realistic picture.

Underlying dynamics of predator–prey interactions: a tendency to cycle

We begin by consciously oversimplifying—ignoring everything but the predator and the prey, and asking what underlying tendency there might be in the dynamics of their interaction. It turns out that the underlying tendency is to exhibit coupled oscillations—cycles—in abundance. With this established, we can turn to the many other important factors that might modify or override this underlying tendency. Rather than explore each and every one of them, however, we examine just two of the more important ones: crowding and spatial patchiness. These two factors cannot, of course, tell the whole story; but they illustrate how the differences in predator–prey dynamics, from example to example, might be explained by the varying influences of the different factors with a potential impact on those dynamics.

Starting simply then, suppose there is a large population of prey. Predators presented with this population should do well: they should consume many prey and hence increase in abundance themselves. The large population of prey thus gives rise to a large population of predators, albeit after a time delay while the positive effects on the predators' survival and fecundity take effect. But this increasing population of predators increasingly takes its toll of the prey. The large population of predators therefore gives rise to a small population of prey, though also after a time delay, while predators (literally) eat into the abundance of the prey. Now the predators are in trouble: there are large numbers of them and very little food. Their abundance declines (another time delay as they starve and fail to reproduce). But this takes the pressure off the prey: the small population of predators gives rise to a large population of prey—and the populations are back to where they started. There is, in short, an underlying tendency for predators and their prey to undergo coupled oscillations in abundance—population cycles (see below, Figure 7.17)—essentially because of these time delays in the response of predator abundance to that of the prey, and vice versa. A simple mathematical model—the Lotka–Volterra model—conveying essentially this message is described in Box 7.1. Some historical background to the two biologically oriented mathematicians whose names have been attached to the model is provided in Box 7.2.

Quantitative Aspects

The Lotka–Volterra predator–prey model

Here, as in Boxes 5.4 and 6.2, we describe one of the foundation-stone mathematical models of ecology. Like the model of interspecific competition in Box 6.2, this model is known by the name of its originators: Lotka and Volterra (Volterra, 1926; Lotka, 1932). It has two components: P, the numbers present in a predator (or consumer) population, and N, the numbers or biomass present in a prey or plant population.

We assume that in the absence of consumers the prey population increases exponentially (see Box 5.4):

$$dN/dt = rN$$

But now we also need a term signifying that prey individuals are removed from the population by predators. They will do this at a rate that depends on the frequency of predator–prey encounters, which will increase with increasing numbers of predators (P) and prey (N). The exact number encountered and consumed, however, will also increase with the searching and attacking efficiency of the predator, denoted by a. The consumption rate of prey will thus be aPN, and overall:

$$dN/dt = rN - aPN \tag{1}$$

Turning to predator numbers, in the absence of food these are assumed to decline exponentially through starvation:

$$dP/dt = -qP$$

where q is their mortality rate. But this is counteracted by predator birth, the rate of which is assumed to depend on: (i) the rate at which food is consumed, aPN; and (ii) the predator's efficiency, f, at turning this food into predator offspring. Overall:

$$dP/dt = faPN - qP \tag{2}$$

Equations 1 and 2 constitute the Lotka–Volterra model.

The properties of this model can be investigated by finding zero isoclines, as we did previously for the model of interspecific competition in Box 6.1. There are separate predator and prey zero isoclines, both of which are drawn on a graph of prey density (x-axis) against predator density (y-axis) (Figure 7.15). The prey zero isocline joins combinations of predator and prey densities that lead to an unchanging prey population, $dN/dt = 0$, while the predator zero isocline joins combinations of predator and prey densities that lead to an unchanging predator population, $dP/dt = 0$.

In the case of the prey, this amounts to saying that $dN/dt = 0$ in equation 1, which means that the other side of the equation also equals zero, and so:

$$P = r/a$$

Thus, since r and a are constants, the prey zero isocline is a line for which P itself is a constant (Figure 7.15a): prey increase when predator abundance is low ($P < r/a$) but decrease when it is high ($P > r/a$).

Similarly, for the predators, we solve for $dP/dt = 0$ in equation 2, giving the equation of the isocline as:

$$N = q/fa$$

The predator zero isocline is therefore a line along which N is constant (Figure 7.15b): predators decrease when prey abundance is low ($N < q/fa$) but increase when it is high ($N > q/fa$).

Putting the two isoclines (and two sets of arrows) together in Figure 7.16 shows the behavior of joint populations. The various combinations of increases and decreases, listed above, mean that the populations undergo 'coupled oscillations' or 'coupled cycles' in abundance; 'coupled' in the sense that the rises and falls of the predators and prey are linked, with predator abundance tracking that of the prey (discussed biologically in the main text). The world is much more complex than imagined here. But the model does capture the essential tendency for coupled cycles in predator–prey interactions.

Note, however, that the model does not suggest that the patterns it generates are exactly what we should observe.

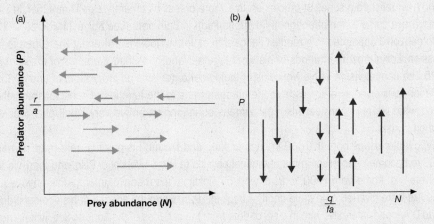

FIGURE 7.15   See box text for details.

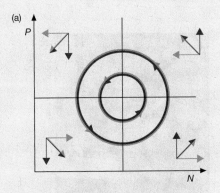
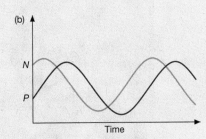

FIGURE 7.16   See box text for details.

7.2 Historical Landmarks

Lotka and Volterra

The two men whose names we associate with the baseline models of both interspecific competition (Box 6.2) and predator–prey interactions (Box 7.1) make up a semi-detached partnership because they never worked together in any real sense. Neither was an ecologist nor, arguably, had much deep-seated interest in ecology, yet each, in his own way, is an unexpected hero in the development of theoretical ecology.

Alfred James Lotka was born in 1880 to American parents in what is now Lviv in Ukraine. After an education in physics and chemistry that took him to England and Germany, he got employment as an assistant chemist with the General Chemical Company in New York in 1902. He took a year out getting a master's degree in physics from Cornell University in 1909, but most of the years until the early 1920s saw him not in any academic establishment, but commercially employed as either a physicist or chemist. His passion, however, was the application of the principles of physics to questions in biology, and he published a series of articles on this, more or less as an enthusiastic amateur, in scientific journals. These came to the attention of Raymond Pearl in Baltimore (see Box 6.1) who by 1922 managed to get Lotka appointed to a modest Fellowship at Johns Hopkins University. There, free to indulge his passion, Lotka drew the threads of his work together and published *Elements of Physical Biology* in 1925. The central vision of his work was of biological organisms as processors of energy. Touching on an enormous diversity of topics from the nitrogen cycle to the function of consciousness, the book included what we now consider his most famous equations as only a very small part, receiving no special emphasis.

Vito Volterra was born in 1860 in Ancona, Italy, and though his parents were poor, he had risen by 1883 to become a professor in mathematical physics at The University of Pisa, and then, via Turin, at the University of Rome from 1900. Years of work on elasticity, the theory of integro-differential equations and the like were diverted onto quite another track in 1925, when the fiancée of his ecologist daughter, Umberto D'Ancona, interested him in a set of data from the fisheries of the Adriatic Sea, where the virtual suspension of fishing around the time of the First World War had led to a marked increase in the numbers of predatory fish species. The following year he published an extended initial analysis in Italian of the dynamics of interacting species, and a much abbreviated English version in the journal *Nature*. His focus

Alfred James Lotka (left) and Vito Volterra

was not so much on energy as on evolution: species interactions as part of the more general struggle for existence.

When Lotka read the article in 1926, his response appears to have been an understandable combination of pride that such an eminent and established mathematician was traveling on so similar a path to his own (albeit with a rather different emphasis) and concern that he would be denied the credit that he deserved. Volterra in his turn seems to have been perfectly happy to concede priority but at pains to point up important differences. Nonetheless, there followed a number of years during which Volterra was much more fully engaged with his new biological project than Lotka, who had become a demographer working for the Metropolitan Life Insurance Company in New York, such that Lotka's contribution was often (far too often for Lotka's liking) relegated to footnotes in much longer descriptions of Volterra's work. Neither, though, seems positively to have sought preeminence, and by the late 1930s it had become common practice, as it has been ever since, to join their names together when referring to the models.

Kingsland (1985), again, provides a much fuller account of Lotka's and Volterra's histories and their contributions to the development of the science of ecology in her book, *Modeling Nature*.

Predator–prey cycles in practice

The underlying tendency for predator–prey interactions to generate coupled oscillations in abundance could lead us to expect such cycles in real populations, but many aspects of predator and prey ecology had to be ignored in order to demonstrate it, and these omissions can greatly modify expectations. It is no surprise, then, that there are few good examples of clear predator–prey cycles—though they have received a great deal of attention from ecologists. Nonetheless, as we try to make sense of predator–prey population dynamics, the underlying tendency to cycle is a good place to start.

the 'expectation' of cycles is only rarely fulfilled

They do occur sometimes. It has been possible in several cases, for example, to generate coupled predator–prey oscillations, several generations in length, in the laboratory (Figure 7.17a; see also Figure 7.21c). Among field populations, there are a number of examples in which regular cycles of prey and predator abundance can be discerned. Cycles in hare populations, in particular, have been discussed by ecologists since the 1920s, and were recognized by fur trappers more than 100 years earlier. Most famous of all is the snowshoe hare, *Lepus americanus*, which in the boreal forests of North America follows a 10-year cycle (although in reality this varies in length between 8 and 11 years; see Figure 7.17b). The snowshoe hare is the dominant herbivore of the region, feeding on the terminal twigs of numerous shrubs and small trees. A number of predators, including the Canada lynx (*Lynx canadensis*), have associated cycles of similar length. The hare cycles often involve 10–30-fold changes in abundance, and 100-fold changes can occur in some habitats. They are made all the more spectacular by being virtually synchronous over a vast area from Alaska to Newfoundland.

But are the hare and lynx participants in a predator–prey cycle? This immediately seems less likely once we note the number of other species with which both interact. Their food web (see Section 9.5) is shown in Figure 7.18. In fact, both experimental studies (Krebs et al., 2001) and statistical analyses of the population dynamics data (Stenseth et al., 1997) suggest that whereas the dynamics of the hares are driven by their interactions with both their food and their predators (especially lynx), the dynamics of the lynx are driven largely by their interaction with their hare prey, much as the food web might suggest. Both the hare–plant and the predator–hare interactions have some propensity to cycle on their own—but, in practice, the cycle seems normally to be generated by the interaction between the two. This warns us that even when we have a predator–prey pair both exhibiting cycles, we may still not be observing simple predator–prey oscillations.

. . . but how are the cycles generated?

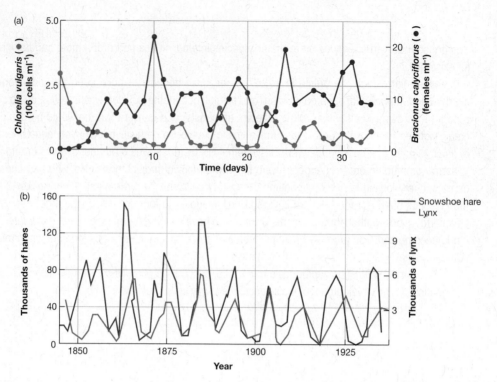

FIGURE 7.17 Coupled oscillations in the abundance of predators and prey. (a) Parthenogenetic female rotifers, *Bracionus calyciflorus* (predators, red circles), and unicellular green algae, *Chlorella vulgaris* (prey, blue circles), in laboratory cultures. (After Yoshida et al., 2003.) (b) The snowshoe hare (*Lepus americanus*) and the Canada lynx (*Lynx canadensis*) as determined by the number of pelts lodged with the Hudson Bay Company. (After Maclulick, 1937.)

The Canada lynx and the snowshoe hare – a predator and prey that may show coupled oscillations.

Cycles of predators and prey can often appear as 'outbreaks' of the predator, and as such they may make the news. An example is described in Box 7.3.

Disease dynamics and cycles

Cycles are also apparent in the dynamics of many parasites, especially microparasites (bacteria, viruses, etc.). To understand the dynamics of any parasite, the best starting point is its basic reproductive number, conventionally called 'R nought', R_0. For microparasites, R_0 is the average number of new infected hosts that would arise from a single infectious host in a population of susceptible hosts. An infection cannot spread for $R_0 < 1$ (each present infection leads to less than one infection in the future), but an infection will spread for $R_0 > 1$. There is therefore a 'transmission threshold' when $R_0 = 1$, which must be crossed if a disease is to spread. A derivation of R_0 for microparasites with direct transmission (see Figure 7.12c) is given in Box 7.4.

Box 7.4 provides us with a crucial insight into disease dynamics—for each directly transmitted microparasite, there is a critical threshold population size that needs to be exceeded for a parasite population to be able to sustain

basic reproductive number and the transmission threshold

threshold population sizes and microparasite cycles

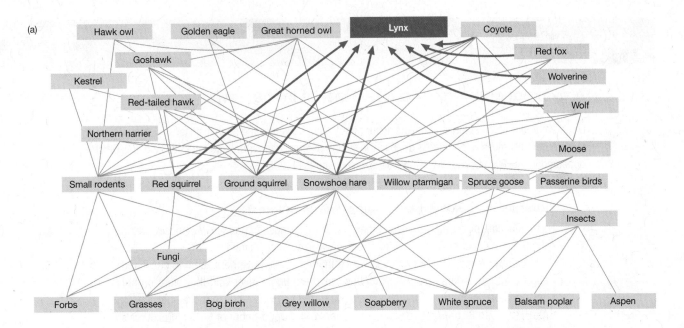

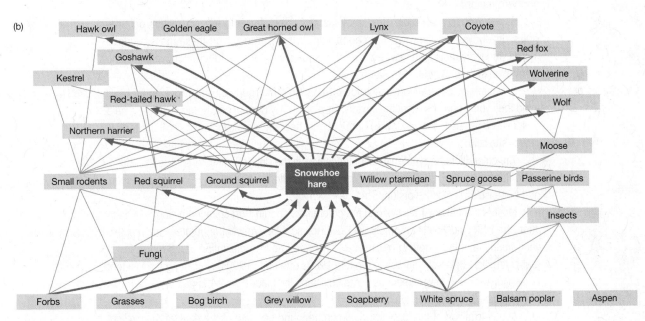

FIGURE 7.18 (a) The main species and groups of species in the boreal forest community of North America, with trophic interactions (who eats who) indicated by lines joining the species, and those affecting the Canada lynx shown as blue arrows, pointing toward the consumer. (b) The same community, but with the interactions of the snowshoe hare shown as arrows. (After Stenseth et al., 1997.)

itself. For example, measles has been calculated to have a threshold population size of around 300,000 individuals and is unlikely to have been of great importance until quite recently in human biology. However, it has generated major epidemics in the growing cities of the industrialized world in the 18th and 19th centuries, and in the growing concentrations of population in the developing world in the 20th century. Current estimates suggest that there were around 750,000 deaths each year from measles infection in the developing world in 2000, but control programs

(mainly immunization) had reduced numbers to around 150,000 per year by the end of that decade (Dabbagh et al., 2009).

Moreover, the immunity induced by many bacterial and viral infections, combined with death from the infection, reduces the number of susceptible hosts in a population, reduces R_0, and therefore tends to lead to a decline in the incidence of the disease itself. In due course, though, there will be an influx of new susceptible hosts into the population (as a result of new births or perhaps immigration), an increase in R_0, an increase

7.3 ECOncerns

A cyclical outbreak of a forest insect in the news

Large outbreaks of forest tent caterpillars occur about every 10 years, and each lasts for 2–4 years. During these outbreaks, massive damage is done to the foliage of forest trees over large tracks of land. This article appeared in the *Telegraph Herald* (Dubuque, Iowa) on June 11, 2001.

Caterpillars making a meal out of northern forests

Forest tent caterpillars have munched their way through much of northern Wisconsin, eating aspen, sugar maple, birch and oak from Tomahawk to southern Canada.

The insects move across roads in waves that make the pavement seem to crawl and hang from trees in large clumps. . . . 'One lady from Eagle River said they were on her house and on her driveway and on her sidewalk, and she was ready to move back to Oak Creek', said Jim Bishop, public affairs manager for the Department of Natural Resource's northern region.

Shane Weber, a DNR forest entomologist from Spooner, said the caterpillars on sidewalks, driveways and highways are a good sign. 'Whenever they start these mass overland moves, suddenly moving in waves across the ground, it means that they're starving, looking for another source of food', he said.

In Superior, customers have inundated Dan's Feed Bin [general store], looking for ways to rid their yards and homes of the insects. Employee Amy Connor said some customers held their telephones up to the window so Connor could hear the worms falling like hail. 'It's terribly gross', she said.

The caterpillars have eaten most of the leaves in the Upper Peninsula, said Jeff Forslund, of Hartland, who drove to Ramsey, Michigan. 'My grandfather has about 500 acres of aspen, and there isn't a leaf left', Forslund said.

Most of the trees will survive and the caterpillars should start spinning cocoons by mid-June, the DNR said. Forest entomologist Dave Hall said he expects the outbreak to peak this year. 'I can't imagine it getting much worse', he said. The last infestation of the native forest tent caterpillars in Wisconsin was in the late 1980s and early 1990s. . . . During the last tent caterpillar outbreak, several serious traffic collisions in Canada were blamed on slick roads from squashed tent caterpillars.

About 4 million of the fuzzy crawlers can be found per acre at the peak of the cyclical infestation, the DNR said.

(All content © 2001 *Telegraph Herald* (Dubuque, IA) and may not be republished without permission.)

1 *From what you have learned about population cycles in this chapter, suggest an ecological scenario to account for the periodic outbreaks of these caterpillars.*

2 *Do you believe the comment attributed to a Department of Natural Resources (DNR) employee that the mass movement of the caterpillars is a good sign? How would you determine whether this behavior heralds an end to the peak phase of the cycle?*

7.4 Quantitative Aspects

Transmission thresholds for microparasites

Putting it simply, for microparasites with direct transmission, the basic reproductive number, R_0, measures the average number of new infections arising from a single infected individual in a population of susceptible hosts. It increases with the average period of time over which an infected host remains infectious, L, since a long infectious period means plenty of opportunity to transmit to new hosts; it increases with the number of susceptible individuals in the host population, S, because more susceptible hosts offer more opportunities ('targets') for transmission of the parasite; and it increases with the transmission rate of the infection, β, because this itself increases first with the infectiousness of the parasite—the probability that contact leads to transmission – but also with the likelihood of infectious and susceptible hosts coming into contact as a reflection of the pattern of host behavior (Anderson, 1982). Thus, overall:

$$R_0 = S\,\beta L$$

We know that $R_0 = 1$ is a transmission threshold, in that below this the infection will fail to spread but above it the infection will spread. But we can also assume that for a particular host–parasite combination, L and β are constant, in that they are characteristic of that pair of species. This then allows us to define a critical *threshold population size S_T*: the number of susceptible hosts that give rise to $R_0 < 1$. At that threshold, making $R_0 = 1$ in the equation means:

$$S_T = 1/\beta L$$

In populations with fewer susceptible hosts than this, the infection will die out ($R_0 < 1$), but with more than this, the infection will spread ($R_0 > 1$). We can immediately see, therefore, that the threshold population size is larger (more individuals are required to sustain an infection) when infectiousness (β) is low and/or infections themselves are short-lived (small L).

---

in incidence, and so on. We are once again looking at a marked tendency to cycle. Such diseases tend to generate a sequence from high incidence, to few susceptibles, to low incidence, to many susceptibles, to high incidence, etc. This undoubtedly underlies the observed cyclic incidence of many human diseases (especially prior to modern immunization programs), with the differing lengths of cycle reflecting the differing characteristics of the diseases: measles with peaks every 2 years, approximately, pertussis (whooping cough) every 3–4 years, diphtheria every 4–6 years, and so on (Figure 7.19).

Crowding

One fundamental feature that we have ignored so far is the fact that no predator lives in isolation: all are affected by other predators. The most obvious effects are competitive. When predator numbers

mutual interference among predators reduces the predation rate

are high, there may be less food than the predators overall require, and individual predators will suffer as a consequence (see Chapter 3). However, even when food is not limited, the consumption rate per individual can be reduced with increases in predator density by a number of processes known collectively as *mutual interference*. For example, many predators interact behaviorally with other members of their population, leaving less time for feeding. Hummingbirds actively and aggressively defend rich sources of nectar; parasitoid wasps will threaten and, if need be, fiercely drive away an intruder from their own area of tree trunk. Alternatively, an increase in consumer density may lead to an increased rate of emigration, or of consumers stealing food from one another (as do many gulls), or the prey themselves may respond to the presence of consumers and become less available for capture.

In all such cases, the underlying pattern is the same: the consumption rate per individual predator

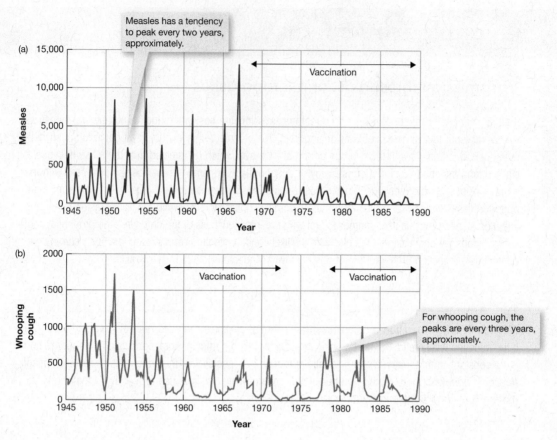

FIGURE 7.19 Reported monthly cases of (a) measles and (b) pertussis (whooping cough) in London from 1945 to 1990. Notice the damping effect of vaccination on the patterns, since this reduces the number and flow of susceptible hosts into the populations. (After Keeling et al., 2001.)

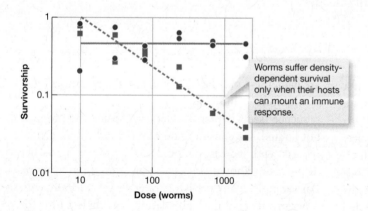

FIGURE 7.20 Host immune responses are necessary for density dependence in infections of the rat with the nematode *Strongyloides ratti*. Survivorship is independent of initial dose in mutant rats without an immune response (●; slope not significantly different from 0), but with an immune response (■) it declines (slope = −0.62, significantly less than 0; $P < 0.001$). (After Paterson & Viney, 2002.)

declines with increasing predator density. This reduction is itself likely to have an adverse effect on the fecundity, growth, and mortality of individual predators, which intensifies as predator density increases. The predator population is thus subject to density-dependent regulation (see Chapters 3 and 5).

With parasites, too, it is to be expected that individuals will often interfere with each other's activities.

competition or the immune response in parasites?

There may, in the first place, be intraspecific competition between parasites for host resources and density dependence in their growth, birth, and/or death rates. However, for vertebrate hosts at least, the intensity of the immune reaction elicited from a host also typically depends on the abundance of parasites. A rare attempt to disentangle these two effects utilized the availability of mutant rats lacking an effective immune response (Figure 7.20). Both these and normal rats were subjected to experimental infection with a nematode, *Strongyloides ratti*, at a range of doses. Any reduction in parasite fitness with dose in the normal rats could be due to either intraspecific competition or an immune response increasing with dose (or both). But clearly, in the mutant rats only the first of these is possible. In fact, there was no observable response in the mutant rats (Figure 7.20), indicating that at these doses, which were themselves similar to those observed naturally, there was no evidence of intraspecific competition. Hence, the pattern observed in the normal rats is entirely the result of a density-dependent immune response. Of course, this does not mean that there is never intraspecific competition among parasites within hosts, but it does emphasize the particular subtleties that arise when an organism's habitat is its reactive host.

Prey, too, are likely to suffer reductions in growth, birth, and survival rates as their abundance increases and their individual intake of resources declines; and the effect

crowding tends to dampen or eliminate predator–prey cycles

of either predator or prey crowding on their dynamics is, in a general sense, fairly easy to predict. Prey crowding prevents their abundance from reaching as high a level as it would otherwise do, which means in turn that predator abundance is also unlikely to reach the same peaks. Predator crowding, similarly, prevents predator abundance from rising so high, but it also tends therefore to prevent them from reducing prey abundance as much as they would otherwise do. Overall, therefore, crowding is likely to have a damping effect on any predator–prey cycles, reducing their amplitude or removing them altogether. This occurs not only because crowding chops off the peaks and troughs, but also because each peak in a cycle tends itself to generate the next trough (e.g., high prey abundance → high predator abundance → *low* prey abundance), so that the lowering of peaks in itself tends to raise troughs.

There are certainly examples that appear to confirm the stabilizing effects of crowding in predator–prey interactions. For instance, there are two groups of primarily herbivorous rodents that are widespread in the Arctic: the microtine rodents (lemmings and voles) and the ground squirrels. The microtines are renowned for their dramatic, cyclic fluctuations in abundance, but the ground squirrels have populations that remain remarkably constant from year to year, especially in open meadow and tundra habitats. There, significantly, they appear to be strongly self-limited by food availability, suitable burrowing habitat and their own spacing behavior (Karels & Boonstra, 2000).

Predators and prey in patches

The second feature that was ignored initially but will be examined here is the fact that many populations of predators and prey exist not as a single, homogeneous mass, but as a **metapopulation**. The term is used when the overall population is divided, by the patchiness of the environment, into a series of subpopulations, each with its own internal dynamics but linked to other subpopulations by movement (dispersal) between patches (a topic developed further in Section 9.3).

We get a good idea of the general effect of this spatial structure on predator–prey dynamics by considering the simplest imaginable metapopulation: one consisting of just two subpopulations. If the patches are displaying the same dynamics, and dispersal is the same in both directions, then the dynamics are unaffected: every 'lost' individual is counteracted by an equivalent gain. To put it simply, patchiness and dispersal have no effect in their own right. Differences between the patches, however, either in the dynamics within subpopulations or in the dispersal between them, tend, in themselves, to stabilize the interaction: to dampen any cycles that might exist. The reason is that any difference leads to asynchrony in the fluctuations in the patches, meaning that a population at the peak of its cycle tends to lose more by dispersal than it gains, whereas a population at a trough gains more than it loses. Also, even with just two patches, if one subpopulation goes extinct, the other (asynchronous) subpopulation is unlikely to do so at the same time. Dispersers from it may therefore 'rescue' the first, allowing the population as a whole, the metapopulation, to persist. Dispersal and asynchrony together, therefore—and some degree of asynchrony is likely to be the general rule—tend to dampen fluctuations in predator–prey dynamics and make population persistence more likely.

Is it possible, though, to see the stabilizing influence of this type of metapopulation structure in practice? One example is provided by experimental work on a laboratory system in which a predatory mite *Phytoseiulus persimilis* fed on a phyophagous mite, its prey, *Tetranychus urticae*. This latter fed on small bean plants collected together on styrofoam 'islands' separated by water, but with bridges connecting them across which the mites could walk: either a single large island containing 90 bean plants, or eight smaller islands each containing 10 plants (Figure 7.21a). Islands were thus connected, but their dynamics were at least semi-independent since the rate of transfer of mites between islands was low. When all the bean plants were collected

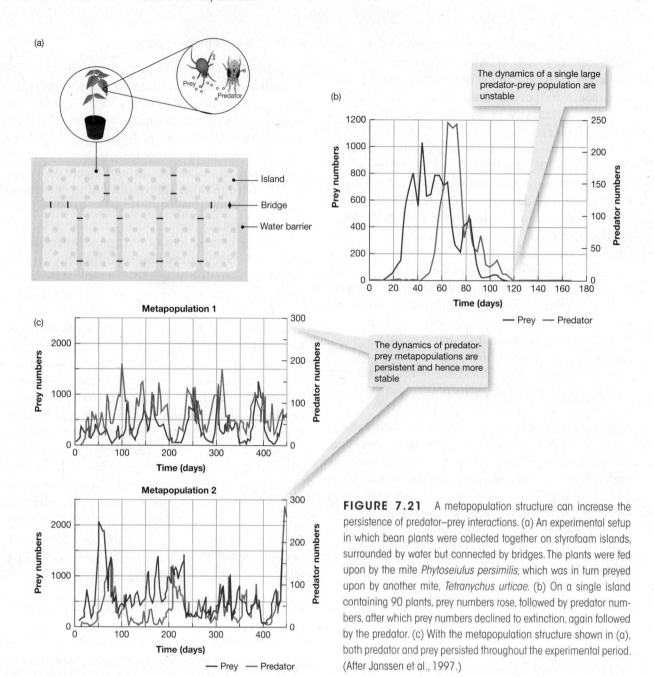

FIGURE 7.21 A metapopulation structure can increase the persistence of predator–prey interactions. (a) An experimental setup in which bean plants were collected together on styrofoam islands, surrounded by water but connected by bridges. The plants were fed upon by the mite *Phytoseiulus persimilis*, which was in turn preyed upon by another mite, *Tetranychus urticae*. (b) On a single island containing 90 plants, prey numbers rose, followed by predator numbers, after which prey numbers declined to extinction, again followed by the predator. (c) With the metapopulation structure shown in (a), both predator and prey persisted throughout the experimental period. (After Janssen et al., 1997.)

together on a single island, the abundance of the prey mite increased, followed, after a delay, by a corresponding rise in the numbers of predators, which then led to a crash in the prey population, followed by a crash of the predators (Figure 7.21b). There was therefore only a single 'cycle' of predator and prey abundance, with the whole system lasting for just 120 days. The underlying predator–prey dynamics were unstable.

When the habitat was broken into eight small islands, however, in two separate runs of the system, both predator and prey persisted for more than a year—indeed until the experiments were terminated. In

one case, there was clear evidence of persistent cycles in both predator and prey; in the other, the dynamics were much more erratic but still persistent (Figure 7.21c). Notably, though, on the individual small islands, there was not one example of the predator and prey populations persisting. Each went extinct at least once, requiring colonizing mites walking over from an occupied island to rescue that particular population. Predators and prey were therefore ultimately doomed to extinction in each patch (island)—that is, the patch dynamics were unstable. But overall, at any one time, there was a mosaic of unoccupied patches, prey–predator patches

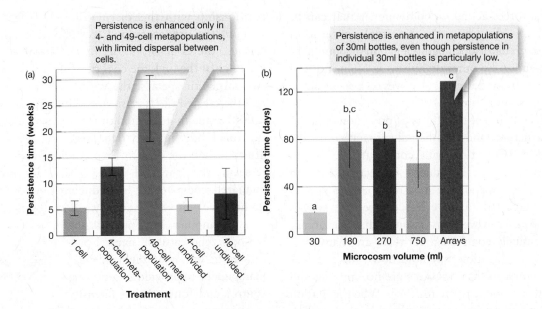

FIGURE 7.22 A metapopulation structure can increase the persistence of predator–prey interactions. (a) The parasitoid, *Anisopteromalus calandrae*, attacking its bruchid beetle host, *Callosobruchus chinensis*, living on beans either in small single 'cells' (short persistence time, left), or in combinations of cells (4 or 49), which either had free access between them so that they effectively constituted a single population (persistence time not significantly increased, right), or had limited (infrequent) movement between cells so that they constituted a metapopulation of separate subpopulations (increased persistence time, center). Bars are standard errors. (After Bonsall et al., 2002.) (b) The predatory ciliate, *Didinium nasutum*, feeding on the bacterivorous ciliate, *Colpidium striatum*, in bottles of various volumes (30–750 mL), where persistence time varied little, except in the smallest populations (30 mL) where times were shorter. But in 'arrays' of 9 or 25 linked 30 mL bottles (metapopulations), persistence was greatly prolonged: all populations persisted until the end of the experiment (130 days). Bars are standard errors; different letters above bars indicate treatments that were significantly different from one another ($P < 0.05$). (After Holyoak & Lawler, 1996.)

heading for extinction, and thriving prey patches; and this mosaic was capable of maintaining persistent populations of both predators and prey, providing stability to the metapopulation as a whole (Figure 7.21c).

A similar example, from a natural population, is provided by work off the coast of southern California on the predation by starfish of clumps of mussels (Murdoch & Stewart-Oaten, 1975). Clumps that are heavily preyed upon are liable to be dislodged by heavy seas so that the mussels die; the starfish are continually driving patches of their mussel prey to extinction. The mussels, however, have planktonic larvae that colonize new locations and initiate new clumps, whereas the starfish disperse much less readily. Thus, patches of mussels are continually becoming extinct, but other clumps are growing prior to the arrival of the starfish. As with the mites, patchiness, and a lack of synchrony between the behavior of different patches, appear capable of stabilizing the dynamics of a predator–prey interaction.

Others, too, have demonstrated the power of a metapopulation structure in promoting the persistence of coupled predator and prey populations when their dynamics in individual subpopulations are unstable. Figure 7.22a, for example, shows this for a parasitoid

attacking its beetle host. Figure 7.22b shows similar results for prey and predatory ciliates (protists), where, in support of the role of a metapopulation structure, it was also possible to demonstrate the asynchrony in the dynamics of individual subpopulations and frequent local prey extinctions and recolonizations (Holyoak & Lawler, 1996).

A metapopulation structure, then, like crowding, can have an important influence on predator–prey dynamics. More generally, however, the message from this section is that predator–prey dynamics can take a wide variety of forms, but we can begin to make sense of this variety through seeing it as a reflection of the way in which the different aspects of predator–prey interactions combine to play out variations on an underlying theme.

7.6 PREDATION AND COMMUNITY STRUCTURE

What roles can predation play when we broaden our perspective from populations to whole ecological communities? Central to this is the notion that predation is just

predation as an interrupter of competitive exclusion: predator-mediated coexistence

one of the forces acting on communities that can be described as a 'disturbance.' For example, in terms of its effect, a predator opening up a gap in a community for colonization by other organisms is often indistinguishable from the battering of waves on a rocky shore or a hurricane in a forest.

In fact, many of the effects of predation on community structure (and of other disturbances) are the result of its interaction with competitive exclusion (taking up a theme introduced in Section 6.2). In an undisturbed world, the most competitive species might be expected to drive less competitive species to extinction. However, this assumes that the organisms are actually competing. Yet there are many situations where predation may hold down the densities of competitors, so that resources are not limiting and individuals do not compete for them. When predation promotes the coexistence of species that might otherwise exclude one another, this is known as **predator-mediated coexistence**.

For example, in a study of nine Scandinavian islands, pigmy owls (*Glaucidium passerinum*) occurred on only four of the islands, and the pattern of occurrence of three species of tit had a striking

relationship with this distribution. The five islands without the predatory owl were home to only one species, the coal tit (*Parus ater*). However, in the presence of the owl, the coal tit was always joined by two larger tit species, the willow tit (*P. montanus*) and the crested tit (*P. cristatus*). Kullberg and Ekman (2000) argue that the coal tit is the superior competitor for food; but the two larger species are less affected than the coal tit by predation from the owl. It seems that the owl is responsible for predator-mediated coexistence, by reducing the competitive dominance enjoyed by coat tits in its absence.

In another example, grazing by local zebu cattle in natural pasture in the Ethiopian highlands was manipulated to provide a no-grazing control and four grazing intensity treatments at two sites. Figure 7.23 shows how the mean number of plant species varied in the sites in October, the period when plant productivity was highest. Significantly more species occurred at intermediate levels of grazing than where there was no grazing or heavier grazing ($P < 0.05$). In the ungrazed plots, several highly competitive plant species, including the

do most species occur at intermediate levels of predation?

(a) Pigmy owls (*Glaucidium passerinum*), (b) coal tit (*Parus ater*), (c) willow tit (*P. montanus*) (d) crested tit (*P. cristatus*).

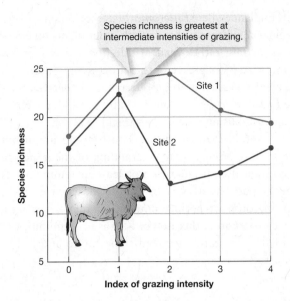

FIGURE 7.23 Mean species richness of pasture vegetation in plots subjected to different levels of cattle grazing in two sites in the Ethiopian highlands in October. 0, no grazing; 1, light grazing; 2, moderate grazing; 3, heavy grazing; 4, very heavy grazing (estimated according to cattle stocking rates). (After Mwendera et al., 1997.)

grass *Bothriochloa insculpta*, accounted for 75–90% of ground cover. At intermediate levels of grazing, however, the cattle kept the dominant grasses in check and allowed a greater number of plant species to persist. But at very high intensities of grazing, cattle were forced to turn to less preferred species, driving some to extinction and allowing grazing-tolerant species such as *Cynodon dactylon* to become dominant, so that plant species numbers were again reduced (Figure 7.23).

Overall, the number of species was greatest at intermediate levels of predation.

This suggests that broadly, *selective* predation should favor an increase in species numbers in a community as long as the preferred prey are competitively dominant. This is supported by a study that took place along the rocky shores of New England, where the most abundant and important herbivore in mid and low intertidal zones is the periwinkle snail *Littorina littorea*. The snail will feed on a wide range of algal (seaweed) species but is relatively selective: it shows a strong preference for small, tender species and in particular for the green alga *Enteromorpha intestinalis*. The least-preferred foods are much tougher (for example, the perennial red alga *Chondrus crispus* and brown algae).

Enteromorpha, the snails' preferred food, is a competitive dominant in their absence. If snails are artificially removed from a rock pool dominated by *Chondrus*, *Enteromorpha* and several other algae settle, grow, and become abundant, and *Enteromorpha* ultimately dominates; whereas naturally, snails feed on the young stages of many ephemeral algae, including *Enteromorpha*. Conversely, *adding* snails to *Enteromorpha* pools leads, in a year, to a decline in the percentage cover of *Enteromorpha* from almost 100% to less than 5%. Clearly, snails are responsible for the dominance of *Chondrus* in *Chondrus* pools.

The natural composition of rock pools varies from almost pure stands of *Enteromorpha* to almost pure stands of *Chondrus*, and it seems that grazing by these periwinkle snails is responsible (Figure 7.24a).

selective predation on a rocky shore

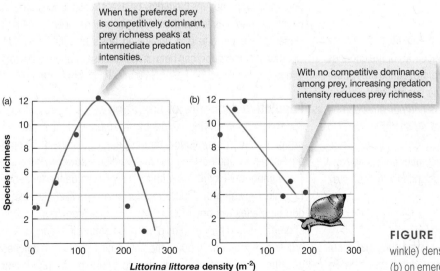

FIGURE 7.24 The effect of *Littorina littorea* (periwinkle) density on species richness (a) in tide pools and (b) on emergent substrata. ((a,b) After Lubchenco, 1978.)

When periwinkles were absent or rare, *Enteromorpha* appeared to competitively exclude other species and the number of algal species was low. At very high densities of periwinkles, all palatable algal species were consumed to extinction, leaving almost pure stands of *Chondrus*. Again, therefore, it was at intermediate predation intensities that the abundance of the preferred *Enteromorpha* (and other ephemeral algal species) was reduced, competitive exclusion was prevented, and many species, both palatable and unpalatable, coexisted.

The picture is quite different, though, when the preferred prey species is not competitively dominant. Here, increased predation pressure should simply reduce the number of prey species in the community. This can also be illustrated on the rocky shores of New England, where *Enteromorpha* loses its competitive dominance when it interacts with other plants on emergent substrata rather than in tide pools. Any increase in the predation pressure, therefore, simply decreases the algal diversity, as the preferred, ephemeral species like *Enteromorpha* are consumed totally and prevented from reestablishing themselves (Figure 7.24b).

Overall, then, predation can have an important role in developing our understanding of the structure of ecological communities, not least in reminding us that the patterns we saw in Chapter 6, when we were focusing on interspecific competition, may never get a chance to express themselves because communities in the real world rarely proceed smoothly to an equilibrium state.

SUMMARY

Predation, true predators, grazers, and parasites

A predator may be defined as any organism that consumes all or part of another living organism (its 'prey' or 'host') thereby benefiting itself, but, under at least some circumstances, reducing the growth, fecundity, or survival of the prey. 'True' predators invariably kill their prey and do so more or less immediately after attacking them, and consume several or many prey items in the course of their life. Grazers also attack several or many prey items in the course of their life but consume only part of each prey item and do not usually kill their prey. Parasites also consume only part of each host and also do not usually kill their host, especially in the short term, but attack one or very few hosts in the course of their life.

Prey fitness and abundance

The fundamental similarity between true predators, grazers, and parasites is that each reduces either the fecundity or the chances of survival of individual prey and may therefore reduce prey abundance. The effects of true predators on the survival of individual prey are obvious. But the effects of grazers and parasites can be equally profound, if more subtle. It is not so straightforward to demonstrate that reductions in the survival or fecundity of individual prey translate into reductions in prey abundance. All sorts of predators can cause reductions in the abundance of their prey, but they do not *necessarily* do so.

The subtleties of predation

Grazers and parasites, in particular, often exert their harm not by killing their prey immediately like true predators, but by making the prey more vulnerable to some other form of mortality.

The effects of grazers and parasites on the organisms they attack are often *less* profound than they first seem because individual plants or host can compensate for these attacks. The effects of predation on a population of prey are complex to predict, because the surviving prey may experience reduced competition for a limiting resource, or produce more offspring, or other predators may take fewer of the prey.

Predator behavior

Many predators and grazers typically forage, moving around within their habitat in search of their prey. Other predators sit and wait for their prey, though almost always in a selected location. With parasites and pathogens there may be direct transmission between infectious and uninfected hosts, or contact between free-living stages of the parasite and uninfected hosts may be important. Optimal foraging theory can help us make sense of where predators forage and how long they spend in particular patches of habitat.

Population dynamics of predation

There is an underlying tendency for predators and prey to exhibit cycles in abundance, and cycles are observed in some predator–prey and host–parasite interactions. However, there are many important factors that can modify or override the tendency to cycle. Crowding of either predator or prey is likely to have a damping effect on any predator–prey cycles. Many populations of predators and prey exist as a metapopulation. In theory, and in practice, asynchrony in population dynamics in different patches and the process of dispersal tend to dampen any underlying population cycles.

Predation and community structure

When predation promotes the coexistence of species among which there would otherwise be competitive exclusion, this is known as 'predator-mediated coexistence.' The effects of predation generally on a group of competing species depend on which species suffers most. If it is a subordinate species, then this may be driven to extinction and the total number of species in the community will decline. If it is the competitive dominants that suffer most, however, the results of heavy predation will usually be to free space and resources for other species, and species numbers may then increase. It is not unusual for the number of species in a community to be greatest at intermediate levels of predation.

REVIEW QUESTIONS

1 With the aid of examples, explain the feeding characteristics of true predators, grazers, parasites, and parasitoids.

2 Discuss the various ways that plants may 'compensate' for the effects of herbivory.

3 Predation is 'bad' for the prey that get eaten. Explain why it may be good for those that do not get eaten.

4 In simple terms, explain why there is an underlying tendency for populations of predators and prey to cycle.

5 Define *mutual interference* and give examples for true predators and parasites. Explain how mutual interference may dampen inherent population cycles.

6 Discuss the evidence presented in this chapter that suggests environmental patchiness has an important influence on predator–prey population dynamics.

7 With the help of an example, explain why most prey species may be found in communities subject to an intermediate intensity of predation.

Challenge Questions

1 True predators, grazers, and parasites can alter the outcome of competitive interactions that involve their 'prey' populations: discuss this assertion using one example from each category.

2 Discuss the pros and cons, in energetic terms, of (i) being a generalist as opposed to a specialist predator, and (ii) being a sit-and-wait predator as opposed to an active forager.

3 You have data that show cycles in nature among interacting populations of a true predator, a grazer and a plant. Describe an experimental protocol to determine whether this is a grazer–plant cycle or a predator–grazer cycle.

Chapter 8

Molecular and evolutionary ecology

CHAPTER CONTENTS

KEY CONCEPTS

After reading this chapter you will be able to:

- explain how molecular markers help identify the extent of subdivision within, and the degree of separation between, species

- describe the coevolutionary "arms race" between plants and their insect herbivores, and between parasites and their hosts

- explain how a range of mutualistic interactions affect both the species concerned and almost all communities on the planet

We have noted earlier that nothing in ecology makes sense, except in the light of evolution. But some areas of ecology are even more evolutionary than others. We may need to look within individuals to examine the details of the genes they carry, or to acknowledge explicitly the crucial and reciprocal role that species play in one another's evolution.

8.1 MOLECULAR ECOLOGY: DIFFERENTIATION WITHIN AND BETWEEN SPECIES

In Chapter 2, we set the scene for the remainder of this book by illustrating how to modify slightly Dobzhansky's famous phrase, 'nothing in ecology makes sense, except in the light of evolution.' But evolution does more than underpin ecology (and the whole of the rest of biology). There are many areas in ecology where evolutionary adaptation by natural selection takes center stage to the extent that the term 'evolutionary ecology' is often used to describe them.

In several earlier chapters, therefore, we dealt with topics in evolutionary ecology as integral parts of broader ecological questions. Of course, this has not been an exhaustive survey of topics in evolutionary ecology. In the present chapter, therefore, we deal with a number of others, though the final list will still be far from exhaustive. We focus especially on pairs of species that act as reciprocal driving forces in one another's evolution. In particular, in Section 8.3, we take up the coevolutionary relationship between predators and their prey, with a particular emphasis on insect–plant and host–pathogen interactions. Here, each adaptation in the prey that fends off or avoids the attacks of a predator provokes a corresponding adaptation in the predator that improves its ability to overcome those defenses, which in turn provokes a response from the prey, and so on.

However, not all coevolutionary interactions are antagonistic. In many interactions between species-pairs, such as in pollination, farming, and **nitrogen fixation**—converting nitrogen into molecules useful for their form and function—for example, both parties benefit, on balance at least, from the interactions in which they take part, as we shall see in Section 8.4.

We begin, though, not with species interactions, but with aspects of evolutionary differentiation within and between species, especially those detectable by modern techniques developed in molecular genetics, which are thus often described as aspects of molecular ecology. We shall see the value of these techniques again when we turn to species interactions in Sections 8.3 and 8.4.

Often, it is entirely appropriate for ecologists to talk about populations or species as if they were singular, homogeneous entities. For example, we may talk about the distribution of Asian elephants, saying nothing about whether the species might be genetically differentiated into distinct **races**, also called **clades**, where relatedness is much greater within these groups than between them, suggesting some evolutionary split between them in the past. But for some purposes, being able to determine who is most closely related to whom (and who is quite distinct from whom) may be critical for understanding whether a species is stable, or declining, or even increasing in abundance, and what might account for that. Is a particular population derived largely from offspring born

the need to know who is most closely related to whom

Helen E. Grose/Shutterstock

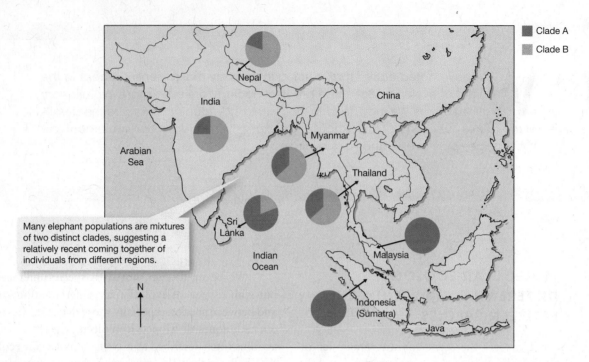

FIGURE 8.1 Distribution of two distinct clades of the Asian elephant, *Elephas maximus* (groups with distinct evolutionary histories following their common origin), revealed only by an analysis of molecular markers (mitochondrial DNA, see Box 8.1). (After Fleischer et al., 2001.)

locally, or from immigrants from a different population? Where exactly does the distribution of one species end and that of another, closely related species begin? In fact, in the case of the Asian elephant there *are* distinct clades, and many populations are mixtures of the two (Figure 8.1). This in turn hints at an important human influence in the past, in bringing together individuals from quite distinct regions.

Our ability to measure levels of relatedness depends on the resolution with which we can differentiate individuals from one another and even determine where they came from or who their parents were. In the past, this was difficult and frequently impossible: reliance on simple, visual markers meant that all individuals within a species often looked the same, and even

members of closely related species could often only be distinguished by experienced taxonomists looking down a microscope at, say, details of a male's genitalia. Now, molecular genetic markers—though they still require experts, and equipment that is expensive (though getting cheaper all the time)—have massively increased the resolution at which we can differentiate between populations and even between individuals, and hence, they have vastly improved our ability to address these types of questions. We begin, therefore, in Box 8.1, with a short survey of some of the most important of these molecular markers and their uses, so readers wanting a brief explanation of some of the terms mentioned in the main text have somewhere convenient to turn.

8.1 Quantitative Aspects

Molecular markers

This is not the place for crash courses in either molecular biology or the laboratory methods used to extract and analyze molecular markers, but it will nonetheless be useful to have some appreciation of their nature and key properties.

Choosing a molecular marker

The basis for all uses of molecular markers in ecology is that we can differentiate one individual from another because of molecular variation among them. Most recent studies (certainly of animals and plants) have used DNA of one type or other for molecular identification. Each portion of the DNA molecule is characterized by the sequence of nucleotide bases of which it is composed—adenine (A), cytosine (C), guanine (G), and thymine (T).

The ultimate source of this variation is mutation in the sequence of bases. What happens to the mutation, and the mutated organism, then depends on the balance between natural selection and **genetic drift**, which is random, undirected changes in gene frequency from generation to generation. If the mutation occurs in a region of DNA that is important because, say, it codes for a crucial part of an essential enzyme, then selection is likely to determine the outcome. An unfavorable mutation (the vast majority in important regions of DNA) will quickly be lost, because the mutated organism is less fit than its counterparts. Individuals will therefore differ relatively little in such regions. If there *is* differentiation in such a region, it is most likely to reflect **adaptive variation**, which occurs when different variants are favored by natural selection in different individuals, perhaps because of where they live.

But there are also regions of DNA that appear not to code for important parts of enzymes or to perform any other function. Here the precise sequence is far less crucial. Variation in these regions is therefore said to be 'neutral,' and mutations can accumulate there over time. Imagine two offspring of a single mating. They will be genetically very similar. But imagine now that each migrates to a different location and starts a lineage of offspring generations there. As each generation passes, in those regions of their genome where variation is neutral, the lineages will become increasingly divergent as mutations accumulate. A snapshot taken in the future should therefore allow us to determine how long the period of divergence has been: how long ago it was that the two groups shared a common ancestor.

However, our ability to do this will depend on the rate of mutation in the DNA region concerned, and indeed, mutation rates in different regions do differ. If the rates are too slow, individuals will tend all to look the same. But if they are too fast, each individual will tend to be so unique that its relationships to others will be hard to discern. Molecular markers should therefore be chosen such that the mutation rate matches the question being addressed, as we see below.

Polymerase chain reaction (PCR)

As a practical point, most studies in molecular ecology, having extracted the DNA from the organism concerned, use the **polymerase chain reaction** (PCR). Very simply, PCR requires the identification of informative target regions of DNA and also flanking regions, either side, that can be used to identify the target region. It then uses molecular primers that match the flanking regions, pick up the target region, and repeatedly copy that target region. Thus, an originally small amount of target DNA in the midst of other, unwanted sequences becomes large enough to be subjected to analysis. Being able to make use of small samples, this technique has revolutionized our ability to sample individuals noninvasively, using small amounts of blood, hair, feces, wing clips, and so on.

Nuclear and mitochondrial DNA

Nuclear DNA is usually inherited equally from both parents and codes for most of an organism's functions, but in the past especially, many studies have instead used the relatively small lengths of mitochondrial DNA (mtDNA) found in the mitochondria of each cell. A major advantage of mtDNA is that lineages can be more clearly traced from generation to generation, because it is almost always inherited only from the mother, who contributes the cytoplasm to the fused egg, and it is not subject to the recombination of genetic material during the process of meiosis. Also, the mutation rate is higher than for coding regions of nuclear DNA, allowing finer resolution differentiation. On the other hand, mtDNA offers only a small number of targets, and its maternal inheritance makes it worthless for following matings since offspring can only be linked to their mothers. Increasingly, therefore, studies focus on regions of nuclear DNA, though often in parallel with analyses of mtDNA genes, combining the advantages of both.

Microsatellites

Within the nuclear genome, sequences coding for proteins (genes) are by no means the only regions molecular biologists have used. **Microsatellites**, for example, are regions of DNA in which the same two, three, or four bases are repeated many times (Figure 8.2a). The variability among individuals, which may be considerable within a population, comes from the precise number of repeats, and we can differentiate the resulting lengths of microsatellite DNA by the speed at which they move through a semisolid medium (a gel) under the influence of an electric current, in electrophoresis. Thus, an appropriately chosen panel of microsatellites for a species may effectively allow each individual in a population to be uniquely identified via a DNA fingerprint, making microsatellites especially appropriate at the finer scales of differentiation.

Sequencing

As far as nuclear or mitochondrial genes are concerned, having chosen, extracted, and amplified the target region from a sample of individuals, we must have some basis for differentiating individuals from one another, determining who is most similar to whom, and so on. Increasingly, the whole sequences of genes are being determined. Initially, the method of gene sequencing developed in the 1970s (and often referred to as Sanger sequencing after its inventor, Nobel Prize winner Frederick Sanger) was carried out manually and thus was slow. This method later gave way to faster, automated methods, but these in their turn are now being replaced by a range of alternative sequencing methods collectively known as **next-generation sequencing**. The speed and lower costs of these methods are making the sequencing not just of genes but of whole genomes increasingly possible.

We have seen that regions of the same gene differ in terms of their functional importance (Figure 8.2b). Some regions are conserved from individual to individual, from population to population, and often from species to species. These are (or are presumed to be) the regions of greatest functional importance, and they play effectively no part in differentiation. But there are other regions where far more variation is observed, and that we presume, therefore, to be neutral or at least subject to weaker selective constraints. It is on the basis of this variation that individuals and populations can be differentiated.

SNPs

With the wider availability of gene sequencing, differentiation based on **single nucleotide polymorphisms** or SNPs (pronounced 'snips') is increasingly possible. A **polymorphism** is the occurrence together within a single population of two or more genetically distinct forms or **morphs**. The melanic and typical forms of the peppered moth, described in Section 2.3, are a good example. A SNP is then simply the occurrence

(a)

Allele 1 which has 10 repeats
```
...GCATTGCGATAACGTGTGTGTGTGTGTGTGTGCCATGCCGGATGA...
...CGTAACGCTATTGCACACACACACACACACACGGTACGGCCTACT...
```
    Flanking region   Microsatellite   Flanking region

Allele 2 which has 8 repeats
```
...GCATTGCGATAACGTGTGTGTGTGTGTGCCATGCCGGATGA...
...CGTAACGCTATTGCACACACACACACACACGGTACGGCCTACT...
```

Note the contrast between the conserved (unvarying) regions in black and a variable region in red.

(b)
```
Individual 1..CGTAACGCTATTGCGCATTGTGATAACACCATGCCGGATGA..
Individual 2..CGTAACGCTATTGCGCCATCCGATCATATCATGCCGGATGA..
Individual 3..CGTAACGCTATTGCGCCTAGTCCTAGTGCCATGCCGGATGA..
Individual 4..CGTAACGCTATTGCGCCTAGCGAGAAAGTCATGCCGGATGA..
Individual 5..CGTAACGCTATTGCGCCTTACGATAACGTCATGCCGGATGA..
```

FIGURE 8.2 The term 'locus' refers to the location of a region in the overall DNA sequence. The term 'allele' refers to a particular variant of the sequence at a particular locus. In animals and plants, that sequence is of two paired strands of DNA. Between the two strands, the bases guanine (G), cytosine (C), adenine (A), and thymine (T) are themselves always paired: G with C, and A with T. (a) Two contrasting alleles at a microsatellite locus. As a microsatellite, its sequence is one of repeated bases, but the lengths of the repeats differ in the two alleles (red), whereas the flanking regions at either end (black) are exactly the same. (b) Here, by contrast, the base sequences are from just one of the DNA strands of a hypothetical gene (a sequence of DNA coding for a protein) from five different individuals, each with a different allele. Differentiation between individuals clearly depends on the variable region towards the center.

together of individuals differing from one another at a single base pair position in the DNA sequence. A single SNP, of course, can only divide a collection of individuals into a very limited number of groups. But SNPs at a larger number of loci (Figure 8.2) allow individuals to be grouped together on the basis of varying degrees of relatedness. Some may be identical at the loci examined. The individuals most closely related to this group would be those that differ from them at a single SNP, and so on. A variety of statistical methods exist for creating relatedness trees on this basis, which in turn suggest the paths by which their gene sequences have diverged in the past over evolutionary time.

DNA barcoding

In cases where we know how much differentiation there is within and between a group of species, we may wish to use that information in order to rapidly and efficiently determine the species identity of any newly acquired samples. DNA or genetic barcoding is an approach to doing this, based on one or just a few DNA sequences. It relies on there being a 'barcoding gap,' such that interspecific divergences are clearly greater than divergences between individuals within a species.

Differentiation within species: albatrosses

Albatrosses are wide-ranging sea birds with the largest wingspans of any birds alive today. Sadly, though, of the 21 species normally recognized, 19 are regarded as threatened with extinction and the other two as near threatened. Taxonomists have recently split the black-browed albatross into two species: *Thalassarche impavida*, found only on Campbell Island, between New Zealand and Antarctica, and *T. melanophris*, with breeding populations elsewhere in the sub-Antarctic, including the Falkland Islands, South Georgia and Chile (Figure 8.3a). The similarly sized gray-headed albatross, *T. chrysostoma*, also breeds on a number of sub-Antarctic islands, including South Georgia. The black-browed species usually remain associated with coastal shelf systems, whereas gray-headed albatrosses feed much more often far from land, but both, like all albatross species, are thought to exhibit **natal philopatry**; that is, they return very close to their place of birth to breed. Numbers at all sites are declining each year. Hence, the questions arise: How connected or separate

are these populations? Should conservation efforts be directed at what we currently perceive to be whole species, or at particular breeding populations?

These questions were addressed for both species by a study that used both mtDNA sequences and a panel of seven microsatellites (Burg & Croxall, 2001; see Box 8.1). The results were clearest for mtDNA (Figure 8.3b, c), but those for the microsatellites told the same story. For the black-browed species (Figure 8.3b), the molecular data confirmed the taxonomists' view that *T. impavida* was a separate species. The data also demonstrated breeding between this species and *T. melanophris* on Campbell Island, and indeed the production of hybrids between the two species there. More surprisingly, it seems that the Falkland Islands support a breeding population of *T. melanophris* that is quite separate from a population shared by Diego Ramirez (Chile), South Georgia, and Kerguelen Island, which is indivisible in spite of the natal philopatry to these three sites. By contrast, the wider ranging gray-headed albatrosses, from all five of their sites, seemed to

The Campbell black-browed albatross

A black-browed albatross from the Falklands

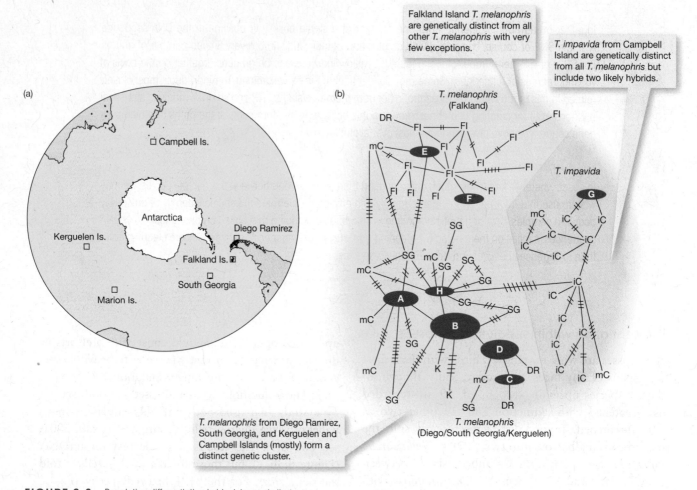

Falkland Island *T. melanophris* are genetically distinct from all other *T. melanophris* with very few exceptions.

T. impavida from Campbell Island are genetically distinct from all *T. melanophris* but include two likely hybrids.

T. melanophris from Diego Ramirez, South Georgia, and Kerguelen and Campbell Islands (mostly) form a distinct genetic cluster.

FIGURE 8.3 Population differentiation in black-browed albatrosses, *Thalassarche melanophris* and *T. impavida*, and the gray-headed albatross, *T. chrysostoma*. (a) Distribution of sites in the sub-Antarctic from which samples were taken. (b) The relationships among the sequences of bases of 73 black-browed albatrosses at a variable region in their mtDNA. The origin of each sample, with species identified morphologically, is as follows. *T. melanophris*: SG, South Georgia; DR, Diego Ramirez (Chile); FI, Falkland Islands; K, Kerguelen Island; mC, Campbell Island. *T. impavida*: iC, Campbell Island. Groups of individuals sharing exactly the same sequence have been assigned a letter code (A, B, etc.) and placed in an oval proportional in size to the number of individuals. The cross-hatches on lines represent the number of base differences between the individuals (or groups) they join. The samples fall into three 'clusters,' though the clustering is not perfect. Some of the *T. melanophris* found on Campbell Island (identified morphologically) had *T. impavida* mtDNA and hence were likely *T. melanophris–T. impavida* hybrids. (c) The relationships among 50 gray-headed albatrosses in the base sequence at a variable region in their mtDNA. Coding is the same as in (b) except that M is Marion Island and C is Campbell Island. (After Burg & Croxall, 2001.)

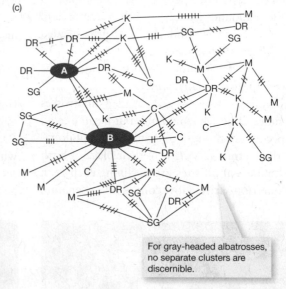

For gray-headed albatrosses, no separate clusters are discernible.

represent a single breeding population (Figure 8.3c)— again in spite of their natal philopatry.

From a conservation point of view, therefore, the most important conclusion relates to the black-browed species. Previously, the relative stability of the large Falkland Islands population was taken as insurance against any real vulnerability to extinction of the species generally. Now, however, in the light of these molecular data, we must consider the Falkland Islands population as somewhat separate from the rest of the species, which itself is far more seriously threatened with extinction than we thought.

A more active and immediate role for molecular markers in practical matters of conservation is described in Box 8.2.

Differentiation between species: the red wolf—species or hybrid?

Issues in conservation surface again when we shift our focus from differentiation within to differentiation between species. The red wolf, *Canis rufus*, once had a widespread distribution in the southeastern United

---

8.2   ECOncerns

The forensic analysis of the origins of the fish we eat

As we shall discuss more fully in Chapter 14, there is an increasingly frequent conflict between exploiting natural populations, for example fish, as a necessary source of food, and conserving those same populations, both as an end in itself and so that future generations will have something to eat. Getting the balance right is not easy. But even with a policy in place, there may still be difficulties in implementing that policy if, somewhere between the sea and the plate, the fish are mislabeled.

Dana Miller and Stefano Mariani (2010) looked at fish for sale in Dublin, Ireland. In their words, "Seafood labels are industry management tools that serve to convey important product information to consumers. They also promote the improvement of industry organization by requiring that valid information is transmitted throughout the chain of production. Provided they are adequately enforced, policies governing seafood labeling . . . should theoretically assist in the monitoring of industry operation, encourage high food-safety standards, help prevent fraud for economic gain, facilitate conservation efforts, and help avoid illegal, unreported, and unregulated fishing products from reaching consumers."

The researchers went out and collected their material (see Figure 8.4a): "We obtained samples labeled as either 'cod' or 'haddock' from randomly chosen retail outlets within 10 postal districts in Dublin. Within each district, we sampled six retailers: two 'fish-and-chips' shops, two fishmongers, and two supermarkets. We collected four samples (two frozen, two packaged fresh) from each supermarket and two samples from each fish-and-chips shop and fishmonger." Each sample was then DNA barcoded (Box 8.1) using the cytochrome c oxidase subunit 1 gene from the mitochondrial DNA. The results are summarized in Figure 8.4b.

ARCTIC IMAGES/Alamy

FIGURE 8.4   (a) The real thing: Atlantic cod, *Gadus morhua*, in a fishmonger's. (b) The percentages of tested samples that were mislabeled. (After Miller & Mariani, 2010.)

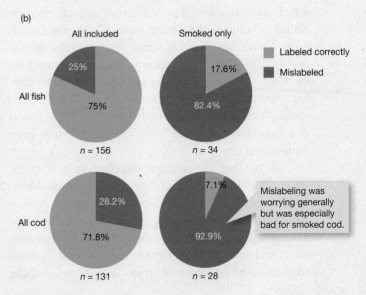

Mislabeling was worrying generally but was especially bad for smoked cod.

Mislabeling was clearly widespread, and as the authors point out: "The levels of mislabeling revealed in this study raise concerns about the functioning of the seafood industry in Europe. The gravity of this problem is amplified if we acknowledge that cod is both a historically and currently popular seafood . . . The high levels of cod mislabeling found in Ireland could be . . . creating a false perception of market availability, and leading consumers to believe that because cod is so widely available, the stocks must be healthy."

Perhaps unsurprisingly, they conclude: "The application of tools for the accurate and cost-effective genetic identification of seafood at all stages in the chain of production should be promoted. Here, the mitochondrial DNA barcode region was successfully amplified, through standard protocols . . . in all samples tested, including many that had been battered, breaded, smoked, deep-fried, and degraded to varying extents. Rapid advances in molecular genetics and bioinformatics will make this, and even more sophisticated methodologies, more available and affordable in the near future."

How can fraudulent mislabeling harm conservation efforts? How does the seriousness of this crime compare to street robbery or the possession of illegal drugs? Should those convicted be punished in proportion to the economic harm they may be doing to these particular fisheries, or should their fines be higher, to send a signal those who ignore the need to restrain activity in vulnerable populations and to conserve them for future generations?

States (Figure 8.5a), but when, by the mid-1970s, that distribution had shrunk to a single population in eastern Texas, the U.S. Fish and Wildlife Service instituted an emergency program to save the wolf from extinction. Fourteen individuals were rescued from its final refuge and bred in captivity with a view to subsequent reintroduction in the wild.

In the United States as a whole, the red wolf coexists with two other, closely related species, the gray wolf, *C. lupus* (once common throughout much of the United States but now confined to the far north and to Canada), and the coyote, *C. latrans*, which has expanded its distribution and can now be found in much of the United States. Traditional analyses, based on morphological features, placed the red wolf as a genuine, separate species, intermediate in many ways between the gray wolf and the coyote (Nowak, 1979). However, as we shall see below, molecular markers suggest strongly that the red wolf is a hybrid arising from interbreeding between gray wolves and coyotes.

A number of questions therefore suggest themselves (Wayne, 1996), including: Should the conservation status of the red wolf, and the amount of money spent on its conservation, be downgraded if it is acknowledged that it is a hybrid and not a full species? And will attempts to save the red wolf by reintroduction be undermined, in any case, because of the movement of genes from gray wolves or coyotes into the red wolf gene pool as a result of interbreeding? In other words, will 'red wolf-ness' be diluted as fast as it is introduced?

The red wolf question has been examined in the past using a variety of techniques and DNA sources, including mtDNA and microsatellites in the nuclear DNA (see Box 8.1). More recently, however, scientists have used the production of commercial SNP arrays (see also Box 8.1) for genotyping individuals of domesticated species. So in this case, given the high degree of relatedness between dogs and the wolves and coyotes, a Canine SNP Genome Mapping Array was directed towards the question.

Dogs themselves are genetically quite separate from all the nondomesticated species (Figure 8.5b); coyotes are also clearly distinct from the wolves; and the wolves of Eurasia are not much less distinct, in their turn, from those of North America. It is also clear, however, that red wolves are much more closely related to coyotes than they are to gray wolves in the United States.

In fact, attribution of the various portions of the red wolf genome to its most likely ancestors (Figure 8.5c) indicates that red wolves are the result of interbreeding between coyotes and gray wolves that began around 114 generations (290–430 years) ago, when the southeastern United States was being

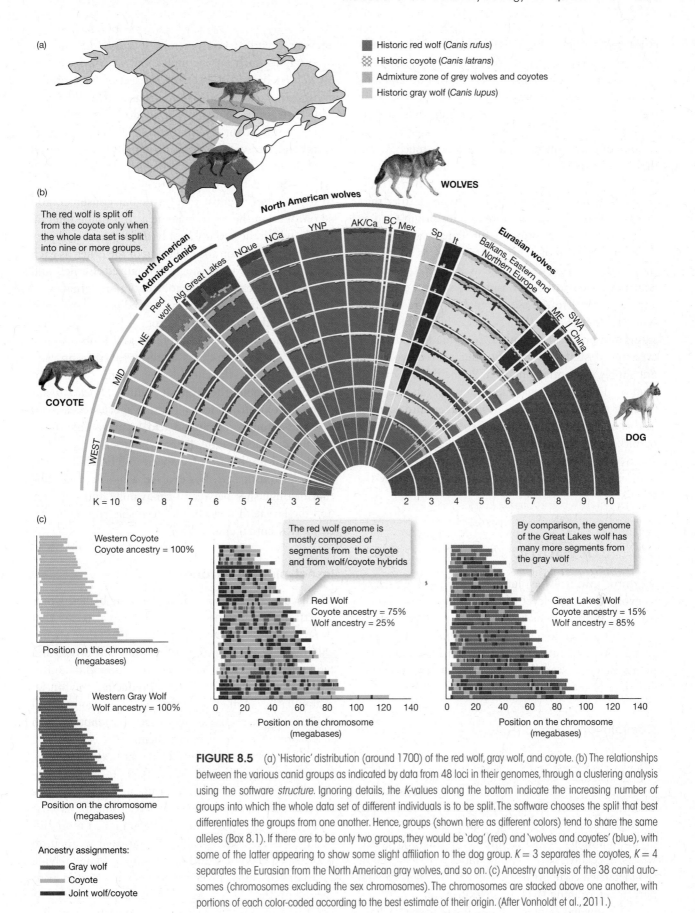

FIGURE 8.5 (a) 'Historic' distribution (around 1700) of the red wolf, gray wolf, and coyote. (b) The relationships between the various canid groups as indicated by data from 48 loci in their genomes, through a clustering analysis using the software *structure*. Ignoring details, the *K*-values along the bottom indicate the increasing number of groups into which the whole data set of different individuals is to be split. The software chooses the split that best differentiates the groups from one another. Hence, groups (shown here as different colors) tend to share the same alleles (Box 8.1). If there are to be only two groups, they would be 'dog' (red) and 'wolves and coyotes' (blue), with some of the latter appearing to show some slight affiliation to the dog group. *K* = 3 separates the coyotes, *K* = 4 separates the Eurasian from the North American gray wolves, and so on. (c) Ancestry analysis of the 38 canid autosomes (chromosomes excluding the sex chromosomes). The chromosomes are stacked above one another, with portions of each color-coded according to the best estimate of their origin. (After Vonholdt et al., 2011.)

converted to agriculture, followed by further interbreeding between these and coyotes as the gray wolves retreated northwestwards. Interestingly, interbreeding between coyotes and gray wolves is still going on in the Great Lakes region (Figure 8.5 a,b,c), giving rise to separate mixed-ancestry populations, though the proportion of gray wolf genes in these is very much higher than for the red wolf.

In answer to our original questions, then, (i) the red wolf seems, ultimately, to be a hybrid rather than a separate species with a more ancient origin (and so arguably less of a priority for conservation), and (ii) if any program of reintroduction has, as a key aim, the avoidance of gene exchange with coyotes, then that goal may need to be reconsidered, since the process appears central to the genetic makeup of red wolves (von Holdt et al., 2011). However, whether biological status and practical difficulties combine to undermine even the desirability of reintroducing red wolves is not simply a scientific question. Public perception and opinion (in this case regarding the conservation importance of the red wolf) must also be taken into account in this and most conservation issues, especially when public funds are involved. A molecular ecology perspective has been immensely informative—but information can sometimes muddy rather than clarify the waters.

8.2   COEVOLUTIONARY ARMS RACES

We turn now from evolution at the molecular level to evolution at the level of species interactions, starting with those in which species are in opposition to one another. Following some general background, we turn first to interactions between insects and the plants they eat, and then to those between parasites and their hosts.

Coevolution

We can link the dynamics of consumer-resource pairs, described in Chapter 7, to the dynamics of whole webs of interacting species, which we will describe in Chapter 9, by noting how specialized or generalized particular consumers are. Generalists draw the species of a community together into large interactive networks. Specialists divide communities into detached or semidetached compartments. Coevolution plays a vital part in determining levels of specialization.

We saw in Chapter 3 that many prey species have evolved defenses that either reduce the chances of an encounter with a consumer or increase the chances of surviving such an encounter. But of course, a better defended prey itself exerts a selection pressure on consumers to overcome that defense. If a consumer evolves features that allow it to do so, it will have stolen a march on its competitors. But diverting its metabolism towards overcoming the prey's defenses is likely to make it more of a specialist on that prey type—which is then under particular selective pressure to defend itself against that particular consumer, and so on. We can therefore imagine a continuing interaction in which the evolution of both consumer and prey depends crucially on the evolution of the other, as each tends to become increasingly specialized. Ehrlich and Raven (1964) called this a coevolutionary 'arms race.' In its most extreme form, a coadapted pair of species is locked together in perpetual struggle.

In fact, what is unacceptable as a food resource to most animals may, as a result of this process, evolve to become the chosen, even unique diet of others. For example, the tropical legume *Dioclea metacarpa* is toxic to almost all insect species because it contains a nonprotein amino acid, l-canavanine, which the insects incorporate (lethally) into their proteins in place of arginine. But a species of bruchid beetle, *Caryedes brasiliensis*, has evolved a modified enzyme that distinguishes between l-canavanine and arginine, and the larvae of these beetles feed solely on *D. metacarpa* (Rosenthal et al., 1982).

Insect–plant arms races

We discussed in Section 3.3 how attacks by herbivores select for plant-defensive chemicals. We also saw that these chemicals can be divided into qualitative chemicals that are poisonous, can kill in small doses, and tend to be induced by herbivore attacks, and quantitative chemicals that are digestion-reducing, rely on an accumulation of ill effects, and tend to be produced all the time. These chemicals will select for adaptations in herbivores that can overcome them.

It seems probable, however, that toxic chemicals, by virtue of their specificity, are likely to be the foundation of an arms race, requiring an equally specific response from a herbivore, whereas chemicals that make plants generally indigestible are much more difficult to overcome through any targeted adaptation (Cornell & Hawkins, 2003). Put simply, plants relying on toxins are more prone to become involved in arms races with their herbivores (like the beetle and legume described above) than those relying on more quantitative chemicals.

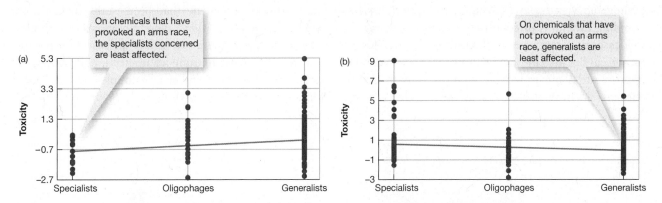

FIGURE 8.6 Following a survey of 892 insect–plant combinations, insect herbivores were split into three groups: specialists (feeding from one or two plant families), 'oligophages' (3–9 families), and generalists (more than nine families). Chemicals were split into two groups: those that are found in the normal hosts of the specialists and oligophages, and those that are not. Toxicity was measured from the mortality rates of insects on a standardized scale. (a) On chemicals found in the normal hosts of the specialists, the specialists suffered lower mortality. (b) On chemicals *not* found in the normal hosts of the specialists, the generalists suffered lower mortality. $P < 0.005$ in both cases. (After Cornell & Hawkins, 2003.)

Is it the case, then, that specialist herbivores, locked in their coevolutionary arms races, tend to perform better when faced with their plants' toxic chemicals than do generalists? And conversely, do generalists, having invested in overcoming a wide range of chemicals, perform better than specialists when faced with chemicals that have not provoked coevolutionary responses? An analysis of a wide range of data sets for insect herbivores fed on artificial diets with added chemicals suggests that this is indeed the case, in spite of a great deal of variation around these general trends (Figure 8.6).

specialists are more prone to arms races

Part of the reason for the large amount of scatter around the relationships in Figure 8.6 may of course be classification errors. For example, a series of specialist species, each feeding on different host plants, may be classified as a single generalist species if we are unable to tell them apart. This is certainly likely to have been the case in the past when we had to rely on tiny anatomical distinctions, but the molecular methods described in Section 8.2 are increasingly helping us to overcome these difficulties. Two examples are shown in Figure 8.7.

one generalist or many specialists? molecular ecology to the rescue

In the first, specimens of the skipper butterfly, *Astraptes fulgerator*, from a 110,000-hectare conservation area in Costa Rica were DNA barcoded, using the mitochondrial cytochrome oxidase 1 (mtCO1) gene (Figure 8.7a). First described in 1775, *A. fulgerator* was, until relatively recently, considered a single variable species ranging in its distribution from the southern United States to northern Argentina.

Astraptes fulgerator

However, even within the boundaries of the study area, evidence before the barcoding study from a combination of caterpillar markings, the food plants from which they were collected, and subtle morphological differences among adults had suggested six or seven different species.

The molecular approach, entirely consistent with the earlier data, raised that number to at least 10. Most of these were food specialists and some were extremely specialized, feeding on just one host plant species. Expanding the study to cover wider geographical areas (if that were practical) would no doubt increase both the number of different species and the strength of the impression that *A. fulgerator* is not one generalist species but many much more specialized ones.

Similarly, a second example examined the mtCO1 gene in the whitefly, *Bemisia tabaci* (Figure 8.7b). This is a globally distributed pest that feeds on vegetables, grain, legumes, cotton, and ornamental plants, causing damage both by feeding and by transmitting plant pathogenic viruses, and giving rise to economically crippling invasions in countries around the world. In this case, therefore, species delineation is not only of ecological but also of practical interest; we can choose appropriate plant protection measures only if we can correctly identify the species. It seems (Figure 8.7b)

that although a tendency had been established to consider variants of *B. tabaci* merely biotypes of a single, variable species, there are in fact at least 24 separate species, many with quite characteristic host-species preferences. As the number of examples like this continues to grow, it seems likely that our view of the

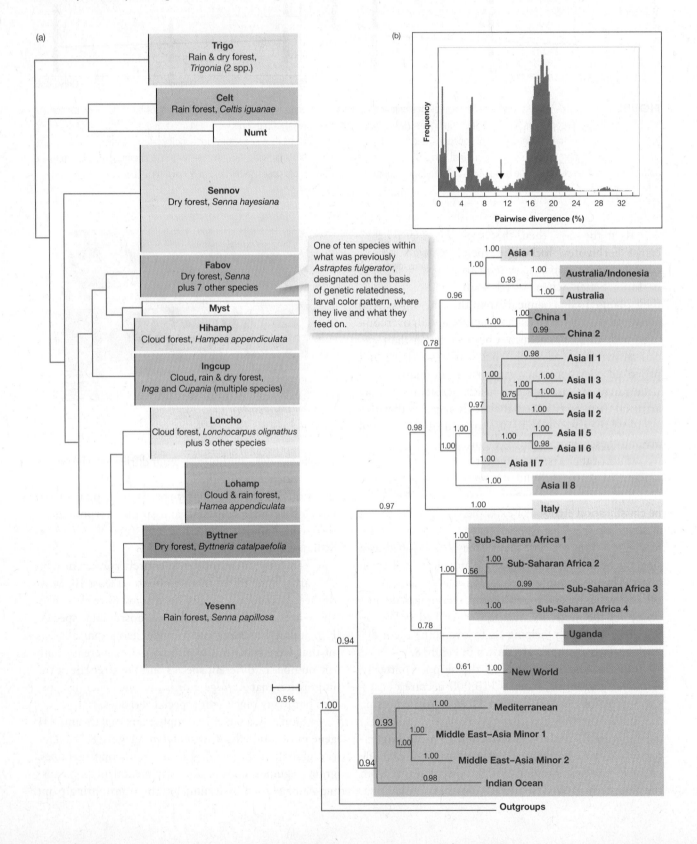

coevolutionary relationships between insects and their host plants will itself continue to evolve.

Coevolution of parasites and their hosts

The intimate association between parasites and their hosts makes them especially prone to coevolutionary arms races. Indeed, the specialization may go further than that between species. Within species, it is common to find a high degree of genetic variation in the virulence of parasites and/or in the resistance or immunity of hosts. Every few years, for example, as we are perhaps more aware than ever, a new strain of the influenza virus evolves of sufficient virulence and novelty to generate a widespread epidemic and mortality in human populations that had been relatively resistant to previously circulating strains. No strain has been more devastating—at the time of writing—than the worldwide epidemic (*pandemic*) of Spanish flu that followed World War I in 1918/19 and killed 20 million people—many more than died in the war itself.

Human diseases can also provide examples of variation in host resistance. When the Native Americans of the Canadian Plains were forcibly settled onto reservations in the 1880s, their death rate due to tuberculosis (TB) initially exploded but then gradually declined (Figure 8.8). Environmental factors (inadequate diet, overcrowding, spiritual demoralization) undoubtedly played some part in these fatalities, but variations in resistance are also likely to have been significant. The mortality rate among the Native Americans was often 20 times that of the surrounding European colonist population, who were living in similar conditions but had already been exposed to TB. Some native families had a particularly low mortality rate in the 1880s epidemic, and many of the survivors in the 1930s were descendants of those families (Ferguson, 1933; Dobson & Carper, 1996).

It may seem straightforward that parasites in a population select for the evolution of more resistant hosts, which in turn select for more infective parasites: a classic arms race. In fact, the process is not necessarily so straightforward, though there are certainly examples where host and parasite drive one another's evolution. A most dramatic and classic example involves the rabbit and the myxoma virus, which causes myxomatosis. The virus originated in the South American jungle rabbit *Sylvilagus brasiliensis*, where it causes a mild disease that only rarely kills the host. The South American virus, however, is usually fatal when it infects the European rabbit *Oryctolagus cuniculus*.

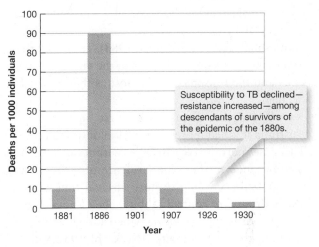

> Susceptibility to TB declined—resistance increased—among descendants of survivors of the epidemic of the 1880s.

FIGURE 8.8   The mortality rate due to tuberculosis in three generations of Canadian Plains Native Americans after their forced settlement onto reservations. (After Ferguson, 1933; Dobson & Carper, 1996.)

myxomatosis

Rabbit with myxomatosis lesion on its eye

FIGURE 8.7   (a) A neighbor-joining tree, linking the mtCO1 gene sequences of 466 individuals (the lengths of vertical lines represent the number of individuals) from the *Astraptes fulgerator* species complex, collected in the Area de Conservación Guanacaste, Costa Rica. Each colored block represents what is likely a separate species, designated on the basis of gene sequence but also larval color pattern, habitat, and sole or main host plant (all indicated). For two groups (not colored), species identity was not certain. In the tree, vertical lines join individuals closely related by their gene sequence. Horizontal lines from individuals to a vertical line represent the degree of difference between each individual and the rest of the group (the length for a 0.5% difference is indicated below). (After Hebert et al., 2004.) (b) Relatedness estimated from the mtCO1 gene sequences of 201 individuals of *Bemisia tabaci* from around the world. The differences between all pairs of individuals were estimated (inset) and 'natural gaps' occurred at differences of 3.5% and 11%. The former was used as a criterion to split the individuals into 24 species; the latter to collect these into 11 groups of species (colored blocks). 'Outgroups' are more distantly related individuals used to 'root' the relatedness tree. (After Dinsdale et al., 2010; de Barro et al., 2011.)

In one of the greatest examples of biological pest control, the myxoma virus was introduced into Australia in the 1950s to control the European rabbit, which had become a pest of grazing lands. The disease spread rapidly in 1950–1951, and rabbit populations were greatly reduced – by more than 90% in some places. At the same time, the virus was introduced to England and France, and there too it resulted in huge reductions in the rabbit populations. The evolutionary changes that then occurred in Australia were followed in detail by Fenner and his associates, who had the brilliant foresight to establish baseline genetic strains of both rabbits and virus (Fenner, 1983). They used these to measure subsequent changes in the virulence of the virus and the resistance of the host as they evolved in the field.

When the disease was first introduced to Australia it killed more than 99% of infected rabbits. This case mortality fell to 90% within one year and then declined further. The virulence of virus isolates was graded according to host survival time and the case mortality of control rabbits. The original, highly virulent virus was grade I, which killed > 99% of infected laboratory rabbits. Already by 1952, most of the virus isolates from the field were the less-virulent grades III and IV. At the same time, the rabbit population in the field was increasing in resistance. When injected with a standard grade III strain, field samples of rabbits in 1950–1951 had a case mortality of nearly 90%. This had declined to less than 30% only 8 years later (Figure 8.9).

This evolution of resistance is easy to understand: resistant rabbits are obviously favored by natural selection in the presence of the myxoma virus. The case of the virus, however, is subtler. The contrast between the virulence of the virus in the European rabbit and its lack of virulence in the South American host with which it had coevolved, combined with the attenuation of its virulence in Australia and Europe after its introduction, fit a commonly held view that parasites evolve toward becoming benign to their hosts in order to prevent eliminating their host and thus their habitat. This view, however, is wrong. The parasites favored by natural selection are those with the greatest fitness (broadly, the greatest reproductive rate). Sometimes this is achieved through a decline in virulence, but sometimes it is not. In the myxoma virus, as we next explain, an initial decline in virulence was indeed favored—but further declines were not.

The myxoma virus is blood-borne and is transmitted from host to host by blood-feeding insect vectors. In the first 20 years after its introduction to Australia, the main vectors were mosquitoes, which

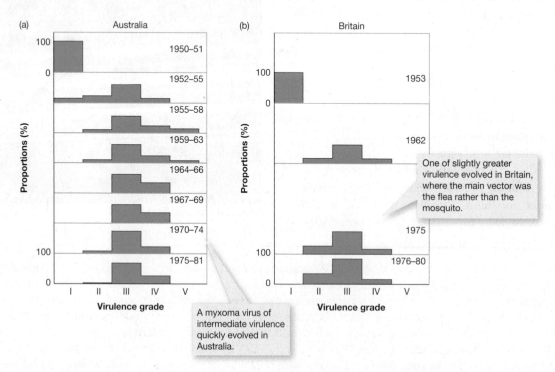

FIGURE 8.9 (a) The percentages in which various grades of myxoma virus have been found in wild populations of rabbits in Australia at different times from 1951 to 1981. Grade I is the most virulent. (After Fenner, 1983.) (b) Similar data for wild populations of rabbits in Great Britain from 1953 to 1980. (From Fenner, 1983; after May & Anderson, 1983.)

feed only on live hosts. The problem for grade I and II viruses is that they kill the host so quickly that there is only a very short time in which the mosquito can transmit them. Effective transmission may be possible at very high host densities, but as soon as densities decline, so does transmission. Hence, following the initial drop in rabbit densities, there was selection against grades I and II and in favor of less virulent grades, giving rise to longer periods of host infectiousness.

At the other end of the virulence scale, however, the mosquitoes are unlikely to transmit grade V of the virus because it produces very few infective particles. Hence the majority were grade III. The situation was complicated in the late 1960s when an alternative vector of the disease, the rabbit flea *Spilopsyllus cuniculi* (the main vector in England), was introduced to Australia, apparently favoring more virulent strains than the mosquitoes had done. Overall, however, there has been selection in the rabbit–myxomatosis system not for decreased virulence as such, but for increased transmissibility and hence increased fitness, which happens in this system to be maximized at intermediate grades of virulence.

In other cases, host–parasite coevolution is more definitely antagonistic; we find increased resistance in the host and increased infectivity in the parasite. A classic example is the interaction between agricultural plants and their pathogens (Burdon, 1987), though in

bacteria and bacteriophage

this case the resistant hosts are often introduced by human intervention. There may even be gene-for-gene matching, with a particular virulence allele in the pathogen selecting for a resistant allele in the host, which in turn selects for alleles other than the original allele in the pathogen, and so on. In fact, these detailed processes have often proved difficult to observe, but this has been done with a laboratory system consisting of the bacterium *Pseudomonas fluorescens* and its viral parasite, the bacteriophage (or phage) SBW25Φ2, where such evolution is relatively easy to observe because generation times are so short.

Changes in both host and parasite were monitored as 18 replicate coexisting populations of bacterium and phage were transferred from culture bottle to culture bottle every two days for 16 transfers (about 120 bacterial generations in all). Crucially, when the proportion of bacteria resistant to the phage was estimated at transfers 2, 4, 6, 8, 10, 12, and 14, the bacteria were tested not only against their contemporary phage. They were also tested against the phage from two transfers previously, against the original phage (retained for this purpose), and against the phage from two transfers in the future (for which purpose the bacteria were also retained). It is apparent that, as the generations passed, the bacteria became generally more resistant to the phage at the same time as the phage became generally more infective to the bacteria; each was being driven by the directional selection of an arms race (Figure 8.10).

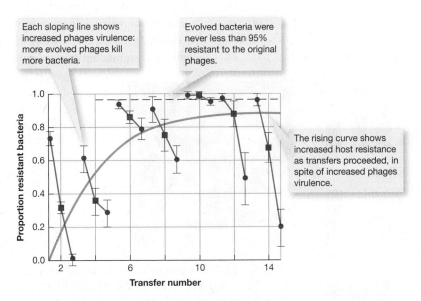

FIGURE 8.10   Laboratory coevolution of the bacterium, *Pseudomonas fluorescens*, and its pathogen, the phage SBW25φ2. Each set of three points shows the proportion of resistant bacteria when tested against contemporary phage (large squares) and also against phage from two transfers earlier and later. Bars are SEs. The horizontal dashed line shows the minimum level of resistance observed against the original phage. The curved line (drawn by eye) shows the trend in increasing bacterial resistance as coevolution proceeded. (After Brockhurst et al., 2003.)

8.3 MUTUALISTIC INTERACTIONS

No species lives in isolation, but often the association with other species is especially close, such as when organisms occupy a habitat that is an individual of another species. Parasites live within the body cavities or even the cells of their hosts, many nitrogen-fixing bacteria live in nodules on the roots of leguminous plants, and so on. **Symbiosis** (meaning 'living together') is the term that has been coined for such close physical associations between species, in which a **symbiont** occupies a habitat provided by a **host**. In fact, though, parasites are usually excluded from the category of symbionts, which is reserved instead for interactions where there is at least the suggestion of **mutualism**. A mutualistic relationship is one in which organisms of different species interact to their mutual benefit. Mutualism, therefore, need not include close physical association: mutualists need not be symbionts. For example, many plants gain dispersal of their seeds by offering a reward to birds or mammals in the form of edible fleshy fruits, and many plants ensure effective pollination by offering a resource of nectar in their flowers to visiting insects. These are mutualistic interactions but they are not symbioses.

It would be wrong, moreover, to see mutualistic interactions simply as conflict-free relationships from which nothing but good things flow | *mutualism: reciprocal exploitation not a cozy partnership* |
for both partners. Rather, current evolutionary thinking views many mutualisms as cases of reciprocal exploitation, in which each partner gains something at the expense of the other, but overall each is a *net* beneficiary (Herre & West, 1997).

Mutualists compose most of the world's biomass. Almost all the plants that dominate grasslands, heaths, and forests have **mycorrhizas**, or roots that have an intimate mutualistic association with fungi. Most corals depend on the unicellular algae within their cells. Many flowering plants need their insect pollinators. Most animals carry communities of microorganisms within their guts that they require for effective digestion.

In the rest of this section we start with mutualisms without intimate symbiosis; rather, the association is largely behavioral. Then we discuss mutualisms between animals and the microbiota living in their guts, where the associations are much closer. Finally we examine still more intimate symbioses in which one partner enters between or within another's cells: mycorrhizas, and the fixation of atmospheric nitrogen in mutualistic plants.

Mutualistic protectors

"Cleaner fish," of which at least 45 species have been recognized, feed | *cleaner and client fish* |
on parasites, bacteria, and necrotic tissue from the body surface of client fish. Indeed, the cleaners often hold territories with cleaning stations that their clients visit—and visit more often when they carry many parasites. The cleaners gain a food source and the clients are protected from infection.

It is not always easy to establish that clients benefit, but experiments off Lizard Island on Australia's Great Barrier Reef were able to do this for the cleaner fish, *Labroides dimidiatus*, which eats parasitic isopods from its client fish, *Hemigymnus melapterus*. Clients had significantly more parasites 12 days after cleaners were excluded from caged enclosures (Figure 8.11a). But even in the short term, up to 1 day, although removing cleaners, which feed only during daylight, had no effect when a check was made at dawn (Figure 8.11b), following a further day's feeding the uncleaned clients had significantly more parasites (Figure 8.11c).

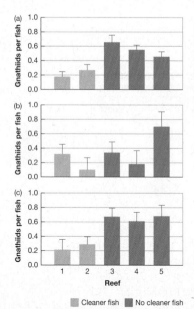

Nature Production/AUSCAPE

FIGURE 8.11 Cleaner fish really do clean their clients. The mean number of isopod parasites per client, *Hemigymnus melapterus*, at five reefs, from three of which cleaners, *Labroides dimidiatus*, were experimentally removed. (a) In a long-term experiment, clients without cleaners had more parasites after 12 days ($F = 17.6$, $P = 0.02$). (b) In a short-term experiment, clients without cleaners did not have significantly more parasites at dawn after 12 hours ($F = 1.8$, $P = 0.21$), presumably because cleaners do not feed at night. (c) However, the difference was significant after a further 12 hours of daylight ($F = 11.6$, $P = 0.04$). Bars are standard errors. (After Grutter, 1999.)

(a)

(b)

© Bob Gibbons/Alamy Limited

Gilbert S. Grant/Photo Researchers, Inc.

FIGURE 8.12   Structures of the Bull's horn acacia (*Acacia cornigera*) that attract its ant mutualist. (a) Hollow thorns used by the ants as nesting sites. (b) Protein-rich Beltian bodies at the tips of the leaflets.

The idea that there are mutualistic, protective relationships between plants and ants was put forward by Belt (1874) after observing the behavior of aggressive ants on species of *Acacia* with swollen thorns in Central America. For example, the Bull's horn acacia (*Acacia cornigera*) bears hollow thorns that are used by its associated ant, *Pseudomyrmex ferruginea*, as nesting sites (Figure 8.12a); its leaves have protein-rich 'Beltian bodies' at their tips (Figure 8.12b) that the ants collect and use for food, and it has sugar-secreting organs called **nectaries** on its vegetative parts that also attract the ants. The ants, for their part, protect these small trees from competitors by actively snipping off shoots of other species, and they also protect the plant from herbivores. Even large vertebrate herbivores may be deterred.

> ant–plant mutualisms: but do the plants benefit?

In fact, ant–plant mutualisms appear to have evolved many times, even repeatedly in the same family of plants; nectaries are present on the vegetative parts of plants of at least 39 families and in many communities throughout the world. The benefits to the plants are not always easy to establish, but in the case of the Amazonian rain forest tree, *Duroia hirsuta*, they are at least twofold. *D. hirsuta* exists naturally as large single-species stands that can last for as long as 800 years. They are known locally as Devil's gardens because of the traditional belief that they are tended by evil spirits that exclude other species. It seems, though, from a series of experiments, that those evil spirits are ants, especially *Myrmelachista schumanni*. On the one hand, these ants removed herbivores from the trees (Figure 8.13a), though they were not as effective in this as some other ant species (*Azteca* spp.). But in addition, *M. schumanni* attacked competing plants within a Devil's garden, injecting poison (formic acid) into their leaves, such that the leaves showed signs of dying within a day and had mostly been shed within five days (Figure 8.13b). The homes and nectaries the plants provide for their devils seem to be a price well worth paying.

Farming crops or livestock

At least in terms of geographic extent, some of the most dramatic mutualisms are those of human agriculture. The numbers of individual plants of wheat, barley, oats, corn, and rice, and the areas these crops occupy, vastly exceed what they would have been if not for cultivation. The increase in the human population since the time of hunter–gatherers is some measure of

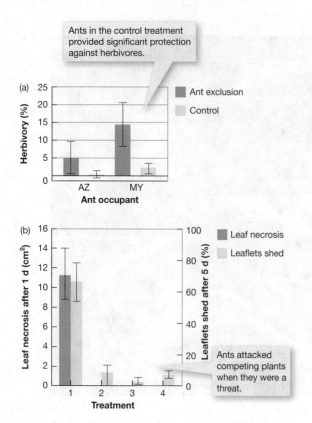

FIGURE 8.13 (a) Levels of herbivory (percentage of leaf area loss) on *Duroia hirsuta* leaves, naturally occupied by ants or with ants artificially excluded. Ants: AZ, *Azteca* spp., MY, *Myrmelachista schumanni*. Bars are 95% confidence intervals. (After Fredrickson, 2005.) (b) Saplings of the competitor plant *Clidemia heterophylla* subjected to different treatments: 1, planted among *D. hirsuta*, ants (*M. schumanni*) present; 2, planted among *D. hirsuta*, ants excluded; 3, planted away from *D. hirsuta*, ants present; 4, planted away from *D. hirsuta*, ants excluded. Bars are SEs. Treatments 2, 3, and 4 were all significantly different from treatment 1 ($P < 0.001$). (After Fredrickson et al., 2005.)

the reciprocal advantage to *Homo sapiens*. Even without doing the experiment, we can easily imagine the effect the extinction of humans would have on the world population of rice plants, or the effect of the extinction of rice plants on the population of humans. Similar comments apply if we consider the domestication of cattle, sheep, and other mammals.

Similar farming mutualisms have developed in termite and especially ant societies, where the insect farmers may protect individuals they exploit from competitors and predators and may even move or tend them. Ants, for example, farm many species of aphids (homopterans) in return for sugar-rich secretions of honeydew. The flocks of aphids benefit through suffering lower mortality rates caused by predators, showing increased feeding and excretion rates, and forming larger colonies.

> aphids farmed by ants: do they pay a price?

But this is not a cozy relationship with nothing but benefits; the aphids are being manipulated and they may also be paying a price. This cost is apparent from a study of colonies of the aphid *Tuberculatus quercicola* attended by the red wood ant *Formica yessensis* on the island of Hokkaido, northern Japan. As expected, in the presence of predators, aphid colonies survived significantly longer when attended by ants than when ants were excluded (Figure 8.14a). However, there *were* also costs for the aphids. When the predators were kept away and the effects of ant attendance on aphids could thus be viewed in isolation, the ant-attended aphids clearly suffered; they grew less well and were less fecund than those where ants were excluded as well as predators (Figure 8.14b).

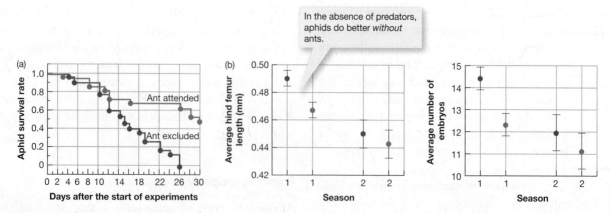

FIGURE 8.14 (a) Colonies of the aphid *Tuberculatus quercicola* from which ants had been excluded (by means of ant repellent smeared at the base of the oak trees on which the aphids lived) were more likely to become extinct than those attended by ants ($\chi^2 = 15.9$, $P < 0.0001$). (b) But in the absence of predators (experimentally removed), ant-excluded colonies performed better than those attended by ants. Shown are averages for aphid body size (hind femur length; $F = 6.75$, $P = 0.013$) and numbers of embryos ($F = 7.25$, $P = 0.010$), $\pm$ SE, for two seasons (1: July 23 to August 11, 1998; 2: August 12 to August 31, 1998). Color key: red circles, predator-free and ant-excluded treatment; blue circles, predator-free and ant-attended treatment. (After Yao et al., 2000.)

The dispersal of seeds and pollen

Very many plant species use animals to disperse their seeds and pollen. About 10% of all flowering plants possess seeds or fruits that bear hooks, barbs, or glues that become attached to the hairs, bristles, or feathers of any animal that comes into contact with them. They are frequently an irritation to the animal, which often cleans itself and removes them if it can, but usually after carrying them some distance. In these cases the benefit is only to the plant (which has invested resources in attachment mechanisms); there is no reward to the animal and no mutualism as such.

Quite different are the true mutualisms between higher plants and the birds and other animals that feed on fleshy fruits, dispersing the seeds. Of course, for the relationship to be mutualistic, it is essential that the animal digests only the fleshy fruit and not the seeds, which must remain viable when regurgitated or defecated. Thick strong defenses that protect plant embryos from digestion are usually part of the price paid by the plant for dispersal by fruit-eaters.

In addition, many different kinds of animals have entered into pollination liaisons with flowering plants, including hummingbirds, bats, and even small rodents and marsupials. Most animal-pollinated flowers offer nectar, pollen, or both as a reward to their visitors. The evolution of specialized flowers, and of specialized animal pollinators, will be favored because a specialist pollinator may be able to recognize and discriminate between different flowers, moving pollen between different flowers of the same species but not to flowers of other species. Thus, where the vectors and flowers are highly specialized, as is the case in many orchids, virtually no pollen is wasted. By contrast, passive transfer of pollen, for example by wind or water, does not discriminate in this way and is therefore much more wasteful.

The pollinators par excellence are, without doubt, the insects. Pollen is a nutritionally rich food resource, and in the simplest insect-pollinated flowers, it is offered freely and in abundance to all and sundry. The plants rely for pollination on the insects being less than wholly efficient in their pollen consumption, carrying their spilled food with them from plant to plant. In more complex flowers, nectar (a solution of sugars) is produced as an additional or alternative reward. The nectar seems to have no value to the plant other than as an attractant to animals, and it has a cost to the plant, because the carbohydrates in nectar might have been used in growth or some other activity.

> insect pollinators: from generalists to ultraspecialists

In the simplest of these plants, the nectaries are unprotected, but, with increasing specialization, nectaries are enclosed in structures that restrict access to the nectar to just a few visitor species. This range can be seen within the family Ranunculaceae (Figure 8.15). In the simple flower of *Ranunculus ficaria*, the nectaries are exposed to all visitors, but in the more specialized flower of *R. bulbosus*, there is a flap over the nectary, and in *Aquilegia*, the nectaries have developed into long tubes and only visitors with long probosces (tongues) can reach the nectar.

Not all insect-pollinated plants (and not all pollinating insects) are specialists, of course; there is an evolutionary balance to be struck. Unprotected nectaries have the advantage of a ready supply of pollinators, but because these pollinators are unspecialized, they transfer much of the pollen to the flowers of other species. Protected nectaries have the advantage of efficient transfer of pollen by specialists to other flowers of the same species, but they rely on there being sufficient numbers of these specialists. Specialist insects have the advantage of an exclusive resource; generalists have the advantage of a more reliable supply, though they share it.

FIGURE 8.15 Increasing protection of nectar among the buttercups and columbines (Ranunculaceae). Left to right: *Ranuculus ficaria, R. bulbosus, Aquilegia vulgaris*.

Mutualistic gut inhabitants

Most of the mutualisms discussed so far have depended on patterns of behavior, where neither species lives entirely within its partner. In many other mutualisms, one of the partners is integrated more or less permanently into the body cavity or even the cells of its partner. The microorganisms occupying parts of various animals' alimentary canals, called the **gut microbiota**, are the best-known extracellular symbionts.

The crucial role of microbes in the digestion of cellulose by vertebrate herbivores has long been appreciated, but it now appears that the gastrointestinal tracts of all vertebrates are populated by a mutualistic microbiota. Protozoa and fungi are usually present, but the major contributors to these fermentation processes are bacteria. Their diversity is greatest in regions of the gut where the pH is relatively neutral and food retention times are relatively long.

In small mammals like rodents, rabbits, and hares, the cecum is the main fermentation chamber, whereas in larger nonruminant mammals such as horses, the colon is the main site. In ruminants, like cattle and sheep, and in kangaroos and other marsupials, fermentation occurs in specialized stomachs. The basis of the mutualism is straightforward. The microbes receive a steady flow of substrates for growth in the form of food that has been eaten, chewed, and partly homogenized. They live within a chamber in which the pH and, in endotherms, temperature are regulated and anaerobic conditions are maintained. On the other side, the vertebrate hosts, especially the herbivores, receive nutrition from food that they would otherwise find literally indigestible.

These mutualistic relationships are by no means confined to vertebrates. Many insects, for example, are equally reliant on their gut microbiota for digesting their food, as we saw in Chapter 1 for termites, which are incapable of digesting their woody food without their bacterial symbionts (see Figure 1.3). In fact, the benefits to the host may themselves not be confined to improved digestion. Increasingly, studies are pointing to an effect of the gut microbiota on the host's resistance to pathogens.

Figure 8.16, for example, shows a case where it had been known that different colonies of bumble bee, *Bombus terrestris*, show different sensitivities to the gut parasite *Crithidia bombi*. It was natural to assume this was the result of the differing genotypes of the colonies. But that assumption was not borne out when the gut microbiota were transferred among the different colonies by feeding emerging worker bees feces, either from their own or from other colonies, and then test-infecting

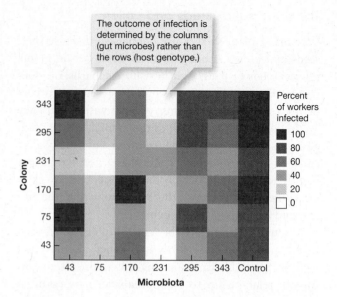

FIGURE 8.16 Gut microbes were transferred among bee hosts in a fully factorial design: each colony of hosts received its own and every other colony's gut microbiota. *Control* refers to a dummy transfer of microbes: no microbes at all, only sugar water. The depth of color indicates sensitivity to infection with *Crithidia bombi* (in terms of percentage of worker hosts dying). The effect of the origin of the microbiota on sensitivity to infection was significant ($P < 0.001$); that of the origin of the hosts was not ($P = 0.66$). (After Koch & Scmid-Hempel, 2012.)

the workers with *C. bombi*. The outcome of infection was determined not by the colony of origin of the bees (their genotype), but by the colony of origin of their gut microbes. As we become ever more able to manipulate even our own gut microbes, interest in what they provide is certain to increase.

Mycorrhizas

Most higher plants do not have roots we can view in isolation; they have mycorrhizas—intimate mutualisms between fungi and root tissue. Plants of only a few families, such as the Cruciferae, are exceptions. Broadly, the fungal networks in mycorrhizas capture nutrients from the soil, which they transport to the plants in exchange for carbon and energy in return. Many plant species can live without their mycorrhizal fungi in soils where neither nutrients nor water are ever limiting, but in the harsh world of natural plant communities, the symbioses, if not strictly necessary, are nonetheless needed if the individuals are to survive in nature (Buscot et al., 2000).

Three major types of mycorrhiza are recognized. Arbuscular mycorrhizas are found in about two-thirds of all plant species, including most nonwoody species and tropical trees (Figure 8.17a). Ectomycorrhizal fungi form symbioses with many trees and shrubs, dominating boreal and temperate forests and also

(a)

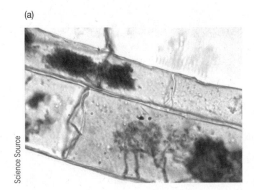

Science Source

(b)

Dr. Jeremy Burgess/Science Source

FIGURE 8.17 (a) An arbuscular mycorrhiza. (b) An ectomycorrhiza.

some tropical rain forests (Figure 8.17b). Finally, ericoid mycorrhizas are found in the dominant plant species of heathland.

In ectomycorrhizas (ECMs), infected roots are usually concentrated in the litter layer of the soil. Fungi form a sheath of varying thickness around the roots. From there, hyphae radiate into the litter layer, extracting nutrients and water. The fungal mycelium also extends inward from the sheath, penetrating between the cells of the root cortex to give intimate cell-to-cell contact with the host and establishing an interface with a large surface area for the exchange of the products of photosynthesis, soil, water, and nutrients, between the host plant and its fungal partner. The ECM fungi are effective in extracting the sparse and patchy supplies of phosphorus and especially nitrogen from the forest litter layer. In the other direction, carbon flows from the plant to the fungus, very largely in the form of simple sugars: glucose and fructose.

The plants pay a price. Fungal consumption of these sugars may represent up to 30% of the plants' net rate of photosynthate production. The plants, though, are often nitrogen-limited, and much of that nitrogen is available as ammonia. It is therefore crucial for forest trees, on the other side of the balance sheet, that ECM fungi can access organic nitrogen directly through enzymic degradation and utilize ammonium as a preferred source of inorganic nitrogen. Nonetheless, the idea that there is a balance of costs and benefits for the plants is emphasized by their responsiveness to changing circumstances. ECM growth rate is high when the rate of flow of sugars from the plant is high. But when nitrate is plentifully available to the plants, plant metabolism is directed away from sugar production and towards amino acid synthesis. As a result, the ECM degrades: the plants seem to support just as much ECM as they need.

Arbuscular mycorrhizas (AMs) do not form a sheath but penetrate *within* the roots of the host. Roots become infected

from mycelium present in the soil or from germ tubes that develop from asexual spores. Initially, the fungus grows between host cells, but eventually it enters them and forms a finely branched intracellular cluster called an **arbuscule**, giving these mycorrhizas their name. Studies generally have tended to emphasize facilitation of phosphorus uptake as the main benefit to plants of AM symbioses (phosphorus is a highly immobile element in the soil, and it therefore frequently limits plants' growth if they do not have the assistance of the mycorrhizas). The truth, however, appears to be more complex than this, with benefits demonstrated, too, in nitrogen uptake, pathogen and herbivore protection, and resistance to toxic metals (Newsham et al., 1995).

To continue a theme that pervades this section, there is also clear evidence that the balance between benefits and costs changes with the ecological circumstances. One example is shown in Figure 8.18. Cilantro, *Coriandrum sativum*, was grown in experimental pots in Costa Rica with and without an inoculation of AM, at a range of phosphorus concentrations, and at high and low densities. When phosphorus concentrations were low, the benefits to the plant of having an AM root system to aid in accessing that phosphorus were clear, especially when resources were in relatively short supply at high densities. But with a relatively plentiful supply of phosphorus, and therefore less need among the plants for aid in accessing it, the plants performed best in the absence of an AM, whether growth was faster (low density) or slower (high density). Removing the need for the benefits that an AM can bring revealed its costs.

Fixation of atmospheric nitrogen in mutualistic plants

The inability of the plants and animals to undertake the fixation of atmospheric nitrogen is one of the great puzzles in the process of evolution, since nitrogen is

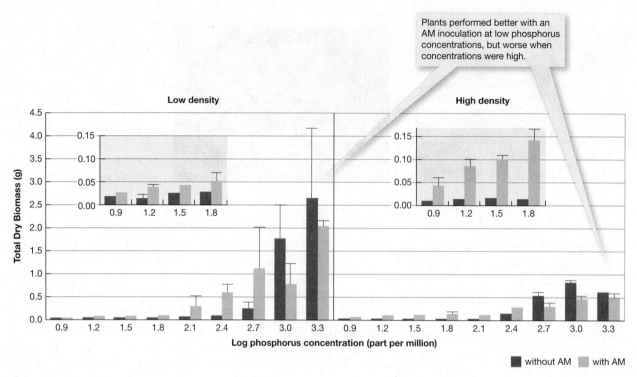

FIGURE 8.18 Mean dry weights after 74 days growth of cilantro plants at a range of phosphorus concentrations, at low and high densities, and both with an inoculation of arbuscular mycorrhiza (orange bars) and without (red bars). Insets show results at low phosphorus concentrations on an expanded scale. Bars are SEs. (After Schroeder & Janos, 2004.)

in limiting supply in many habitats (see Chapter 11). However, the ability to fix nitrogen is widely though irregularly distributed among both the eubacteria ('true' bacteria) and the archaea (Archaebacteria), and many of these have been caught up in tight mutualisms with distinct groups of eukaryotes. The best known, because of the huge agricultural importance of legume crops (peas, beans, and their relatives), are the **rhizobia**, bacteria that fix nitrogen in the root nodules of most leguminous plants and just one nonlegume, *Parasponia*.

The establishment of the liaison between rhizobia and legume plants proceeds by a series of reciprocating steps. The bacteria occur in a free-living state in the soil and are stimulated to multiply by root exudates and cells that have been sloughed from roots as they develop. In a typical case, a bacterial colony develops on the root hair, which then begins to curl and is penetrated by the bacteria. The host responds by laying down a wall that encloses the bacteria and forms an *infection thread*. This grows within the root, and the rhizobia proliferate within it. Rhizobia cannot fix nitrogen, but some are released into host cells in a developing nodule where, surrounded by a host-derived membrane, they differentiate into a particular form, called a bacteroid, that can fix nitrogen. Meanwhile, a special vascular system develops in the host, supplying the products of photosynthesis to the nodule tissue and

carrying away fixed-nitrogen compounds to other parts of the plant.

The costs and benefits of this mutualism need to be considered carefully. From the plant's point of view, we need to compare the energetic costs of alternative ways of obtaining fixed nitrogen. The route for most plants is direct from the soil as nitrate or ammonium ions. The metabolically cheapest route is the use of ammonium ions, but in many soils most nitrogen is in the form of nitrates. For the plants alone, the unit energetic cost of reducing nitrate from the soil to ammonia is about 12 mol of adenosine triphosphate (ATP, the cell's energy currency). The mutualistic process with rhizobia is energetically slightly more expensive to the plant: about 13.5 mol of ATP.

However, we must also add the costs of forming and maintaining the nodules, which may be about 12% of the plant's total photosynthetic output. Hence, mutualistic nitrogen fixation is energetically inefficient. Energy, though, may be much more readily available for green plants than nitrogen. A rare and valuable commodity (fixed nitrogen), bought when you have more currency (energy) than you need, may be quite a bargain. On the other hand, in a pattern analogous to what we saw with the mycorrhizas, when a nodulated legume is provided with nitrates (that is, when nitrate

costs and benefits of rhizobial mutualisms

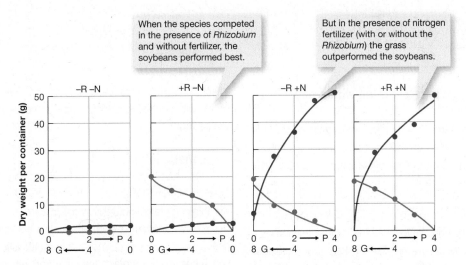

FIGURE 8.19   The growth of soybeans (*Glycine soja*, G, ●) and a grass (*Paspalum*, P, ●) grown alone and in mixtures with and without nitrogen fertilizer (N) and with and without inoculation with nitrogen-fixing *Rhizobium* (R). The plants were grown in pots containing 0–4 plants of the grass together with 0–8 plants of *Glycine*. Thus, moving left to right on the horizontal axis, the treatments are zero *Paspalum* (0P) and 8 *Glycine* (8G), 1P with 6G, 2P with 4G, 3P with 2G, and, finally, 4P with 0G. The vertical scale on each figure shows the mass of plants of the two species in each container. −R −N, no *Rhizobium* and no fertilizer; +R −N, inoculated with *Rhizobium* but no fertilizer; −R +N, no *Rhizobium* but nitrate fertilizer was applied; +R +N, inoculated with *Rhizobium* and nitrate fertilizer was supplied. (After de Wit et al., 1966.)

is *not* a rare commodity), nitrogen fixation declines rapidly.

Of course, we must not think of the mutualisms of rhizobia and legumes (and other nitrogen-fixing mutualisms) as isolated interactions between bacteria and their own host plants. In nature, legumes normally form mixed stands in association with nonlegumes. These are potential competitors with the legumes for nitrogen. The nodulated legume sidesteps this competition by its access to its own unique source. It is in this ecological context that nitrogen-fixing mutualisms gain their main advantage. Where nitrogen is plentiful, however, the energetic costs of nitrogen fixation often put the plants at a competitive *dis*advantage.

Figure 8.19, for example, shows the results of a classic experiment in which soybeans (*Glycine soja*, a legume) were grown in mixtures with *Paspalum*, a grass. The mixtures either received mineral nitrogen, or were inoculated with *Rhizobium*, or both. The experiment was designed as a replacement series, which allows us to compare the growth of pure populations of the grass and the legume with their performances in the presence of each other, at a range of ratios. In the pure stands of soybean, yield was increased very substantially *either* by inoculation with *Rhizobium*, *or* by application of fertilizer nitrogen, or both. The legumes can use either source of nitrogen as a substitute for the other. The grass, however, responded only to the fertilizer. Hence, when the species competed in the presence of *Rhizobium* but with no fertilizer, the legume contributed far more to the overall yield than

did the grass; over a succession of generations, the legume would have outcompeted the grass. When they competed in soils supplemented with fertilizer nitrogen, however, it was the grass that made the major contribution, whether or not *Rhizobium* was also present; long term, it would have outcompeted the legume.

Quite clearly, then, it is in environments deficient in nitrogen that nodulated legumes have a great advantage over other species. But their activity raises the level of fixed nitrogen in the environment. After death, legumes augment the level of soil nitrogen on a very local scale, with a delay of weeks to months and even longer as they decompose. Thus, their advantage is lost; they have improved the environment of their competitors, and the growth of associated grasses will be favored in these local patches. In an important sense, organisms that can fix atmospheric nitrogen are locally suicidal. This is one reason why it is very difficult to grow repeated crops of pure legumes in agricultural practice without aggressive grass weeds invading the nitrogen-enriched environment. It may also explain why leguminous herbs or trees usually fail to form dominant stands in nature. Grazing animals, on the other hand, continually remove grass foliage, and the nitrogen status of a grass patch may again decline to a level at which the legume is at a competitive advantage.

We end this section, then, on a theme that has recurred repeatedly. To understand the ecology of mutualistic pairs, we must look beyond those species to the wider community and ecosystem of which they are a part.

SUMMARY

Molecular ecology: differentiation within and between species

For some purposes, knowing how much differentiation there is within species or between one species and another is critical. Molecular genetic markers, of a variety of types, have massively increased the resolution at which we can differentiate between populations and even individuals. Studies on albatrosses illustrate how even within a species of conservation importance, separate populations even more threatened with extinction may be hidden. Studies on commercial fish illustrate how molecular markers can be used to detect, and to prosecute, illegal fishermen. A threatened 'species,' the red wolf, may in fact be a hybrid between two other, relatively common species, with implications for both the desirability and the practicality of its conservation.

Coevolutionary arms races

Consumers and their living resources may become involved in coevolutionary 'arms races.' Plants relying on toxins are more prone to becoming involved in arms races with their herbivores than those relying on more 'quantitative' (digestion-reducing) chemicals. The intimate association between parasites and their hosts makes them especially prone to arms races. The evolution of resistance to the myxoma virus in the European rabbit is easy to understand, but the parasites favored by natural selection are those with the greatest reproductive rate, which occurs at intermediate levels of virulence because of increased transmissibility. In other cases, host–parasite coevolution is more definitely antagonistic—increasing resistance in the host and increasing

infectivity in the parasite—a process that can be observed in action with bacteria and their viruses, because generation times are so short.

Mutualistic interactions

A mutualistic relationship is one in which organisms of different species interact to their mutual benefit, though current evolutionary thinking views mutualisms as cases of reciprocal exploitation where nonetheless each partner is a *net* beneficiary. Pairs of species from many taxa take part in mutualistic associations in which one species protects the other from predators or competitors but gains privileged access to a food resource on the protected species. 'Farming' mutualisms have developed not only in human agriculture, but also in termite and especially ant societies, where many species farm aphids in return for sugar-rich secretions of honeydew. Very many plant species use animals to disperse their seeds and pollen, and many different kinds of animals (but especially insects) have entered into pollination liaisons with flowering plants. The gastrointestinal tracts of many animals are populated by a mutualistic microbiota, essential for digestion. Most higher plants do not so much have roots as mycorrhizas—intimate mutualisms between fungi and root tissue—where the fungi play crucial roles especially in extracting nutrients from the soil. Finally, the ability to fix nitrogen is widely distributed among bacteria, and many of these have been caught up in tight mutualisms with distinct groups of eukaryotes, the best known of which are the rhizobia, which fix nitrogen in the root nodules of most leguminous plants.

REVIEW QUESTIONS

1 Explain why molecular (DNA) markers have improved the ability of ecologists to study degrees of differentiation within and between species.

2 Explain the similarities between evolution and coevolution.

3 Why are some plants more likely than others to be involved in arms races with their insect herbivores?

4 Account for the decline in virulence of the myxomatosis virus in European rabbits after its initial introductions in Australia and Europe.

5 Compare and contrast the mutualistic associations of ants with plants they protect and aphids they farm.

6 Compare the roles of fruits and nectar in the interactions between plants and the animals that visit them.

7 What are mycorrhizas and what is their significance?

Challenge Questions

1   Should the red wolf be conserved, or would that be a misguided waste of public money? Weigh up carefully the two sides of the argument.

2   The following propositions may seem radical but discuss whether they have any validity: 'Most herbivores are not really herbivores but consumers of the by-products of the mutualists living in their gut,' and 'Most gut parasites are not really parasites but competitors with their hosts for food the host has captured.'

3   Explain why leguminous plants may be considered to be a perfect example of a mutualistic association that can be understood only in the context of the ecological community within which it normally exists.

Part 4

Communities and Ecosystems

Andrea Gingerich/Getty Images

Andrea Gingerich/Getty Images

Chapter 9

From populations to communities

CHAPTER CONTENTS

KEY CONCEPTS

After reading this chapter you will be able to:

- differentiate between the roles of interacting abiotic and biotic factors in the determination and the regulation of population abundance

- explain how patchiness and dispersal between patches influence the dynamics of both populations and communities

- describe how disturbances influence community patterns, and their role in community succession

- explain the importance of direct and indirect interactions in food webs, the distinction between bottom-up and top-down control, and the relationship between food webs' structure and stability

n preceding chapters, we generally dealt with individual species or pairs of species in isolation, as ecologists often do. Ultimately, however, we must recognize that every population exists within a web of interactions with myriad other populations, across several trophic levels. Each population must be viewed in the context of the whole community. We also need to understand that populations occur in patchy and inconstant environments in which disturbance and local extinction may be common.

9.1 MULTIPLE DETERMINANTS OF THE DYNAMICS OF POPULATIONS

In preceding chapters, single-species populations have often been our main focus. In attempting to understand what determines species' abundances and distributions, we have dealt more or less separately with the roles of conditions and resources, of migration, of competition (both intra- and interspecific), of mutualism, and of predation and parasitism. In reality, though, the dynamics of any population reflect a combination of these effects, with their relative importance varying from case to case. Now, therefore, we need to view the population in the context of the whole community, since each exists within a whole web of interactions (Figure 9.1).

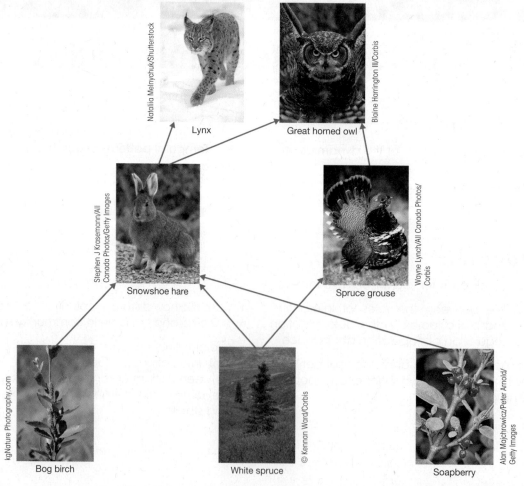

FIGURE 9.1 A simple community matrix, itself abstracted from a fuller, more realistic community from the North America boreal forest (see Figure 7.18), illustrating how each species may interact with several others in competitive interactions (among the plant species—bog birch, white spruce, and soapberry; or between the grazers—snowshoe hares and spruce grouse; or between predators—lynx and great horned owls) and predator–prey interactions (such as between lynx and hares, or between spruce grouse and white spruce).

In this section we consider how abiotic and biotic factors combine to determine the dynamics of species populations. Then, in Section 9.2, we revisit one of the major themes of this book—the importance of patchiness and dispersal between patches—and discuss especially the importance of the concept of the metapopulation. In Section 9.3, we deal with temporal patterns in community composition, including community succession. We acknowledge that disturbances, such as forest fires and the storm-battering of seashores, play an important role in the dynamics of many populations and the composition of most communities. We trace the patterns of reestablishment of species that are played out after each disturbance against a background of changing conditions, resources, and population interactions. Finally, in Section 9.4 we broaden our view further to examine food webs, like the one illustrated in Figure 9.1, with usually at least three trophic levels (plant–herbivore–predator). We emphasize the importance not only of direct but also of indirect effects that a species may have on others on the same trophic level or on levels below or above it.

Why are some species rare and others common? Why does a species occur at low population densities in some places and at high densities in others? What factors cause fluctuations in a species' abundance? These are crucial questions when we wish to conserve rare species, or control pests, or manage natural living resources, or when we wish simply to understand the patterns and dynamics of the natural world. To provide complete answers for even a single species in a single location, we need to know the physicochemical conditions, the level of resources available, the organism's life cycle, and the influence of competitors, predators, parasites, and so on—and how all these factors influence abundance through effects on birth, death, dispersal, and migration. We now bring these factors together and consider how we might discover which actually matter in particular examples.

The raw material for the study of abundance is usually some estimate of the numbers of individuals in a population. However, a record of numbers alone can hide vital information. Picture three human populations, shown to contain identical numbers of individuals. One is an old people's residential area, the second is a population of young children, and the third is a population of mixed age and sex. In the absence of information beyond mere numbers, it would not be clear that the first population was doomed to extinction (unless maintained by immigration), the second would grow fast but only

what total numbers can and cannot tell us

after a delay, and the third would continue to grow steadily. The most satisfactory studies, therefore, estimate not only the numbers of individuals but also partition these into different age classes, sexes, and size groups.

We can use the data that accumulate from estimates of abundance to establish correlations with external factors like food or weather. These correlations in turn can help predict the future. For example, high intensities of the late blight disease in potato crops usually occur 15–22 days after a period in which the minimum temperature is above 10°C and relative humidity is more than 75% for two consecutive days. Such a correlation may helpfully alert the potato grower to the need for protective spraying.

what correlations can and cannot tell us

Correlations may also suggest—but not prove—causal relationships. Figure 9.2, for example, shows four examples in which population growth rate declines as the availability of food declines. But ultimately, certifying a cause requires us to identify a mechanism. When the population is large, and hence little food is available, many individuals may starve to death, fail to reproduce, or become aggressive and drive out the weaker members. The correlations in Figure 9.2 cannot tell us which. Nonetheless, correlations can be informative. Figure 9.2 also suggests, for example, that such relationships are likely to be flat at the highest food levels where some other factor or factors place an upper limit on the growth rate.

Fluctuation or stability?

Some populations appear to change very little in size. One study of birds that covered an extended time span—though it was not necessarily the most scientific study—examined swifts (*Micropus apus*) in the village of Selborne in southern England over more than 200 years. In one of the earliest published works on ecology, Gilbert White, who lived in the village, wrote in 1778 (see White, 1789):

> I am now confirmed in the opinion that we have every year the same number of pairs invariably. . . . The number that I constantly find are eight pairs, about half of which reside in the church, and the rest in some of the lowest and meanest thatched cottages.

More than 200 years later, Lawton and May (1984) visited the village and, not surprisingly, found major changes. Swifts are unlikely to have nested in the

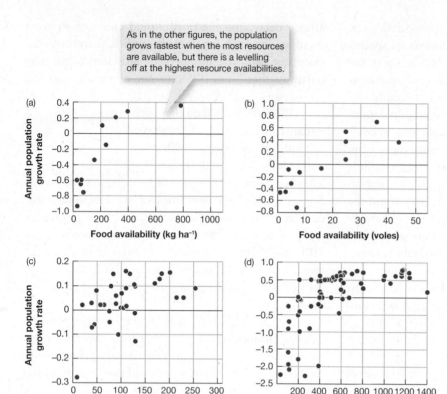

As in the other figures, the population grows fastest when the most resources are available, but there is a levelling off at the highest resource availabilities.

FIGURE 9.2 Increases in annual population growth rate with the availability of food, measured as pasture biomass (kg ha⁻¹) in (a) and (c), as vole abundance in (b), and as availability of food per capita in (d). (a) Red kangaroo (Bayliss, 1987). (b) Barn owl (modified from Taylor, 1994). (c) Wildebeest (Krebs et al., 1999). (d) Feral pig (Choquenot, 1998). Positive growth rates indicate increasing abundance; negative growth rates decreasing abundance. (After Sibly & Hone, 2002.)

church for 50 years, and the thatched cottages have either disappeared or had their roofs covered with wire. Yet the number of breeding pairs of swifts regularly to be found in the village is now 12. In view of the many changes that have taken place in the intervening centuries, this number is remarkably close to the eight pairs so consistently found by White.

But the overall stability of a population may be combined with more complex underlying patterns. For example, the Australian mice *stability need not mean 'nothing changes'* (*Mus domesticus*) in Figure 9.3a have extended periods with low numbers, interrupted by sporadic and dramatic irruptions. Nonetheless, these seem always to be short-lived and followed by a return to relatively stable low abundances. Similarly, steady trends in numbers may conceal fluctuations that seem, temporarily, to be bucking those trends. The French voles (*Microtus arvalis*) in Figure 9.3b could be said to have declined steadily between 1968 and 1996, but throughout this period there were many years in which their numbers increased dramatically.

Determination and regulation of abundance

Is the similarity between 8 and 12 pairs of swifts over 200 years most important, or the difference between them? Is it the irruptions in the abundance of Australian mice that are most interesting, or the predictable return to much lower levels? Should the overall decline in vole numbers be the focus of attention, or the change in numbers from one year to the next? Some investigators have emphasized the apparent constancy of populations; others have emphasized the fluctuations.

Those who have emphasized constancy argue that we need to look for stabilizing forces *within* populations to explain why they do not usually exhibit unfettered increase or a decline to extinction (even the decline in vole numbers in Figure 9.3b appears to be leveling off). Those who have emphasized fluctuations often look to external factors, weather or disturbance, to explain the changes. Can the two sides be brought together to form a consensus?

To do so, it is important to understand clearly the difference between questions about the ways abundance is determined and questions about the ways abundance is regulated. **Regulation** is the tendency of a population to decrease in size when it is above a particular level, but to increase in size when below that level. In other words, regulation of a population can, by definition, occur only as a result of one or more density-dependent processes (see Chapters 3 and 5)

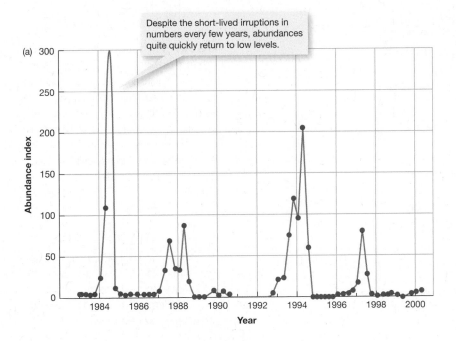

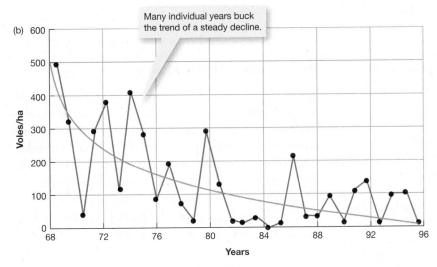

FIGURE 9.3 (a) Irregular irruptions in the abundance of house mice (*Mus domesticus*) in an agricultural habitat in Victoria, Australia, where the mice, when they irrupt, are serious pests. The abundance index is the number caught per 100 trap-nights. In the fall of 1984 the index exceeded 300. (After Singleton et al., 2001.) (b) Estimated abundance of voles (*Microtus arvalis*) in a drained marshland area in southwest France. (After Butet & Leroux, 2001.)

that act on rates of birth and/or death and/or movement (Figure 9.4a). We have discussed various potentially density-dependent processes in earlier chapters on competition, predation, and parasitism. We must look at regulation, therefore, to understand how it is that a population tends to remain within defined upper and lower limits.

On the other hand, the **determination** of the precise abundance of individuals will reflect the combined effects of all the factors and all the processes that affect a population, whether they are dependent or independent of density (Figure 9.4b). We must look at the determination of abundance, therefore, to understand how it is that a particular population exhibits a particular abundance at a particular time, and not some other abundance.

Some have taken extreme stances, believing that density-dependent, biotic interactions play the main role not only in regulating but also in determining population size. Those on the other side feel equally strongly that we should view most natural populations as passing through a repeated sequence of density-independent setbacks and recovery. But there is no need for any sharp polarization of views. What regulates population size and what determines population size are both interesting and important questions. No population can be absolutely free of regulation—long-term unrestrained population

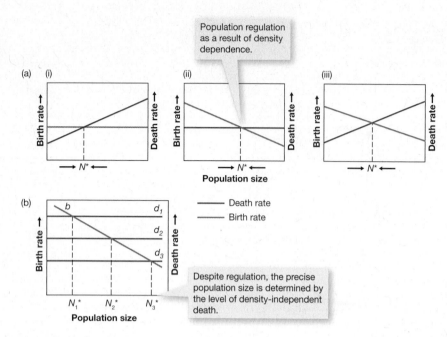

Population regulation as a result of density dependence.

Despite regulation, the precise population size is determined by the level of density-independent death.

FIGURE 9.4 (a) Population regulation with: (i) density-independent birth and density-dependent death; (ii) density-dependent birth and density-independent death; and (iii) density-dependent birth and death. Population size increases when birth rate exceeds death rate and decreases when death rate exceeds birth rate. N^* is therefore a stable equilibrium population size. The actual value of the equilibrium population size is seen to depend on both the magnitude of the density-independent rate and the magnitude and slope of any density-dependent processes. (b) Population regulation with density-dependent birth, b, and density-independent death, d. Death rates are determined by physical conditions that differ in three sites (death rates d_1, d_2, and d_3). Equilibrium population size varies as a result.

growth is unknown, and unrestrained declines to extinction are rare—and density-dependent processes are neither uncommon nor unimportant. For instance, density dependence was detected in more than 80% of studies of insect dynamics, as long as these studies lasted for more than 10 years (Hassell et al., 1989; Woiwod & Hanski, 1992).

On the other hand, for many populations, weather is the major determinant of abundance, and other factors are of relatively minor importance. For instance, in one classic study of a pest, the apple thrips, weather accounted for 78% of the annual variation in numbers and was therefore of paramount importance in predicting the thrips' potential for crop damage (Davidson & Andrewartha, 1948).

So, what regulates the size of a population need not determine its size for most of the time. It would be wrong to give regulation or density dependence some kind of preeminence. Active regulation may be occurring only infrequently or intermittently, and it is likely that no natural population is ever truly at equilibrium: even when regulation is occurring, it may be drawing abundance toward a level that is itself changing in response to changing levels of resources. Thus, there are a range of possibilities: some populations in nature are almost always recovering from the last disaster, others are usually limited by an abundant or a scarce resource and are therefore constantly common or rare, and others are usually in decline after sudden episodes of colonization.

Using *k*-value analysis

We can distinguish clearly between what regulates and what determines the abundance of a population, and see how regulation and determination relate to one another, by examining an approach known as **k-value analysis** or, originally, as key factor analysis. In fact, as key factor analysis it was poorly named, since the technique actually identifies key *phases* (rather than key factors) in the life of a study organism. Shortcomings in the method have been identified and alternatives proposed (Coulson et al., 2005), but these alternatives are far less straightforward, and *k*-value analysis remains valuable as a tool in disentangling the determination and regulation of abundance. Details are described in Box 9.1, but we can understand the approach simply by appreciating that the *k*-values can be computed for each phase of the life cycle, and these measure the amount of mortality in that phase. The higher the *k*-value, the greater the mortality (*k* stands for 'killing power').

To carry out a *k*-value analysis, we compile data each year or season in the form of a life table (see Chapter 5). For the Canadian population of the Colorado potato beetle (*Leptinotarsa decemlineata*) in Box 9.1, the sampling program provided estimates of the population at seven stages: eggs, early larvae, late larvae, pupae, summer adults, hibernating adults, and spring adults. The first question we can ask is: "How much of the total 'mortality' tends to occur in each

9.1   Quantitative Aspects

k-value analysis

Table 9.1 sets out a typical set of life table data, collected by Harcourt (1971) over a singe season for the Colorado potato beetle, *Leptinotarsa decemlineata*, in Canada. The first column lists the various phases of the life cycle. *Spring adults* emerge from hibernation around the middle of June, when potato plants are breaking through the ground. Within 3 or 4 days, egg laying begins, and it continues for about 1 month. *The eggs* are laid in clusters of approximately 34 on the lower leaf surface, and *the larvae* crawl to the top of the plant, where they feed throughout their development, passing through four stages (divided here into *early and late larvae*). When mature, they drop to the ground and form *pupal cells* in the soil. *Summer adults* emerge in early August, feed, and then reenter the soil at the beginning of September to *hibernate* and become the next season's spring adults.

TABLE 9.1   Life table data for the Canadian Colorado potato beetle.

Age Interval	Numbers per 96 Potato Hills	Numbers Dying	Mortality Factor	Factor $\text{Log}_{10}n$	*k*-Value	
Eggs	11,799	2,531	Not deposited	4.072	0.105	(k_{1a})
	9,268	445	Infertile	3.967	0.021	(k_{1b})
	8,823	408	Rainfall	3.946	0.021	(k_{1c})
	8,415	1,147	Cannibalism	3.925	0.064	(k_{1d})
	7,268	376	Predators	3.861	0.023	(k_{1e})
Early larvae	6,892	0	Rainfall	3.838	0	(k_2)
Late larvae	6,892	3,722	Starvation	3.838	0.337	(k_3)
Pupal cells	3,170	16	Parasitism	3.501	0.002	(k_4)
Summer adults	3,154	−126	Sex (52% ♀)	3.499	−0.017	(k_5)
Summer females X 2	3,280	3,264	Emigration	3.516	2.312	(k_6)
Hibernating adults	16	2	Frost	1.204	0.058	(k_7)
Spring adults	14			1.146		
					2.926	(k_{total})

The next column lists the estimated numbers (per 96 potato hills) at the start of each phase, and the third column then lists the numbers leaving the population in each phase, before the start of the next. This is followed, in the fourth column, by what were believed to be the main causes of 'deaths' in each stage of the life cycle, though defined loosely in this case. For example, the very first row is labeled 'Not deposited' to describe eggs that were not so much killed as not laid by the adults carrying them in the first place. The entry below summer adults is labeled 'Sex (52% ♀)' to take account of the unequal sex ratio.

The fifth and sixth columns then show how *k*-values are calculated. In the fifth column, the logarithms of the numbers at the start of each phase are listed. The *k*-values in the sixth column are then simply the differences between successive values in column 5. Thus, each value refers to deaths in one of the phases on a logarithmic scale, and, as in column 3, the total of the column refers to the total deaths throughout the life cycle. Moreover, each *k*-value, because it is logarithmic, measures the rate or intensity of mortality in its own phase, whereas this is not true for the values in column 3. There, values tend to be higher earlier in the life cycle, simply because there are more individuals 'available' to die. These useful characteristics of logarithms are put to use in *k*-value *analysis*.

An adult Colorado potato beetle (*Leptinotarsa decemlineata*) taking off from its host plant. Emigration by summer adults represents the key phase in the population dynamics of potato beetles.

TABLE 9.2 Summary of the life table analysis for Canadian Colorado beetle populations (see Box 9.1). (After Harcourt, 1971.)

		Mean	Coefficient of Regression on k_{total}
Eggs not deposited	k_{1a}	0.095	−0.020
Eggs infertile	k_{1b}	0.026	−0.005
Rainfall on eggs	k_{1c}	0.006	0.000
Eggs cannibalized	k_{1d}	0.090	−0.002
Egg predation	k_{1e}	0.036	−0.011
Larvae 1 (rainfall)	k_2	0.091	0.010
Larvae 2 (starvation)	k_3	0.185	0.136
Pupae (parasitism)	k_4	0.033	−0.029
Unequal sex ratio	k_5	−0.012	0.004
Emigration	k_6	1.543	0.906
Frost	k_7	0.170	0.010
	k_{total}	2.263	

of the phases?" (Mortality here refers to all losses from the population.) We can answer the question by calculating the mean *k*-values for each phase, in this case determined over 10 seasons (that is, from 10 tables like the one in Box 9.1). These are presented in the third column of Table 9.2. Here, most loss occurred among summer adults—in fact, mostly through emigration rather than mortality as such. There was also substantial loss of older larvae (starvation), hibernating adults

(frost-induced mortality), young larvae (rainfall), and eggs (cannibalization and not being laid).

It is also valuable, however, to ask a second question: "What is the relative importance of these phases as determinants of year-to-year *fluctuations* in mortality, and hence of year-to-year fluctuations in abundance?" This is rather different. For instance, a phase might repeatedly witness a significant toll being taken from a population (a high mean *k*-value), but if that toll is always roughly the same, it may play little part in determining the particular population size in any particular year.

We can address this second question in the following way. Mortality during a phase of the life cycle that is important in determining population change will vary in line with total mortality in terms of both size and direction. It will be a key phase if, when mortality during it is high, total mortality tends to be high and the population declines, whereas when phase mortality is low, total mortality tends to be low, and the population tends to remain large. By contrast, a phase with a k-value that varies quite randomly with respect to total k will have little influence on changes in mortality and population size.

We need therefore to measure the relationship between phase mortality and total mortality, and this is achieved by the regression coefficient of the former on the latter. The largest regression coefficient will be associated with the key phase causing population change, whereas phase mortality that varies at random with total mortality will generate a regression coefficient close to zero. In Table 9.2, the summer adults, with a regression coefficient of 0.906, are the key phase. Other phases (with the possible exception of older larvae) have a negligible effect on the changes in generation mortality. These first two questions, then, tell us something about the *determination* of abundance in the population.

What about the possible role of these phases in the *regulation* of the Colorado beetle population, however? In other words, which, if any, act in a density-dependent way? We can answer this most easily by plotting k-values for each phase against the numbers present at the start of the phase. For density dependence, the k-value should be highest (that is, mortality should be greatest) when density is highest.

For the beetle population, two phases are notable in this respect: for both summer adults (the key phase) and older larvae, there is evidence that losses are density-dependent (Figure 9.5). Hence, both have a possible role in regulating the size of the beetle population. In this case, therefore, the phases with the largest role in determining abundance are also those that seem likely to play the largest part in regulating abundance. But as we see next, this is by no means a general rule.

A k-value analysis has been applied to a great many insect populations, but to far fewer vertebrate or plant populations. Examples of these, though, are shown in Table 9.3 and Figure 9.6. We start with populations of the wood frog (*Rana sylvatica*) in three regions of the United States (Table 9.3). The larval period was the key phase determining abundance in all regions, largely as a result of year-to-year variations in rainfall. In low-rainfall years, the ponds often dry out, reducing larval survival to catastrophic levels.

two further examples of k-value analysis

Such mortality, however, was inconsistently related to the size of the larval population (only one of two ponds in Maryland, and only approaching significance in Virginia), and hence it played an inconsistent part in regulating the sizes of the populations. Rather, in two regions it was during the adult phase that mortality was clearly density-dependent (apparently as a result of competition for food), and, indeed, in two regions mortality was also most intense in the adult phase (first data column). The key phase determining abundance in a Polish population of the sand-dune annual plant *Androsace septentrionalis* (Figure 9.6) were the seeds in the soil. Once again, however, mortality there did not operate in a density-dependent manner, whereas mortality of seedlings (not the key phase) was density-dependent.

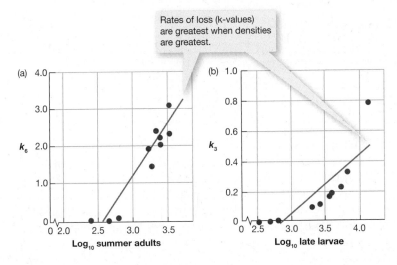

FIGURE 9.5 (a) Density-dependent emigration by Colorado beetle summer adults. (b) Density-dependent starvation of larvae. (After Harcourt, 1971.)

TABLE 9.3 *k*-value analysis for wood frog populations in the United States: Maryland (two ponds, 1977–1982), Virginia (seven ponds, 1976–1982), and Michigan (one pond, 1980–1993). In each area, the phase with the highest mean *k*-value, the key phase, and any phase showing density dependence are highlighted in bold. (After Berven, 1995.)

Age Interval	Mean *k*-Value	Coefficient of Regression on k_{total}	Coefficient of Regression on Log (Population Size)
Maryland			
Larval period	1.94	**0.85**	**Pond 1 : 1.03 (*P* = 0.04)**
			Pond 2 : 0.39 (*P* = 0.50)
Juvenile: up to 1 year	0.49	0.05	0.12 (*P* = 0.50)
Adult: 1–3 years	**2.35**	0.10	0.11 (*P* = 0.46)
Total	4.78		
Virginia			
Larval period	**2.35**	**0.73**	0.58 (*P* = 0.09)
Juvenile: up to 1 year	1.10	0.05	−0.20 (*P* = 0.46)
Adult: 1–3 years	1.14	0.22	**0.26 (*P* = 0.05)**
Total	4.59		
Michigan			
Larval period	1.12	**1.40**	1.18 (*P* = 0.33)
Juvenile: up to 1 year	0.64	1.02	0.01 (*P* = 0.96)
Adult: 1–3 years	**3.45**	−1.42	**0.18 (*P* = 0.005)**
Total	5.21		

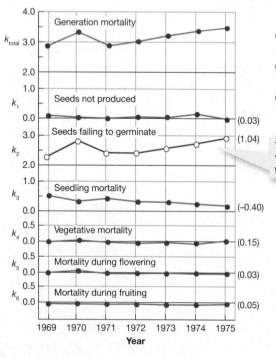

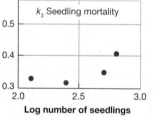

The key phase since *k*-values vary in line with overall generation mortality.

FIGURE 9.6 A *k*-value analysis of the sand-dune annual plant *Androsace septentrionalis*, consisting of a graph of total generation mortality (k_{total}) and of various *k*-factors. The values of the regression coefficients of each individual *k*-value on k_{total} are given in brackets. The largest regression coefficient signifies the key phase and is shown as a red line. Alongside is shown the one *k*-value that varies in a density-dependent manner. (After Symonides, 1979; analysis in Silvertown, 1982.)

Overall, therefore, *k*-value analysis is useful in identifying important phases in the life cycles of study organisms, and useful too in distinguishing the variety of ways in which phases may be important: in contributing significantly to the overall sum of mortality, in contributing significantly to variations in mortality, and in contributing significantly to the regulation of abundance by virtue of the density dependence of the mortality.

9.2   DISPERSAL, PATCHES, AND METAPOPULATION DYNAMICS

In many studies of abundance, researchers have assumed that the major events all occur within the study area, and that immigrants and emigrants can safely be ignored. But migration can be a vital factor in determining and/or regulating abundance. We have already seen, for example, that emigration was the predominant reason for the loss of summer adults of the Colorado potato beetle, which was both the key phase in determining population fluctuations and one in which loss was strongly density-dependent.

Dispersal has a particularly important role to play when populations are fragmented and patchy, | habitable sites and dispersal distance | as many are. We can think of the abundance of patchily distributed organisms as being determined by two things: first, the size of, and distance between, the species' habitable sites, and second, the species' typical dispersal distance. Thus, a population may be small if its habitable sites are small or short-lived, or only few in number. But it may also be small if the dispersal distance between habitable sites is great relative to the dispersibility of the species, such that habitable sites that go extinct are unlikely to be recolonized (Gadgil, 1971).

We can illustrate this for a number of butterfly species, because their larvae feed on only one or a few species of patchily distributed plants, and it is therefore possible to identify habitable sites (containing these plant species) that are not inhabited. For example, the violet copper butterfly, *Lycaena helle*, one of the rarest butterfly species in Central Europe, was studied in the Westerwald region in Germany, where it is confined to patches of habitat containing adderwort, *Bistorta officinalis*, its only larval food plant in the area. This was one of several features used to classify a habitat patch as suitable for the violet copper, giving 230 patches in total. These patches were then surveyed to determine whether they were occupied by

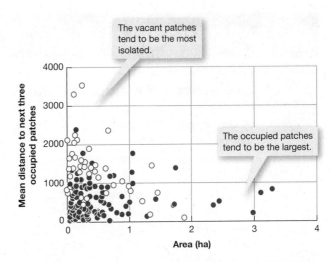

FIGURE 9.7 *The sizes of vacant patches (open circles) and those occupied by the violet copper butterfly (shaded circles) in the Westerwald, Germany, and also their level of isolation from other occupied patches. (After Bauerfeind et al., 2009.)*

the butterfly, and what features distinguished occupied from vacant patches. Some of the results are shown in Figure 9.7.

Three factors had a significant influence on occupancy. Patches were more likely to be occupied the higher the percentage of the ground covered by adderwort—an indicator of patch quality from the butterflies' point of view. But occupancy was also more likely in larger patches and in patches that were less isolated from other occupied sites. As with many rare species, overall abundance was being limited, at least in part, not simply by a lack of habitat but by difficulties in reaching it.

More often than not, the roles of patchiness and dispersal in driv- | metapopulations | ing the dynamics of populations are brought together in the concept of a **metapopulation**, the origins and definition of which are described in Box 9.2.

We can describe a population as a metapopulation when it consists of a collection of subpopulations, each one of which has a realistic chance both of going extinct and of appearing again through recolonization. Thus, compared to more conventional views of population dynamics, in thinking in terms of metapopulations we give less emphasis to the birth, death, and movement processes going on within a single subpopulation, but much more to the colonization (= birth) and extinction (= death) of whole subpopulations. Put simply, metapopulations may persist, stably, as a result of the balance between extinctions and recolonizations, even though none of the local subpopulations is stable in its own right.

The genesis of metapopulation theory

A classic book, *The Theory of Island Biogeography*, written by Robert MacArthur and Edward Wilson and published in 1967, was an important catalyst in radically changing ecological theory. MacArthur and Wilson showed how the distribution of species on islands could be interpreted as a balance between the opposing forces of extinction and colonization (see Chapter 10), and they focused attention especially on situations where those species available for colonization all derived from a common source—the mainland. They developed their ideas in the context of the communities inhabiting oceanic islands, but their thinking has been rapidly assimilated into much wider contexts, with the realization that patches everywhere have many of the properties of true islands—ponds as islands of water in a sea of land, trees as islands in a sea of grass, and so on.

At about the same time as MacArthur and Wilson's book was published, a simple model of metapopulation dynamics was proposed by Richard Levins (1969). Levins introduced the concept of a metapopulation to refer to a subdivided and patchy population in which the population dynamics operate at two levels:

1 The dynamics of individuals within patches (determined by the usual demographic rates of birth, death, and movement).

2 The dynamics of the patches or subpopulations within the overall metapopulation (determined by the rates of colonization of empty patches and extinction of occupied patches).

Richard Levins (left), Ilkka Hanski (right)

Both Levins's and MacArthur and Wilson's theories embraced the idea of patchiness, and both focused on colonization and extinction rather than on the details of local dynamics. But whereas MacArthur and Wilson's theory was based on a vision of mainlands as rich sources of colonists for whole archipelagos of islands, in a metapopulation there is a collection of patches but no such dominating mainland.

Levins introduced the variable $p(t)$, the fraction of habitat patches occupied at time t, which immediately introduces the notion that not all habitable patches are always inhabited. The rate of change in $p(t)$ depends on the rate of local extinction of patches and the rate of colonization of empty patches. It is not necessary to go into the details of Levins's model; suffice to say that as long as the intrinsic rate of colonization exceeds the intrinsic rate of extinction within patches, the total metapopulation will reach a stable, equilibrium fraction of occupied patches, even if none of the local populations is stable in its own right.

Perhaps because of the powerful influence on ecology of MacArthur and Wilson's theory, the idea of metapopulations was largely neglected during the 20 years after Levins's initial work. The 1990s, however, saw a great flowering of interest, both in the underlying theory and in populations in nature that might conform to the metapopulation concept. Much of this interest was driven by the Finnish ecologist Ilkka Hanski, who in 2011 was awarded the prestigious international Crafoord Prize for his work by the Royal Swedish Academy of Sciences.

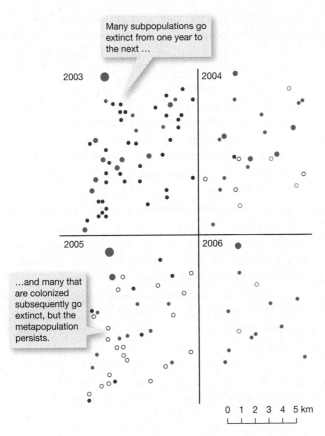

FIGURE 9.8 The spatial distribution of subpopulations in a meta-population of the bee, *Andrena hattorfiana*, in southern Sweden over four years. Filled blue dots represent occupied patches of increasing subpopulation size, in four classes: 1-10, 11-50, 51-100, and > 100. Red dots indicate subpopulations that had gone extinct by the subsequent year. Green open dots indicate newly colonized patches. Red open dots indicate either patches that were colonized one year but had gone extinct by the next, or subpopulations that went extinct one year but had been recolonized by the next. (After Franzen & Nilsson, 2010.)

An example of this is shown in Figure 9.8, where a metapopulation of solitary bees, *Andrena hattorfiana*, in southern Sweden was surveyed several times each year from 2003 to 2006. Habitat patches (potential subpopulations) were those containing the bee's main pollen source, *Knautia arvensis*, and it is clear that from one year to the next, the chance for an individual subpopulation to go extinct (or a vacant patch to be colonized) was extremely high. The metapopulation was stable, the subpopulations were not; only 16% of the subpopulations remained occupied throughout the study. Those with the smallest bee populations were most likely to go extinct; those covering the largest area were most likely to be colonized.

We can also appreciate, in a similar vein, that in what we can now call the violet copper metapopulation in Figure 9.7, the smaller patches were more likely to be vacant because they were less likely to be colonized (smaller target) and more likely to go extinct (smaller, more vulnerable subpopulation). This emphasizes, in turn, that there is likely to be a continuum of types of metapopulation: from collections of nearly identical local populations, all equally prone to extinction, to metapopulations in which there is great inequality between local populations, some of which are effectively stable in their own right. This contrast is further illustrated in Figure 9.9 for the silver-studded blue butterfly (*Plejebus argus*) in North Wales, UK.

Additional aspects of the dynamics of meta-populations can be illustrated in a study of a small mammal, the American pika, *Ochotona princeps*, in

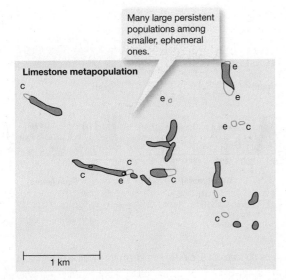

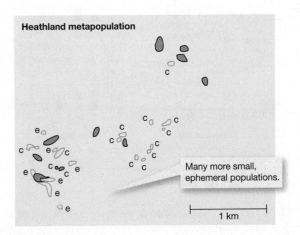

FIGURE 9.9 Two metapopulations of the silver-studded blue butterfly (*Plejebus argus*) in North Wales: filled outlines, present in both 1983 and 1990 (persistent); open outlines, not present at both times; e, present only in 1983 (presumed extinct); c, present only in 1990 (presumed colonization). (After Thomas & Harrison, 1992.)

California (Figure 9.10). The overall metapopulation could itself be divided into northern, middle, and southern networks of patches, and the patch occupancy in each was determined on four occasions between 1972 and 1991. These data (Figure 9.10a) show that the northern network maintained a high occupancy throughout the study period, the middle network maintained a more variable and much lower occupancy, while the southern network suffered a steady and substantial decline.

The dynamics of individual subpopulations were not monitored, but these were simulated using models based on the principles of metapopulation dynamics and on general information about pika biology. When each of the three networks was simulated in isolation (Figure 9.10b), the northern network remained at a stable high occupancy (as observed in the data), but the middle network rapidly and predictably crashed, and the southern network eventually suffered the same fate. However, when the entire metapopulation was simulated as a single entity (Figure 9.10c), the northern network again achieved stable high occupancy, but this time

the middle network was also stable, although at a much lower occupancy (again as observed), while the southern network suffered periodic collapses (also consistent with the real data).

This all suggests that within the metapopulation as a whole, the northern network acts as a net source of colonizers that prevent the middle network from suffering overall extinction. These in turn delay extinction in, and allow recolonization of, the southern network. The study therefore illustrates how whole metapopulations can be stable when their individual subpopulations are not. Moreover, the comparison of the northern and middle networks, both stable but at very different occupancies, shows how occupancy may depend on the size of the pool of dispersers, which itself may depend on the size and number of the subpopulations.

Finally, the southern network in particular emphasizes that the observable dynamics of a metapopulation may have more to do with | transient dynamics may be as important as equilibria |

transient behavior, far from any equilibrium, than with the theoretical equilibrium state itself. To take another example, the silver-spotted skipper butterfly (*Hesperia*

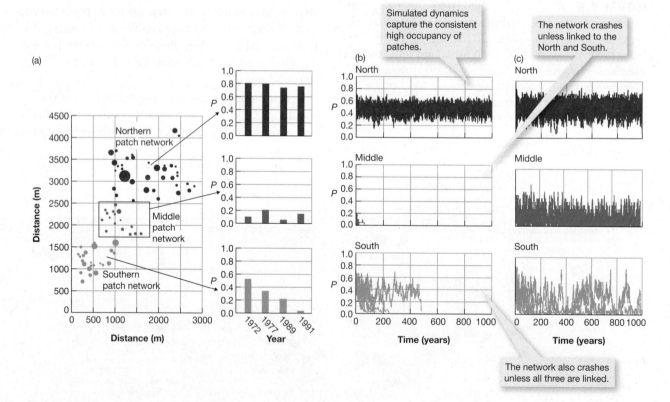

FIGURE 9.10 The metapopulation dynamics of the American pika, *Ochotona princeps*, in Bodie, California. (a) The relative positions (distance from a point southwest of the study area) and approximate sizes (as indicated by the size of the dots) of the habitable patches, and the occupancies (as proportions, P) in the northern, middle, and southern networks of patches in 1972, 1977, 1989, and 1991. (b) The simulated temporal dynamics of the three networks, with each of the networks simulated in isolation. Ten replicate simulations are shown, overlaid on one another, each starting with the actual data in 1972. (c) Equivalent simulations to (b) but with the entire metapopulation treated as a single entity. (After Moilanen et al., 1998.)

comma) declined steadily in Great Britain from a widespread distribution over most calcareous hills in 1900, to 46 or fewer refuge localities (local populations) in 10 regions by the early 1960s (Thomas & Jones, 1993). The probable reasons were changes in land use—increased plowing of grasslands, reduced stocking with grazing animals—and the virtual elimination of rabbits by myxomatosis with its consequent profound vegetational changes. Throughout this nonequilibrium period, rates of local extinction generally exceeded those of recolonization.

In the 1970s and 1980s, however, reintroduction of livestock and recovery of the rabbits led to increased grazing, and suitable habitats increased again. This time, recolonization exceeded local extinction, but the spread of the skipper remained slow, especially into localities isolated from the 1960s refuges. Even in southeast England, where the density of refuges was greatest, it is predicted that the abundance of the butterfly will increase only slowly—and remain far from equilibrium—for at least 100 years. Thus, it seems that around a century of transient decline in the dynamics of the metapopulation is to be followed by another century of transient increase – except that the environment will no doubt alter again before the transient phase ends and, most likely, a new transient phase begins.

Finally, we must be wary of assuming that all patchy populations are truly metapopulations, that is, that every subpopulation

> metapopulations of plants? remember the seed bank

has a measurable probability of going extinct or being recolonized. In particular, there is no doubt that many plants inhabit patchy environments, and apparent extinctions of local populations may be common, but the suggestion that recolonizations following a genuine extinction are common is nonetheless questionable in any species that has a buried seed bank (see Section 5.2). Recolonizations may often simply be the result of the germination of seeds following habitat restoration. Recolonization by dispersal, a prerequisite for a true metapopulation, may be extremely rare.

There are, though, at least some examples of plant metapopulations, including where the application of molecular methods (see Section 8.2) has proved invaluable. One such study looked at an annual plant, the treacle mustard, *Erisymum cheiranthoides*, that occupies sites on stony banks of the River Meuse in Belgium (Figure 9.11a). As an annual, the species clearly shows no continuity of adults at a site from one year to the next, but individual sites are subject to flooding each winter, which may be powerful enough to also flush away all seeds from a site, leading to

extinction of a subpopulation. The river, though, may also bring new seeds (colonists) to a site.

Samples were taken from sites, and DNA extracted and analyzed, allowing each site to be genetically characterized and differentiated from other sites. This in turn allowed at least a proportion of individuals sampled in 2006 to be assigned to populations from 2005, and likewise for individuals in 2007 to be assigned to 2006 populations. It was apparent not only that seeds moved between the subpopulations, but also that in some cases they were carried to vacant sites and recolonized them (Figure 9.11b). In this case at least, the plant formed a true metapopulation; its dynamics had at least as much to do with extinction and colonization as with local birth and death.

9.3 TEMPORAL PATTERNS IN COMMUNITY COMPOSITION

Founder-controlled and dominance-controlled communities

The metapopulation concept is important for our understanding of population dynamics, because it acknowledges that a combination of patchiness and dispersal between patches can give rise to dynamics quite different from those that would be observed if there were just one, homogeneous patch. We can apply the same thinking to community organization.

Disturbances that open up gaps are common in many kinds of community. Gaps are simply patches within which many species suffer local extinction simultaneously. In forests, high winds, elephants, or simply the death of a tree through old age may all create gaps. In grassland, agents include frost, burrowing animals, and cattle dung. On rocky shores, gaps may be formed as a result of severe wave action during hurricanes, battering by moored boats, or the action of predators. Gaps in a community are ripe for colonization, and two fundamentally different kinds of community organization can be recognized (Yodzis, 1986). When all species are good colonists and essentially equal competitors, communities are described as founder controlled. When some species are strongly superior competitively, communities can be described as dominance controlled. The dynamics of the two are quite different, and we deal with them in turn.

In **founder-controlled communities**, species are approximately equivalent in their ability to invade gaps and can hold the gaps against all comers during their lifetime. On each occasion that a population goes locally extinct, a gap is opened

> founder-controlled communities: competitive lotteries and priority effects

(a)

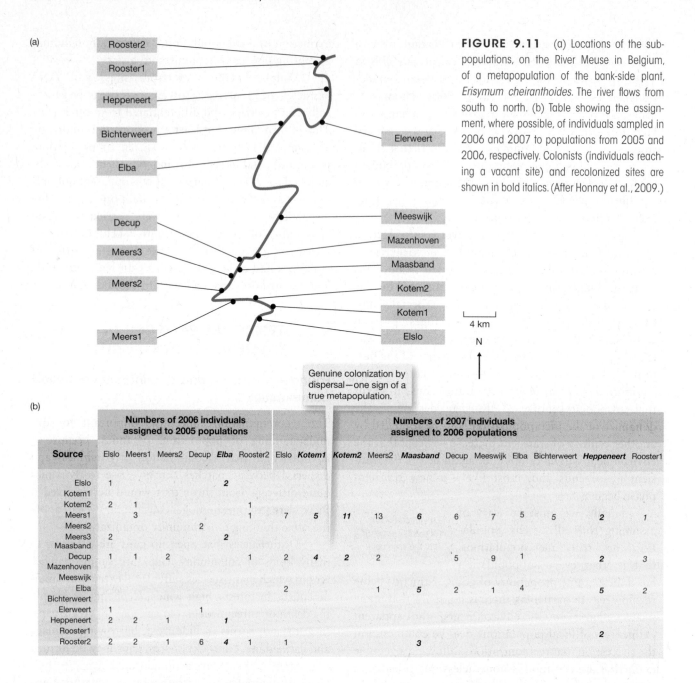

FIGURE 9.11 (a) Locations of the sub-populations, on the River Meuse in Belgium, of a metapopulation of the bank-side plant, *Erisymum cheiranthoides*. The river flows from south to north. (b) Table showing the assignment, where possible, of individuals sampled in 2006 and 2007 to populations from 2005 and 2006, respectively. Colonists (individuals reaching a vacant site) and recolonized sites are shown in bold italics. (After Honnay et al., 2009.)

Genuine colonization by dispersal—one sign of a true metapopulation.

(b)

Source	Numbers of 2006 individuals assigned to 2005 populations						Numbers of 2007 individuals assigned to 2006 populations											
	Elslo	Meers1	Meers2	Decup	*Elba*	Rooster2	Elslo	*Kotem1*	*Kotem2*	Meers2	*Maasband*	Decup	Meeswijk	Elba	Bichterweert	*Heppeneert*	Rooster1
Elslo	1				2												
Kotem1																	
Kotem2	2	1				1											
Meers1		3				1	7	5	11	13	6	6	7	5	5	2	1
Meers2				2													
Meers3	2				2												
Maasband																	
Decup	1	1	2			1	2	4	2	2		5	9	1		2	3
Mazenhoven																	
Meeswijk																	
Elba							2			1	5	2	1	4		5	2
Bichterweert																	
Elerweert	1			1													
Heppeneert	2	2	1		*1*												
Rooster1																2	
Rooster2	2	2	1	6	4	1	1			3							

up for invasion. Hence, if gaps are appearing continually and randomly, and all conceivable replacements are possible, **species richness** – simply the total number of species in a community – is maintained at a high level in the system as a whole. Such competitive lotteries have been hypothesized, especially for communities of tropical reef fish (Sale & Douglas, 1984). These are extremely rich in species.

The number of fish species on the Great Barrier Reef, for example, ranges from 900 in the south to 1500 in the north, and more than 50 resident species may be recorded on a single patch of reef 3 m in diameter. Only a proportion of this species richness is likely to be attributable to resource partitioning of food and space, since the diets of many of the coexisting species

are very similar. Rather, it has been proposed that the crucial limiting factor is vacant living space, generated unpredictably in space and time when a resident dies or is killed. The fish species, then, breed often, sometimes year-round, and produce numerous clutches of dispersive eggs or larvae that compete in a lottery for living space in which larvae are the tickets. The first arrival at the vacant space wins the site, matures quickly, and holds the space for its lifetime.

While the idea is plausible, there is little direct evidence for such competitive lotteries, certainly at the whole community level. There are, though, good examples of one key element of the process: a **priority effect** between competing species, in which the species that arrives first at a site is able to hold it against competing invaders, whatever the outcome would be if they competed as simultaneously arriving equals. This is shown for two reef fish in Figure 9.12a and among species of ectomycorrhizal fungi (Section 8.4) in Figure 9.12b.

In **dominance-controlled communities**, by contrast, some species are competitively superior to others, and an initial colonizer of a patch cannot necessarily maintain its presence

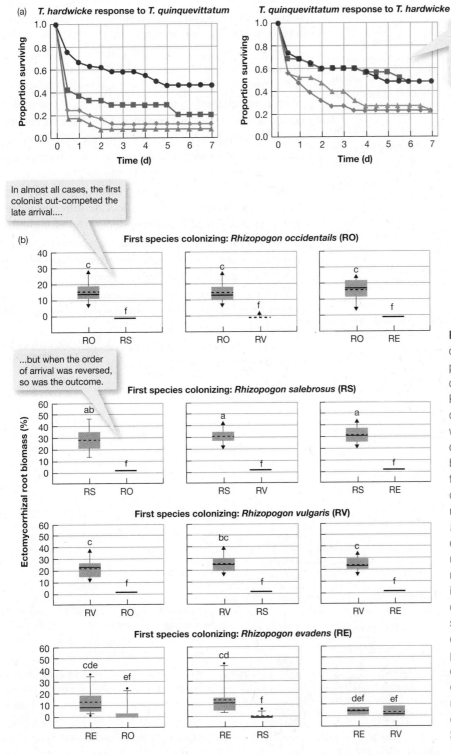

FIGURE 9.12   Priority effects among competitors. (a) Survival time in two competing reef fish, *Thalassoma hardwicke* and *T. quinquevittatum* in Moorea, French Polynesia, when alone (●), when they arrived at the same time at a site (■) and when one arrived either 5 days (♦) or 12 days (▲) after the other. (b) Percentage biomass of the roots of seedlings of the pine tree, *Pinus muricata*, from the California coast, made up by species of ectomycorrhizal fungi, *Rhizopogon evadens, occidentalis, salebrosus,* and *vulgaris* (RE, RO, RS, and RV, respectively) when roots were inoculated with the fungi, taken in pairs. One month after germination, seedlings were inoculated with the first fungus, grown for a further three months, inoculated with the second species, grown three months more, and then harvested. Vertical boxes in each panel of the figure cover 25–75% of the data (the central line is the median), vertical bars 10–90%, and dots 0–100%. Bars not sharing a letter above them are significantly different ($P < 0.05$). (After Geange & Stier, 2009; Kennedy et al., 2009.)

there. In these cases, disturbances that open up gaps lead to reasonably predictable sequences of species, because different species have different strategies for exploiting resources—early species are good colonizers and fast growers, whereas later species can tolerate lower resource levels and grow to maturity in the presence of early species, eventually outcompeting them. Such sequences are examples of **community successions** (see Section 1.3).

An idealized view of a succession is shown in Figure 9.13. Open space is colonized by one or more of a group of opportunistic, early-succession species (p_1, p_2, etc., in the figure). As time passes, more species invade, often with poorer powers of dispersal. These eventually reach maturity, dominating midsuccession (m_1, m_2, etc.), and many or all of the pioneer species are outcompeted to extinction. Later still, the community reaches a **climax stage** when the most efficient competitors (c_1, c_2, etc.) oust their neighbors. In this sequence, if it runs its full course, the number of species first increases (because of colonization) then decreases (because of competition).

Some disturbances are synchronized over extensive areas. A forest fire may destroy a huge tract of a climax community. The whole area then proceeds through a more or less synchronous succession. Other disturbances are much smaller and produce a patchwork of habitats. If these disturbances are out of phase with one another, the resulting community comprises a mosaic of patches at different stages of succession.

Community succession

If an opened-up gap has not previously been influenced by a community, we refer to the sequence of species as a **primary succession**. Examples include lava flows caused by volcanic eruptions, substrate exposed by the retreat of a glacier, and freshly formed sand dunes. But where the species of an area has been partially or completely removed but seeds and spores remain, we call the subsequent sequence a **secondary succession**. The loss of trees locally as a result of high winds may lead to secondary successions, as can cultivation followed by the abandonment of farmland (so-called old-field successions).

Primary successions often take several hundreds of years to run their course, though on recently denuded rocks on a seashore, for instance, a primary succession may take only a decade or so. The research life of an ecologist is long enough to follow the latter but not, for example, the succession following a glacial retreat. Fortunately, however, we can sometimes gain information over the longer time scale, because different stages of the succession may be represented by community gradients in space. For example, as a glacier retreats, the first stage of the succession may be observed just beyond its tip, with later stages strung out further down the glacial valley. We can then use this series of communities currently in existence – or **chronosequence**—to infer what the succession must have been.

An extensive chronosequence of dune ridges occurs on the coast of

> primary and secondary successions

> a primary succession in duneland

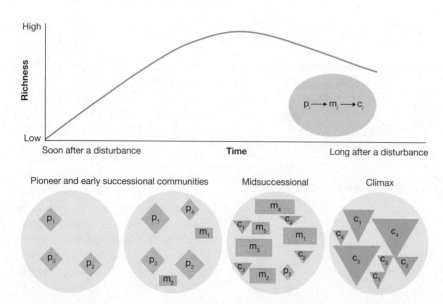

High

Richness

Low

Soon after a disturbance   **Time**   Long after a disturbance

$p_i \longrightarrow m_i \longrightarrow c_i$

Pioneer and early successional communities   Midsuccessional   Climax

FIGURE 9.13 Hypothetical succession in a gap – an example of dominance control. The occupancy of gaps is reasonably predictable. Richness begins at a low level as a few pioneer (p_i) species arrive; reaches a maximum in midsuccession when a mixture of pioneer, midsuccessional (m_i) and climax (c_i) species occur together; and drops again as competitive exclusion by climax species takes place.

Lake Michigan, where 13 ridges of known age (30–440 years old) show a clear pattern of primary succession to forest. The dune grass *Ammophila breviligulata* dominates the youngest, still mobile, dune ridge. Within 100 years, these are replaced by evergreen shrubs such as *Juniperus communis* and by prairie bunch grass *Schizachyrium scoparium*. Conifers begin colonizing the dune ridges after 150 years, and a mixed forest of pine species develops between 225 and 400 years. Deciduous trees such as the oak and maple do not become important components of the forest until 440 years have passed.

Experimental seed addition and seedling transplants have shown that later species are nonetheless capable of germinating in young dunes (Figure 9.14a). The more developed soil of older dunes may improve the performance of late successional species, but their successful colonization of young dunes is mainly prevented by limited seed dispersal, together with seed predation by rodents (Figure 9.14b). Eventually, however, the early species are competitively excluded by later ones as trees establish and grow.

Successions on old fields have been studied primarily in the eastern and northern United States, where many farms were abandoned by farmers who moved west after the frontier was opened up in the 19th century, and where indeed farmland continued to be abandoned through much of the 20th century in some places. Most of the precolonial mixed forest had previously been destroyed by the farmers, but regeneration was swift after the 'disturbance' caused by farmers came to an end. The early pioneers of the U.S. West left behind exposed land

> secondary succession on fields abandoned by farmers

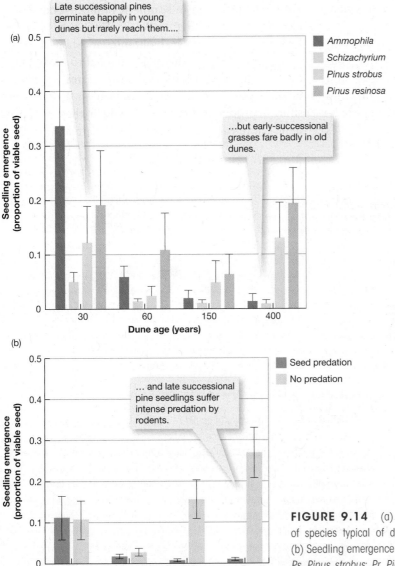

FIGURE 9.14 (a) Seedling emergence (means + SE) from added seeds of species typical of different successional stages on dunes of four ages. (b) Seedling emergence of the four species (*Ab, Ammophila; Ss, Schizachyrium; Ps, Pinus strobus; Pr, Pinus resinosa*) in the presence and absence of rodent predators of seeds. (After Lichter, 2000.)

that was colonized by pioneers of a very different kind. The typical sequence of dominant vegetation is: annual weeds → herbaceous perennials → shrubs → early successional trees → late successional trees. A particularly detailed study of old-field succession performed at the Cedar Creek Natural History Area in Minnesota was discussed in detail in Section 1.3.

Old-field succession has also been studied at a site in Lower Saxony, Germany, going back to 1968. We can divide the communities there into four broad types. The earliest, lasting only around two years, is dominated by short-lived annuals. The second, dominated by perennial herbs and grasses, lasts to around the eighth year and is followed by a shrub-dominated stage. Then, from around 14 years, the community is pioneer forest. The first stage is initiated by the germination of seeds, either from the seed bank or blown in. Thereafter, new colonizers also drive the transition from the first to the second stage (Figure 9.15a), but colonization declines in importance over subsequent transitions.

By contrast, few species are lost from the community over the first transition (Figure 9.15a), but many more go extinct as the third stage passes into the fourth. This shifting balance between colonization and extinction gives rise, as in the idealized succession of Figure 9.13, to an initial increase in species number, followed by a decline between the third and fourth stages (Figure 9.15b). Moreover, the species-rich third stage is also characterized by the widest range from common to rare species (Figure 9.15c), indicative perhaps of the combination of stage 3 species themselves, early successional species that have not yet gone extinct, and late successional species that have only just arrived, again as described in Figure 9.13.

Early succession plants have a fugitive lifestyle. Their continued survival depends on dispersal to other disturbed sites. They cannot persist in competition with later species, and thus they must grow and consume the available resources rapidly. High growth and photosynthetic rates are crucial properties of the fugitive. Rates among later successional plants are much lower. These and other trends were apparent in a study of old-field succession in southern France (some examples are shown in Figure 9.16). With stages ranging from 0–5 years after abandonment (stage 1) to > 100 years (stage 5), the plants tended to have higher rates of photosynthesis and faster rates of growth in the earlier stages, but they lived short lives with short reproductive periods in which they produced light (more easily dispersed) seeds. Late arrivals tend

early and late successional species have different properties

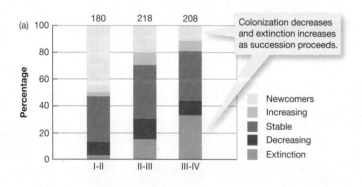

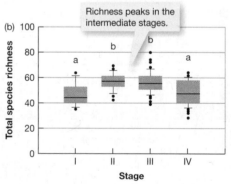

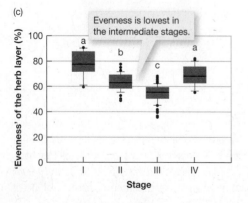

FIGURE 9.15 The dynamics of an old-field succession in Germany studied in a total of 175 plots. (a) The dynamics of the transitions between the four stages of the succession (I–IV). The cumulative numbers of species in the pairs of stages are shown above the bars. (b) Box and whisker plots of the species richness in the different stages. Boxes extend from the first to the third quartile of data with the median per plot in between. Whiskers extend to plots twice the interquartile range from the median. Dots represent plots outside this range. Boxes not sharing the same letter differ significantly. (c) Similar box and whisker plots of the 'evenness' of different species' contributions to the herb layer in the different stages. Evenness is at its maximum when all species are equally common. (After Dölle et al., 2008.)

to be those with heavier seeds, poorer dispersal, and long juvenile phases. The contrast is between two lifestyles: 'Quickly come, quickly gone,' and 'What I have, I hold.'

Figure 9.16 also makes another, arguably more subtle, point that applies also to the other succession data sets we have described, and much more broadly, too. The relationships in the figure, while clear, are only general trends. In the latest successional stage, some species had thicker leaves or lighter seeds than the thinnest leaves or heaviest seeds in the earliest stage, just as in Figure 9.15 some plots at the earliest stage had greater richness or lower evenness than many in the peak/trough stage 3. It is important that we characterize communities along the successional gradient by their defining features; but equally important that we recognize that all communities support the coexistence of species living their lives in very different ways.

Primary producers provide the starting point for all food webs and determine much of the character of the physical environment in which animals live. It is perhaps not surprising, then, that the changing animal community

| animals are often affected by, but may also affect, plant successions |

over the course of a succession is often a reflection of the changing plant community. This is shown for the Cedar Creek succession in Figure 9.17. The number of arthropod species—the species richness—increased significantly with successional age, whereas plant species richness showed only a nonsignificant upward trend. However, arthropod species richness also increased with plant species richness, and if the latter was accounted for in the arthropod regression, the relationship disappeared. For the most part, at least, the arthropods simply followed the plants. In some cases, though, animals play key roles in determining the nature and/or the speed of a plant succession, for instance, through heavy grazing or trampling. An example is described in Box 9.3.

Figure 9.13 was described as an idealized succession, and one respect in which it was so is that it

| the concept of a climax community |

arrived at a climax community at the end. Do real successions reach a climax? Some may. The succession of seaweeds on an overturned boulder may reach a climax in only a few years. Old-field successions, on the other hand, might take 100–300 years to reach a climax, but in that time the probabilities of fire or severe hurricanes,

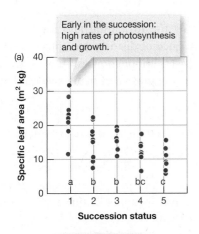

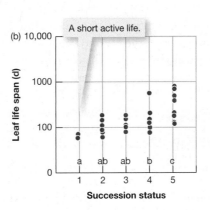

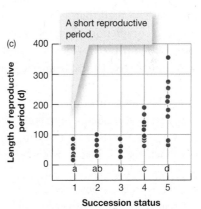

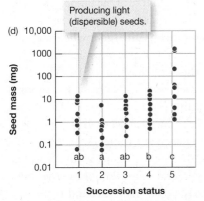

FIGURE 9.16 Changes in selected plant traits over the course of old-field succession in southern France, divided into five stages (1 is 0–5 years after abandonment, 2 is 5–15 years, 3 is 15–50 years, 4 is 50–100 years, 5 is > 100 years). In each case, stages not sharing the same letter differ significantly. (a) Specific leaf area is the ratio of leaf area to leaf weight. Plants with thinner leaves (greater specific leaf area) with the ability to capture light effectively in chloroplasts spread over a wider area, tend to be those that photosynthesize, and grow fastest. (b) Leaf life span. (c) Length of the reproductive period. (d) Seed mass. (After Navas et al., 2010.)

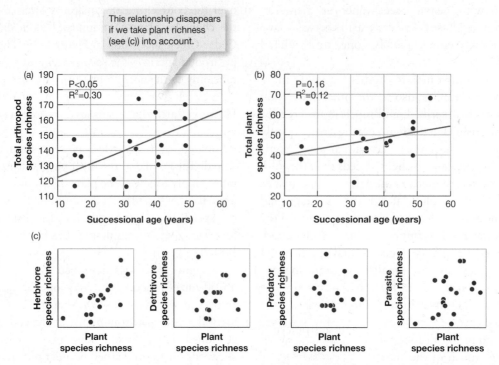

FIGURE 9.17 Arthropod and plant species richness in the study of old-field succession at the Cedar Creek site in Minnesota. (a) Arthropod richness increased significantly with successional age. (b) However, plant richness showed only a nonsignificant upward trend. (c) Also, the richness of arthropods that were herbivores or parasites, especially, increased with plant richness. If this was accounted for in the arthropod regression in (a), the relationship disappeared. (After Sieman et al., 1999.)

which occur every 70 or so years in New England, for example, are so high that a process of succession may never go to completion. Given that forest communities in northern temperate regions, and probably also in the tropics, are still recovering from the last glaciation, it is questionable whether the idealized climax vegetation is often reached in nature.

In fact, our decision about whether a climax has been reached, like so much else in ecology, is likely to be a matter of scale. As we saw above, many successions take place in a mosaic of patches, with each patch, having been disturbed independently, at a different successional stage. Boulders on a rocky shore are a good example. Climax communities in such cases can then occur, at best, only on a very local scale. Moreover, when successions occur in a patchwork, the nature of the succession, both locally and overall, is likely to depend on the size and shape of the patches (gaps). The centers of very large gaps are most likely to be colonized by individuals that have traveled relatively great distances. But such mobility is less important in small gaps, since much of the recolonization will be by established individuals from around the periphery.

> successions in a patchwork—the size and shape of gaps

Intertidal beds of mussels provide excellent opportunities to study the formation and the filling-in of gaps. In the absence of disturbance, mussel beds may persist as extensive monocultures. More often, they are an ever-changing mosaic of many species that inhabit gaps formed by the action of waves. The size of these gaps at the time of formation ranges from a single mussel space to hundreds of square meters. Gaps begin to fill as soon as they are formed. An experimental study of mussel beds of *Brachidontes solisianus* and *B. darwinianus* in Brazil aimed to determine the effects of patch size and location within a patch on the dynamics of succession (Figure 9.18).

Some of the colonizers were of the limpet *Collisella subrugosa*. These are particularly vulnerable to visually hunting predators and most vulnerable when they are most isolated. No doubt for this reason, peak rates of colonization by the limpet occurred in the smallest gaps in the mussel bed in the first six months after gap formation, but not in medium or large gaps (Figure 9.18a), and the limpets were also much better at colonizing the periphery than the center of the large gaps (Figure 9.18b). Small gaps were also most quickly colonized by lateral migration of the two mussel species (Figure 9.18a), but from around 6 months,

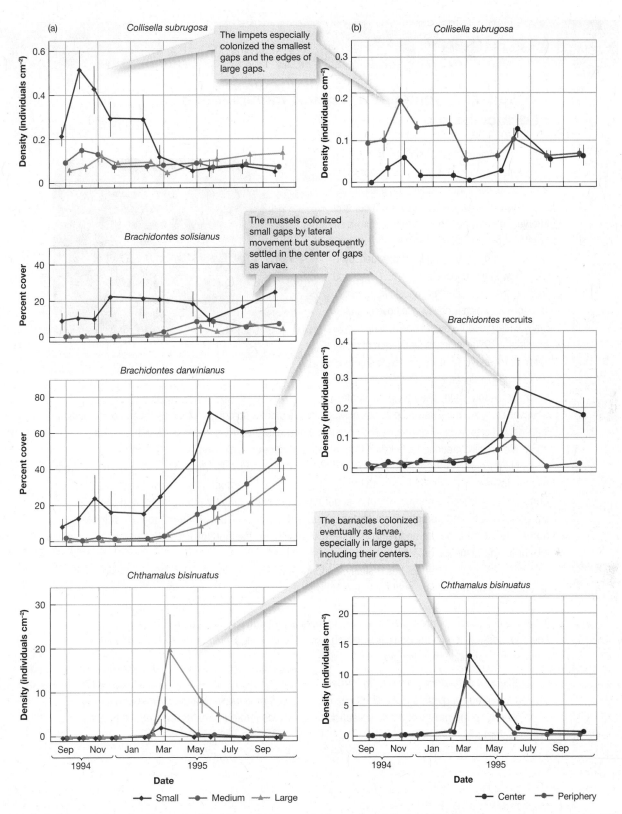

FIGURE 9.18   (a) Mean abundances (± SE) of four colonizing species in experimentally cleared small, medium, and large gaps in intertidal mussel beds. (b) Recruitment of three species at the periphery (within 5 cm of the gap edge) and in the center of large gaps. (After Tanaka & Magalhaes, 2002.)

9.3 ECOncerns

Conservation sometimes requires manipulation of a succession

Some endangered animal species are associated with particular stages of a succession. Their conservation then depends on a full understanding of the successional sequence, and intervention may be required to maintain their habitat at an appropriate successional stage.

An intriguing example is provided by a giant New Zealand insect, the weta, *Deinacrida mahoenuiensis* (Orthoptera, Anostostomatidae). This species, which is believed to have been formerly widespread in forest habitat, was discovered in the 1970s in an isolated patch of gorse (*Ulex europaeus*). Ironically, in New Zealand gorse is an introduced weed that farmers spend much time and effort attempting to control.

The giant weta. The world's largest cricket and one of the world's largest insects.

However, its dense, prickly sward provides a refuge for the giant weta against other introduced pests, particularly rats but also hedgehogs, stoats, and possums, which could readily capture wetas in their original forest home. Mammalian predation is believed to be responsible for weta extinction elsewhere.

New Zealand's Department of Conservation purchased this important patch of gorse from the landowner, who insisted that his cattle should still be permitted to overwinter in the reserve. Conservationists were unhappy about this, but the cattle subsequently proved to be part of the weta's salvation. By opening up paths through the gorse, cattle provided entry for feral goats that browse the gorse, producing a dense hedge-like sward and preventing the habitat from succeeding to a stage inappropriate to the wetas.

This story therefore features a single, endangered, endemic insect together with a whole suite of introduced pests (gorse, rats, goats) and introduced domestic animals (cattle). Before the arrival of people in New Zealand, the island's only land mammals were bats, and New Zealand's endemic fauna have proven to be extraordinarily vulnerable to the mammals that arrived with people. However, by maintaining gorse succession at an early stage, the grazing goats provide a habitat in which the wetas can escape the attentions of rats and other predators.

Because of its economic cost to farmers, ecologists have been trying to find an appropriate biological control agent for gorse, ideally one that would eradicate it. How would you weigh up the needs of a rare insect against the economic losses associated with gorse on farms?

B. darwinianus increasingly predominated and also built up its numbers in the medium and large gaps. In the absence of further disturbance, *B. darwinianus* would seem likely to outcompete *B. solisianus*. After around 6 months, too, the *Brachiodontes* mussels, which cannot be identified to species when they are small, recruited

significantly from larvae that settled into the central areas of the large gaps (Figure 9.18b). Finally, the barnacle *Chthamalus bisinuatus* also recruited from settled larvae, largely as a pulse after around 6 months, especially in the largest gaps (Figure 9.18a) and more in the center than at the periphery of the large gaps (Figure 9.18b).

Thus, the smaller the gap, the more the succession within it was dominated by lateral movement than by true migration, and even within a large gap, succession proceeded differently at the center and at the periphery. On the shore as a whole, therefore, as in any patchy and disturbed habitat, there was a mosaic of patches in different successional states, with those states being determined by patch size, the time since the last disturbance, and even location within a patch.

9.4   FOOD WEBS

No predator–prey, parasite–host, or grazer–plant (grazer–phytoplankton, in aquatic communities) pair exists in isolation. Each is part of a complex web of interactions with other predators, parasites, food sources, and competitors within its community. Ultimately, it is these food webs that ecologists wish to understand. However, it has been useful to isolate groups of competitors as we did in Chapter 6, of predator–prey and parasite–host pairs as in Chapter 7, and of mutualists as in Chapter 8, simply because we have little of understanding the whole unless we have some understanding of the component parts. Toward the end of Chapter 7 (Section 7.6), we expanded our field of view to include the effects of predators on groups of competitors and to show, for example, the importance of predator-mediated coexistence.

We now take this approach a stage further to focus on systems with at least three trophic levels (primary producer–herbivore–predator) and consider not only direct but also indirect effects that a species may have on others at the same or at other trophic levels. The effects of a predator on populations of its herbivorous prey, for example, are direct and relatively straightforward. But effects may also be felt by any plant population on which the herbivore feeds, other predators of the herbivore, other consumers of the plant, competitors of the herbivore, or the myriad species linked even more remotely in the food web.

Indirect and direct effects

The deliberate removal of a species from a community can be a powerful tool in unraveling the workings of a food web. We might expect such removal to lead to an increase in the abundance of a competitor, or, if the species removed is a predator, to an increase in the abundance of its prey. Sometimes, however, when a species is removed, a competitor may actually decrease in abundance, and the removal of a predator can lead to a decrease in a prey population. Such unexpected effects arise when direct effects are less important than effects that occur through indirect pathways. For example, removal of a species might increase the density of one competitor, which in turn causes another competitor to decline.

These indirect effects are brought especially into focus when the initial removal is carried out for some managerial reason, since the deliberate aim is to solve a problem, not create further, unexpected problems. For example, there are many islands on which feral cats have been allowed to escape domestication and now threaten native prey, especially birds, with extinction. The seemingly obvious response is to eliminate the cats and conserve their island prey, but as a simple model shows (Figure 9.19), such programs may not have the desired effect, especially where, as is often the case, rats have also been allowed to colonize the island.

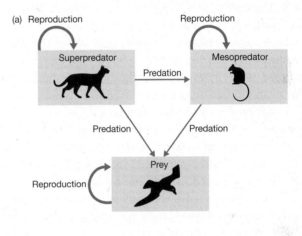

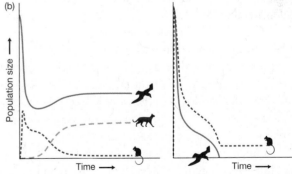

FIGURE 9.19   (a) Schematic representation of a model of an interaction in which a superpredator such as a cat preys both on mesopredators such as rats, for which the cat shows a preference, and on prey such as birds, while the mesopredator also attacks the prey. Each species also recruits to its own population via reproduction. (b) The output of the model with realistic values for rates of predation and reproduction. With all three species present, the superpredator keeps the mesopredator in check and the three coexist (left); but in the absence of the superpredator, the mesopredator drives the prey to extinction (right). (After Courchamp et al., 1999.)

Robin Bush/Oxford Scientific/Getty Images

The kakapo, a flightless parrot from New Zealand

The cats normally prey upon the rats as well as the birds, while the rats typically both compete with and prey upon the birds. Hence, removal of the cats will increase the abundance of rats and may thus increase rather than decrease the threat to the birds. For example, cats introduced on Stewart Island, New Zealand, preyed upon an endangered flightless parrot, the kakapo, *Strigops habroptilus* (Karl & Best, 1982). But controlling cats alone would have been risky, since their preferred prey are three species of introduced rats that, unchecked, could pose far more of a threat to the kakapo. To solve the problem, conservationists translocated Stewart Island's kakapo population to smaller offshore islands where exotic predators like rats were absent.

The indirect effect within a food web that has received most ⬛ trophic cascades attention is the so-called **trophic cascade**, which occurs when a predator reduces the abundance of its prey, and this cascades down to the trophic level below, such that the prey's own resources (typically plants or phytoplankton) increase in abundance.

Of course, it need not stop there. In a four-trophic-level system, if it is subject to trophic cascade, we might expect that as the abundance of a top carnivore increases, the abundances of primary carnivores in the trophic level below decrease, those of the herbivores in the level below that therefore increase, and plant abundance at the lowest level decreases. This is what researchers found in a study in the tropical lowland forests of Costa Rica. *Tarsobaenus* beetles preyed on *Pheidole* ants that preyed on a variety of herbivores that attacked ant-plants, *Piper cenocladum* (Figure 9.20a). These showed precisely the alternation of abundances expected in a four-trophic-level cascade: relatively high abundances of plants and ants associated with low levels of herbivory and beetle abundance at three sites, but low abundances of plants and ants associated with high levels of herbivory and beetle abundance at a fourth (Figure 9.20b).

However, results were more subtle in another four-trophic-level community, this one in the Bahamas (Figure 9.20c) and consisting of sea grape shrubs fed on by herbivorous arthropods, which were fed on by web spiders (primary carnivores). There were also lizards (top carnivores) that fed on both the spiders and the herbivores. When lizards were removed from experimental plots and results were compared to controls, the subtlety arose because there were different types of herbivores, and because the lizards ate large numbers of plant-sucking homopterans but very few gall-forming midges (which were mostly eaten by the spiders). Therefore, for the damage caused to the plants by the midges, the expected four-level alternation was observed. Removing lizards benefitted the plants. However, removing the lizards also directly and significantly benefitted the homopterans, which in turn harmed the plants, so in this regard at least, the four trophic–level community functioned as if it had only three levels.

In rare cases, a cascade can be observed to extend through five levels. Figure 9.21 shows the effects of the commercial fishing of cod (*Gadus morhua*) and other fish off the coast of Nova Scotia, Canada. As the cod were steadily overfished from the 1960s through the 1990s, the abundance of their own food (small fish and bottom-dwelling invertebrates) increased, the zooplankton on which these feed decreased, the phytoplankton on which the zooplankton feed increased,

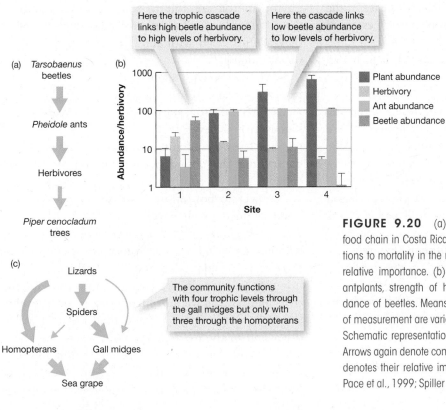

FIGURE 9.20 (a) Schematic representation of a four-level food chain in Costa Rica. Direction of the arrows denotes contributions to mortality in the resource; width of the arrows denotes their relative importance. (b) At four sites, the relative abundance of antplants, strength of herbivory, abundance of ants, and abundance of beetles. Means and standard errors are shown; the units of measurement are various and given in the original references. (c) Schematic representation of a four-level food chain in Costa Rica. Arrows again denote contributions to mortality in the resource; width denotes their relative importance. (After Letourneau & Dyer, 1998; Pace et al., 1999; Spiller & Schoener, 1990.)

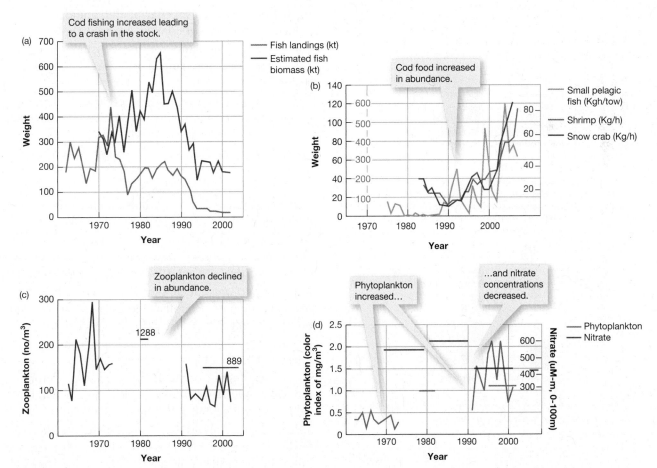

FIGURE 9.21 The effects of cod fishing (and that of other fish) off Nova Scotia from the 1960s through the 1990s (a) on the cod themselves, (b) on their food, estimated from standard samples from the environment, (c) on the zooplankton (horizontal lines are averages for separate samples from the periods concerned), and (d) on the phytoplankton and on nitrate concentrations. (After Frank et al., 2005.)

and it was even possible to monitor a decrease in the concentration of one of the phytoplankton's essential resources, nitrate.

A final example of a trophic cascade is provided by another aquatic example—a comparison of lakes in Wisconsin with either fish-eating bass or plankton-eating minnows at the top trophic level. When the bass are present, they effectively eliminate these plankton-eaters. This has the potential to reduce the grazing pressure on the zooplankton, allowing them to increase their own grazing pressure on the phytoplankton, cascading down to reduce phytoplankton abundance. In fact, however, this full cascade was observed only if the communities were also subjected to an experimental addition of nutrients (Figure 9.22). Without added nutrients, the effect on the zooplankton was apparent, but this did not cascade down to the trophic level below. It appears that the

phytoplankton are able to respond differentially to the different levels of zooplankton grazing only when they are provided with sufficient nutrients to allow them to do so.

We have seen, therefore, that trophic cascades are generally viewed from the top, starting at the highest trophic level. So, in a three-trophic–level community, we think of the predators controlling the abundance of the herbivores—so-called top-down control. Reciprocally, the predators are subject to bottom-up control: their abundance is determined by their resources. With these three levels, the plants are also subject to bottom-up control, having been released from top-down control by the effects of the predators on the herbivores. Thus, in a trophic cascade, top-down and bottom-up controls alternate as we move from one trophic level to the next. But suppose instead that we start at the other end of the food chain, and assume the plants are controlled bottom-up by competition for their resources. It is still possible for the herbivores to be limited by competition for plants—*their* resources— and for the predators to be limited by competition for herbivores. In this scenario, all trophic levels are subject to bottom-up control, because the resource controls the abundance of the consumer but the consumer does not control the abundance of the resource.

> top-down or bottom-up control of food webs?

Of course, it must often be the case that top-down and bottom-up effects combine. We saw an example in Figure 9.22, where the top-down effects of the fish-eating bass on the phytoplankton were apparent only when the phytoplankton were supplied, bottom up, with sufficient nutrients. Nonetheless, it is possible to find examples of bottom-up control to set alongside the trophic cascades described above. In one study in Brazil, for example, the perennial shrub, *Chromolaena pungens*, was subjected to two levels of clipping—either around 50% of leaves were removed or the plant was cut at the base – and allowed to regrow. Clipping generally is known to give rise to poorly defended, nutrient-rich foliage in the rapidly (re-)growing plants, which were then compared to unclipped control plants, in terms of the plants' own traits, and in terms of the herbivores and predators in the two trophic levels above. Partial clipping had relatively little effect, but fully clipped plants grew faster and had larger leaves and were much less likely to flower (Figure 9.23a). This led more seed eaters to infest flower heads in the clipped plants, and the level of leaf consumption was also higher (Figure 9.23b,c). This increased availability of food led in turn to much

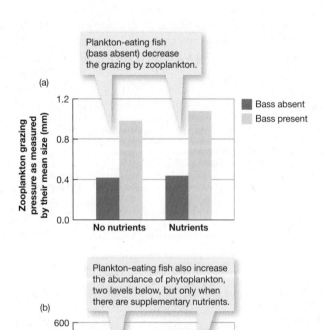

FIGURE 9.22 Responses in Wisconsin lakes of (a) zooplankton and (b) phytoplankton to either one trophic level ('Bass absent') or two trophic levels ('Bass present') above them, with and without the addition of nutrients. In (a), zooplankton grazing pressure is strongly correlated with their mean size, which is much lower when plankton-eating fish are abundant because they selectively prey on the larger, more voracious zooplankton. (After Carpenter et al., 1995: Pace et al., 1999.)

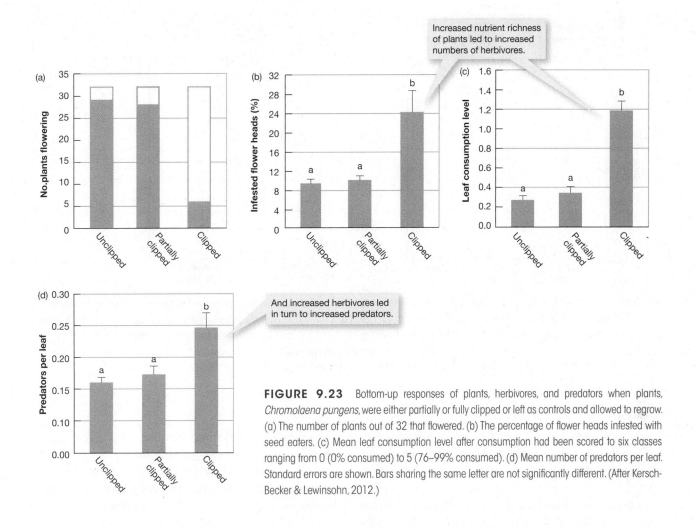

Increased nutrient richness of plants led to increased numbers of herbivores.

And increased herbivores led in turn to increased predators.

FIGURE 9.23 Bottom-up responses of plants, herbivores, and predators when plants, *Chromolaena pungens*, were either partially or fully clipped or left as controls and allowed to regrow. (a) The number of plants out of 32 that flowered. (b) The percentage of flower heads infested with seed eaters. (c) Mean leaf consumption level after consumption had been scored to six classes ranging from 0 (0% consumed) to 5 (76–99% consumed). (d) Mean number of predators per leaf. Standard errors are shown. Bars sharing the same letter are not significantly different. (After Kersch-Becker & Lewinsohn, 2012.)

larger numbers of predators on the clipped plants (Figure 9.23d).

There are examples, then, of both top-down and bottom-up control, but the question nonetheless arises – are food webs, or are particular *types* of food web, dominated by either one or the other? The widespread importance of top-down control was first advocated in a famous paper by Hairston et al. (1960), which asked 'Why is the world green?' The authors answered, in effect, that the world is green because top-down control predominates: green plant biomass accumulates because predators keep herbivores in check. Murdoch (1966), in particular, challenged these ideas. His view, described by Pimm (1991) as 'the world is prickly and tastes bad,' emphasized that even if the world is green (assuming it is), it does not necessarily follow that the herbivores are failing to capitalize on this because they are limited, top down, by their predators. Many plants have evolved physical and chemical defenses that make life difficult for herbivores. The herbivores may therefore be competing fiercely for a

why is the world green?

limited amount of palatable and unprotected plant material, and their predators may, in turn, compete for scarce herbivores. A world controlled from the bottom up may still be green.

The appeal of general rules is understandable, but it remains a challenge for the future to discover whether top-down or bottom-up control predominates in particular types of communities. Indeed, it is likely that in many communities, both forces have a role to play. Nonetheless, there have been several **meta-analyses**—structured analyses of large numbers of data sets with a view to discerning consistent trends—of the large number of studies of possible top-down or bottom-up control, and the results of one of these are shown in Figure 9.24. Predator manipulation experiments consistently reveal a significant negative effect of the predators on herbivores, and a somewhat less profound but nonetheless significant positive, top-down, trophic cascade effect of the predators on the plants (Figure 9.24a). However, while fertilization experiments consistently have a significant bottom-up positive effect on the plants themselves, there is no

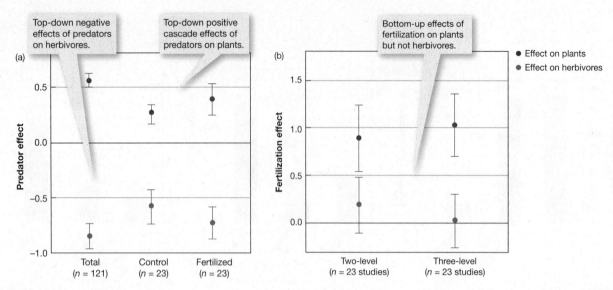

FIGURE 9.24 Results from a meta-analysis of studies of top-down and bottom-up community control. (a) Effects of predator manipulation across 121 studies from terrestrial, freshwater, and marine systems (Total) and the subset of studies that simultaneously manipulated predators and nutrients (Control compared to Fertilized). The predator effect is the log ratio of biomass with and without predators. (b) Effects of fertilization in the presence and absence of predators in the 23 studies in which nutrients were manipulated. The fertilization effect is the log ratio of biomass with and without fertilization. In both cases, zero represents no effect. Standard errors are shown. (After Borer et al., 2006.)

evidence overall for this effect being transmitted further up the food chain to the herbivores, whether or not there are predators present as well (Figure 9.24b). Also, fertilization from below tends to have no attenuating action on the top-down effects of predator manipulation (Figure 9.24a). Hence, while both may be important, it seems that top-down effects are more likely to reverberate throughout the system than bottom-up effects.

Population and community stability and food web structure

Of all the imaginable food webs in nature, are there particular types that we tend to observe repeatedly? Are some food web structures more stable than others? Do we observe particular types of food web *because* they are stable (and hence persist)? Are populations themselves more stable when embedded in some types of food web than in others? These are important practical questions. We require answers if we are to determine whether some communities are more fragile (and more in need of conservation) than others, or whether there are certain natural structures we should aim for when we construct communities ourselves, or whether communities that have been restored are likely to remain so. Before we address these questions, though, we must explore briefly some different meanings and aspects of the word 'stable' (Box 9.4.)

Whatever meaning we use, | keystone species | 'stability' usually means stability in the face of a disturbance or perturbation, and most disturbances are, in practice, the loss of one or more populations from a community. What are the ripple effects of such a loss? How profound are the consequences of the loss of that population for the rest of the community? Consider that in building construction, a keystone is the wedge-shaped block at the highest point of an arch that locks the other pieces together. The removal of the keystone in an arch leads to the collapse of the whole structure. Similarly, some species—strong interactors—are more intimately and tightly woven into the fabric of a food web than others, and the removal of these *keystone species* leads to extinctions or large changes in abundance of several species, producing a community with a very different species composition.

A more precise definition of a **keystone species** is one whose impact on community composition is disproportionately large relative to its abundance (Power et al., 1996). This definition has the advantage of excluding from keystone status what would otherwise be rather trivial examples, especially species at lower trophic levels that may provide the resource on which myriad other species depend—for example, a coral, or the oak trees in an oak woodland.

Although the term was originally applied only to predators, it is now widely accepted that keystone species can occur at any trophic level. For example, lesser

Quantitative Aspects

What do we mean by 'stability'?

In trying to decide what we mean by stability, we should first draw a distinction between the resilience of a community and its resistance. A **resilient community** is one that returns rapidly to something like its former structure after that structure has been altered. A **resistant community** is one that undergoes relatively little change in its structure in the face of a disturbance.

A second distinction is between fragile and robust stability. A community has only **fragile stability** if it remains essentially unchanged in the face of a small disturbance but alters utterly when subjected to a larger disturbance. By contrast, a community is **dynamically robust** if it stays roughly the same in the face of much larger disturbances.

To illustrate these distinctions by analogy, consider the following:

- a pool or billiard ball balanced carefully on the end of a cue

- the same ball resting on the table

- the ball sitting snugly in its pocket

The ball on the cue is stable in the narrow sense that it will stay there forever as long as it is not disturbed—but its stability is fragile, and both its resistance and its resilience are low: the slightest touch will send the ball to the ground, far from its former state (low resistance), and it has not the slightest tendency to return to its former position (low resilience).

The same ball resting on the table has a similarly low resilience. It has no tendency to return to exactly its former state (assuming the table is level), but its resistance is far higher; pushing or hitting it moves it relatively little. And its stability is also relatively robust. It remains 'a ball on the table' in the face of all sorts and all strengths of assault with the cue.

The ball in the pocket, finally, is not only resistant but resilient too—it moves little and then returns—and its stability is highly robust. It will remain where it is in the face of almost everything, other than a hand that carefully plucks it away.

snow geese (*Chen caerulescens caerulescens*) are herbivores that breed in large colonies in coastal marshes along the west coast of Hudson Bay in Canada. At their nesting sites in spring, before growth of aboveground foliage begins, adult geese grub for the roots in dry areas and eat the swollen bases of shoots in wet areas. Their activity creates bare areas (1–5 m²) of peat and sediment. Few pioneer plant species are able to recolonize these patches, and recovery is very slow. Furthermore, in areas of intense summer grazing, lawns of *Carex* and *Puccinellia* spp. have become established. Overall, therefore, high densities of grazing geese are essential to maintain the species composition of the vegetation and its above-ground production (Kerbes et al., 1990). The lesser snow goose is a keystone species; the whole structure and composition of these communities are drastically altered by its presence.

A keystone species may be critical to the stability of a community, but are there characteristics of the community as a whole that stabilize it? For a long time, the conventional wisdom, arrived at largely through apparently logical argument, was that increased complexity within a community leads to increased stability (MacArthur, 1955; Elton, 1958). For example, it was thought that in more complex communities, with more species and more interactions, there were more possible pathways by which energy passed through the community. Hence, if there was a perturbation to the community—a change in the density of one of the species—this would affect only a small proportion of the energy pathways and would have relatively little effect on the densities of other species: the complex community would be resistant to change (Box 9.4). However, when ecologists came to

> complexity leads to stability? what models suggest

formalize their seemingly logical arguments into mathematical models of food webs, the conventional wisdom was by no means always supported (May, 1981; Tilman, 1999). Indeed, conclusions differ depending on whether we focus on individual populations within a community or on aggregate properties of the community such as its biomass or productivity.

Briefly, the model food webs have been characterized by one or more of the following: (i) the number of species they contain; (ii) the **connectance** of the web, meaning the fraction of all possible pairs of species that interact directly with one another; and (iii) the average interaction strength between pairs of species. At the level of the individual population, models overall have tended to suggest that increases in the number of species, increases in connectance, and increases in average interaction strength—each representing an increase in complexity—all tend to *decrease* the resilience of individual populations within the community, that is, their tendency to return to their former state following a disturbance (Figure 9.25). Thus, these models suggest, if anything, that community complexity leads to population *instability*.

However, the effects of complexity, especially species richness, on the stability of aggregate properties of model communities have been more consistent. Broadly, in richer communities, the dynamics of these aggregate properties are *more* stable (Figure 9.25). The reason is largely that, as long as the fluctuations in different populations are not perfectly correlated, there is an inevitable statistical averaging effect when populations are added together—when one goes up, another is going down—and this tends to increase in

effectiveness as richness (the number of populations) increases.

In fact, though, our preoccupation with the effect on stability of *overall* complexity, *average* interaction strength, and so on may itself be misleading. Models suggest that it may be the distribution of complexity and interaction strengths within a food web that are key to its stability. For example, a good case can be made for the critical importance of the **compartmentalization** of a food web, meaning the tendency for subsets of species to interact more strongly and more frequently among themselves than with other species in the community (Figure 9.26a). Results from models suggest that more compartmentalized webs are more resilient to perturbation (the removal of one species), and indeed that the contribution compartmentalization makes to such stability is greater the greater that compartmentalization is (Figure 9.26b). This occurs because most subsequent extinctions occur in the same compartment as the original perturbation, rather than spreading evenly through the whole web (Figure 9.26c).

What is the evidence from real communities? The prediction that populations in richer communities are less stable when disturbed was investigated by Tilman (1996), who pooled data for 39 common plant species from 207 grassland plots in Cedar Creek Natural History Area, Minnesota, collected over an 11-year period. He found that variation in the biomass of individual species increased significantly, but only very weakly, with the richness of the plots (Figure 9.27a). Thus, like the theoretical studies, this empirical study hints at decreased population stability and increased variability in more complex communities, but the effect seems to be weak and inconsistent.

> complexity and stability in practice

Turning to the aggregate, whole-community level, we find the evidence is largely consistent in supporting the prediction that increased richness in a community increases stability and decreases variability. For example, in Tilman's Minnesota grasslands study, in contrast to the weak negative effect found at the population level, there was a strong positive effect of richness on the stability of community biomass (Figure 9.27b). Also, when richness in aquatic microbial communities (producers, herbivores, bacterivores, predators) was manipulated, it was apparent that variation in another community measure, carbon dioxide flux—a measure of community respiration—also declined with richness (Figure 9.28).

On the other hand, in an experimental study of small grassland communities perturbed by an induced drought, Wardle et al. (2000) found detailed

> the devil is in the detail

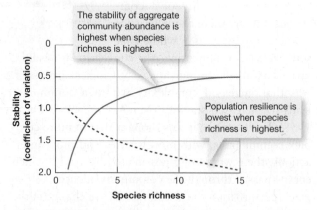

FIGURE 9.25 An example of the effect of species richness (number of species) on the stability of population size and of aggregate community abundance in model communities. Stability is measured as temporal variability (coefficient of variation, CV). Thus, high values for CV correspond to low levels of stability. (After Cottingham et al., 2001.)

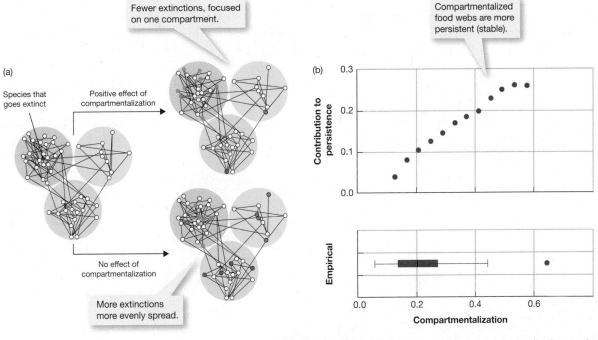

FIGURE 9.26 (a) Potential effects of compartmentalization on the fate of a food web, following the initial extinction of one species (red dot). Subsequent extinctions are also in deep colors. Compartmentalization in such webs is zero when species are connected at random and increases as more connections are confined within compartments. 'Persistence' describes the fraction of the original species remaining at the end of a model simulation. (b) Above, the contribution of compartmentalization to the persistence of species in model food webs varying in compartmentalization and connectance when one species is removed. The contribution is estimated from the proportion of persistence not accounted for by connectance and 'baseline' persistence. Below, box-whisker plots for estimated compartmentalization from 15 actual (empirical) food webs. The mid-line is the median; the box covers the first-third quartiles; the whiskers span 1.5 times the interquartile range; the one dot is an outlier. (c) The effect of compartmentalization on the relative chance of extinction in the same compartment as the original extinction. Values greater than zero indicate that extinctions are more likely in the same compartment. Bars are SEs. (After Stouffer & Bascompte, 2011.)

community composition to be a far better predictor of stability than overall richness. Indeed, the whole concept of a keystone species (see above) recognizes that the effects of a disturbance on structure or function are likely to depend very much on the precise nature of the disturbance, and in particular on *which* species are lost.

The theme of the fine detail being more important than the overall average is repeated, too, in the case of compartmentalization. Real food webs certainly seem to be more compartmentalized than we would expect from making merely random connections between even connectable species (Figure 9.26b). It seems, however, that the compartmentalization itself may emerge naturally or even inevitably. For example, five major compartments were identified

in an analysis of a Caribbean marine food web covering 3,313 trophic interactions between 249 species, and running from the surface to a depth of 100 m. It seems, though, that much of the subdivision between these compartments occurs either because members of different compartments simply occupy different habitats (shore or off-shore; Figure 9.29a), or because the morphological constraints on each species confine them to feeding on prey in a particular size-range (Figure 9.29b), or because the combinations of trophic levels occupied by each species bring them into closer connection with other species occupying similar levels (Figure 9.29c).

Moreover, the facts that real food webs are compartmentalized, and that compartmentalization promotes stability, do not imply that food webs are

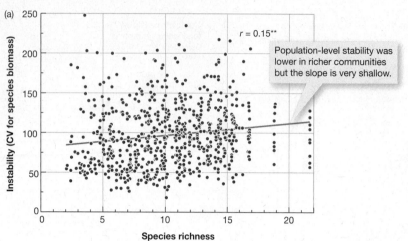

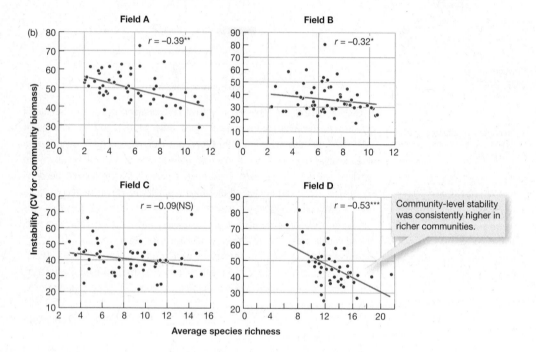

FIGURE 9.27 (a) The coefficient of variation (CV) of population biomass for 39 plant species from plots in four fields in Minnesota over 11 years (1984–1994), plotted against species richness in the plots. (b) The CV for community biomass in each plot plotted against species richness for each of the four fields (A–D). In both cases, regression lines and correlation coefficients are shown (*, $P < 0.05$; **, $P < 0.01$; ***, $P < 0.001$). (After Tilman, 1996.)

compartmentalized *because* of the stability this imparts. Rather, it helps explain why we can frequently observe complex food webs in nature, despite the fact that such complexity tends to promote instability. One reason, at least, is that they are compartmentalized.

In fact, even if complexity and instability are connected in models, this does not necessarily mean we should expect to see an association between complexity and instability in real communities. In a stable and predictable environment, a community that is dynamically fragile may nevertheless persist. But in a variable and unpredictable environment, only a community that is dynamically robust will be able to persist. Hence, we might expect to see: (i) complex and relatively fragile communities in stable and predictable environments, with simple and robust communities in variable and unpredictable environments,

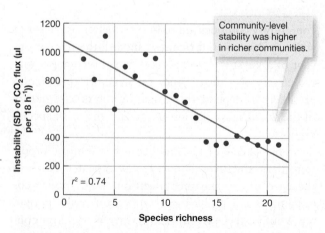

FIGURE 9.28 Variation in community productivity [standard deviation (SD) of carbon dioxide flux] declined with species richness in microbial communities observed over a 6-week period. (After McGrady-Steed et al., 1997.)

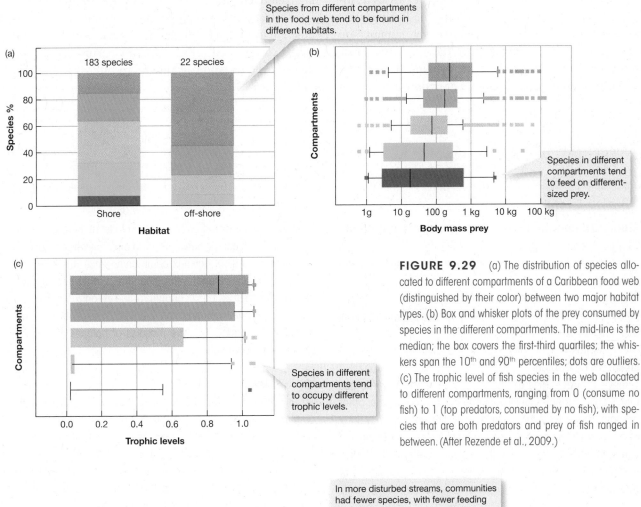

FIGURE 9.29 (a) The distribution of species allocated to different compartments of a Caribbean food web (distinguished by their color) between two major habitat types. (b) Box and whisker plots of the prey consumed by species in the different compartments. The mid-line is the median; the box covers the first-third quartiles; the whiskers span the 10th and 90th percentiles; dots are outliers. (c) The trophic level of fish species in the web allocated to different compartments, ranging from 0 (consume no fish) to 1 (top predators, consumed by no fish), with species that are both predators and prey of fish ranged in between. (After Rezende et al., 2009.)

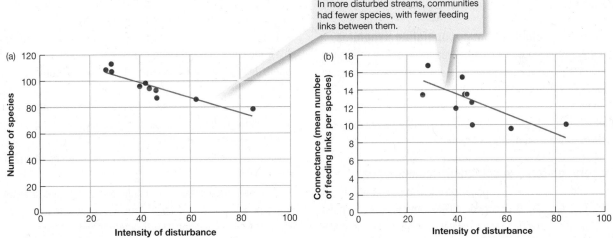

FIGURE 9.30 In New Zealand streams, the relationship between the intensity of disturbance (percentage of standard stream-bed particles moving in a defined period) and (a) the number of species in the community, and (b) the connectance between species in the community. (After Townsend et al., 1998.)

but (ii) approximately the same *observed* stability (in terms of population fluctuations, and so forth) in all communities, since this will depend both on the inherent stability of the community itself and on the variability of the environment. One study supporting at least the first of these points investigated 10 small

streams in New Zealand that differed in the intensity and frequency of flow-related disturbances to their beds (Figure 9.30). Food webs in the more disturbed, unstable streams were characterized by less complex communities, with fewer species and fewer feeding links between them.

Finally, then, this line of argument carries a further, very important implication for the likely effects of unnatural perturbations caused by humans on communities. We might expect these to have their most profound effects on the dynamically fragile, complex communities of stable environments, which are relatively unaccustomed to perturbations, and least effect on the simple, robust communities of variable environments, which have previously been subjected to repeated (albeit natural) perturbations. We pick up the question of threats to biodiversity again in Chapter 13.

SUMMARY

Multiple determinants of the dynamics of populations

To understand the factors responsible for the population dynamics of even a single species in a single location, we need knowledge of physicochemical conditions, available resources, the organism's life cycle, and the influence of competitors, predators, and parasites on rates of birth, death, immigration, and emigration.

Contrasting theories explain the abundance of populations. At one extreme, researchers emphasize the apparent stability of populations and point to the importance of forces that stabilize them (density-dependent factors). At the other extreme, those who focus on density fluctuations may look at external (often density-independent) factors to explain the changes. We can apply k-value analysis to life table studies to throw light on both the determination and the regulation of abundance.

Dispersal, patches, and metapopulation dynamics

Movement can be a vital factor in determining and/or regulating abundance. A radical change in the way ecologists think about populations has focused attention less on processes occurring within populations and more on patchiness, the colonization and extinction of subpopulations within an overall metapopulation, and dispersal between subpopulations.

Temporal patterns in community composition

Disturbances that open up gaps (patches) are common in all kinds of community. Founder-controlled communities are those in which all species are approximately equivalent in their ability to invade gaps and are equal competitors that can hold the gaps against all comers during their lifetime. Dominance-controlled communities are those in which some species are competitively superior to others so that an initial colonizer of a patch cannot necessarily maintain its presence there.

The phenomenon of dominance control is responsible for many examples of community succession. Primary successions occur in habitats where no seeds or spores remain from previous occupants of the site; all colonization must be from outside the patch. Secondary successions occur when existing communities are disturbed but some at least of their seeds and so on remain. It can be very difficult to pinpoint when a succession reaches a stable climax community, since this may take centuries to achieve and in the meantime further disturbances are likely to occur. The exact nature of the colonization process in an empty patch depends on the size and location of that patch. Many communities are mosaics of patches at different stages in a succession.

Food webs

No predator–prey, parasite–host, or grazer–plant pair exists in isolation. Each is part of a complex food web including other predators, parasites, food sources, and competitors within the various trophic levels of a community. One of the most common indirect effects of one species on another is a trophic cascade, in which, say, a predator reduces the abundance of a herbivore, thus increasing the abundance of plants.

Top-down control of a food web occurs in situations in which the structure (abundance, species number) of lower trophic levels depends on the effects of consumers from higher trophic levels. Bottom-up control occurs in a community dependent on factors, such as nutrient concentration and prey availability, that influence a trophic level from below. The relative importance of these forces varies according to the trophic level under investigation and the number of trophic levels present.

Some species are more tightly woven into the food web than others. A species whose removal would produce a significant effect like extinction or a large change in density in at least one other species may be thought of as a strong interactor. Removal of such keystone species leads to significant changes that spread throughout the food web.

The relationship between food web complexity and stability is uncertain (and we need to define stability carefully). Mathematical and empirical studies agree in suggesting that, if anything, population stability decreases with complexity, whereas the stability of aggregate properties of whole communities increases with complexity, especially species richness.

REVIEW QUESTIONS

1   We can use population census data to establish correlations between abundance and external factors such as weather. Why can't we use such correlations to prove a causal relationship that explains the dynamics of the population?

2   Distinguish between the determination and regulation of population abundance.

3   What is meant by a metapopulation and how does it differ from a simple population?

4   Define founder control and dominance control as they apply to community organization. In a mosaic of habitat patches, how would you expect communities to differ if they were dominated by founder or dominance control?

5   What factors are responsible for changes in species composition during an old-field succession?

6   What are bottom-up and top-down control? How is the importance of each likely to vary with the number of trophic levels in a community?

7   Discuss what we know about the relationship between the complexity and stability of food webs.

Challenge Questions

1   Use boxes and arrows to construct a flow diagram with a named population at its center to illustrate the wide range of abiotic and biotic factors that influence its pattern of abundance.

2   Imagine a number of species with patchy distributions, including a plant, an insect, and a mammal, or consider examples of such species with which you are familiar. How would you identify habitable patches of these species that they do not currently occupy?

3   Draw up a food web of, say, six or seven species with which you are familiar and that spans at least three trophic levels. Take each species in turn and suggest the kind of community organization necessary for this to be a keystone species.

Chapter 10

Patterns in species richness

CHAPTER CONTENTS

KEY CONCEPTS

After reading this chapter you will be able to:

- explain how species richness may be limited by available resources

- describe how species richness can vary with productivity, predation intensity, energy inputs, spatial heterogeneity, and environmental harshness

- describe how species richness varies with levels of unpredictable disturbance, with variation in conditions over the annual cycle, and over evolutionary time

- explain the importance of habitat area and remoteness in determining species richness

- describe and explain how species richness changes with latitude, altitude, and depth, and during community succession

An accurate appreciation of the world's biological diversity is becoming increasingly important. For our conservation efforts to be effective, we must understand why species richness varies widely across the face of the Earth. Why do some communities contain more species than others? Are there patterns or gradients in this biodiversity? If so, what are the reasons for these patterns?

10.1   QUANTIFYING SPECIES RICHNESS AND DIVERSITY

Why does the number of species vary from place to place? And from time to time? These are questions that present themselves not only to ecologists but to anybody who observes and ponders the natural world. They are interesting questions in their own right—but they are also questions of practical importance. If we wish to conserve or restore the planet's biological diversity, then we must understand how species numbers are determined and how it comes about that they vary. We will see that there are plausible answers to the questions we ask, but these answers are not always clear-cut. This may be disappointing, but it is also a challenge to ecologists of the future. Much of the fascination of ecology lies in the fact that many of the problems are blatant, whereas the solutions can be difficult to find. We will see that a full understanding of patterns in species richness must draw on our knowledge of all the areas of ecology discussed so far in this book.

The number of species in a community is referred to as its *species richness*. Counting or list- | determining species richness | ing the species present in a community may sound a straightforward procedure, but in practice it is often surprisingly difficult, partly because of taxonomic inadequacies, but also because we can usually count only a proportion of the organisms in an area. The number of species recorded then depends on the number of samples we have taken, or on the volume of the habitat we have explored. The most common species are likely to be represented in the first few samples, and as we take more samples, we begin adding rarer species to the list.

At what point do we stop taking further samples? Ideally, we should continue to sample until the number of species stops increasing. For one thing, we can compare the species richness of different communities only if they are based on the same sample sizes, in terms of area of habitat explored, time devoted to sampling, or, best of all, number of individuals included in the samples.

When we simply state the number of species present, however, we overlook an important aspect of a community's structure—namely, that some species are rare and oth- | diversity indices and rank–abundance diagrams | ers common. Intuitively, a community of 10 species with equal numbers in each seems more diverse than another with 10 species in which 91% of the individuals belong to the most common species and just 1% to each of the other nine. Yet, each community has the same species richness. **Diversity indices** are measures that combine both species richness and the evenness or equitability of the distribution of individuals among those species (Box 10.1). Moreover, attempts to describe a complex community structure by one single attribute, such as richness, or even diversity, can still be criticized because so much valuable information is lost. A more complete picture of the distribution of species abundance in a community is therefore sometimes provided in a **rank–abundance diagram** (Box 10.1).

Nonetheless, for many purposes, the simplest measure, species richness, suffices. In the following sections, therefore, we examine the relationships between species richness and a variety of factors that may, in theory, influence richness in ecological communities. It will become clear that it is not always easy to come up with unambiguous predictions and clean tests of hypotheses.

To try to understand the determinants of species richness, we begin with a simple model (Figure 10.3). Assume, for simplicity, that we can depict the resources available to a community as a one-dimensional continuum, R units long. Each species uses only a portion of this resource continuum, and the lengths of these portions define the **niche breadth** (n) of each species; the average niche breadth within the community is then $\bar{n}$. Some of these niches overlap, and we can measure the overlap between adjacent species by the value o. The average

Quantitative Aspects

Diversity indices and rank–abundance diagrams

The measure most often used to combine both species richness and the relative abundances of those species in a community is the **Shannon** or **Shannon–Weaver diversity index** (denoted by *H*). We calculate this by determining, for each species, the proportion of individuals or biomass (P_i for the *i*th species) that the species contributes to the total in the sample. Then, if *S* is the total number of species in the community (the richness), diversity (*H*) is:

$$H = -\Sigma P_i \ln P_i$$

where the summation sign Σ indicates that the product ($P_i \ln P_i$) is calculated for each of the *S* species in turn and these products are then summed. As required, the value of the index depends on the species richness, but also on the evenness (equitability) with which individuals are distributed among the species. Thus, for a given richness, *H* increases with equitability, and for a given equitability, *H* increases with richness.

An example of an analysis using diversity indices is provided by the unusually long-term study begun in 1856 in an area of pasture at Rothamsted in the UK. Experimental plots received a fertilizer treatment once every year, and control plots did not. Figure 10.1 shows how species diversity (*H*) of grass species changed between 1856 and 1949. While the unfertilized area remained essentially unchanged, the fertilized area showed a progressive decline in diversity. This 'paradox of enrichment' is discussed in Section 10.3.

Rank–abundance diagrams, on the other hand, make use of the full array of P_i values by plotting P_i against rank. That is, the most abundant species takes rank 1, the second most abundant rank 2, and so on, until the array is completed by the rarest species of all. The steeper the slope of a rank–abundance diagram, the greater the dominance of common species over rare species in the community (a steep slope means a sharp drop in relative abundance, P_i, for a given drop in rank). Thus, in the case of the Rothamsted experiment, Figure 10.2 shows how the dominance of commoner species steadily increased (steeper slope) while species richness decreased over time.

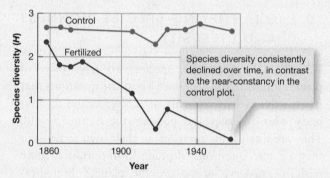

FIGURE 10.1 Species diversity (*H*) over almost 100 years on two plots of pasture in an experiment commencing in 1856 at Rothamsted in the UK. One plot regularly received fertilizer; the other control plot received no fertilizer. (After Tilman, 1982.)

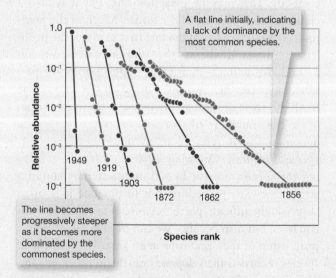

FIGURE 10.2 Change in the rank–abundance pattern of plant species in the Rothamstead fertilized plot from 1856 to 1949. Note how the slope of the regression line becomes progressively steeper with time since commencement of fertilizer addition. A steeper plot indicates that the commoner species comprise a greater proportion of the total community – in other words, this pasture community gradually became dominated by just a few species. (After Tokeshi, 1993.)

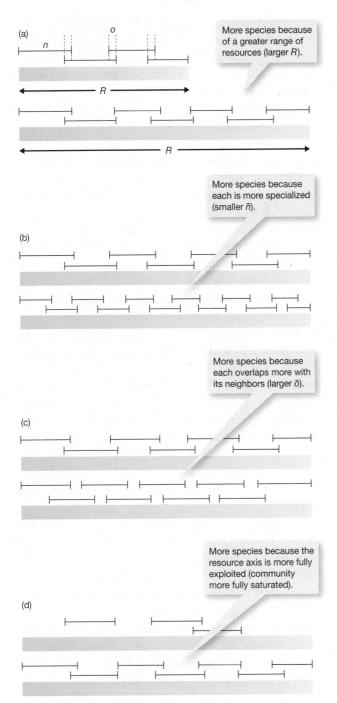

FIGURE 10.3   A simple model of species richness. Each species utilizes a portion n of the available resources (R), overlapping with adjacent species by an amount o. More species may occur in one community than in another because: (a) a greater range of resources is present (larger R), (b) each species is more specialized (smaller average n), (c) each species overlaps more with its neighbors (larger average o), or (d) the resource dimension is more fully exploited. (After MacArthur, 1972.)

niche overlap within the community is then $\bar{o}$. With this simple background, it is possible to consider why some communities should contain more species than others.

First, for given values of $\bar{n}$ and $\bar{o}$, a community will contain more species the larger the value of R, that is, the greater the range of resources (Figure 10.3a). Second, for a given range of resources, more species will be accommodated if $\bar{n}$ is smaller, that is, if the species are more specialized in their use of resources (Figure 10.3b). Alternatively, if species overlap to a greater extent in their use of resources (greater $\bar{o}$), then more may coexist along the same resource continuum (Figure 10.3c). Finally, a community will contain more species the more fully saturated it is; conversely, it will contain fewer species when more of the resource continuum is unexploited (Figure 10.3d).

We can now consider the relationship between this model and two important kinds of species interactions described in previous chapters: interspecific competition

> competition and predation may influence species richness

and predation. If a community is dominated by interspecific competition (see Chapter 6), the resources are likely to be fully exploited. Species richness will then depend on the range of available resources, the extent to which species are specialists, and the permitted extent of niche overlap (Figure 10.3a–c). As the chapter proceeds, we will examine a range of influences on each of these three.

Predation, on the other hand, is capable of exerting contrasting effects (see Chapter 7). First, we know that predators can exclude certain prey species; in the absence of these species the community may then be less than fully saturated (Figure 10.3d), and predation may therefore reduce species richness. On the other hand, predation may tend to keep species below their carrying capacities for much of the time, reducing the intensity and importance of direct interspecific competition for resources. This may then permit much more niche overlap and a *greater* richness of species than in a community dominated by competition (Figure 10.3c).

The next two sections examine a variety of factors that influence species richness. To organize these, Section 10.2 focuses on factors that often vary spatially (from place to place): productivity, energy, predation intensity, spatial heterogeneity, and environmental harshness. Section 10.3 then focuses on factors reflecting temporal variation: climatic variation, disturbance, and evolutionary age.

10.2 SPATIALLY VARYING FACTORS INFLUENCING SPECIES RICHNESS

We can subdivide into two separate hypotheses the general idea that spatial (geographic) variations in climate play a critical role in determining species richness. The first is the **productivity hypothesis**. This emphasizes the importance of climate in determining productivity at the lowest trophic level – the plants and microbes – and the resources these then provide for herbivores and then carnivores further up the food chain. Of course, we must remember that productivity at a particular location can also be determined by other factors unrelated to climate.

We deal with this productivity hypothesis first. The alternative, the **energy hypothesis**, emphasizes the direct role of energy (often measured by environmental temperature) on organisms throughout the community. We discuss this hypothesis second.

Productivity and resource richness

For plants, the productivity of the environment can depend on whichever nutrient or condition is most limiting to growth (see Chapter 11). Broadly speaking, the productivity of the environment for animals follows the same trends as for plants, mainly as a result of the changes in resource levels at the base of the food chain. If higher productivity is correlated with a wider *range* of available resources, then this is likely to lead to an increase in species richness (Figure 10.3a). However, a more productive environment may have a higher rate of supply of resources but not a greater variety of resources. This might lead to more individuals per species, rather than more species.

Again, however, it is possible, even if the overall variety of resources is unaffected, that rare resources in an unproductive environment may become abundant enough in a productive environment for extra species to be added, because more specialized species can be accommodated (Figure 10.3b). In general, then, there are rather straightforward grounds for expecting that species richness should increase with productivity.

Examples do appear to support this contention (Figure 10.4). The species richness of fish in North American lakes, for example, increases with an increase

in productivity of the lake's phytoplankton (Figure 10.4a). There are also strong positive correlations between species richness and precipitation for both seed-eating ants and seed-eating rodents in the southwestern deserts of the United States (Figure 10.4b), where it is well established that mean annual precipitation is closely related to plant productivity and thus to the amount of seed resource available. In this latter case, it is particularly noteworthy that in species-rich sites the communities contain more species of very large ants (which consume large seeds) and more species of very small ants (which take small seeds). It seems that either the range of seed sizes is greater in the more productive environments (Figure 10.3a), or the abundance of seeds becomes sufficient to support extra consumer species with narrower niches (Figure 10.3b).

On the other hand, an increase in diversity with productivity is by no means universal. We saw in Box 10.1, for example, in the long-term experiment at Rothamsted, that fertilized areas showed a progressive decline in species richness and diversity, and we had previously seen the same pattern for Minnesota grasslands in Figure 1.6. Similarly, a survey of plants in rich-fen sites in England and Wales (wetland sites where the water table was at or slightly above the substratum and with a high proportion of plants that thrive in lime-rich soils) showed clear evidence that species richness was lowest where productivity was greatest (Figure 10.5).

Indeed, an association between high productivity and low species richness has often been found in plant communities. We also see it where human activities lead to an increased input of nutrient resources like nitrates and phosphates into lakes, rivers, estuaries, and coastal marine regions. When such cultural eutrophication is severe, we consistently see a decline in the species richness of phytoplankton (despite an increase in their productivity). Rosenzweig (1971) referred to such declines as illustrating 'the paradox of

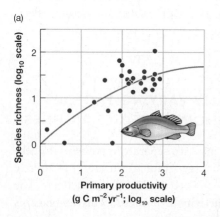

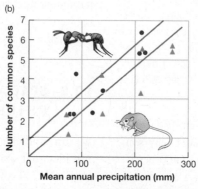

FIGURE 10.4 Positive relationships between species richness and productivity. The best-fit lines are statistically significant. (a) Species richness of fish increased with primary productivity of phytoplankton in a series of North American lakes. (After Dodson et al., 2000.) (b) The species richness of seed-eating rodents (triangles) and ants (circles) inhabiting sandy soils increased along a geographic gradient of increasing precipitation and, therefore, of increasing productivity. (After Brown & Davidson, 1977.)

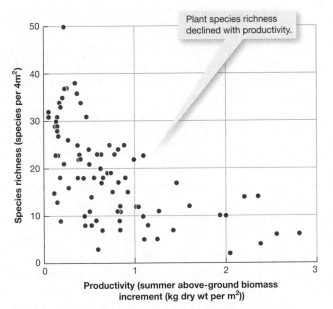

Plant species richness declined with productivity.

FIGURE 10.5 The relationship species richness and productivity for rich-fen communities of plants in England and Wales, as evidenced by samples from 4m² quadrats. (After Wheeler & Shaw, 1991.)

enrichment.' One possible resolution of the paradox is that high productivity leads to high rates of population growth, bringing about the extinction of some of the species present because of a speedy conclusion to any potential competitive exclusion (see Section 6.2). With

lower productivity, the environment is more likely to have changed before competitive exclusion has been achieved.

It is perhaps not surprising, then, that several studies have demonstrated both an increase and a decrease in richness with increasing productivity – that is, that species richness may be highest at intermediate levels of productivity. Species richness declines at the lowest productivities because of a shortage of resources, but it also declines at the highest productivities where competitive exclusions speed rapidly to their conclusion. We see such humped curves, for instance, when we plot the number of lake phytoplankton species against overall phytoplankton productivity (Figure 10.6a) for the same lakes we have already seen in Figure 10.4a; we see it when we plot the species richness of desert rodents against precipitation (and thus productivity) along a geographic gradient in Israel (Figure 10.6b), in contrast to the relationship in Figure 10.4b; and we see it when we plot the diversities of both gastropod and bivalve mollusc species living on the sea floor in deep-sea communities against the level of **particulate organic carbon** (POC), the rain of chemical energy falling as dead organic matter from the sea surface (described in Section 4.5).

Indeed, a number of meta-analyses have been carried out of the wide range of studies of the richness–productivity relationship, from both terrestrial and

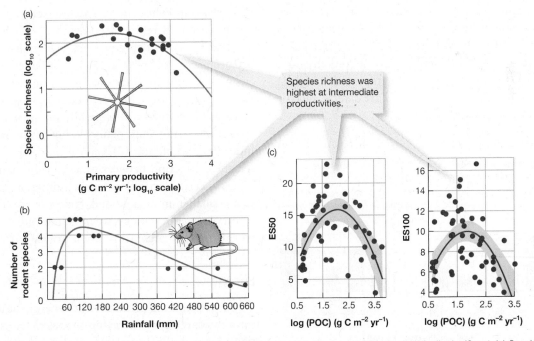

Species richness was highest at intermediate productivities.

FIGURE 10.6 Humped relationships between species richness and productivity. The best-fit lines are statistically significant. (a) Species richness of phytoplankton at a range of productivities for the same lakes as for the fish data in Figure 10.4a. (After Dodson et al., 2000.) (b) Species richness of desert rodents in Israeli deserts plotted against annual rainfall. (After Abramsky & Rosenzweig, 1983.) (c) Species diversity of gastropod molluscs, left, and bivalve molluscs, right (ES50/ES100, the expected number of species normalized to a sample size of 50 or 100 individuals, respectively) in deep-sea communities, plotted against the flux of particulate organic carbon (POC) falling from above. (After Tittensor et al., 2011.)

aquatic systems, in a search for some general rules. These have been heavily criticized for the arguably uncritical way in which the original data sets have been accepted at face value and combined (Whittaker, 2010) – a problem that besets all such analyses, despite their attractions. Nonetheless, one conclusion that survives such criticism is that a very wide range of relationships has been found: some positive, some negative, some humped, some with no detectable relationship at all, and even some U-shaped curves, cause unknown (Figure 10.7). Clearly, increased productivity can and does lead to increased or decreased species richness, or both, or neither.

We can account for some of this variation among studies simply by assuming that in particular

α-, β-, and γ-diversity

cases, only a proportion of the productivity spectrum was sampled, so that, for example, only the ascending or descending limb of a humped-shaped curve was observed. And where no relationship was observed, the reason may of course be shortcomings in the data collection. But we can make a good case for the nature of the relationship changing, depending on the spatial scale at which it is observed. To make this overall pattern clearer, ecologists have often distinguished between α-, β-, and γ-diversity. Here, **α-diversity** refers to diversity at a local scale (*within* a community), so it is high when there are many species within each community. Then, **β-diversity** refers to the differences among communities within a region, so it is high when different communities in a region differ in the species they contain. Finally, **γ-diversity** refers to diversity at the whole regional scale (collections of communities). It therefore combines α- and β-diversity and is highest when both individual communities are diverse and the communities in a region differ.

In particular, it has often been suggested that at the local scale, either the relationship between α-diversity declines with productivity or there is a hump-shaped relationship, whereas at the regional scale γ-diversity tends to increase with productivity (Whittaker, 2010). This suggestion is borne out by a study of ponds in southwestern Michigan and northeastern Pennsylvania, where α-diversity focused on the relationship within ponds, while γ-diversity focused on whole watersheds, each of which was a collection of ponds. For both plants and animals, there was a hump-shaped relationship between α-diversity and pond productivity, but γ-diversity increased with the productivity of whole watersheds (Figure 10.8a, b). This in turn suggests that the differences between communities within a region (β-diversity) must increase with productivity, and this, too, was borne out by the data (Figure 10.8c). The implication, clearly, is that the drivers of the richness–productivity relationship are different at different scales, and this, too, is a safe conclusion to accept.

One plausible, more detailed suggestion is that at the very smallest scales, increased productivity leads to a decline in richness as competition excludes rare species for which there is no space (as in the small quadrats in Figure 10.5). At somewhat larger scales (the ponds in Figure 10.8), there is sufficient space for initial rises in productivity to enhance richness prior to the subsequent decline, and hence the whole hump-shaped relationship is apparent. But at larger scales still (the watersheds in Figure 10.8), species excluded from one community in a region are likely to survive in other, somewhat different communities, such that we observe no decline in richness at higher productivities, and richness simply continues to rise as more exploitable niches become available for the regional species pool. Work in the future will hopefully tell us whether this suggestion is not only plausible but also correct – or indeed, the whole story. Whatever the truth of the matter, the distinction between small, medium, and large scales is certain to differ between different communities – smaller in a grassland, for example, than in a forest. As ever in ecology, we must remember to see the world from the organisms' point of view.

Energy

When we turn from productivity to energy, the argument for a direct effect of energy on species richness is essentially a metabolic argument – higher temperatures for longer periods mean more time for species to be active, higher rates of metabolism (more growth, more reproduction, more individuals) and less chance of cold- and

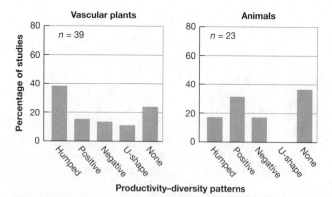

FIGURE 10.7 Percentage of published studies on plants and animals showing various patterns in the relationship between species richness and productivity. (After Mittelbach et al., 2001.)

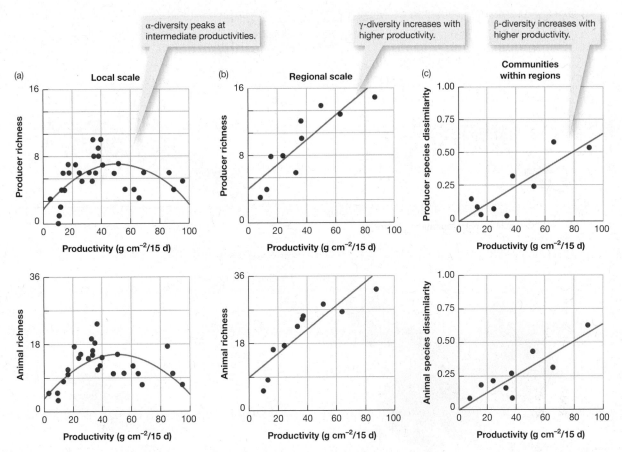

FIGURE 10.8 Diversity–productivity relationships for ponds in southwestern Michigan and northeastern Pennsylvania. Producers are above (vascular plants and macroalgae); animals below (insects, crustaceans, amphibians, etc). All lines are statistically significant. (a) α-diversity – the diversity within individual ponds. (b) γ-diversity – the diversity within regions: watersheds containing several ponds. (c) β-diversity – the dissimilarity among ponds within watersheds. Each point is the average dissimilarity (1 – Jaccard's similarity index) among ponds in one watershed. (After Chase & Leibold., 2002.)

frost-damage (Hawkins et al., 2003). Certainly there are examples in which species richness for whole groups is correlated positively either with indicators of energy such as annual **potential evapotranspiration** (PET – the amount of water that would evaporate or be transpired from a saturated surface, and hence a measure of atmospheric energy; see the American birds and butterflies in Figure 10.9a,b), or simply with annual mean temperature itself (the Chinese vertebrates in Figure 10.9c,d).

On the other hand, we must be careful, as always, not to mistake association for causation, especially given that energy is itself bound to be one of the drivers of the productivity variations discussed above. We can see, for example, that the species richness of vertebrates in China is also correlated with annual rainfall and with an index of vegetation productivity (Figure 10.9e,f). Is only one of these the true driver of variations in species richness? Or do all three have some part to play? We can also see that the association between species richness and PET for U.S. birds and

butterflies applied only at the lower end of the scale – it is clearly not the whole story (Figure 10.9a,b).

To try to disentangle the separate effects of energy and productivity, each in its own right, Belmaker and Jetz (2011) compiled large data sets on amphibians, birds, and mammals. They took the statistical approach of determining the effect of net primary productivity (NPP) after having taken account of variations in energy (mean annual temperature) (Figure 10.10a), and conversely determining the effect of energy after having taken account of variations in NPP (Figure 10.10b). They also studied these relationships at a range of scales. The largest compared assemblages encompassed within circles 2000 km in diameter; the smallest, local assemblages (median area 488 km², equivalent to a circle 22 km in diameter). Between these were assemblages encompassed within circles 200 km in diameter.

It is apparent, first, that both energy and productivity retain their associations with species richness even after the other has been taken into account.

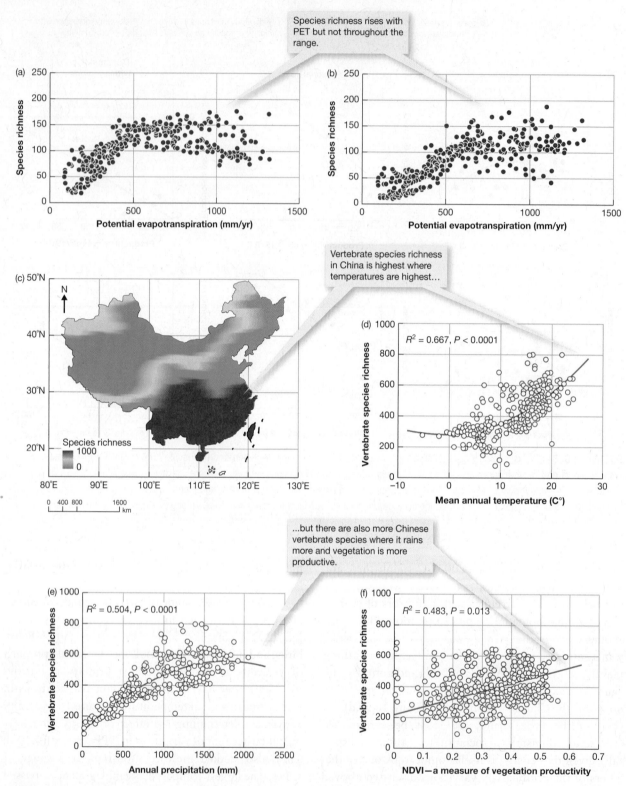

FIGURE 10.9 Species richness and environmental energy. (a) The relationship between the species richness of birds in North America and potential evapotranspiration (PET), a measure of environmental energy. (b) A similar relationship for North American butterflies. (c) Variation in the species richness of vertebrates over the whole of China. (d) The relationship between the species richness of vertebrates in China and mean annual temperature. (e) The relationship between the species richness of vertebrates in China and mean annual precipitation. (f) The relationship between the species richness of vertebrates in China and NDVI, a measure of vegetation productivity. ((a), (b) After Hawkins et al., 2003; (c)-(f) After Luo et al., 2012.)

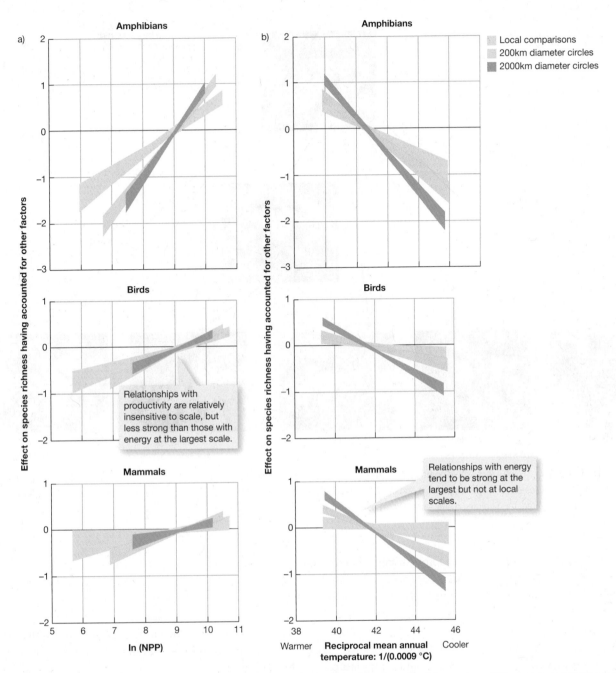

FIGURE 10.10 (a) Relationships and 95% confidence intervals for the regressions of species richness of amphibians, birds, and mammals on net primary productivity, NPP [kg C m⁻² year⁻¹ x 0.0001 (log scale)], having taken other factors into account. Data are collected from throughout the world and calculated on the basis of comparisons at a range of scales, as indicated. (b) Similar, but for energy, measured as a reciprocal mean annual temperature so scales are comparable. (After Belmaker & Jetz, 2011.)

It is also apparent, though, that the strengths of these relationships are strongly dependent on the scales at which we study them. In particular, the associations for energy were strongest of all at the largest scale, where energy seemed, if anything, to be more important than productivity, whereas at smaller scales, associations with energy were generally much weaker and in some cases nonexistent (Figure 10.10b). By contrast, the relationships with productivity were much less sensitive to changes of scale. While they were less strong than the energy relationships at the largest scale, they were not much altered at the local scale (Figure 10.10b).

Thus, energy, beyond its effects on productivity, seems to have an important role to play in helping us understand patterns of species richness at the global scale. Perhaps it sets an upper limit on richness,

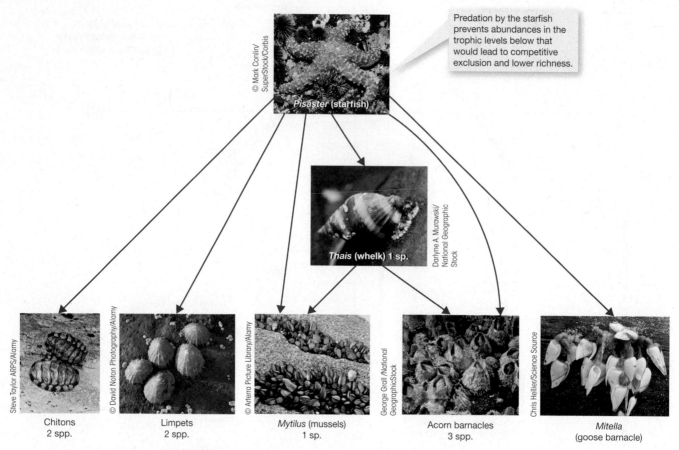

Predation by the starfish prevents abundances in the trophic levels below that would lead to competitive exclusion and lower richness.

Pisaster (starfish)

Thais (whelk) 1 sp.

Chitons
2 spp.

Limpets
2 spp.

Mytilus (mussels)
1 sp.

Acorn barnacles
3 spp.

Mitella
(goose barnacle)

FIGURE 10.11 Paine's rocky shore community. It was possible to detect the profound influence of the predatory starfish only by removing them. In the absence of *Pisaster*, other species became dominant (first barnacles and then mussels) leading to an overall reduction in species richness. This is a classic case of predator-mediated coexistence. (After Paine, 1966.)

regardless of variations in other factors discussed in this chapter. But the role of productivity, as we discussed earlier, reflecting many things other than energy, seems also to be important at a whole range of scales. We should remember, for example, that there are highly species-rich communities at the bottom of the deep oceans (see Section 4.5) where there is no light and temperatures are low. Here, clearly, energy-input and productivity are very largely decoupled from one another.

Predation intensity

We examined the possible effects of predation on the species richness of a community in Chapter 7: predation may increase richness by allowing otherwise competitively inferior species to coexist with their superiors (predator-mediated coexistence), but intense predation may reduce richness by driving prey species to extinction (whether they are strong competitors or not). Overall, therefore, there may also be a humped relationship between predation intensity and species richness in a community, with greatest richness

at intermediate intensities, such as that shown by the effects of cattle grazing (see Figure 7.23). A classic example of predator-mediated coexistence is provided by a study that established the concept in the first place: the work of Paine (1966) on the influence of a top carnivore on community structure on a rocky shore (Figure 10.11).

The starfish *Pisaster ochraceus* preys on sessile filter-feeding barnacles and mussels, and also on browsing limpets and chitons and a small carnivorous whelk. These species, together with a sponge and four macroscopic algae (seaweeds), form a typical community on rocky shores of the Pacific coast of North America. Paine removed all starfish from a stretch of shoreline about 8 m long and 2 m wide and continued to exclude them for several years. The structure of the community in nearby control areas remained unchanged during the study, but the removal of *Pisaster* had dramatic consequences.

Within a few months, the barnacle *Balanus glandula* settled successfully. Later mussels (*Mytilus californicus*) crowded it out, and eventually the site became dominated by these. All but one of the species of alga disappeared, apparently through lack of space, and the browsers tended to move away, partly because space was limited and partly because there was a lack of suitable food. The main influence of the starfish *Pisaster* appears to be to make space available for competitively subordinate species. It cuts a swathe free of barnacles and, most importantly, free of the dominant mussels that would otherwise outcompete other invertebrates and algae. Overall, the natural situation is for there to be predator (starfish)-mediated coexistence, but the removal of starfish led to a reduction in number of species from 15 to eight.

The concept of predator-mediated coexistence also finds a surprising application in the field of restoration ecology (Box 10.2).

10.2   ECOncerns

Using exploiter-mediated coexistence to assist grassland restoration

Species-rich meadows are now uncommon in agricultural landscapes in Europe because decades of intensive fertilizer application have allowed a few species to competitively exclude others, a pattern that echoes the results of the remarkable century-long Rothamsted experiment (see Figure 10.1). Attempts to restore the lost species richness of these pasture settings are now common. One approach is to use what we know about predator-mediated coexistence or, more generally, exploiter-mediated coexistence. This occurs when one species exploits as food a number of species in the community, reducing the dominance of the most competitively superior species and allowing less competitive species to maintain a foothold.

Parasites, too, are capable of promoting such exploiter-mediated coexistence. One such species is *Rhinanthus minor*, an annual plant capable of its own limited photosynthesis but known as a *hemiparasite*, because it typically steals the photosynthetic products of other plants by building connections with their roots. Researchers reasoned that the presence of the hemiparasite might facilitate recovery from an agriculturally impoverished to a species-rich grassland via exploiter-mediated coexistence (Pywell et al., 2004). To test this hypothesis, they established experimental plots with various densities of *Rhinanthus minor* and sowed a mixture of seeds of 10 native wildflower species that had been lost from the grassland as a result of intensive agriculture. After 2 years, the hemiparasite was found to have suppressed the growth of those plants that it had parasitized, and this led, the following year, to the desired increase in grassland species richness because competitive exclusion had been circumvented (Figure 10.12).

A species-rich flower meadow

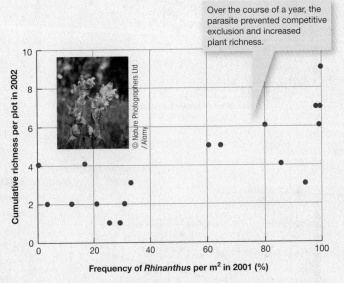

Over the course of a year, the parasite prevented competitive exclusion and increased plant richness.

FIGURE 10.12   Relationship between frequency of occurrence of the hemiparasite *Rhinanthus minor* (see photo inset) and species richness of plants per experimental plot of grassland at a study site in Oxfordshire, UK. ((left) © alamy images a02y49; (right) After Pywell et al., 2004.)

An understanding of exploiter-mediated coexistence holds promise for future meadow restoration efforts. Can you think of any other aspects of the theory of species richness that could be applied to the benefit of impoverished grasslands? (Clue – check out the 'intermediate disturbance hypothesis,' described in Section 10.3. These intensively farmed landscapes have also been subject to regular and intensive disturbances caused by heavy mowing or grazing. What might the intermediate disturbance hypothesis have to offer in restoring grassland species richness?)

Spatial heterogeneity

We can expect environments that are more spatially heterogeneous to accommodate extra species because they provide a greater variety of microhabitats, a greater range of microclimates, more types of places to hide from predators, and so on. In effect, the extent of the resource spectrum is increased (see Figure 10.3a). Confirming this expectation, a study of plant species growing in 51 plots alongside the Hood River, Canada, revealed a positive relationship between species richness and an index of spatial heterogeneity

based, among other things, on the number of categories of substrate, slope, drainage regimes, and soil pH present (Figure 10.13a).

Most studies of spatial heterogeneity, however, have related the species richness of animals to the structural diversity of the plants in their environment. Occasionally, this has been the result of experimental manipulation of the plants, as with the spiders in Figure 10.13b. In other cases, though, the species richness of the plants has been used as a surrogate for the spatial heterogeneity they provide (e.g., Figure 10.13c),

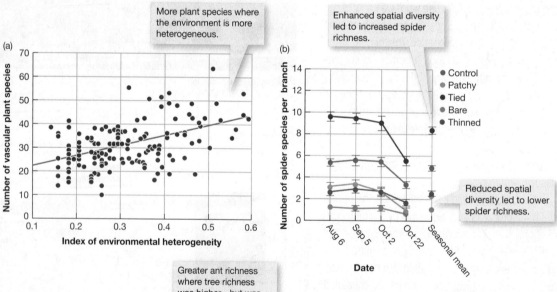

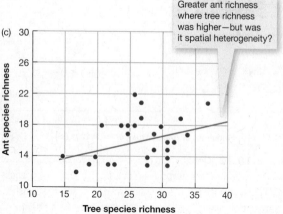

FIGURE 10.13 (a) Relationship between the number of plants per 300 m² plot beside the Hood River, Northwest Territories, Canada, and an index (ranging from 0 to 1) of spatial heterogeneity in abiotic factors associated with topography and soil. (After Gould & Walker, 1997.) (b) In an experimental study, the number of spider species living on Douglas fir branches increased with their structural diversity. Those bare, patchy, or thinned were less diverse than normal ('control') by virtue of having needles removed; those tied were more diverse because their twigs were entwined together. (After Halaj et al., 2000.) (c) Relationship between arboreal ant species richness in Brazilian savanna and the species richness of trees (a surrogate for spatial heterogeneity). (After Ribas et al., 2003.)

inviting the question of whether the driving force is spatial heterogeneity or one of the many other things associated with plant richness and so correlated with spatial heterogeneity. Nonetheless, whether spatial heterogeneity arises from the abiotic environment or is provided by biological components of the community, it is clearly capable of promoting an increase in species richness.

Environmental harshness

Environments dominated by an extreme abiotic factor – often called *harsh environments* (see Section 3.2) – are more difficult to recognize than might be immediately apparent. From our own point of view, we might describe as extreme both very cold and very hot habitats, unusually alkaline lakes and grossly polluted rivers. However, species have evolved and live in all such environments, and what is very cold and extreme for us must seem benign and unremarkable to a penguin.

We might try to get around the problem of defining environmental harshness by letting the organisms decide. Thus, we can classify an environment as extreme if organisms, by their failure to live there, show it to be so. But if the claim is to be made – as it often is – that species richness is lower in extreme environments, then this definition is circular, since it is designed to prove the very claim we wish to test.

Perhaps the most reasonable definition of an extreme condition is one that requires, of any organism tolerating it, a morphological structure or biochemical mechanism that is not found in most related species and is costly, either in energetic terms or in terms of the compensatory changes in the organism's biological processes that are needed to accommodate it. For example, plants living in highly acidic soils (low pH) may be affected directly through injury by hydrogen ions or indirectly via deficiencies in the availability and uptake of important resources such as phosphorus, magnesium, and calcium. In addition, the solubility of aluminium, manganese, and heavy metals may increase to toxic levels. Moreover, the activity of mycorrhizas, enhancing uptake of dissolved nutrients, or bacteria, fixing atmospheric nitrogen (Section 8.4) may be impaired. Plants can tolerate low pH only if they have specific structures or mechanisms allowing them to avoid or counteract these effects.

Environments that experience low pHs can thus be considered harsh, and the mean number of plant species recorded per sampling unit in a study in the Alaskan Arctic tundra was indeed lowest in soils of low pH (Figure 10.14a). Similarly, the species richness of *benthic* (that is, bottom-dwelling) invertebrates in streams in southern England was markedly lower in the more acidic streams (Figure 10.14b).

Other extreme environments associated with low species richness include hot springs, caves, and highly saline water bodies such as the Dead Sea. The problem with these examples, however, is that each is also characterized by other features associated with low species richness, such as low productivity and low spatial heterogeneity. In addition, many occupy small areas (caves, hot springs) or areas that are rare compared to other types of habitat (only a small proportion of the streams in southern England are acidic). Hence, we often see extreme environments as small and isolated islands. We will see in Section 10.4 that these features, too, are usually associated with low species richness. Although it appears reasonable that intrinsically extreme environments should as a consequence support few species, this has proved an extremely difficult proposition to establish.

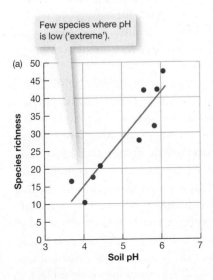

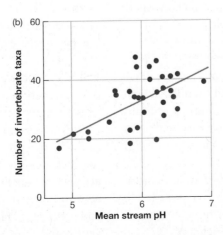

FIGURE 10.14 (a) The number of plant species in the Alaskan Arctic tundra increases with soil pH. (After Gough et al., 2000.) (b) The number of taxa of invertebrates in streams in southern England increases with the pH of stream water. (After Townsend et al., 1983.)

10.3 TEMPORALLY VARYING FACTORS INFLUENCING SPECIES RICHNESS

Temporal variation in conditions and resources can be predictable or unpredictable. The distinction is an important one. Predictable variation often occurs on a time scale similar to the generation time of the species in a community themselves. This is the case for seasonal climatic variation, discussed next. The organisms can adapt to such variations, usually by adopting patterns in their own metabolism that parallel the seasonal cycle, while the predictable *absence* of seasonal variation may allow specialized adaptation, without the threat of local extinction that a marked change in conditions would provoke. By contrast, unpredictable changes in the habitat are most naturally viewed as a disturbance to a community that changes its composition, following which the community may gradually revert to its pre-disturbance state. We discuss this below.

Climatic variation and its absence

In a predictable, seasonally changing environment, different species may be suited to conditions at different times of the year. We might therefore expect more species to coexist in a seasonal environment than in a completely constant one (see Figure 10.3a). Different annual plants in temperate regions, for instance, germinate, grow, flower, and produce seeds at different times during a seasonal cycle. Similarly, phytoplankton and zooplankton pass through a seasonal succession in large, temperate lakes, with a variety of species dominating in turn as changing conditions and resources become suitable for each.

> niche differentiation in seasonal and nonseasonal environments

On the other hand, opportunities exist for specialization in a nonseasonal environment that do not exist in a seasonal environment. For example, it would be difficult for a specialist fruit-eater to persist in a seasonal environment when fruit is available for only a very limited portion of the year. But such specialization is found repeatedly in nonseasonal, tropical environments where fruit of one type or another is available continuously.

Broadly, studies tend to support the second of these two suggestions – that species richness increases as climatic variation *decreases*. For example, as we move along the west coast of North America from Panama in the south to Alaska in the north, we find a significant decrease in the species richness of birds, mammals, and gastropods as the range of monthly mean temperatures steadily increases (MacArthur, 1975). However, this correlation does not of course prove causation, since many other things also change between Panama and Alaska. Thus, there is no established relationship between climatic constancy, as such, and species richness.

More supportive, perhaps, is evidence from species-rich biomes in especially unvarying environments, where neither high productivity, nor high energy inputs, nor any of the other factors usually proposed provide a ready explanation. Two examples were described in Section 4.5. Communities on the abyssal seafloor, 3000–6000m below the sea surface, are characterized by an absence of light and low temperatures (–0.5 – 3.0°C) and dependent for a food resource on detritus falling from the surface communities above. Nonetheless, various studies have reported abyssal sediments containing around 50 different species of polychaete worm for every 150 individuals sampled, 100 nonmicrobial species per 0.25 m², and frequently very high levels of β-diversity – different species found in different local communities (Smith et al., 2008). The unvarying nature of these communities, buffered from the fluctuations in the habitats high above them, are likely to have favored both narrow specialization and local speciation.

Similarly, the subtropical oceanic gyres have low productivity but high species richness; a single trawl may bring up 50 different species of fish from their open waters (Barnett, 1983). The biome, though, is characterized by remarkable spatial and temporal invariance. The species are also ancient and likely to have existed in the same positions in the oceans for tens of millions of years.

Disturbance

We discussed the influence of disturbance on community structure in Section 9.4 and demonstrated that when a disturbance opens up a gap, and the community is dominance-controlled (strong competitors can replace residents), a community succession tends to occur in which richness initially increases as a result of colonization, but then declines as a result of competitive exclusion.

> the intermediate disturbance hypothesis

If we now superimpose the frequency of disturbance on this picture, it seems likely that very frequent disturbances will keep most patches in the early stages of succession (where there are few species), but also that very rare disturbances will allow most patches to become dominated by the best competitors (where there are also few species). This suggests an **intermediate disturbance hypothesis**, in which communities are expected to contain most species when the frequency of disturbance is neither too high nor too low (Connell, 1978). The intermediate disturbance hypothesis was originally proposed to account for patterns of richness in tropical rain forests and coral reefs. However, it has become apparent that very many (arguably all) communities are subject to disturbances – the differences being only in their frequencies and intensities, and the sizes of the gaps created. The hypothesis has therefore come to occupy an important place in our attempts to understand patterns in species richness.

A number of studies have provided support for this hypothesis. Disturbances in small streams often take the form of bed movements during periods of high discharge, and because stream beds differ in their make up and their flow regimes, some stream communities are disturbed more frequently than others. This variation was assessed in 54 stream sites in the Taieri River in New Zealand. The pattern of richness of macroinvertebrate species conformed to the intermediate disturbance hypothesis (Figure 10.15a).

Another supportive example comes from the plants and animals living on rocky shores in north-central Chile that were subjected, in an experiment, to one of seven disturbance treatments, clearing patches in the experimental plots between zero and twelve times during the course of the experiment (Figure 10.15b). Finally, in controlled field experiments, researchers disturbed natural phytoplankton communities in Lake Plußsee (north Germany) at intervals of 2–12 days by disrupting the normal stratification in the water column with bubbles of compressed air. Again, both species richness and Shannon's diversity index were highest at intermediate frequencies of disturbance (Figure 10.15c).

Once again, however, we must be careful not to jump from supportive examples to claims that there is a general rule. Surveys, whether of observational or experimental studies, have shown that peaks of richness at intermediate disturbance are far from being a universal finding (Figure 10.16). As with the productivity–richness relationships, a partial explanation may be that only one limb of a humped-shaped curve was sampled. But again, it is also likely that reality is too complex to be captured in one catchall explanation. One particular complication is that as well as disturbance affecting the diversity of a community, the reverse may also be true: diversity can affect the nature and extent of the disturbance. Once we include these reciprocal effects, we can predict a much wider range of richness–disturbance relationships (Hughes et al., 2006).

Environmental age: evolutionary time

If we consider disturbances acting on much longer timescales, it has also often been suggested that communities may differ in species richness because some are closer to equilibrium and are therefore more saturated than others (see Figure 10.3d). For example, many have argued that

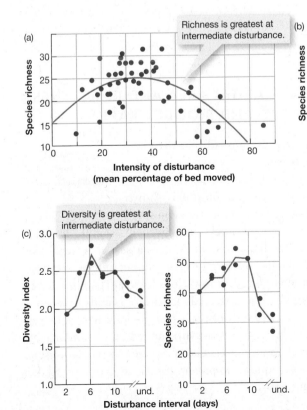

FIGURE 10.15 (a) Relationship between insect species richness and the intensity of disturbance, measured as the average percentage of the bed that moves during successive 2-month periods, in each of 54 stream sites in the Taieri River, New Zealand. (After Townsend et al., 1997.) (b) Species richness in benthic communities on the coast of Chile subjected to disturbances (clearances of 20% of the substratum) at a range of frequencies. (After Valdivia et al., 2005.) (c) Species diversity (Shannon index) and species richness of phytoplankton communities in controlled field experiments in Lake Plußsee in north Germany. 'und' represents species richness in the undisturbed state. (After Flödder & Sommer, 1999.)

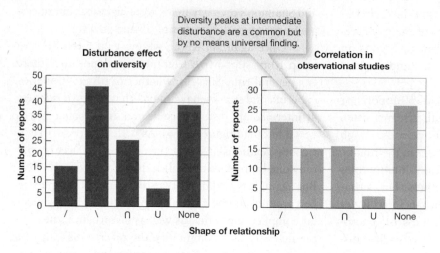

FIGURE 10.16 Results of a survey of the findings from studies of the diversity–disturbance relationship, either through experimental manipulation of disturbance (left) or direct observation of associations between the two (right). (After Hughes et al., 2006.)

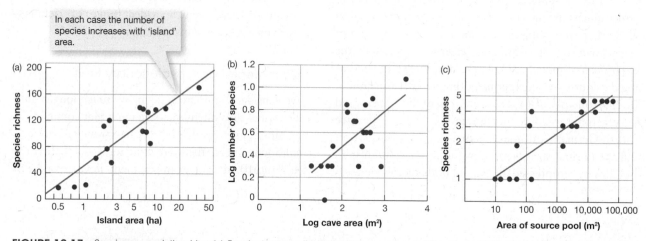

FIGURE 10.17 Species–area relationships. (a) For plants on small islands off the east coast of Sweden in 1999. (After Lofgren & Jerling, 2002.) (b) For bats inhabiting different-sized caves in Mexico. (After Brunet & Medellín, 2001.) (c) For fish living in Australian desert springs connected to pools of different sizes. All regression lines are significant at $P < 0.05$. (After Kodric-Brown & Brown, 1993.)

the tropics are richer in species than temperate regions at least in part because the tropics have existed over long and uninterrupted periods of evolutionary time, whereas the temperate regions are still recovering from the Pleistocene glaciations when temperate biotic zones shifted in the direction of the tropics.

It now seems, however, that tropical areas were also disturbed during the ice ages, not directly by ice but by associated climatic changes that saw tropical forest contracting to a limited number of small refuges surrounded by grassland. We saw in Section 2.5 (Figure 2.18) that the communities of animals in forests in northeast Australia were richest where the forest itself had been most consistently present over the past 18,000 years. Thus, it seems likely that some communities, by virtue of disturbances in their distant past, are less saturated than others. However, it is only rarely that we can pinpoint these communities with confidence.

10.4 HABITAT AREA AND REMOTENESS: ISLAND BIOGEOGRAPHY

It is well established that the number of species on islands decreases as island area decreases. Such a **species–area relationship** is shown in Figure 10.17a for plants on small islands east of Stockholm, Sweden. 'Islands,' however, need not be islands of land in a sea of water. Lakes are islands in a sea of land, mountaintops are high-altitude islands in a low-altitude ocean, gaps in a forest canopy where a tree has fallen are islands in a sea of trees, and islands of particular geological, soil, or vegetation types can exist surrounded by dissimilar types of rock, soil, or vegetation. Species–area relationships can be equally apparent for these types of islands (Figure 10.17b,c).

species–area relationships on oceanic and habitat islands

The relationship between species richness and habitat area is one of the most consistent of all ecological patterns. But is this impoverishment of species on islands more than we would expect in comparably small areas of mainland? Or to look at it the other way round, does the characteristic isolation of islands contribute to their impoverishment of species? These are important questions for an understanding of community structure, since there are many oceanic islands, many lakes, many mountaintops, many woodlands surrounded by fields, many isolated trees, and so on.

Probably the most obvious reason why larger areas should contain more species is that larger areas typically encompass more different types of habitat. However, MacArthur and Wilson (1967), in their *equilibrium theory of island biogeography* (Box 10.3), argued that this explanation was too simple, and that island size and isolation themselves played important roles. They proposed that the number of species on an island is determined by a balance between immigration and extinction; that this balance is dynamic, with species continually going extinct and being replaced (through immigration) by the same or by different species; and that immigration and extinction rates may vary with island size and isolation.

10.3 **Historical Landmarks**

MacArthur and Wilson's equilibrium theory of island biogeography

When Robert H. MacArthur and Edward O. Wilson published their book *The Theory of Island Biogeography* in 1967, they said in their Preface, 'We do not seriously believe that the particular formulations advanced in the chapters to follow will fit for very long the exacting results of future empirical investigation.' For a book that has so evidently stood every test that time might reasonably set, this is a remarkably modest and, in a sense, a remarkably inaccurate statement. As they also point out in that Preface, the authors came together as an ecologist (MacArthur) and a taxonomist and zoogeographer (Wilson) but with faith in 'the ultimate unity of population biology.' They had faith, too, in the power of general theory. Not the power to be right – to know the truth, the whole truth, and nothing but the truth – since that is not theory's purpose (and their 'particular formulations' have indeed been found not always to fit the data exactly) – but the power to stimulate thought, stimulate new observations and experiments, and so stimulate further theory as this interplay between data and theory carries our science forward.

MacArthur, a Princeton professor, died only five years later, tragically young at 42, having arguably already done more than anyone to establish modern ecology as a synergy between theory and empiricism. E. O. Wilson, who has been at Harvard since 1953, has gone on, among other things, to be dubbed 'the father of Sociobiology' following publication of his *Sociobiology: The New Synthesis* in 1975 (not without its critics), and to be widely acknowledged as the world's greatest authority on social insects, especially ants.

Robert MacArthur (left) and Edward (E.O.) Wilson (right)

Their basic theory of island biogeography runs as follows. Taking immigration first, imagine an island that as yet contains no species at all. The rate of immigration of *species* will be high, because any colonizing individual represents a species new to that island. However, as the number of resident species rises, the rate of immigration of new, unrepresented species diminishes. The immigration rate reaches zero when all species from the *source pool* (from the mainland or from other nearby islands) are present on the island in question (Figure 10.18a).

The immigration graph is drawn as a curve, because immigration rate is likely to be particularly high when there are low numbers of residents and many of the species with the greatest powers of dispersal are yet to arrive. In fact, the curve should really be a blur rather than a single line, since the precise data will depend on the exact sequence in which species arrive, and this will vary by chance. In this sense, we can think of the immigration curve as the most probable curve.

The exact immigration curve will depend on the degree of remoteness of the island from its pool of potential colonizers (Figure 10.18a). The curve will always reach zero at the same point (when all members of the pool are resident), but it will generally have higher values on islands close to the source of immigration than on more remote islands, since colonizers have a greater chance of reaching an island the closer it is to the source. It is also likely that immigration rates will generally be higher on a large island than on a small island, since the larger island represents a larger target for the colonizers (Figure 10.18a).

The rate of species extinction on an island (Figure 10.18b) is bound to be zero when there are no species there, and it will generally be low when there are few species. However, as the number of resident species rises, the theory assumes the extinction rate increases. This increase is itself likely to occur at an accelerating rate, since with more species, competitive exclusion becomes more likely, and the population size of each species is on average smaller, making it more vulnerable to chance extinction. Similar reasoning (population sizes will typically be smaller on small islands) suggests that extinction rates should be higher on small than on large islands (Figure 10.18b). As with immigration, the extinction curves are best seen as 'most probable' curves.

To see the net effect of immigration and extinction, we can superimpose their two curves (Figure 10.18c). The number of species where the curves cross (S^*) is a dynamic equilibrium and should be the characteristic species richness for the island in question. Below S^*, richness increases (immigration rate exceeds extinction rate); above S^*, richness decreases (extinction exceeds immigration). The theory, then, makes a number of predictions, described in the text.

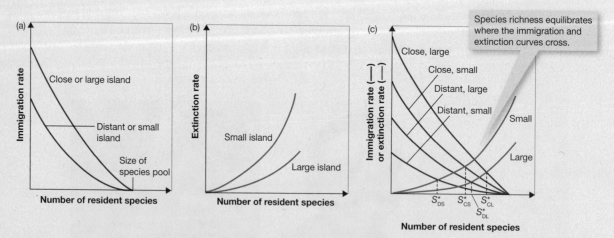

FIGURE 10.18 MacArthur and Wilson's (1967) equilibrium theory of island biogeography. (a) The rate of species immigration to an island, plotted against the number of resident species on the island, for large and small and for close and distant islands. (b) The rate of species extinction on an island, plotted against the number of resident species on the island for large and small islands. (c) The balance between immigration and extinction on small and large and on close and distant islands. In each case, S^* is the equilibrium species richness; C, close; D, distant; L, large; S, small.

MacArthur and Wilson's theory makes several predictions:

1 The number of species on an island should eventually become roughly constant through time.

2 This should be a result of a continual **turnover** of species, that is, the result of some becoming locally extinct and others immigrating.

3 Large islands should support more species than small islands.

4 Species number should decline with the increasing remoteness of an island.

On the other hand, we would expect a higher richness on larger islands, simply as a consequence of larger islands having more habitat types. Does richness increase with area at a rate *greater* than could be accounted for by increases in habitat diversity alone? Some studies have attempted to partition species–area variation on

> partitioning variation between habitat diversity and area itself

islands into these effects of habitat diversity and island area. For beetles on the Canary Islands, for example, the relationship between species richness and habitat diversity (as measured by plant species richness) is much stronger than between richness and island area, and this is particularly marked for the herbivorous beetles, presumably because of their particular food plant requirements (Figure 10.19a).

By contrast, a study of a variety of animal groups living on the Lesser Antilles islands in the West Indies partitioned the variation in species richness from island to island into that attributable to island area alone, that attributable to habitat diversity alone, that attributable to correlated variation between area and habitat diversity (and hence not attributable to either alone), and that attributable to neither (Figure 10.19b). For reptiles and amphibians, as for the beetles of the Canary Islands, habitat diversity was far more important than island area. But for bats, the reverse was the case, and for birds and butterflies, both area itself and habitat diversity had important roles to play. Overall, therefore, studies like this suggest a separate area effect

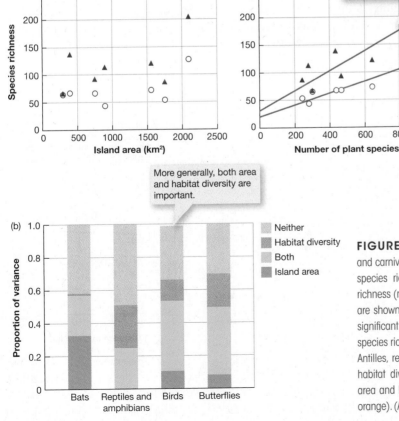

> Beetle richness increases with habitat diversity but not with island area (left).

> More generally, both area and habitat diversity are important.

FIGURE 10.19 (a) The relationships – for herbivorous (circles) and carnivorous (triangles) beetles of the Canary Islands – between species richness and both island area (left) and plant species richness (right). Regression lines are significant at $P < 0.05$; no lines are shown in the left panel of (a) because the regressions are not significant. (After Becker, 1992.) (b) The proportion of variance in species richness, for four animal groups among islands in the Lesser Antilles, related uniquely to island area (blue), related uniquely to habitat diversity (orange), related to correlated variation between area and habitat diversity (green), and unexplained by either (pale orange). (After Ricklefs & Lovette, 1999.)

(larger islands are larger targets for colonization; populations on larger islands have a lower risk of extinction) beyond a simple correlation between area and habitat diversity.

Species richness on islands can certainly be lower the farther the islands are from a mainland source | richness and island isolation | of immigrants, illustrated for example by the fungi and fish in Figure 10.20a and b. However, there are also many examples where the relationship between richness and isolation is weak or nonexistent. For the trees and lichens in Figure 10.20c, sampled on islands in lakes within the Boundary Waters Canoe Area Wilderness in Minnesota, there is some indication that the most isolated islands have the fewest species, but there are no clear overall trends. In these and similar cases, at least part of the explanation is likely to be simply that the effects of the many other factors that determine species richness at a site are too strong for us to observe any effects of isolation.

A more transient but nonetheless important reason for the species impoverishment of islands, especially remote ones, is the fact that | species missing because of insufficient time for colonization | many lack species they could potentially support, simply because there has been insufficient time for the species to colonize. An example is the island of Surtsey, which emerged in 1963 as a result of a volcanic eruption. The new island, 40 km southwest of Iceland, was reached by bacteria and fungi, some sea birds, a fly, and seeds of several beach plants within 6 months of the start of the eruption. Its first

established vascular plant was recorded in 1965, the first moss colony in 1967, and the first bush (a dwarf willow, *Salix herbacea*) in 1998. An earthworm was found in 1993 and slugs in 1998, probably carried in by birds (Hermannsson, 2000). By 2004, more than 50 species of vascular plant, 53 mosses, 45 lichens, and 300 species of invertebrate had been recorded, though not all persisted (Surtsey Research Society, website www.surtsey.is). Colonization by new species occurred both above and below the water line, with marine invertebrates, which disperse as larval stages in the ocean, accumulating faster than terrestrial plants (Figure 10.21).

Finally, no aspect of ecology can be fully grasped without reference to evolutionary processes (see Chapters 2 and 8), and this is particularly true for an understanding | evolution rates on islands may be faster than colonization rates | of island communities. On isolated islands, the rate at which new species evolve may be comparable to or even faster than the rate at which they arrive as new colonists. Clearly, the communities of these islands will be incompletely understood by reference only to ecological processes. Take the remarkable numbers of *Drosophila* species (fruitflies) found on the remote volcanic islands of Hawaii. There are probably about 1,500 *Drosophila* species worldwide, but at least 500 of these have evolved almost entirely on the Hawaiian Islands. The communities of which they are a part are clearly much more strongly affected by local evolution and speciation than by the processes of invasion and extinction.

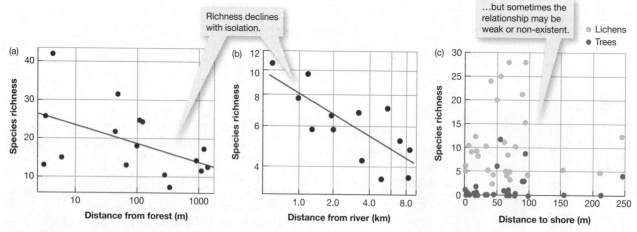

FIGURE 10.20 Relationships between species richness and island isolation from a mainland species pool. (a) Ectomycorrhizal fungi on the roots of single trees in tree clumps isolated from contiguous forest at Point Reyes National Seashore, California, USA; *p* = 0.048. (After Peay et al., 2010.) (b) Fish in spring-fed ponds connected to the Shigenobu River in southwestern Japan; *P* = 0.012. (After Uchida & Inoue, 2010.) (c) Trees and lichens on islands in lakes in the Boundary Waters Canoe Area Wilderness, Minnesota; no significant linear relationships. (After Ames et al., 2012.)

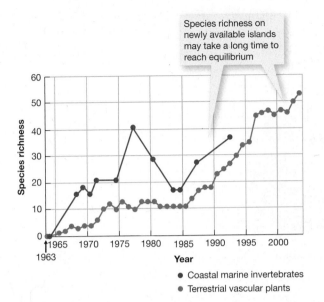

FIGURE 10.21 Regular surveys of species richness of animals and plants have occurred since the emergence in 1963 of volcanic Surtsey Island, near Iceland. Shown here are the results of standard surveys of coastal marine invertebrates up to 1992 (barnacles, isopods, decapods, molluscs, starfish, brittlestars, sea urchins, and sea squirts; red circles) and of terrestrial vascular plants up to 2004 (blue circles). (After Hermannson, 2000; Surtsey Research Society, website www.surtsey.is.)

10.5 GRADIENTS OF SPECIES RICHNESS

In the preceding sections we have been looking at the possible effects of various factors – productivity, island isolation, and so on – on species richness and biodiversity generally. Here we turn to patterns in species richness, either in space or in time. We seek to use what we have learned about the effects of the various factors to account for the patterns we observe. Even in the strongest of these patterns, however, there is variation around any general trend. As in any statistical relationship, we can attribute some variation to either sampling or observer error; data collected from nature are never likely to reflect perfectly the underlying reality. But variation will also arise simply because the observed species richness will have been co-determined by several of the factors we have discussed, and not all of them will vary in parallel with the gradient we are following. So, for example, when we turn next to latitudinal gradients, the question we will ask will be 'what accounts for the general trend?' not 'what accounts for each individual value within the trend?'

Latitudinal gradients

One of the most widely recognized patterns in species richness, then, is the increase that occurs from the poles to the tropics. This can be seen in a wide variety of groups, including trees, marine invertebrates, butterflies, and lizards (Figure 10.22). The pattern can be seen, moreover, in terrestrial, marine and freshwater habitats.

No doubt the most commonly proposed explanations for the latitudinal patterns have been variations in productivity and energy, which, as we discussed in Section 10.2, may often be difficult to disentangle. It was apparent there, though (see Figure 10.10), that variations in energy input, in their own right, are particularly important at global scales, and the latitudinal trends certainly come into that category. Indeed, given what we know about global climatic patterns (Chapter 4), it is hard to see why consistent latitudinal trends in productivity would *not* be related to energy inputs. Remember, too, also from Section 10.2, that while richness may often peak at intermediate productivities, this is least likely to be observed at larger spatial scales, which again certainly applies to global latitudinal trends.

We can safely propose, therefore, that the core explanation for the latitudinal gradient in species richness is the underlying variation in energy input, acting directly on metabolism, and the consequences of metabolism at all trophic levels, but acting too through gradients in primary productivity. On the other hand, to repeat the general point made above, this should not blind us to the glaring exceptions to this trend, where richness is no doubt more dependent on factors other than energy input. The high diversity in the low productivity subtropical ocean gyres is a prime example.

A number of other explanations have also been put forward for the overall latitudinal trend. Some have pointed out that equatorial regions are generally less seasonal than temperate regions, and this may allow species to be more specialized (that is, to have narrower niches; see Figure 10.3b). The greater evolutionary age of the tropics has also been proposed as a reason for their greater species richness, and another line of argument suggests that the repeated fragmentation of tropical forest into patches and the subsequent mergers of these promoted genetic differentiation and speciation, accounting for much of the high richness in tropical regions. Each of these answers may contribute to the greater species richness of the tropics; there is no need to imagine that there must be only a single contributory factor.

The richness of tropical communities has also been attributed to a greater intensity of predation and to more specialized predators. More intense predation could reduce the importance of competition, permitting greater niche overlap and promoting higher richness (see Figure 10.3c). Of course, predation cannot be the root cause of tropical richness, since this assumption

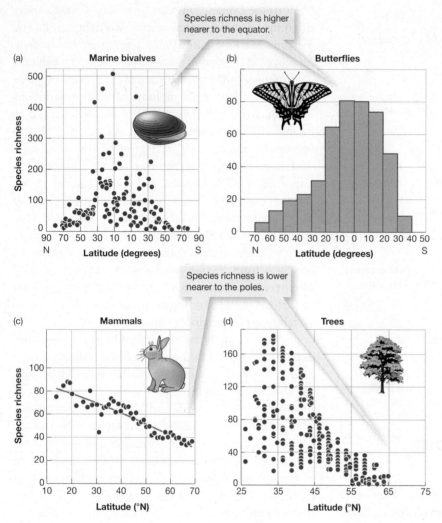

FIGURE 10.22 Latitudinal patterns in species richness for: (a) marine bivalves, (After Flessa & Jablonski, 1995.) (b) swallowtail butterflies, (After Sutton & Collins, 1991.) (c) mammals in North America, (After Rosenzweig & Sandlin, 1997.) and (d) trees in North America. In each case there is a decline from low latitudes (the equator is at 0°) to high latitudes (the poles are at 90°). (After Currie & Paquin, 1987.)

simply invites us to ask what gives rise to the richness of the predators themselves. But a focus on predation raises a point we return to later: the various factors acting on a trend may interact with and even reinforce one another. For example, greater energy may lead to higher species richness, to more predation, and hence to even higher species richness.

Gradients with altitude and depth

A decrease in species richness with altitude, reminiscent of that observed with latitude, has frequently been reported in terrestrial environments (Figure 10.23a, b). On the other hand, some studies, especially those of particular groups of species, have reported an increase with altitude (Figure 10.24c), while about half the studies of altitudinal species richness have described hump-shaped patterns (Figure 10.24d) (Rahbek, 1995).

At least some of the factors affecting the latitudinal trend in richness are also likely to play a part in explaining altitudinal trends. For example, declines in species richness have often been explained in terms of decreasing productivity associated with lower temperatures and shorter growing seasons at higher altitude, or physiological stress associated with climatic extremes near mountaintops. Indeed, the explanation for the converse, positive relationship between ant diversity and altitude, in Figure 10.23c, is that precipitation increased with altitude in this case, resulting in higher productivity at higher altitudes. In addition, high-altitude communities almost invariably occupy smaller areas than lowlands at equivalent latitudes, and they will usually be more isolated from similar communities than lowland sites. Therefore, the effects of area and isolation are likely to contribute to observed decreases in species richness with altitude.

In aquatic environments, the change in species richness with depth shows important similarities to the terrestrial gradient with altitude, but there is also

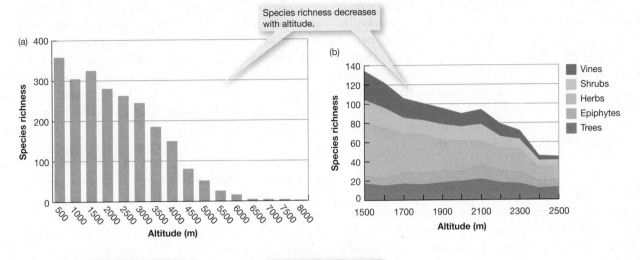

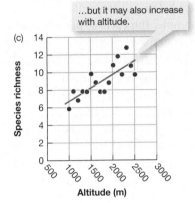

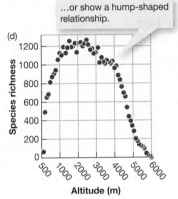

FIGURE 10.23 Relationships between species richness and altitude for: (a) breeding birds in the Nepalese Himalayas, (After Hunter & Yonzon, 1992.) (b) plants in the Sierra Manantlán, Mexico, (After Vázquez & Givnish, 1998.) (c) ants in Lee Canyon in the Spring Mountains of Nevada, (After Sanders et al., 2003.) and (d) flowering plants in the Nepalese Himalayas. (After Grytnes & Vetaas, 2002.)

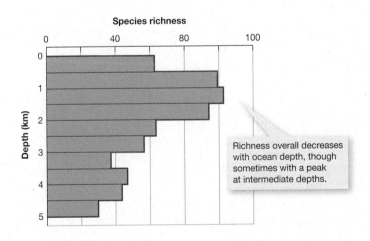

FIGURE 10.24 Depth gradient in species richness of bottom-dwelling vertebrates and invertebrates (fish, decapods, holothurians, asteroids) in the ocean southwest of Ireland. (After Angel, 1994.)

a fundamental difference. With increasing depth in the water column, conditions change but the structural nature of the habitat remains essentially the same. On the bottom, however, the structure of the environment is quite different. The bottom acts, too, as a sink for detritus falling from the communities in the water column above. Thus, in larger lakes, the cold, dark, oxygen-poor abyssal depths contain fewer species than the shallow surface waters, and likewise, in marine habitats, plants and algae are confined to the photic zone (where light penetrates and they can photosynthesize), which rarely extends below 30 m.

In the open ocean, though, while there is a rapid decrease in richness with depth, this is markedly reversed on the ocean floor where, as we have noted previously, diversity may be very high. Interestingly, however, in coastal regions, the effect of depth on the species richness of bottom-dwelling animals is to produce not a single gradient, but a peak of richness at about 1000 m, possibly reflecting higher environmental predictability there (Figure 10.24).

Gradients during community succession

Section 9.4 described how, in community successions, if they run their full course, the number of species first increases (because of colonization) but eventually decreases (because of competition). This is most firmly established for plants, but the few studies that have been carried out on animals in successions indicate, at the least, a parallel increase in species richness and biodiversity in the early stages of succession. Figure 10.25 illustrates this for birds following the reclamation of land used for open-cast mining in the Czech Republic, for benthic invertebrates living on a submersed cliff off the coat of Greece, and for ants colonizing new housing developments in Puerto Rico.

To a certain extent, the successional gradient is a necessary consequence of the gradual colonization of an area by species from surrounding communities that are at later successional stages; that is, later stages are more fully saturated with species (see Figure 10.3d). However, this is a small part of the story, since succession includes the replacement of species and not just the addition of new ones.

Indeed, as with the other gradients in species richness, there is something of a cascade effect with succession: one process that increases richness kick-starts a second, which feeds into a third, and so on. The earliest species will be those that are the best colonizers and the best competitors for open space. They immediately provide resources (and introduce heterogeneities) that were not present before. For example, the earliest plants generate resource depletion zones (see Section 3.3) in the soil that inevitably increase the spatial heterogeneity of plant nutrients. The plants themselves provide a new variety of microhabitats, and for the animals that might feed on them, they provide a much greater range of food resources (see Figure 10.3a). The increase in herbivory and predation may then feed back to promote further increases in species richness (predator-mediated

a cascade effect?

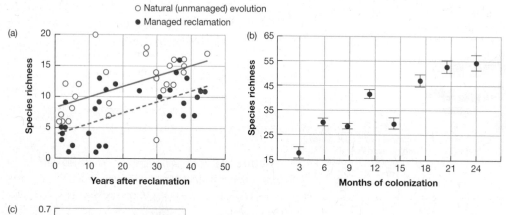

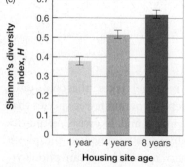

FIGURE 10.25 Increases in animal biodiversity during succession. (a) Bird species richness following reclamation of abandoned open-cast mines in the Czech Republic. (After Salek, 2012.) (b) Species richness of marine invertebrates following colonization of artificial tiles placed off the coast of northern Greece. Bars are SEs. (After Antoniadou et al., 2010.) (c) Shannon's diversity index (see Box 10.1) for ants collected from housing developments in Puerto Rico, 1, 4, and 8 years after they had been built. (After Brown et al., 2012.)

coexistence, Figure 10.3c), which provides further resources and more heterogeneity, and so on.

In addition, temperature, humidity, and wind speed show much less temporal variation within a forest than in an exposed early successional stage, and the enhanced constancy of the environment may provide a stability of conditions and resources that permits specialist species to build up populations and persist (Figure 10.3b). Within all gradients of species richness, the interaction of many factors makes it difficult to disentangle cause from effect. But within the successional gradient, this tangled causal web appears to be of the essence.

SUMMARY

Quantifying species richness and diversity
The number of species in a community is referred to as its species richness. Diversity indices are designed to combine species richness and the evenness of the distribution of individuals among those species. We can draw a still more complete picture with a rank–abundance diagram. A simple model can help us understand the determinants of species richness. Within it, a community will contain more species the greater the range of resources, when the species are more specialized in their use of resources, when species overlap to a greater extent in their use of resources, or when the community is more fully saturated.

Spatially varying factors influencing species richness
Species richness often increases with the richness of available resources and productivity, but in many cases too, the reverse has been observed – the paradox of enrichment – and other studies have found species richness to be highest at intermediate levels of productivity. Richness also tends to increase with increases in energy input into a community, acting either directly on organisms' metabolisms or as one of the factors determining variations in productivity. Predation can exclude certain prey species and reduce richness or permit more niche overlap and thus greater richness (predator-mediated coexistence). Overall, therefore, there may be greatest richness at intermediate intensities.

Environments that are more spatially heterogeneous often accommodate extra species because they provide a greater variety of microhabitats, a greater range of microclimates, more types of places to hide from predators, and so on; the resource spectrum is increased. Environments dominated by an extreme abiotic factor are more difficult to recognize than might be immediately apparent. Some apparently harsh environments do support few species, but any overall association has proved extremely difficult to establish.

Temporally varying factors influencing species richness
Seasonally changing environments may allow seasonal specialization and an increase in richness, but opportunities for specialization exist in a nonseasonal environment that are not available in a seasonal environment. On balance, richness seems to be higher in nonseasonal environments. The intermediate disturbance hypothesis suggests that there should be peak richness at intermediate frequencies of disturbance, because very frequent disturbances keep most patches at an early stage of succession (where there are few species), but very rare disturbances allow most patches to become dominated by the best competitors (where there are also few species). Some studies support the hypothesis.

Communities may also differ in richness because some are closer to equilibrium and therefore more saturated than others. Hence, the tropics may be rich in species in part because they have existed over long and uninterrupted periods of evolutionary time.

Habitat area and remoteness: island biogeography
Islands need not be islands of land in a sea of water. The number of species on islands decreases as island area decreases, in part because larger areas typically encompass more different types of habitat. MacArthur and Wilson's equilibrium theory of island biogeography argues for a separate island effect based on a balance between immigration and extinction, and the theory has received much support. On isolated islands especially, the rate at which new species evolve may be comparable to or even faster than the rate at which they arrive as new colonists.

Gradients in species richness
Richness overall increases from the poles to the tropics, with notable exceptions. Predation, climatic variation, and the greater evolutionary age of the tropics have been put forward as partial explanations, but the key factors are productivity and especially energy input. In terrestrial environments, richness often decreases with altitude (but not always). Factors instrumental in the latitudinal trend are important, but so are area and isolation. In aquatic environments, richness in the water column usually decreases with depth for similar reasons, but the ocean floor communities can be highly diverse. In successions, richness first increases (because of colonization) but eventually decreases (because of competition).

REVIEW QUESTIONS

1 Explain species richness, diversity indexes, and rank–abundance diagrams and compare what each measures.

2 What is the paradox of enrichment, and how can it be resolved?

3 Explain the variety of mechanisms by which energy input may influence species richness.

4 Explain, with examples, the contrasting effects that predation can have on species richness.

5 Why is it so difficult to identify harsh environments?

6 Explain the intermediate disturbance hypothesis.

7 Islands need not be islands of land in an ocean of water. Compile a list of other types of habitat islands over as wide a range of spatial scales as possible.

Challenge Questions

1 Researchers have reported a variety of hump-shaped patterns in species richness, with peaks of richness occurring at intermediate levels of productivity, predation pressure, disturbance, and depth in the ocean. Review the evidence and consider whether these patterns have any underlying mechanisms in common.

2 An experiment was carried out to try to separate the effects of habitat diversity and area on arthropod species richness on some small mangrove islands in the Bay of Florida. These consisted of pure stands of the mangrove species *Rhizophora mangle*, which support communities of insects, spiders, scorpions, and isopods. After a preliminary faunal survey, some islands were reduced in size by means of a power saw and brute force! Habitat diversity was not affected, but arthropod species richness on three islands nonetheless diminished over a period of 2 years (Figure 10.26). A control island, the size of which was unchanged, showed a slight increase in richness over the same period. Which of the predictions of island biogeography theory are supported by the results in the figure? What further data would you require to test the other predictions? How would you account for the slight increase in species richness on the control island?

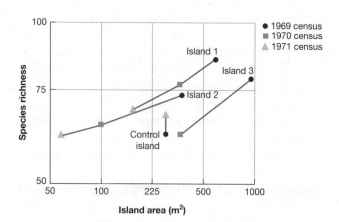

FIGURE 10.26 The effect on the number of arthropod species of artificially reducing the size of three mangrove islands. Islands 1 and 2 were reduced in size after both the 1969 and the 1970 censuses. Island 3 was reduced only after the 1969 census. The control island was not reduced, and the change in its species richness was attributable to random fluctuations. (After Simberloff, 1976.)

3 A cascade effect is sometimes proposed to explain the increase in species richness during a community succession. How might a similar cascade concept apply to the commonly observed gradient of species richness with latitude?

Andrea Gingerich/Getty Images

Chapter 11

The flux of energy and matter through ecosystems

KEY CONCEPTS

After reading this chapter, you will be able to:

- explain how communities and their abiotic environment are linked by the fluxes of energy and matter through the ecosystem

- describe the variation in the rates of net primary productivity both on land and in oceans across the Earth

- explain how net primary productivity in terrestrial ecosystems is controlled largely by water and temperature

- describe why nutrient limitation is a far more important control on net primary productivity in aquatic ecosystems, where productivity is controlled largely by the balance between light and nutrient availability

- show how the flow of energy through grazing food webs is related to the efficiency of energy transfers at each tropic level

- describe how much of an ecosystem's energy and matter passes through the decomposer system, particularly in terrestrial ecosystems

- contrast the cycles of more geologically controlled nutrients such as phosphorus and the cycle of nitrogen, which is so intimately regulated by bacteria

- describe how the nutrient inputs to ecosystems come from a variety of sources, including weathering of soil and bedrock, deposition from the atmosphere, nitrogen fixation, and fluxes in flowing waters

Like all other biological entities, ecosystems require matter for their construction and energy for their activities. We need to understand the routes by which matter and energy enter and leave ecosystems, how they are transformed into the biomass of primary producers, and how this fuels the rest of the community—herbivores, detritivores, decomposers, and their consumers.

11.1 THE ROLE OF ENERGY IN ECOLOGY

The laws of thermodynamics tell us that the universe is steadily moving towards greater entropy and less order. However, biological systems are sites of localized order maintained by continuous inputs of energy, working against the general trend. All biological entities require this continuous input of energy to support their activities and structure, as well as inputs of matter for their construction. This is true not only for individual organisms, but also for populations, communities, and ecosystems. In this chapter, we explore the intricate interactions of energy and material flows that occur in ecosystems, and the ways in which these flows and their interactions vary across biomes.

> the standing crop and primary and secondary productivity

Before examining ecosystem processes, here are some key terms we'll be using.

- **Standing crop.** The bodies of the living organisms within a unit area constitute a standing crop of biomass.

- **Biomass.** By biomass we mean the mass of organisms per unit area of ground (or water), usually expressed in units of energy (such as joules per square meter), dry organic matter (grams per square meter), or mass of carbon (grams of carbon per square meter).

- **Primary productivity.** Primary productivity is the rate at which biomass is produced per unit area or volume through photosynthesis. Like biomass, it can be expressed in many different units, including energy (such as joules per square meter per day, or per year), dry organic matter (grams per square meter per day), or mass of carbon (grams carbon per square meter per day). Ecologists consider both *gross* primary productivity and *net* primary productivity.

- **Gross primary productivity (GPP).** Gross primary productivity is the total fixation of energy by photosynthesis. A proportion of this, however, is respired away by the primary producer organisms (the autotrophs) themselves and is lost from the ecosystem as **respiratory heat** (R_{auto}).

- **Net primary productivity (NPP).** This is the difference between GPP and R_{auto}. It represents the actual rate of production of new biomass that is available for consumption by heterotrophic organisms (bacteria, fungi, and animals).

- **Secondary productivity.** By secondary productivity we mean the rate of production of biomass by heterotrophs.

- **Net ecosystem productivity (NEP).** Net ecosystem productivity is the difference between GPP and the respiration of all organisms in an ecosystem (R_{total}). It measures the net rate of accumulation or loss of organic matter, energy, or organic carbon from the ecosystem and is equivalent to the rate of NPP minus the respiration of all heterotrophic organisms (R_{het}).

A proportion of primary production is consumed by herbivores, which, in turn, are consumed by carnivores. These two groups constitute the **live consumer system**. The fraction of NPP that is not eaten by herbivores passes through the **decomposer system**. We distinguish two groups of organisms responsible for the decomposition of dead organic matter (detritus): bacteria and fungi are called **decomposers**, while animals that consume dead matter are known as **detritivores**.

> live consumer systems and decomposer systems

11.2 GEOGRAPHIC PATTERNS IN PRIMARY PRODUCTIVITY

The communities and ecosystems across the surface of the planet depend crucially for their energy on the levels of NPP that plants, algae, and other primary producers are able to achieve. NPP is also critical to the global carbon cycle, since it removes carbon dioxide from the atmosphere and interacts strongly with global climate change, as we explore in Chapter 12. Of course, the release of carbon dioxide back to the atmosphere through R_{het} is also important, and it is the NEP, or the balance between NPP and R_{het}, that determines the net influence on carbon exchanges of an ecosystem with the atmosphere.

The rate of NPP varies dramatically across the surface of the Earth, both in terrestrial ecosystems and in the oceans (Figure 11.1). We see the very highest rates, approaching 800 g C m^{-2} year^{-1} and more, in the tropical rain forests. Equally high rates of NPP occur in wetlands, estuaries, and giant kelp beds (see Chapter 4), but the spatial area of these ecosystems is too small for them to be apparent in Figure 11.1. Lower but still high rates (300 to 600 g C m^{-2} year^{-1}) typify temperate forests and grasslands, tropical savannas, upwelling zones, and much of the North Atlantic Ocean north of 40° N. The lowest rates of NPP on Earth (50 to 150 g C m^{-2} year^{-1}) occur in deserts and in the subtropical gyres of the world's oceans.

The rates illustrated in Figure 11.1 represent estimates of what the rate of NPP in natural ecosystems would be in the absence of human disturbance. In fact, as we discuss in Box 11.1 and in Chapter 14, humans have had a dramatic influence on land use and on rates of NPP across the land masses of the planet. About one third of the ice-free land area on Earth is now used for agriculture (Ramankutty et al., 2008).

The total potential rate of NPP of the planet's natural ecosystems (if there were no human disturbance) is estimated to be about 100 petagrams of carbon per year (1 Pg = 10^{15} g). Of this, slightly more than half is produced in terrestrial ecosystems and slightly less than half in the oceans (Table 11.1). Of course, the oceans cover 70% of the Earth's surface, so the average rate of NPP per area is higher on land. The single most important biome on the planet in terms of overall NPP is the tropical rain forest, which makes up more than 15% of all NPP globally and more than 30% of

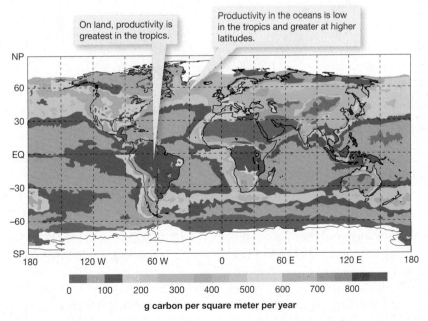

FIGURE 11.1   Annual average rates of net primary productivity across the planet for both the oceans and terrestrial ecosystems (g carbon per square meter per year). (Reprinted from Field et al., 1998.)

ECOncerns

Human appropriation of net primary productivity

Human activity, particularly from agriculture, has had a remarkable influence on net primary productivity (NPP). One measure of this influence is the 'human appropriation' of NPP (Haberl et al., 2007).

Some NPP is directly harvested when agricultural crops are collected and timber is cut from forests. Some of the potential NPP of natural ecosystems is lost due to soil degradation and other land-use changes, including urban and suburban development. And some is lost due to human-caused fires. Cumulatively, almost one-quarter of the potential NPP of the Earth's terrestrial biomes is lost through this human appropriation. Of that amount, 53% is due to harvest of crops and timber, 40% to lost productivity from land-use changes, and 7% to fires. As Haberl et al. (2007) noted, 'this is a remarkable impact on the biosphere caused by just one species!'

The pattern of human appropriation of net primary productivity (HANPP) is far from uniform over the planet (Figure 11.2). A large percentage of the potential rate of NPP by natural ecosystems has now been used directly by humans or diminished by human activity in India, much of Europe, the eastern half of the United States, the eastern third of China, southeast Asia, and the seasonal tropical forests of western Africa. Where the natural rate of NPP is low, as in the tundra biomes and many of the major deserts of the world, humans have had little impact. On the other hand, in some major areas of the planet with very high rates of NPP—notably the rain forests of the Amazon in Latin America and the Congo in Africa, as well as much of the savanna region of Africa—humans have until now appropriated less than 20%.

Is there any inherent reason it is good or bad for humans to appropriate a quarter of the world's NPP by terrestrial ecosystems? Would setting a goal of, say, 10% be better? Is there any reason to fear increasing the human appropriation of NPP to 50% or more? What might the consequences be?

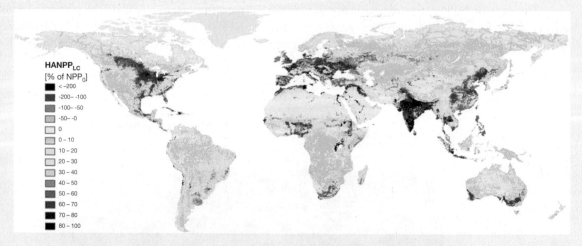

FIGURE 11.2　Extent of human appropriation of net primary productivity (HANPP) as a percentage of the potential natural rate of net primary productivity (NPP_0) across the land surface of the Earth. (Reprinted from Haberl et al., 2007.)

terrestrial NPP. Tropical savannas fall close behind and contribute almost as much to global NPP (Table 11.1). This contribution to global productivity by rain forests and savannas reflects both the high rates of NPP per area in these biomes and the large areas they cover. Remember that the Mercator mapping projection used in Figure 11.1 overemphasizes areas increasingly as we move away from the Equator.

11.3　FACTORS LIMITING TERRESTRIAL PRIMARY PRODUCTIVITY

What, then, regulates primary productivity? In terrestrial communities, solar radiation, carbon dioxide, water, and soil nutrients are the resources required for primary production, while temperature, a condition, also has a strong influence on the rate of NPP.

TABLE 11.1 Net primary production (NPP) per year, summed globally for each of the major biomes and for the planet in total (in units of petragrams of carbon).

Marine	NPP	Terrestrial	NPP
Tropical and sub-tropical oceans	13.0	Tropical rain forests	17.8
Temperate oceans	16.3	Broadleaf deciduous forests	1.5
Polar oceans	6.4	Mixed broad/needleleaf forests	3.1
Coastal	10.7	Needleleaf evergreen forests	3.1
Salt marsh/estuaries/seaweed	1.2	Needleleaf deciduous forests	1.4
Coral reefs	0.7	Savannas	16.8
		Perennial grasslands	2.4
		Broadleaf shrubs with bare soil	1.0
		Tundra	0.8
		Desert	0.5
		Cultivation	8.0
Total	48.3	Total	56.4

From Geider et al., 2001.

Carbon dioxide is normally present at a level of around 0.03% to 0.04% (300 to 400 ppm) of atmospheric gases and seems to play no significant role in determining differences between the productivities of different communities (although global increases in carbon dioxide concentration may bring big changes; see Chapter 12). On the other hand, we see that the intensity of radiation, the availability of water and nutrients, and temperature all vary dramatically from place to place. They are all candidates for the role of limiting factor. Which of them actually sets the limit to primary productivity?

Depending on location, something between 0 and 5 J of solar energy strike each square meter of the Earth's surface every minute. If all this were converted by photosynthesis to plant biomass (that is, if photosynthetic efficiency were 100%), we would see an enormous generation of plant material, ten to a hundred times greater than recorded values. Actual efficiencies of photosynthesis are generally less than a few percent, and often lower still. For example, the conifer communities shown in Figure 11.3 have the highest net photosynthetic efficiencies, but these are only between 1% and 3%. For a similar level of incoming radiation, deciduous forests achieve 0.5–1%, and, despite their greater energy input, deserts manage only 0.01–0.2%. Compare these with short-term peak efficiencies achieved by crop plants under ideal conditions, which can reach 3% to 10%.

> terrestrial communities use radiation inefficiently

There is no doubt that available radiation would be used more efficiently if other resources were in sufficient supply. The much higher efficiencies for primary productivity in agricultural systems bear witness to this. Shortage of water—an essential resource both as a constituent of cells and for photosynthesis—is often the critical factor. We should not be surprised, therefore, that precipitation in a region is quite closely correlated with its productivity, shown in Figure 11.4 for tropical savanna ecosystems. As we discussed in Chapter 3, water availability interacts with the ability of plants to assimilate carbon dioxide: when water stress is great, the plants will close their stomata,

> water and temperature as critical factors

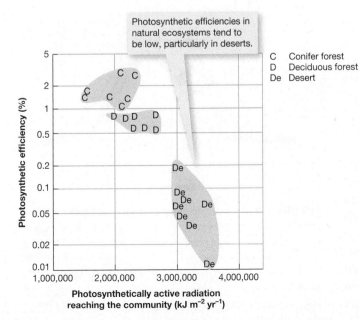

FIGURE 11.3 Photosynthetic efficiency (percentage of incoming photosynthetically active radiation converted to above-ground net primary production) for three sets of terrestrial communities in the United States. Desert ecosystems receive the greatest levels of radiation but are much less efficient than forests at converting it to biomass. Note the log scale for photosynthetic efficiency. (After Webb et al., 1983.)

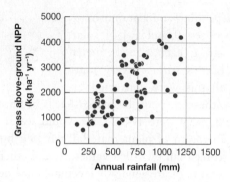

FIGURE 11.4 Above-ground net primary productivity (NPP) of grass in savanna regions of the world in relationship to annual rainfall. (After Higgins et al., 2000.)

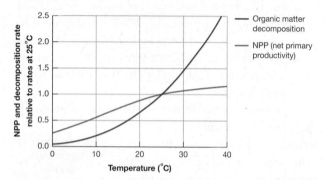

FIGURE 11.5 The effect of temperature on rates of net primary productivity (NPP) and organic matter decomposition in a typical forest ecosystem, relative to the rates at 25° C. Decomposition increases exponentially as the temperature rises, while the influence of temperature on NPP is lower, especially as temperature rises above 20° C. (After Kirschbaum, 2000.)

preventing the flow of carbon dioxide from the atmosphere into their leaves. Much of the global pattern of NPP in terrestrial ecosystems is controlled by precipitation patterns and the availability of water, which is controlled in turn by the balance between precipitation and evaporation.

Precipitation and temperature work together to structure the distribution of terrestrial biomes across the Earth (Chapter 4), and temperature also has an influence on NPP, working through three different mechanisms. First, temperature has a direct effect on the physiological processes of photosynthesis, which tends to lead to higher rates of NPP as temperature increases. The ecological effect of this increase is relatively small, though, and is dwarfed by the other two factors. The second factor is the relationship between temperature and evaporation: evaporation increases greatly with increased temperature, and this can lead to lower water availability to plants, greatly reducing NPP. And third, the rate of decomposition of dead organic matter increases as temperature rises. Decomposition leads to release of inorganic nutrients, so higher rates of decomposition result in a higher rate of release of nutrients, providing a

greater supply to plants, which in turn can increase rates of NPP. In general, the effect of temperature on organic matter decomposition is greater than its effect on NPP (Figure 11.5). The highest rates of NPP for terrestrial ecosystems occur in tropical rain forests (see Figure 11.1 and Chapter 4), where both precipitation and temperature are uniformly high, and input of solar radiation is high throughout the year.

An ecosystem can sustain its productivity only for that part of the year when the plants bear photosynthetically active foliage. Deciduous trees bear foliage for only part of the year, while evergreen trees hold a canopy throughout the year. However, for much of the year a conifer forest may barely photosynthesize at all, particularly in the colder boreal zones where solar radiation input is very low in the winter and water availability becomes extremely low as soils freeze (Figure 11.6).

Plants require many nutrients, including nitrogen, phosphorus, sulfur, potassium, calcium, magnesium, and iron (Chapter 3). In most terrestrial ecosystems,

> NPP increases with the length of the growing season

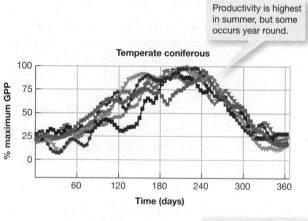

> Productivity is highest in summer, but some occurs year round.

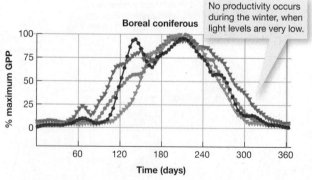

> No productivity occurs during the winter, when light levels are very low.

FIGURE 11.6 Seasonal development of maximum daily gross primary productivity (GPP) for conifer forests in temperate (Europe and North America) and boreal locations (Canada, Scandinavia, and Iceland). The different symbols in each panel relate to different forests. Daily GPP is expressed as the percentage of the maximum achieved in each forest during the 365 days of the year. Note the extended periods with no photosynthesis in the colder boreal locations. (After Falge et al., 2002.)

Historical Landmarks

Justin von Liebig and his Law of the Minimum

Justin von Liebig was one of the leading chemists of the 19th century. He became a professor at the University of Giessen in Germany when he was only 21 years old in 1824 (Figure 11.7), spending half of his 50-year career there before moving to the University of Munich. He edited the leading German chemical science journal of his time and played a large role in the development of the modern method of laboratory-based instruction in chemistry that is still widely used to this day.

Although a chemist, Liebig also had an enduring influence on ecology. In 1816, when he was only 13 years old, a massive volcanic eruption led to the 'year without a summer.' The resulting global famine hit Germany particularly hard and undoubtedly precipitated Liebig's life-long interest in using science to improve agriculture and to understand the growth of plants. From an ecological standpoint, his most important discovery was that plants 'feed' on inorganic nitrogen compounds in soils and carbon dioxide from the atmosphere to create new organic tissue. His work led eventually to the use of inorganic nitrogen fertilizers in agriculture, and he recognized that nitrogen is more important to agricultural crop growth than any other nutrient. The massive explosion in agricultural productivity over the past 150 years – an increase that is absolutely necessary to support the current human population – resulted in large part from this recognition.

Liebig's Law of the Minimum states that the growth of a plant is limited primarily by the one nutrient that is in relatively short supply – relative, that is, to the needs of the plant. This concept remains important in ecological thought in the 21st century.

FIGURE 11.7 Painting of Justin von Liebig at age 18 while a student at the University of Erlangen, Germany (1821). Liebig first demonstrated that plants take carbon dioxide and nitrogen from the environment and convert it into their organic biomass. He also proposed that plant growth is most limited by a single nutrient, such as nitrogen. We now call this idea "Liebig's Law of the Minimum." (Source: http://en.wikipedia.org/wiki/Justus_von_Liebig)

the supply of most of these nutrients is more than sufficient to meet plant needs. However, nitrogen, phosphorus, or both are often in relatively short supply, leading to nutrient limitation of NPP. The fact that one particular element—usually either nitrogen or phosphorus—can limit productivity was first noted by the 19th-century German chemist Justin von Liebig and is still called Liebig's Law of the Minimum (Box 11.2). In some ecosystems, both nitrogen and phosphorus are limiting to production, a phenomenon called **colimitation**. Many agricultural systems are strongly nitrogen limited, and fertilizing

> rates of NPP can be influenced by nutrient resources

with nitrogen is at the root of the green revolution that so dramatically increased global food production during the second half of the 20th century (Chapter 14).

A recent global model (Wang et al., 2012) predicts the spatial patterning of nitrogen and phosphorus limitation across the natural terrestrial biosphere by estimating, for both nutrients, the rate of supply, and the potential demand to support NPP as a function of climate, vegetation type, and soil type (Figure 11.8). According to this model, either limitation by nitrogen or colimitation by nitrogen and phosphorus is likely in most temperate grasslands and forests, phosphorus alone is more critical in many tropical forests and

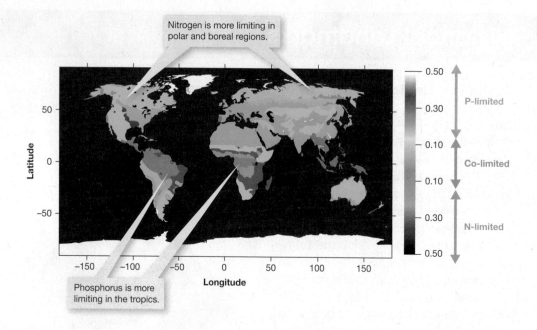

FIGURE 11.8 Model predictions of where terrestrial net primary productivity is most likely to be limited by nitrogen alone, by phosphorus alone, or by nitrogen and phosphorus together. In general, nitrogen is more limiting at higher latitudes and in deserts, while phosphorus is more limiting in tropical rain forests and savannas. (After Wang et al., 2010.)

savannas, and nitrogen alone is the more limiting nutrient in deserts, tundra systems, and boreal forests. This nutrient limitation depresses the rate of NPP by up to 20% in the phosphorus-limited ecosystems, and by up to 40% in the nitrogen-limited ecosystems.

The model predictions of Wang et al. (2010) agree with a large amount of data on nutrient limitation in terrestrial ecosystems collected from across the Earth, and indeed phosphorus limitation is very common in tropical forests. This limitation results at least in part from a very low availability of phosphorus in the old, heavily weathered soils that dominate much of the tropics. Relatively young soils formed from recent weathering of bedrock often contain high levels of available phosphorus, but over geological time scales, weathering leads to less available phosphorus as it is leached away or bound in unavailable mineral forms. The older soils are far more prevalent in the tropics, which have not seen any glacial activity for millions of years.

The importance of soil age on nutrient limitation in tropical forests was well demonstrated by experimentally fertilizing forests at three different sites in Hawaii (Figure 11.9). All the Hawaiian Islands were formed from volcanoes and from the same magma source, so they all have the same type of bedrock. The oldest are more than 4 million years old, while on the youngest island volcanoes are still active, and new rock is still being continuously formed. The islands all have the same climate, so they provide an ideal site for studying just the

> phosphorus is often limiting in tropical forests . . .

effect of soil age. Ecosystems with young soils (300 years) were nitrogen limited, while those on old soils (4.1 million years) were phosphorus limited. The tropical forest ecosystem on soils of intermediate age (20 thousand years) were colimited by nitrogen and phosphorus.

Nitrogen limitation is prevalent in high-latitude ecosystems such as boreal forests and the tundra. In these regions, we find soils that are very young due to recent glaciation: often, the glaciers have retreated only in the past few thousand years or less, leaving exposed bedrock to slowly weather into soil. These new soils provide plenty of plant-available phosphorus, which helps explain the nitrogen limitation.

> . . . while nitrogen is often limiting in boreal forests and tundra ecosystems

But why does nitrogen limitation persist, since nitrogen-fixing bacteria are capable of capturing molecular N_2 from the atmosphere and converting it into plant-available nitrogen (Chapter 8)? Nitrogen fixation occurs at high rates in many ecosystems, which helps alleviate nitrogen limitation, and can lead to phosphorus limitation as phosphorus becomes the scarcer resource. However, high rates of nitrogen fixation are far more prevalent in tropical regions than elsewhere on the planet, and rates of nitrogen fixation in boreal and tundra ecosystems are quite low (Figure 11.10). Ecologists are still searching for an explanation for the low rates of nitrogen fixation in high-latitude regions, but the relative lack of nitrogen fixation clearly allows nitrogen limitation to continue in these regions.

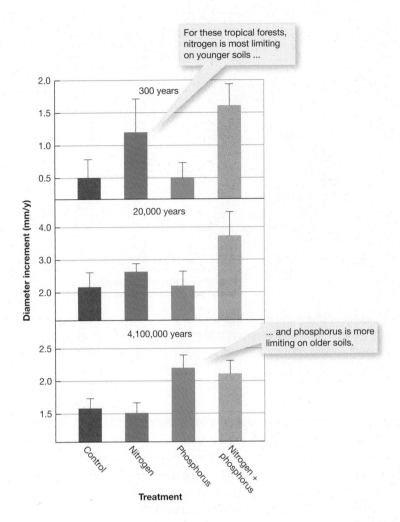

FIGURE 11.9 The increased growth in tree diameters in response to fertlization at three sites in the Hawaiian Islands with soils of very different ages: 300 years (top), 20,000 years (middle), and 4.1 million years (bottom). Compared to unfertilized controls plots, trees on the young soils responded to nitrogen or to nitrogen + phosphorus, but not to phosphorus alone. On the other hand, trees on the oldest soils responded to phosphorus or to nitrogen + phosphorus, but not to nitrogen alone. Trees on the soils of intermediate age showed relatively little response to fertilization with just nitrogen or just phosphorus but had markedly higher rates of growth when fertilized with nitrogen + phosphorus. (After Vitousek & Farringon, 1997.)

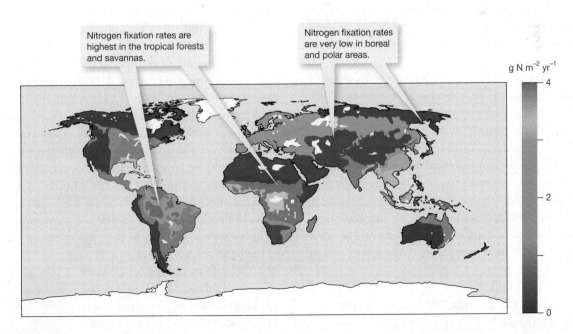

FIGURE 11.10   Rates of nitrogen fixation in natural ecosystems across the land surfaces of the Earth. Note the generally high rates in tropical regions, intermediate rates in many temperate areas, and very low rates in desert ecosystems and in northern latitude regions dominated by boreal forests and tundra. (After Cleveland et al., 1999.)

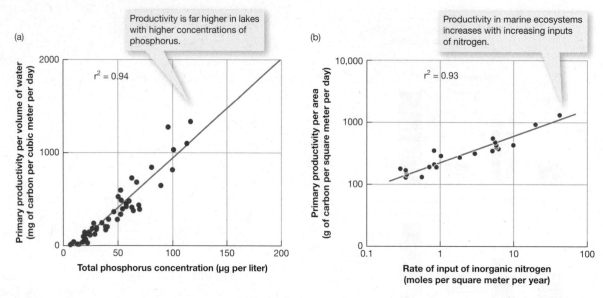

FIGURE 11.11 (a) The relationship between the primary productivity of phytoplankton and phosphorus concentrations across a large number of freshwater lakes in the temperate zone. (After Smith, 1979.) (b) The relationship between primary productivity and the input of dissolved inorganic nitrogen (DIN) across a wide range of marine and estuarine ecosystems. Both the *y*-axis and the *x*-axis are on log scales. (After Nixon et al., 1996.)

In temperate-zone forests and grasslands, we can view the balance between nitrogen limitation and colimitation by nitrogen and phosphorus as a function of soil age and the extent of nitrogen fixation. The soils of the temperate zone range in age from 10,000 years or so in areas that were glaciated during the last Ice Age to 1 million years or more in areas that have not seen recent glaciation, resulting in large differences in phosphorus availability. And rates of nitrogen fixation are variable in the temperate zone, but generally lower than in the tropics and greater than in boreal and tundra ecosystems (Figure 11.10).

The factors that control NPP can vary substantially over the course of a year in those biomes that have pronounced seasonality in precipitation, temperature, or both. For instance, NPP in temperature grasslands may be far below the theoretical maximum because the winters are too cold, with frozen soils and low-intensity radiation; because the summers are too dry; because the rate of nitrogen supply is too low; or because heavy grazing reduces the standing crop of photosynthetic leaves and much of the incident radiation falls on bare ground.

> a succession of factors may limit primary productivity through the year

11.4 FACTORS LIMITING AQUATIC PRIMARY PRODUCTIVITY

The interaction of nutrients and light structures aquatic biomes (Chapter 4), and not surprisingly, the interaction of these two factors also regulates the rates of NPP. Nutrient limitation is particularly important, and in fact it is far more important than in terrestrial ecosystems. While a relative scarcity of nitrogen or phosphorus may depress NPP 1.2- to 1.4-fold (20 to 40%) in terrestrial ecosystems, NPP in aquatic ecosystems is much more responsive to the limiting nutrient. In the examples shown in Figure 11.11, we see that primary productivity in lakes increases more than 20-fold as the concentration of phosphorus increases. And for the marine ecosystems, primary productivity increases 10-fold as the rate of nitrogen inputs increases.

As we noted in Chapter 4, NPP in marine biomes as well as in many lakes is controlled by a balance between light and nutrient availability. At one extreme, NPP in the subtropical ocean gyres is extremely low, even though light penetration through the water column is high, because the nutrient concentrations and rate of nutrient inputs are so low. On the other hand, NPP in upwelling ecosystems is high because the upward-moving water provides a high concentration of nutrients and also keeps the phytoplankton buoyed in the light-rich surface layers.

Why do nutrients play a greater role in regulating NPP in aquatic ecosystems than in terrestrial ones? The reason lies in the huge difference between the relative proportions of nitrogen and phosphorus on the one hand and of the major elements—carbon, hydrogen, and oxygen—on the other

> nutrients are a greater control on NPP in aquatic systems than on land

in the biomass of the primary producers. The proportion of these latter three elements in biomass is relatively constant, driven in part by the molar ratio that characterizes carbohydrates—one carbon atom to two hydrogen atoms to one oxygen atom—and ecologists tend to focus just on carbon. On the other hand, the relative amounts of nitrogen and phosphorus vary tremendously.

For instance, while the ratio of carbon to nitrogen in most marine phytoplankton is approximately 7:1, the ratio in many grasses is 35:1 or more, and for trees 600:1 or more (Figure 11.12). These differences are driven by the fact that structural material is important in land plants but not in algae and cyanobacteria.

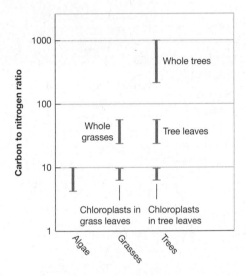

FIGURE 11.12   Algae and the chloroplasts of grasses and trees all have a similarly low ratio of carbon to nitrogen. The ratio in whole grass plants is higher, because much of the structural material needed to support the grasses contains little or no nitrogen. Tree leaves have a ratio similar to that of whole grasses, while whole trees have a substantially higher ratio, due to the large amount of cellulose and other compounds with no nitrogen that provide their structure. Note the log scale.

The marine phytoplankton are composed largely of the essential materials for life and photosynthesis, including chlorophyll and enzyme systems, and these materials are rich in nitrogen (and phosphorus). The chloroplasts of tree and grass leaves are similar in composition to the phytoplankton, but whole trees are made up largely of wood, a necessity as they reach up into the sky to compete with their neighbors for light. The cellulose and lignin of wood contain very little nitrogen. Even the leaves of trees and grasses contain more structural carbon than do marine phytoplankton, since the air around them provides far less support to tissues than does the water of lakes and oceans. The relationship between the abundances of elements in organisms is called **stoichiometry**, a topic we will consider in more detail when we examine the relative importance of nutrients in different aquatic ecosystems below.

Next, we turn to the relative importance of nitrogen and
phosphorus as limiting factors in
three different types of aquatic ecosystems: subtropical ocean gyres, freshwater lakes, and estuaries. NPP in the subtropical gyres is very low and is colimited by nitrogen and phosphorus. Concentrations of both in the water are low (Figure 11.13), and in fact, the concentration of inorganic phosphorus is lower than that of inorganic forms of nitrogen. However, phytoplankton also contain less phosphorus than nitrogen, and the relative proportions of dissolved nutrient concentrations in the water and in phytoplankton biomass are virtually the same throughout much of the world's oceans: an N:P ratio by moles of 15:1, often termed the Redfield ratio (Box 11.3).

At any particular time and place, either nitrogen or phosphorus may be in slightly lower supply relative to the needs of the phytoplankton, and therefore

> colimitation in subtropical ocean gyres

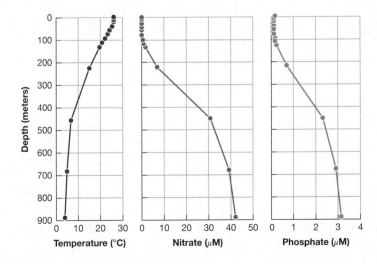

FIGURE 11.13   Concentrations of both nitrate and phosphate are very low in the surface waters of the subtropical gyres, here illustrated for the gyre in the North Pacific Ocean. Concentrations increase rapidly at depth, due to release of decomposing materials that sink from the surface oceans. The ratio of the dissolved inorganic nitrogen (nitrate) to dissolved inorganic phosphorus (phosphate) in these deep-ocean waters is very close to 15:1. The small extent of mixing of the deep-ocean waters to the surface waters of the subtropical gyres provides nutrients in this ratio. (After Smith, 1984.)

Alfred Redfield and the stoichiometry of the world's oceans

A long-time professor at Harvard University and one of the first research scientists at the Woods Hole Oceanographic Institution on the Massachusetts coast, Alfred Redfield was among the great marine ecologists and biogeochemists of the 20th century (Figure 11.14). Originally trained as an animal physiologist, he was fascinated by large-scale patterns in ocean chemistry and how these related to the controls on primary production in the oceans.

By early in the 20th century, marine chemists had documented that the ratio of dissolved nitrogen to phosphorus throughout much of the ocean was relatively fixed at 15:1 (by moles). The ratio of nitrogen to phosphorus in the tissues of marine phytoplankton is also 15:1, and Redfield's creative mind looked for a connection. His hypothesis, developed in a set of papers during the 1930s and 1950s (he spent the 1940s on research related to supporting Allied efforts in World War II), was that the stoichiometric ratio in the phytoplankton is determined by the basic chemical constituents of the machinery necessary for life for a phytoplankton cell, the relative amount of nitrogen and phosphorus in cell membranes, chlorophyll, enzymes, DNA, etc. Over geological time scales, the ratio of dissolved nutrients in the oceans came to reflect this same ratio, due to biogeochemical feedbacks.

Redfield argued that phytoplankton must take up the dissolved nitrogen and phosphorus in the same 15:1 ratio as in their cellular chemistry. If nitrogen were in short supply (that is, available at ratios less than 15:1), then nitrogen-fixing cyanobacteria would respond and create more biologically available nitrogen from the massive supply of molecular N_2 dissolved in seawater. As phytoplankton died and were recycled, this would gradually build up the amount of 'fixed' inorganic nitrogen (nitrate, nitrite, and ammonium) in the oceans. On the other hand, Redfield concluded, if the ratio of inorganic nitrogen to phosphorus in the oceans were greater than 15:1 for any substantial period of time (geologically speaking), bacteria that convert nitrate to molecular N_2 ('denitrifying' bacteria) would gradually consume the excess biologically available nitrogen.

In this model view of the world's oceans, phosphorus is the nutrient that ultimately limits primary productivity. The availability of phosphorus, in turn, is regulated by its inputs from land and by its interaction with the sediments of the deep ocean. The amount of biologically available inorganic nitrogen roughly balances the needs of the phytoplankton—relative to phosphorus—and is regulated by the balance between nitrogen fixation and the work of the denitrifying bacteria. In broad brush, Redfield's model remains the way that marine scientists view the ecology and biogeochemistry of the oceans to this day, and it has played a huge role in the way freshwater and terrestrial ecologists think as well.

Photo by Jan Hahn © Woods Hole Oceanographic Institution

FIGURE 11.14 Alfred Redfield, a marine ecologist who pioneered our understanding of the stoichiometric relationships between carbon, nitrogen, and phosphorus in the oceans. (Source: Woods Hole Oceanographic Institution http://www.whoi .edu/75th/gallery/week10.html.)

slightly more limiting to NPP. But this is a transient feature, with nitrogen and phosphorus being equally important in limiting production in the gyres over periods from years to decades and longer. The very small input of deep-ocean nutrients to the surface waters of the subtropical gyres has an N:P ratio that is also very close to the Redfield ratio of 15:1. This largely explains the colimitation by nitrogen and phosphorus.

In a large number of lakes, particularly in the temperate zone, NPP is tightly controlled by the phosphorus supply. When the input of phosphorus is greater, the concentration of phosphorus and rates of NPP are also greater (see Figure 11.11). Excess inputs of phosphorus can lead

> phosphorus is limiting in many lakes . . .

to great ecological damage, including waters depleted of oxygen by 'blooms' of primary producers. These waters cannot support animal life (particularly in the bottom waters of stratified lakes), and there is a loss of biodiversity. This process of excess nutrient enrichment, called **eutrophication**, accelerated greatly in the 1950s and 1960s across much of North America and Europe. The whole-lake ecosystem experiments by scientists at Canada's Experimental Lakes Area (ELA), described in Chapter 1, established that phosphorus was the central culprit in lake eutrophication. Since the early 1970s, developed countries have worked hard to reduce excess phosphorus inputs, although challenges remain and phosphorus pollution has also become a problem in the developing world.

The experiments at ELA explicitly used Redfield's ocean model (see Box 11.3) as a way to understand lakes. When lakes were fertilized with phosphorus and nitrogen was abundant, NPP was clearly tied to the phosphorus inputs. But the ELA scientists wanted to know what would happen if nitrogen were in relatively scarce supply. Nitrogen-fixing cyanobacteria were not observed in the ELA lakes when nitrogen was plentiful, but when the nitrogen fertilization to a lake was cut—while keeping the phosphorus fertilization level high so the inputs of nitrogen and phosphorus fell below the Redfield ratio of 15:1—nitrogen-fixing cyanobacteria appeared in a matter of weeks (Table 11.2). These came to dominate the phytoplankton community and fixed sufficient nitrogen to balance its relative deficit compared to phosphorus and maintain phosphorus limitation of the lake. Thus, in these lake experiments, the feedbacks that led to increased nitrogen fixation occurred on the time scales of weeks, much faster than the time scale that may balance nitrogen and phosphorus in the oceans.

TABLE 11.2 Response of Lake 227 of the Experimental Lakes Area to nitrogen and phosphorus fertilization at ratios above and below the Redfield ratio.

Year	Nitrogen inputs from watershed and fertilization (moles yr⁻¹)	N:P ratio of inputs from watershed and fertilization	Nitrogen fixation (moles yr⁻¹)	N:P ratio of all inputs including N fixation
1972	27	32:1	0	32:1
1973	26	31:1	0	32:1
1974	27	32:1	0	32:1
1975	12	14:1	1.9	17:1

After Flett et al., 1980.

Efforts by governments to reduce phosphorus inputs to lakes had an immediate, demonstrable effect in improving water quality in the 1970s, on the Great Lakes of North America and on many smaller lakes in the United States, Canada, and Europe. This is one of the great success stories of ecological research leading to immediate benefit in better environmental management. But eutrophication in estuaries and coastal waters continued to worsen for several more decades (see Box 1.4). Why? As it turns out, primary production (and therefore eutrophication) is regulated more by nitrogen than by phosphorus in estuaries and coastal marine ecosystems, and eutrophication in these systems worsened as nitrogen inputs and the resulting NPP grew (Figure 11.15).

> Nitrogen is limiting in many estuaries and coastal marine ecosystems . . .

Many factors make nitrogen limitation more likely in estuaries than in lakes. One of these is the interaction of estuaries with coastal waters: while almost all the nutrient inputs to lakes come from upstream watershed sources, estuaries also receive nutrients from the coastal waters (as we discuss below in Section 11.5). These coastal waters are often relatively depleted in nitrogen compared to phosphorus, so they can serve as a rich source of phosphorus to estuaries, making phosphorus limitation less likely. Another important difference is the lack of nitrogen fixation by planktonic cyanobacteria in most estuaries. When nitrogen deficits start to develop in lakes, nitrogen-fixing cyanobacteria begin to make up this deficit, but no such response occurs in estuaries except at very low (almost lake-like) salinities.

Microbiologists and ecologists have been fascinated by this dichotomy for decades and have proposed and tested many hypotheses to explain the pattern. Simple physiological explanations have proven insufficient. It turns out that the absence of the nitrogen fixers in estuaries results from an interaction of a physiological constraint and grazing by zooplankton. The physiological constraint comes from the very high concentrations of sulfate in seawater, which interferes with and slows the uptake of molybdenum, an element essential for nitrogen fixation. The resulting slow growth of the nitrogen-fixing cyanobacteria makes them very vulnerable to grazing.

> why is there so little nitrogen fixation by phytoplankton in estuaries?

This interaction was first tested in a model in the late 1990s (Figure 11.16). In the model, nitrogen fixation is a function of the population size of the planktonic cyanobacteria under conditions when the ratio of available N:P is low. The population size, then, is controlled by the balance between the growth rate and death rate of the cyanobacteria. Death is through

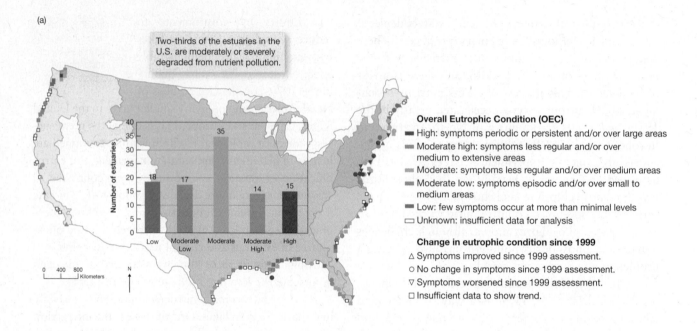

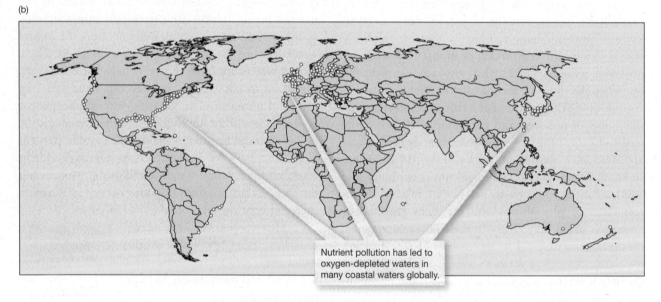

FIGURE 11.15 (a) An assessment of water quality in the estuaries and coastal waters of the United States with regard to eutrophication from excess nutrient inputs. Each symbol represents a different ecosystem. The color of the symbol indicates the extent of degradation from nutrient pollution, and the shape indicates how the condition had changed over the prior decade. Two-thirds of the estuaries and coastal marine ecosystems are moderately to severely degraded, and for most of these, conditions had either remained unchanged or deteriorated over the prior decade. (After Bricker et al., 2008.) (b) A global assessment of the state of low-oxygen conditions in coastal marine ecosystems and estuaries resulting from nutrient pollution. Each circle represents a location where oxygen levels are too low to fully support animal life. (After UNEP, 2008.)

zooplankton grazing, and growth is controlled by the phosphorus and nitrogen supply; the nitrogen supply in turn is a function of both the environmental supply and the rate of nitrogen fixation. This results in a positive feedback, where high growth leading to a high cyanobacteria population size further increases the growth rate. The ultimate growth rate is constrained by phosphorus.

Under conditions of normal zooplankton grazing and sulfate concentrations found in freshwaters, the model correctly predicts that cyanobacterial populations and rates of nitrogen fixation will grow to high levels in around 30 days (Figure 11.16a). However, for the same zooplankton grazing but seawater levels of sulfate (which will slow molybdenum uptake), the model indicates no significant increase in the cyanobacterial population or rates of nitrogen fixation (Figure 11.16b).

Even though we rarely if ever observe nitrogen-fixing cyanobacteria in estuaries except at low salinities,

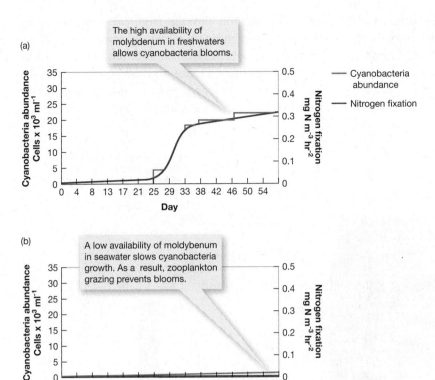

(a)

The high availability of molybdenum in freshwaters allows cyanobacteria blooms.

— Cyanobacteria abundance
— Nitrogen fixation

(b)

A low availability of moldybenum in seawater slows cyanobacteria growth. As a result, zooplankton grazing prevents blooms.

FIGURE 11.16 Model predictions of the change in abundance of cyanobacterial cells (left-hand axes) and rates of nitrogen fixation (right-hand axes) over time in aquatic ecosystems. The top plot (a) represents typical conditions of zooplankton grazing and molybdenum availability for a freshwater lake. For the bottom plot (b), all model conditions are the same as for the freshwater lake, except the molybdenum availability is lowered to reflect that of estuarine conditions. The model correctly predicts the dynamics of cyanobacteria growth and nitrogen fixation in freshwaters, and it illustrates that a slowing of growth in estuaries due to low molybdenum availability can interact with zooplankton grazing to prevent nitrogen fixation. (After Howarth et al., 1999.)

they do in fact occur and can grow if grazing by zooplankton is suppressed. This was demonstrated in mesocosm experiments with water from Narragansett Bay (Rhode Island), where researchers kept zooplankton populations very low by adding fish that consumed the zooplankton (Figure 11.17a). In one experiment (1996), nitrogen-fixing cyanobacteria grew in the mesocosms when zooplankton were largely absent, but not when zooplankton populations were at normal estuarine levels (Figure 11.17b). In a subsequent experiment (1998), a very small population of cyanobacteria grew even in the presence of zooplankton, but the population was significantly larger when zooplankton were suppressed. Even these populations were some 10-fold lower than observed in the ELA freshwater lake experiments, though. This was not simply an artifact of the mesocosm tanks: in the same tanks in a freshwater experiment with no zooplankton grazing, nitrogen-fixing cyanobacteria were almost 30 times higher than in the estuarine experiment and 3 to 5 times higher than in the ELA experiment (Marino et al., 2006).

> a mesocosm test of the importance of grazing in estuaries

If grazing by zooplankton in estuaries can interact with the sulfate-induced slow growth rate of planktonic cyanobacteria to prevent significant rates of nitrogen fixation, how can we

> What about nitrogen fixation in the subtropical ocean gyres?

explain the reasonably high rates of planktonic nitrogen fixation in the subtropical ocean gyres? Presumably the sulfate there also leads to a slow growth rate of the cyanobacteria. The most reasonable hypothesis is that these cyanobacteria do indeed grow slowly, but that slow growth per se does not prevent the eventual establishment of a significant population of cyanobacteria and resulting nitrogen fixation. It is the grazing in estuaries that makes the slow growth problematic. This grazing is by generalist species of zooplankton that are primarily feeding on other species of phytoplankton of similar size as—and far more abundant than—the cyanobacteria. In the subtropical gyres, most phytoplankton are far smaller than the cyanobacteria (see Section 4.4), and there are no generalist zooplankton that feed on phytoplankton that are as large as the cyanobacteria.

While nitrogen and phosphorus are the nutrients most limiting to NPP in most of the world's oceans, iron is more limiting over some 20% of these waters. In these areas, which include much of the Southern Ocean around Antarctica as well as the equatorial region of the Pacific Ocean, inorganic nitrogen and phosphorus are abundant, yet NPP and phytoplankton populations are low (Lalli & Parson, 2004). For many decades, marine ecologists thought the low phytoplankton biomass

> iron can be limiting in some parts of the ocean

resulted from very high grazing on them, but in the late 1980s it became apparent that these areas have unusually low iron levels (Martin & Fitzwater, 1988). This idea was controversial, and it became even more so when Martin proposed to artificially fertilize the oceans with iron to increase productivity and draw carbon dioxide from the atmosphere as a way to help mitigate global warming (Box 11.4). Nonetheless, a series of iron-addition experiments in both the Southern Ocean and the equatorial Pacific have subsequently demonstrated the prevalence of iron limitation in these waters.

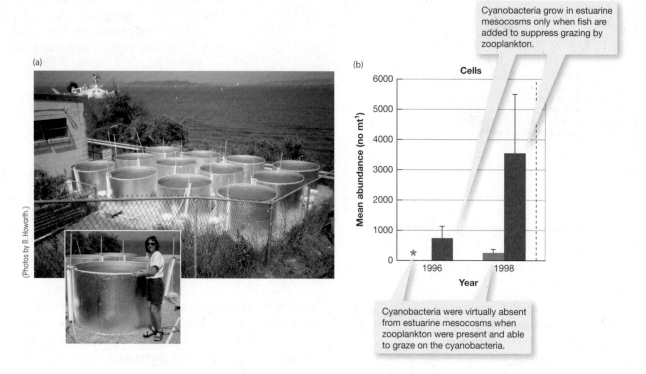

(Photos by R. Howarth.)

Cyanobacteria grow in estuarine mesocosms only when fish are added to suppress grazing by zooplankton.

Cyanobacteria were virtually absent from estuarine mesocosms when zooplankton were present and able to graze on the cyanobacteria.

FIGURE 11.17 (a) Mesocosm tanks used to test the effect of grazing by zooplankton on potential rates of nitrogen fixation by planktonic cyanobacteria in estuaries. The inset photo shows one of the experimenters for scale. (b) The mean abundance of planktonic cyanobacteria cells in the estuarine mesocosms shown in Figure 11.13 over the course of several months in the summer. Two different experiments are shown, one in 1996 and one in 1998. Cyanobacteria grew and were present in both years when zooplankton grazing in the mesocosms was suppressed (red bars). When zooplankton grazing occurred at a normal rate of an estuarine ecosystem, no cyanobacteria were present in 1996 and were present at only very low levels in 1998 (blue bar). (After Marino et al., 2006.)

11.4   ECOncerns

Should we fertilize the oceans with iron?

The oceans take up a significant amount of the carbon dioxide released by industry and human activity to the atmosphere (see Chapter 12). Can the oceans be managed so as to increase this uptake and partially mitigate global warming? Beginning in the late 1980s, the marine chemist John Martin and other scientists proposed that we fertilize parts of the oceans (such as the Southern Ocean and equatorial Pacific) with iron, to increase NPP and therefore the uptake of atmospheric carbon dioxide into the oceans.

From the very start, this idea has drawn strong reactions. Some environmental organizations are appalled at the idea of such a large-scale manipulation of otherwise pristine ocean waters, and many scientists have cautioned that there might be unanticipated consequences such as harmful algal blooms, depletion of oxygen in deep-ocean waters, and release of sulfur-containing gases to the atmosphere. On the other hand, several commercial companies have explored the idea as a way of creating marketable

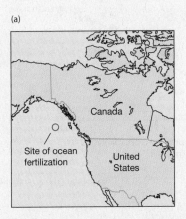

FIGURE 11.18   (a) Map indicating the site of a controversial iron-fertilization of waters in the Pacific Ocean in 2012. (b) The boat releasing the iron sulfate in this controversial fertilization. (Source: Tollefson, 2012.)

carbon credits to sell in the European Union and those countries with established markets for credits for reducing atmospheric carbon dioxide.

The controversy erupted in 2012 after a U.S. business executive working with a native village fertilized the coastal Pacific Ocean waters some 300 km west of Canada with 100 tons of iron sulfate (Figure 11.18). The first press report, in *The Guardian* of Manchester, UK, was headlined "World's Biggest Geoengineering Experiment 'Violates' UN Rules" (October 15, 2012 article by Martin Lukacs). (http://www.guardian.co.uk/environment/2012/oct/15/pacific-iron-fertilisation-geoengineering) Despite the headline, the legal status of the experiment was not clear. Dumping of ocean wastes has long been prohibited by international law, but was this waste? Newer treaties have voluntary restrictions against such geoengineering, but "voluntary" is the operative word; these treaties have no teeth. They allow for scientific experimentation yet give little guidance as to what constitutes legitimate science.

The apparent purpose in this case, according to press accounts, was to stimulate NPP both to increase salmon production and to sell carbon credits. The amount of iron added was five times more than in any previous scientific fertilization with iron, and in this case, no scientists were involved. The business executive claimed the fertilization created a phytoplankton bloom over 10,000 km², but the claim has not been supported by evidence or verified by others.

Should society encourage experimentation with the fertilization of the oceans as a means to help counteract global warming? Or should all experimentation be banned, on the basis that we simply must protect the oceans? If some experimentation is allowed, how can we distinguish between well-conducted science and commercial interests that masquerade as science?

11.5 THE FATE OF PRIMARY PRODUCTIVITY: GRAZING

Animals, fungi, and most bacteria are heterotrophs: they derive their matter and energy either directly by consuming plant material or indirectly from plants by eating other heterotrophs (Chapter 3). Plants, algae, and other autotrophic primary producers make up the first trophic level in an ecosystem; primary consumers (herbivores) occur at the second trophic level; secondary consumers (carnivores) at the third, and so on.

Since secondary productivity depends on primary productivity, we should expect a positive relationship between the two variables. Figure 11.19 illustrates this general relationship for zooplankton and bacterial production across aquatic ecosystems. Secondary productivity by zooplankton, whose main food is phytoplankton cells, is positively related to phytoplankton productivity in a range of lakes in different parts of the world (Figure 11.19a). The productivity of heterotrophic bacteria in lakes and oceans also parallels that of phytoplankton (Figure 11.19b); the bacteria metabolize dissolved organic matter released from intact phytoplankton cells or produced as a

> there is a general positive relationship between primary and secondary productivity

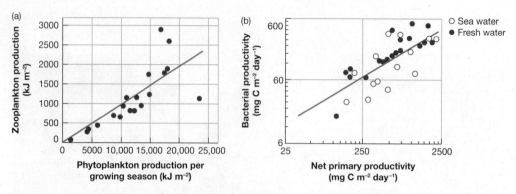

FIGURE 11.19 The relationship between primary and secondary productivity for: (a) zooplankton in lakes (After Brylinsky & Mann, 1973.), and (b) bacteria in fresh and sea water. (After Cole et al., 1988.)

result of messy feeding by animals grazing on the phytoplankton.

Secondary productivity by herbivores is always less, and often far less, than the primary productivity upon which it is based. Where has the missing energy gone? First, not all the primary producer biomass produced (plant, algal, and so on) is consumed alive by herbivores. Much dies without being grazed and supports a community of decomposers (bacteria, fungi, and detritivorous animals). Second, not all the plant biomass eaten by herbivores (nor all the herbivore biomass eaten by carnivores) is assimilated and available for incorporation into consumer biomass. Some is lost in feces, and this also passes to the decomposers. Third, not all the energy that has been assimilated is actually converted to biomass. A proportion is lost as respiratory heat. This occurs both because no energy conversion process is 100% efficient (some energy is lost as unusable random heat, consistent with the second law of thermodynamics), and also because the organisms do work that requires energy, again released as heat. These three energy pathways occur at all trophic levels and are illustrated in Figure 11.20.

> most of the primary productivity does not pass through the grazer system

A unit of energy (a joule) may be consumed and assimilated by an invertebrate herbivore that uses part of it to do work and loses it as respiratory heat. Or it might be consumed by a vertebrate herbivore and later be assimilated by a carnivore that dies and enters the dead organic matter compartment. Here, what remains of the joule may be assimilated by a fungus and consumed by a soil mite, which uses it to do work, dissipating a further part of the joule as heat. At each consumption step, the rest of the joule may fail to be assimilated and pass in the feces to dead organic

> possible pathways of a joule of energy through an ecosystem . . . efficiencies matter!

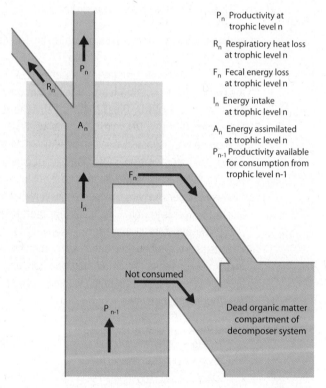

P_n Productivity at trophic level n

R_n Respiratiory heat loss at trophic level n

F_n Fecal energy loss at trophic level n

I_n Energy intake at trophic level n

A_n Energy assimilated at trophic level n

P_{n-1} Productivity available for consumption from trophic level n-1

FIGURE 11.20 The pattern of energy flow through a trophic compartment (represented as the blue box).

matter, or it may be assimilated and respired, or assimilated and incorporated into growth of body tissue (or the production of offspring). The body may die and what remains of the joule enters the dead organic matter compartment, or it may be captured alive by a consumer in the next trophic level where it meets a further set of possible branching pathways. Ultimately we see that each joule will have found its way out of the ecosystem, dissipated as respiratory heat at one or more of the transitions in its path along the food chain. Whereas an element such as nitrogen or iron may cycle endlessly through the food chains of a community, energy passes through just once.

The possible pathways in the herbivore/carnivore (live consumer) and decomposer systems are the same, with one critical exception—feces and dead bodies are lost to the former (and enter the decomposer system), but feces and dead bodies from the decomposer system are simply sent back to the dead organic matter compartment at its base. Thus, the energy available as dead organic matter may finally be completely metabolized—and all the energy lost as respiratory heat—even if this requires several circuits through the decomposer system. The exceptions are situations in which:

1 matter is exported out of the local environment to be metabolized elsewhere, for example, detritus is washed out of a stream or sinks from the surface waters to the deep ocean;

2 local abiotic conditions have inhibited decomposition and left pockets of incompletely metabolized high-energy organic matter, which over geological time can become coal or oil.

The proportions of net primary production flowing along each of the possible energy pathways depend on **transfer efficiencies** from one step to the next. We need to know about just three categories of transfer efficiency to be able to predict the pattern of energy flow (Figure 11.21). These are consumption efficiency, assimilation efficiency, and production efficiency.

> consumption, assimilation, and production efficiencies determine the relative importance of energy pathways

Consumption efficiency (CE) is the percentage of total productivity available at one trophic level that is consumed ('ingested') by the trophic level above. For primary consumers, CE is the percentage of joules (or organic carbon, which is the source of the potential energy) produced per unit time and area as NPP that finds its way into the guts of herbivores. In the case of secondary consumers, it is the percentage of herbivore productivity eaten by carnivores. The remainder dies without being eaten and enters the decomposer system. Reasonable average values for CE by herbivores are approximately 5% in forests, 25% in grasslands, and 50% in phytoplankton-dominated communities. For carnivores, CE varies from 25% to almost 100%.

Assimilation efficiency (AE) is the percentage of food energy taken into the guts of consumers in a trophic level that is assimilated across the gut wall and becomes available for incorporation into growth or to do work. The remainder is lost as feces and enters the decomposer system. It is harder for us to ascribe an assimilation efficiency to microorganisms, in which food does not pass through a gut and feces are not produced. Bacteria and fungi digest dead organic matter externally and, between them, typically absorb almost all the product: they are often said to have AEs of 100%. AEs are typically low for terrestrial herbivores, detritivores, and microbivores (20–50%), somewhat higher for aquatic herbivores, and high for carnivores (around 80% and sometimes higher). The way plants allocate production to roots, wood, leaves, seeds, and fruits also influences their usefulness to herbivores. Seeds and fruits may be assimilated with efficiencies as high as 60–70% and leaves with about 50% efficiency, while the AE for wood may be as low as 15%.

Production efficiency (PE) is the percentage of assimilated energy incorporated into new biomass; the remainder is entirely lost to the community as respiratory heat. PE varies according to the taxonomic class of the organisms concerned. Invertebrates in general have high efficiencies (30–50%), losing relatively little energy in respiratory heat. Among the vertebrates, ectotherms (whose body temperature varies according to environmental temperature; see Chapter 3) have intermediate values for PE (around 10%), while endotherms, which expend considerable energy to maintain a constant temperature, convert only 1% to 5% of assimilated energy at most into production. Microorganisms, including protozoa, tend to have very high PEs, 50% or greater.

The overall **trophic transfer efficiency** from one trophic level to the next is simply CE × AE × PE. This represents the percentage of energy (or organic matter) at one trophic level that is transferred to the next. This

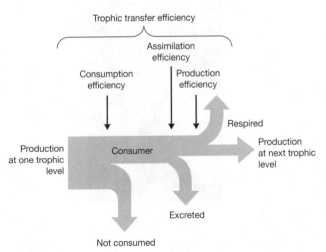

FIGURE 11.21 The relationship of trophic transfer efficiency to its three subcomponent parts: consumption efficiency, assimilation efficiency, and production efficiency.

concept originated with the pioneering work of Raymond Lindeman in 1942, and for many decades after that, it was generally assumed that trophic transfer efficiencies were around 10%. Indeed some ecologists referred to a 10% 'law.' However, there is certainly no law of nature that results in precisely one-tenth of the energy that enters a trophic level transferring to the next. Consider a zebra grazing on grass in a savanna ecosystem. As noted above, CE for herbivores in grasslands is commonly 25%; AE for a herbivore on grass is fairly low, perhaps is in the range of 20%; and PE for an endotherm such as a zebra is in the range of 3%. Therefore, CE × AE × PE = 0.15%, or two orders of magnitude less than 10%. The transfer efficiency from zebra to lion is somewhat greater, but still far below 10%, due to the low PE for endotherms. Even with high values for CE (80%) and AE (80%), the trophic transfer efficiency from the zebra to the lion is in the range of only 1%.

A compilation of 48 different detailed trophic studies from a wide range of aquatic ecosystems, both freshwater and marine, revealed that trophic-level transfer efficiencies were in fact 10% on average (actually 10.13%, with a standard error of 0.49), though they varied between about 2% and 24% (Figure 11.22). The finding is logical for the phytoplankton (primary producer) to zooplankton (herbivore) step, when we consider the component efficiencies. CE for herbivores is often 50% in phytoplankton-based systems, AE is in the range of 50%, and PE for an average invertebrate is 40%. Therefore, CE × AE × PE = 10%.

> trophic transfer efficiencies in aquatic ecosystems are generally greater

However, the analysis also included many studies at higher trophic levels, and the efficiencies there were on average also 10% and indistinguishable from the primary producer to herbivore level (Pauly & Christensen, 1995). These higher-level transfers would include invertebrate carnivores (carnivorous zooplankton) eating invertebrates (herbivorous zooplankton), vertebrate carnivores (fish) eating invertebrates (zooplankton), and vertebrate carnivores (fish) eating other vertebrates (fish). Given that AE for these carnivores is in the range of 80%, and that PE is in the range of 40% for invertebrates and 10% for fish, CE must be in the range of 25% of the invertebrates and 100% for the fish—values that are at the extremes listed above for carnivores. We see, then, that CE for aquatic carnivores is in fact quite high.

Ecologists generally express rates of NPP in terms of units of energy, organic matter, or organic carbon (all per area and per time), as we have done throughout this text

> comparing the rates of production of energy and protein in primary producers

so far. The organic carbon and organic matter themselves represent energy, since energy is released when these are oxidized to carbon dioxide. We can, however, also consider NPP in terms of production of protein per area per time, or nitrogen per area per time, since most of the nitrogen in primary producers is in protein. Table 11.3 compares rates of NPP expressed as organic carbon and as nitrogen for four different ecosystems, using typical values found in these ecosystems.

When we view them in terms of carbon (energy), the tropical forest and estuary are the most productive, and the subtropical gyre looks like a desert. However,

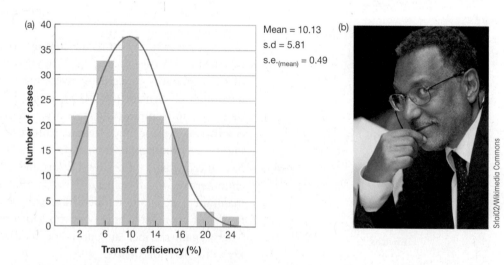

FIGURE 11.22 (a) Frequency distribution of trophic transfer efficiencies for 48 different aquatic ecosystems, includng lakes and marine systems. Different trophic levels are included, such as phytoplankton to herbivore, herbivore to first-level carnivore, and so on, resulting in 140 separate observations. The mean trophic transfer efficiency is slightly over 10%. (After Pauly and Christensen, 1995.) (b) Daniel Pauly, the first author of the plot shown in (a). (Source: http://en.wikipedia.org/wiki/Daniel_Pauly.)

when we think in terms of nitrogen (protein), the rate of production is very similar in the tropical forest, grassland, and subtropical gyre, while the rate is very much greater in the estuary. This reflects the high C:N ratio for the trees in the forest, with much of the production going into structural production of wood, an intermediate ratio for the grasses, and a low ratio for phytoplankton in the subtropical gyre and estuary (see Figure 11.12). From the viewpoint of protein production, the reason for the higher trophic transfer efficiencies in the aquatic ecosystems becomes very clear: it is easier for an herbivore to produce its new protein by eating a protein-rich diet. This also helps us to understand why production of top predators can be as high as it is in the subtropical gyres, even though from the

standpoint of carbon NPP they look like deserts. These subtropical gyres are highly efficient protein machines.

If we know the NPP (reverting back to units of energy or carbon) at a site, and the CE, AE, and PE for all the trophic groupings present (herbivores, carnivores, decomposers, detritivores), we can map out the relative importance of different pathways. Figure 11.23 does this, in a general way, for a forest, a grassland, a plankton community (of the ocean or a large lake), and the community of a small stream or pond. The decomposer system is responsible for the majority of secondary production, and therefore of respiratory heat loss, in many of the ecosystems in the world (Figure 11.24). The live consumers have their greatest role in open-water aquatic communities based on phytoplankton, or in the beds of algae that often dominate in shallow water. In each case, a large proportion of NPP is consumed alive and assimilated at quite a high efficiency (Figure 11.23a). In contrast, the decomposer system plays its greatest role where vegetation is woody—forests, shrublands, and mangroves (Figure 11.23b). Grasslands and aquatic systems based on large primary producers [seagrasses, freshwater weeds, and macroalgae] occupy intermediate positions.

The live consumer system holds little sway in terrestrial communities because of low herbivore consumption efficiencies and assimilation efficiencies, and

> the relative roles of the live consumer and decomposer systems

TABLE 11.3 Comparison of net primary production (NPP) expressed in units of organic carbon with nitrogen for four representative ecosystems.

	g C m^{-2} year^{-1}	g N m^{-2} year^{-1}
Tropical forest	800	10
Temperate grassland	250	5
Subtropical marine gyre	50	7
Productive estuary	800	115

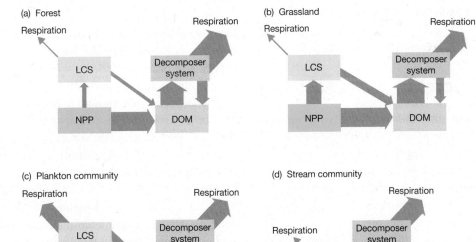

FIGURE 11.23 General patterns of energy flow for: (a) forest, (b) grassland, (c) a plankton community in the sea, and (d) the community of a stream. The sizes of the boxes and arrows are proportional to the relative magnitude of compartments and flows. (DOM refers to dead organic matter; LCS, live consumer system; NPP, net primary production.)

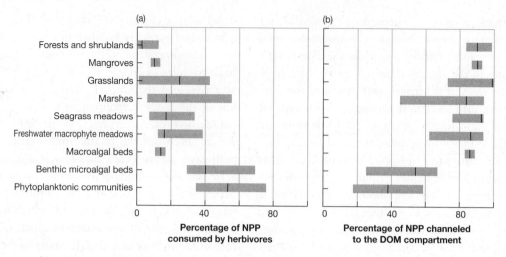

FIGURE 11.24 Box plots for a range of ecosystem types showing: (a) percentage of net primary production (NPP) consumed by herbivores and (b) percentage of NPP entering the dead organic matter (DOM) compartment. Boxes encompass 25% and 75% percentiles of published values and the central lines represent the median values. Phytoplankton and aquatic microalgal communities channel the largest proportions of NPP through herbivores and the smallest proportions through the DOM compartment. (After Cebrian, 1999.)

it is almost nonexistent in many small streams simply because NPP is so low (Figure 11.23d). Small streams often depend for their energy base on dead organic matter that falls or is washed or blown into the water from the surrounding terrestrial environment. The energy flows in the deep-ocean biome are completely supported by an input of organic matter from other areas, in this case the rain of detritus from the surface ocean waters into the dark, deep ocean where no NPP occurs (Chapter 4). In some ways, the ocean bed is equivalent to a forest floor beneath an impenetrable forest canopy.

11.6 THE PROCESS OF DECOMPOSITION

Given the profound importance of the decomposer system, and thus of decomposers (bacteria and fungi) and detritivores, we begin to appreciate the range of organisms and processes active in decomposition.

Immobilization occurs when an inorganic element is incorporated into organic form, often during primary production, for example, when carbon dioxide becomes incorporated into a plant's carbohydrates. Energy is required and comes, in the case of plants, from the sun. Conversely, decomposition includes the release of energy and the **mineralization** of chemical nutrients—the conversion of elements from organic back to an inorganic form. **Decomposition** is the gradual disintegration of dead organic matter (dead bodies, shed parts of bodies, feces) and is brought about by both physical and biological agencies.

| decomposition defined |

Note that some immobilization can occur during decomposition, for instance when a microbe growing on a dead leaf with a high C:N ratio incorporates inorganic nitrogen from the environment into its tissue, which becomes part of the organic nitrogen of the leaf-microbe detritus. In the end, though, decomposition culminates with complex, energy-rich molecules being broken down by their consumers (decomposers and detritivores) into carbon dioxide, water, and inorganic nutrients. Ultimately, the incorporation of solar energy in photosynthesis, and the immobilization of inorganic nutrients into biomass, is balanced by the loss of energy and organic nutrients when the organic matter is mineralized.

| bacteria, archaea, and fungi are early colonists of newly dead material |

If a scavenging animal, a vulture, or a burying beetle, perhaps, does not take a dead resource immediately, the process of decomposition usually starts with colonization by bacteria, archaea, and fungi. Spores for these organisms are always present in the air and the water and are usually on (and often in) organic material before it is dead. The early colonists tend to use soluble materials, mainly amino acids and sugars that are freely diffusible. The residual resources, though, are not diffusible and are more resistant to attack. Subsequent decomposition therefore proceeds more slowly, aided by microbial specialists that can break down structural carbohydrates such as celluloses, lignins, and complex proteins such as suberin (cork) and insect cuticle.

The **microbivores** are a group of animals that operate alongside the detritivores and can be difficult to distinguish from them. The name *microbivore* is

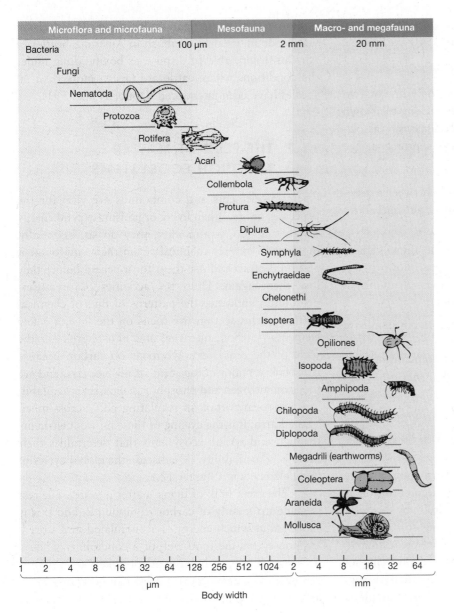

Microflora and microfauna	Mesofauna	Macro- and megafauna

FIGURE 11.25 Size classification by body width of organisms in terrestrial decomposer food webs. Bacteria and fungi are decomposers. Animals that feed on dead organic matter (plus any associated bacteria and fungi) are detritivores. Carnivores that feed on detritivores include Opiliones (harvest spiders), Chilopoda (centipedes), and Araneida (spiders). (After Swift et al., 1979.)

reserved for the minute animals that specialize at feeding on bacteria or fungi but are able to exclude detritus from their guts. In fact, though, the majority of detritivorous animals are generalist consumers, of both the detritus itself and the associated bacterial and fungal populations. The invertebrates that take part in the decomposition of dead plant and animal materials are a taxonomically diverse group. In terrestrial environments they are usually classified according to their size (Figure 11.25). This is not an arbitrary basis for classification, because size is an important feature for organisms that reach their resources by burrowing or crawling among cracks and crevices of litter or soil.

In freshwater ecology, on the other hand, the study of detritivores has been concerned less with the size

> specialist microbivores feed on bacteria and fungi, but most detritivores consume detritus too

of the organisms than with the ways in which they obtain their food. For example, **shredders** are detritivores that feed on coarse particulate organic matter, such as tree leaves fallen into a river; these animals fragment the material into finer particles. On the other hand, **collector–filterers**, such as larvae of blackflies in rivers, consume the fine particulate organic matter that otherwise would be carried downstream. Because of very high densities (sometimes as many as 600,000 blackfly larvae per square meter of riverbed), a very large quantity of fine particulate matter is converted by the larvae into fecal pellets that settle on the bed and provide food for other detritivores.

> aquatic detritivores are usually classified according to their feeding mode

Two of the major organic components of dead leaves and wood are cellulose and lignin. These pose considerable digestive problems for animal consumers. Digesting cellulose requires cellulase enzymes but, surprisingly, cellulases of animal origin have been definitely identified in only one or two species. The majority of detritivores, lacking their own cellulases, rely on the production of cellulases by associated bacteria or fungi or, in some cases, protozoa. The interactions are of a range of types:

the challenge of decomposing wood and dead leaves

1 *obligate mutualisms* (see Chapter 8) between a detritivore and a specific and permanent gut microflora (such as bacteria);

2 *facultative mutualisms*, in which the animals make use of cellulases produced by a microflora that is ingested with detritus as it passes through an unspecialized gut (for example, woodlice);

3 *external rumens*, in which animals simply assimilate the products of the cellulase-producing microflora associated with decomposing plant remains or feces [such as springtails (Collembola)].

The dung of carnivorous vertebrates is relatively poor-quality stuff. Carnivores assimilate their food with high efficiency (usually 80% or more is digested); their feces retain only the least digestible components, whose decomposition is probably accomplished almost entirely by bacteria and fungi. In contrast, herbivore dung still contains an abundance of organic matter and is spread thickly enough in the environment to support its own characteristic fauna, consisting of many occasional visitors and several specific dung-feeders. A good example is elephant dung; within a few minutes of dung deposition, the area is alive with beetles. The adult dung beetles feed on the dung, but they also bury large quantities along with their eggs to provide food for developing larvae.

feces and carrion are quickly consumed

All animals defecate and die, yet feces and dead bodies are not generally very obvious in the environment. The reason is the efficiency of the specialist consumers of these dead organic products. On the other hand, where consumers of feces are absent, a buildup of fecal material may occur. We find a dramatic example in the accumulation of cattle dung where these domestic animals have been introduced to locations lacking appropriate dung beetles. In Australia, for example, during the past 200 years, the cow population increased from just seven individuals (brought over by the first English colonists in 1788) to 30 million or so, producing 300 million cowpats per day. The lack of native dung beetles led to losses of up to 2.5

million hectares per year under dung. The decision was made in 1963 to establish in Australia beetles of African origin, able to dispose of bovine dung under the conditions where cattle are raised; more than 20 species have been introduced (Doube et al., 1991).

11.7 THE FLUX OF MATTER THROUGH ECOSYSTEMS

Chemical elements and compounds are vital for the processes of life. When living organisms expend energy (as they all do, continually), they do so, essentially, in order to extract chemicals from their environment and hold on to and use them for a period before they lose them again. Thus, the activities of organisms profoundly influence the patterns of flux of chemical matter. In this section, we focus on the flux of a few elements—carbon, nitrogen, and phosphorus—at the scale of the ecosystem. We focus on carbon because it is such a major component of all biomass, and we focus on nitrogen and phosphorus since their availabilities are so important in regulating NPP across much of the Earth. It is the cycling of these nutrient elements through and within ecosystems that determines their biological availability. (We discuss the global cycles of these elements in Chapter 12.)

The great bulk of living matter is water. The rest is made up mainly of carbon compounds, and this is the form in which energy is accumulated and stored. Carbon enters the food web of a community when a simple molecule, carbon dioxide, is taken up in photosynthesis. Once incorporated through NPP, it is available for consumption as part of a sugar, a fat, a protein, or, very often, a cellulose molecule. It follows exactly the same route as energy, being successively consumed and either defecated, assimilated, or used in metabolism, during which the energy of its molecule is dissipated as heat while the carbon is released to the atmosphere again as carbon dioxide. Here, though, the tight link between energy and carbon ends.

Once energy has been transformed into heat, it can no longer be used by living organisms to do work or to fuel the synthesis of biomass. The heat is eventually lost to the universe and can never be recycled: life on Earth is possible only because a fresh supply of solar energy is made available every day. In contrast, the carbon in carbon dioxide can be used again in photosynthesis. Carbon and nutrient elements like nitrogen, phosphorus, and others are available to primary producers as simple inorganic molecules and ions (carbon dioxide, ammonium,

energy cannot be cycled and reused—matter can

nitrate, phosphate) in the atmosphere, soils, or water. Each is assimilated and incorporated into complex organic compounds in biomass during primary production. Ultimately, however, when the organic compounds are metabolized to carbon dioxide, the nitrogen, phosphorus, and other nutrients are released to the environment again in simple inorganic form. Another primary producer may then assimilate them, and so an individual atom of a nutrient element may pass repeatedly through one food chain after another. This continued uptake and release is termed *recycling*.

Unlike the energy of solar radiation, nutrients and carbon dioxide are present on the Earth in a finite amount. As they are incorporated into the organic matter of biomass, they are no longer available to other primary producers until the biomass has decomposed and the nutrients have been mineralized. For many of the elements in living matter, such as sodium, potassium, chlorine, and sulfur, the amount tied up in living biomass and organic detritus is quite small relative to the supply in virtually all ecosystems on Earth. On the other hand, the amount of phosphorus in living biomass and detritus is large relative to the inorganic supply in many ecosystems, which is why phosphorus can commonly limit NPP. Recycling of phosphorus is very important to continued primary production in such ecosystems.

Nitrogen is unique in that it is commonly limiting in many types of ecosystems and yet vast supplies are always nearby as molecular N_2 in the atmosphere and dissolved in water. In fact, well over 99% of nitrogen on Earth is present as molecular N_2, and only a tiny fraction is present in biomass and organic detritus. The amount in readily biologically available ions such as ammonium and nitrate is an even tinier fraction. As we have discussed several times earlier in this text, many types of bacteria including cyanobacteria in lakes and subtropical ocean gyres (Chapter 1 and earlier in this chapter) and rhizobium symbionts of plants (Chapter 8) can fix molecular N_2, converting it into ammonium. At the global scale, this process is essential for all life on Earth, for without it the supply of biologically available nitrogen would gradually be depleted: other bacteria are constantly converting nitrate (which together with ammonium makes up most of the biologically available nitrogen on Earth) into molecular N_2 in a process called **denitrification**. Denitrification occurs in environments largely devoid of oxygen and is a type of respiration in which the bacteria use nitrate instead of oxygen as an electron acceptor.

Nitrogen fixation and denitrification are two of many bacterially controlled processes in the

> nitrogen is unique among the nutrients

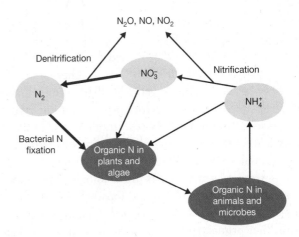

FIGURE 11.26   A simplified depiction of the nitrogen cycle. (After Howarth, 2002.)

nitrogen cycle, the continuing transformation of nitrogen including assimilation into biomass and mineralization back to ammonium (Figure 11.26). Another critical process of the nitrogen cycle is **nitrification**, in which bacteria convert ammonium to nitrate. The bacteria mediating nitrification are autotrophs, using the resulting energy of the nitrogen conversion to fix carbon dioxide into organic biomass (see Chapter 3). During both nitrification and denitrification, there is a leakage of some intermediary compounds from the enzymatic machinery, such as N_2O, NO, and NO_2. All these are gases, and some quantity invariably escapes to the atmosphere. The flux of N_2O (nitrous oxide) is particularly important, because this is a powerful greenhouse gas (see Chapter 12).

We can logically consider N_2 the start of the nitrogen cycle, since it makes up the vast mass of nitrogen on the planet. Nitrogen fixation converts this into organic nitrogen in the biomass of primary producers, which is then fed upon by animals and decomposed by bacteria, becoming organic nitrogen in these heterotrophs (Figure 11.26). The heterotrophs mineralize nitrogen to ammonium. The ammonium then is either assimilated by primary producers again or is nitrified to nitrate, with some loss of the nitrogen gases such as N_2O. Nitrate in turn is either assimilated by primary producers or denitrified back to N_2, closing the cycle. Before the industrial and agricultural revolutions, the amount of biologically available nitrogen on Earth was balanced by its creation through nitrogen fixation and its loss through denitrification. Humans have now dramatically altered this natural cycle, as we explore in Chapters 12 and 14.

The nitrogen cycle is dominated by a variety of bacterially mediated processes—in fact, even more than the main ones described above. The

> the phosphorus cycle is more geologically controlled

interaction of the nitrogen cycle with geology is limited, and nitrogen is found as a component of rock only in sedimentary rocks formed from ancient sediments rich in organic detritus. On the other hand, the phosphorus cycle is more geologically determined and is far less mediated by bacterial processes. Many types of rocks contain phosphorus, though generally in only trace amounts, and the weathering of the rock—its gradual conversion over geological time scales to soils—frees this and starts the biological phosphorus cycle. Inorganic phosphate ions are the only biologically available form of phosphorus, and it is this form that primary producers assimilate into their biomass. As organic matter is consumed or decomposed, the organic phosphorus is mineralized back to inorganic phosphate ions. The phosphate can be strongly adsorbed onto soil and sediment particles, and it can indeed be precipitated into new minerals such as apatite (a mineral that contains calcium as well as phosphorus and other elements) that form in some sediments and soils. Note that the bones and teeth of vertebrate animals are also composed of phosphorus-rich apatite.

11.8 NUTRIENT BUDGETS AND CYCLING AT THE ECOSYSTEM SCALE

We can consider the cycling of nutrients and carbon at many scales, up to the global scale presented in Chapter 12. The ecosystem scale is particularly appropriate for evaluating the factors that regulate the availability of nutrients to primary producers, and for understanding the interactions of different element cycles. Nutrients cycle within the ecosystem but also move across the boundaries of the ecosystem, primarily via exchanges with the atmosphere and fluxes in flowing water. Nutrients can also enter the ecosystem through weathering of the rocks and soil within the ecosystem.

We can characterize an ecosystem as being more open or more closed by comparing the exchanges | closed vs. open ecosystems

of limiting nutrients across its border with the rate of recycling within the ecosystem (Figure 11.27). The subtropical ocean gyres represent ecosystems that are tightly closed: for every new atom of nitrogen or phosphorus that enters the surface waters of the gyre from the small mixing of nutrient-rich deep-ocean waters from below or from the atmosphere, perhaps 50 to 100 atoms of nitrogen and phosphorus are recycled within the gyre and assimilated by phytoplankton during NPP. At the other extreme, for heavily fertilized agricultural ecosystems or coastal salt marshes that have high rates of exchange of water with adjacent estuaries through tidal action, the external supply of nitrogen may match or even exceed the rate of recycling within the ecosystem, and is often sufficient on its own to support NPP.

Colimitation of NPP by nitrogen and phosphorus is more prevalent in relatively closed ecosystems, where recycling of the two nutrients tends to proceed in parallel to the demand in NPP. In an open

(a)

Input to ecosystem of 0.5 mg nitrogen per square meter per day

Uptake by primary producers of 30 mg nitrogen per square meter per day

Julian J. Schanze

(b)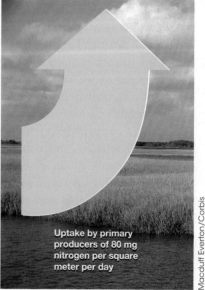

Input to ecosystem of 90 mg nitrogen per square meter per day

Uptake by primary producers of 80 mg nitrogen per square meter per day

Macduff Everton/Corbis

FIGURE 11.27 (a) Subtropical ocean gyres such as the Sargasso Sea are perhaps the most closed ecosystems on the planet, with a rate of nitrogen uptake by primary producers far greater than the external input of nitrogen into the ecosystem. (b) Coastal salt marshes are one of the most open ecosystems on Earth with regard to nutrient cycling, and the external input of nitrogen may even exceed the rate of uptake by the primary producers.

ecosystem, the external inputs of nitrogen and phosphorus are far less likely to meet the balanced needs of primary producers, and limitation by a single element is far more likely, depending upon the relative inputs of nitrogen and phosphorus.

Weathering of parent bedrock and soil, by both physical and chemical processes, is the main | sources of nutrients to terrestrial ecosystems | source of nutrients such as calcium, iron, magnesium, phosphorus, and potassium in many terrestrial ecosystems, which may then be taken up via the roots of plants. Plants themselves can further this process, at the extreme by growing roots into fractures in the rock to make more cracks (Figure 11.28). Plants further weather in more subtle ways as well, for example, by secreting organic acids that can help solubilize minerals. An accumulation of carbon dioxide in soils from respiration also contributes to the acidity and often accelerates weathering. Weathering is particularly important on young soils, such as those created by volcanic action in the past few tens of thousands of years or those glaciated in the last Ice Age. Older soils that may have been weathered for millions of years are very limited in their ability to provide further inputs of nutrients from further weathering.

The other major input of most nutrient elements to terrestrial ecosystems is from atmospheric **deposition**, the net flux of materials from the atmosphere to the ecosystem. Deposition has two components: **dryfall**, the settling of particles during periods without rain, and **wetfall**, flux of materials dissolved in rain and snow. These depositional inputs can be substantial, and in some ecosystems—particularly those with very old weathered soils—they can be the dominant input. For instance, most of the phosphorus and iron inputs to the Amazon River basin come from the atmosphere. Dust containing these elements is eroded into the atmosphere by wind from the Sahel and Sahara regions of Africa, and prevailing easterly trade winds carry some of the dust more than 5,000 km before depositing it into the Amazon (Figure 11.29).

Most bedrocks have little if any nitrogen in them, so weathering is seldom an important source of nitrogen in terrestrial systems. Rather, the major sources are nitrogen fixation and atmospheric inputs in dryfall and wetfall. For nitrogen, dryfall includes not only the nitrogen that settles in particles but also the uptake by vegetation and other moist surfaces of reactive nitrogen gases, such as ammonia and nitric acid vapor. These inputs are particularly important downstream of major pollution sources to the atmosphere.

Some very low rate of nitrogen fixation by free-living cyanobacteria occurs in many ecosystems, but this is seldom a major input to the ecosystem. On the other hand, nitrogen fixation can be very important in tropical rainforests, with fixation both by heterotrophic bacteria that are symbionts associated with tree roots, and by cyanobacteria associated with lichens in the forest canopy. Wetlands also frequently have high inputs from nitrogen fixation, both by free-living bacteria and by symbionts of some tree species such as alder (*Alnus* sp), as discussed in Chapter 8.

Nutrients can circulate within the ecosystem for many years, or a nutrient atom can pass through the system in a matter of minutes, perhaps without interacting with the biota at all. Whatever the case, the atom will eventually be lost through one of the variety of processes that remove nutrients from the system. These processes constitute the debit side of the nutrient budget equation. The rate of loss is greater in more open ecosystems. | nutrient outputs from terrestrial ecosystems |

Release to the atmosphere is one pathway of nutrient loss. Nitrogen may be lost as N_2, in ecosystems where low-oxygen environments encourage denitrification. Some nitrogen can also be lost as N_2O, both through denitrification and nitrification. Further, ammonia gas is released during the decomposition of vertebrate excreta. Other pathways of nutrient loss are important in particular instances. For example, fire (either natural, or when, for instance, agricultural practice includes the burning of stubble) can volatilize large amounts of both nitrogen and sulfur gases to the atmosphere.

For many elements in many ecosystems, the most substantial pathway of loss is in streamflow. The water that drains from the soil of a terrestrial ecosystem into a stream carries a load of nutrients that is partly

Susan Dykstra/Design Pics/Corbis

FIGURE 11.28 Tree roots growing into fractures in a rock, furthering weathering of the rocks.

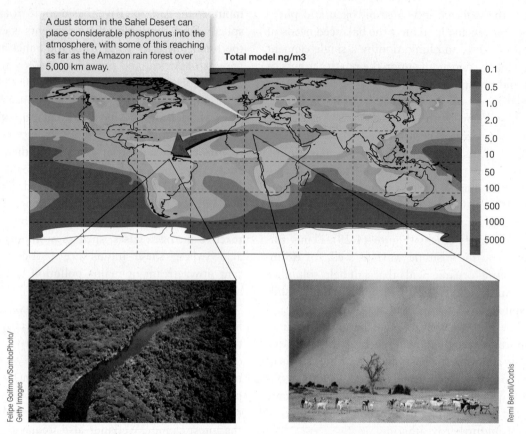

A dust storm in the Sahel Desert can place considerable phosphorus into the atmosphere, with some of this reaching as far as the Amazon rain forest over 5,000 km away.

Total model ng/m3

0.1
0.5
1.0
2.0
5.0
10
50
100
500
1000
5000

Felipe Goitman/SambaPhoto/ Getty Images

Remi Benali/Corbis

FIGURE 11.29 Dust in the atmosphere can contain significant amounts of phosphorus, as illustrated here by a model estimate of the amount of phosphorus in the atmosphere globally. Phosphorus in this dust can move long distances and can be an important input to ecosystems as the dust settles on or rains out to the surface. (After Mahowald et al., 2008.)

dissolved and partly particulate. Phosphorus and iron tend to move largely attached to particles, so losses are related to the overall rate of erosion of soils. Nitrogen, on the other hand, is usually lost primarily as dissolved forms (both nitrate and dissolved organic nitrogen compounds) in flowing water.

Vegetation can have a pronounced influence on the retention or loss of nutrients from terrestrial ecosystems, as we introduced in Chapter 1 (see Figure 1.11). When the forests in a catchment were cut in the Hubbard Brook Experimental Forest, the export of nitrate increased more than 100-fold compared to uncut reference catchments (see Figure 1.12). Losses of potassium and calcium also increased, though less dramatically. Trees and other vegetation help retain nutrients in an ecosystem through two mechanisms. First, they transpire water and thereby reduce the water flow downstream that carries dissolved substances with it and that erodes the soil; second, they take up the nutrients that are being rapidly mineralized by bacteria and other heterotrophs so they are not available to be carried away to the stream water. When undisturbed, forest ecosystems tend to be closed ecosystems with regard to nutrient cycling, but

they become very open systems when cut or otherwise disturbed, such as through fire or wind damage.

The loss of nutrients from terrestrial ecosystems in stream flow becomes the input of nutrients to streams, ponds, and small lakes. The extent of input to these aquatic ecosystems is a function of the hydrologic losses from the upstream terrestrial ecosystem and how open or closed the upstream ecosystem is. Streams and ponds that are downstream from open terrestrial ecosystems will have very large nutrient inputs, while the inputs to streams from undisturbed forests will be much smaller. Nitrogen fixation can be a major input of nitrogen to streams such as those in desert regions, where the stream is not shaded and cyanobacteria thrive.

nutrient inputs to and exports from streams, ponds, and small lakes

Biological fluxes can be important inputs of nutrients in some streams, as evidenced by salmon (Figure 11.30 and Table 11.4). Coho salmon, like many salmon species, migrate back to small streams to spawn and then die within a few weeks of spawning. These salmon are large fish, with large amounts of nitrogen, phosphorus, and other elements in their

FIGURE 11.30   As salmon such as the coho salmon shown here return to streams to spawn, they import huge quantities of nitrogen, phosphorus, and other elements in their biomass. They die soon after spawning, and their decomposing bodies fertilize the streams.

TABLE 11.4   Percentage of nitrogen that originated from ocean sources in various components of the biota of two streams in Washington state, USA. The nitrogen was brought into the streams by migrating salmon, which died after spawning. As their carcasses decomposed, the mineralized nitrogen fertilized the streams. Data come from measurement of the stable isotopic composition of the nitrogen. Note that not all components were measured in both streams.

	Grizzly Creek	East Fork Creek
Algae and organic matter on rocks	20 to 30%	
Grazing insects	25 to 30%	33%
Invertebrate predators	11 to 19%	19%
Cutthroat trout	19 to 45%	45%
Young coho salmon	31%	30%

After Bilby et al., 1996.

biomass that accumulated while they were growing in the Pacific Ocean for many years. The release of these elements upon their death is a large source of nutrient inputs to the streams.

The major losses of nutrients from streams, ponds, and small lakes are the mirror image of the inputs: hydrologic fluxes further downstream to some other ecosystem. Nitrogen losses to the atmosphere as both N_2 and N_2O from denitrification can also be substantial when nitrate inputs to these ecosystems are high, since the sediments in these ecosystems are commonly devoid of oxygen and provide an ideal habitat for denitrifying bacteria.

Nutrient inputs from upstream are also important in lakes, especially when the watershed area that drains to the lake is large or

atmospheric inputs become more important in larger lakes

when the terrestrial ecosystems in the watershed are disturbed and have open nutrient cycles that readily leak downstream. However, as the surface area of the lake itself becomes larger, we find the importance of direct deposition of nutrients to the lake from the atmosphere in both dryfall and wetfall will increase. In some lakes where the ratio of watershed area to lake surface area is relatively small and the upstream terrestrial ecosystems are relatively undisturbed, atmospheric inputs can dominate the nutrient inputs to the lake.

Estuaries, where rivers meet the coastal ocean, are like lakes in many ways, receiving nutrients both from upstream ecosystems and from the atmosphere. Unlike

estuaries receive nutrients from upstream and downstream

lakes, however, estuaries also receive nutrients from downstream, from the coastal ocean waters. Estuaries grade from freshwater at one end to the full salinity of seawater at the other. The mixing that creates this gradient can actually pull salty seawater up into the estuary, particularly along the bottom, in what is called reverse estuarine flow. This salty water carries along with it the nutrients that are in the coastal ocean waters. These inputs can be substantial, particularly for phosphorus. For example, Chesapeake Bay—the largest estuary in the United States and one of the most sensitive to nutrient pollution—gets 25% of its phosphorus inputs but less than 1% of its nitrogen inputs from the coastal ocean.

On the continental shelves, ecosystems obtain nutrients from rivers and estuaries, from atmospheric deposition, but also from mixing up of the nutrient-rich

terrestrial sources and deep-ocean nutrients vie for dominance in the coastal oceans

waters from the deep oceans (Table 11.5; also see Chapter 4). The relative importance of these sources is determined by the physics on the continental shelf in the local area, by the amount of nutrients flowing down rivers in that area, and by the amount of atmospheric pollution from the upstream land sources in the area. Both the northeastern coast of the United States (the continental shelf from Maine south to Virginia) and the North Sea (between the United Kingdom and mainland Europe) receive a substantial amount of nitrogen from deep-ocean waters, while the Gulf of Mexico is more isolated from the influence of the deep ocean. On the other hand, the Gulf of Mexico receives a very large amount of nitrogen from terrestrial sources, particularly from the Mississippi River, which drains over 40% of the lower 48 states of the United States, including the Midwestern corn belt. The North Sea also receives substantial nitrogen from terrestrial sources; its watersheds are smaller than for the Gulf of Mexico,

TABLE 11.5 Sources of nitrogen to three contrasting continental shelf ecosystems (Tg nitrogen per year) and the percent increase in nitrogen inputs as a result of human activities. The estimate for atmospheric deposition is just for the direct deposition onto the water surfaces of these ecosystems. Atmospheric deposition onto the landscape with subsequent runoff to the continental shelves is included in the estimates for terrestrial watersheds.

	Terrestrial watersheds	Atmospheric deposition	Deep oceans	Percent increase due to humans
Gulf of Mexico	2.1	0.3	0.15	275%
NE coast of U.S.	0.3	0.2	1.5	28%
North Sea	1.0	0.6	1.3	98%

After Boyer & Howarth, 2008.

but human activity is intense in those watersheds. The terrestrial inputs to the continental shelf off the northeastern United States are smaller, despite a great deal of human activity, because the watershed area is so much less than for the Gulf of Mexico; when expressed per area of watershed, the fluxes from watersheds are actually higher in the northeastern United States than for the Mississippi River basin (Chapter 12).

All three of these continental shelf ecosystems receive considerable inputs of nitrogen from direct atmospheric deposition onto their water surfaces, but this input is not dominant in any of them. Most of the land and atmosphere source is pollution from human activity, while the deep-ocean sources are natural.

In the rest of the oceans away from the continental shelves, the major nutrient inputs to the surface waters come from the deep ocean and from atmospheric deposition. The relative contribution of these two sources is determined by the physics of the surface waters and by the sources that are upwind in the atmosphere (Chapter 12). In the subtropical gyres, the nutrient-rich waters of the deep ocean—only a few hundred meters below the surface waters!—are largely unavailable (Chapter 4). A slight amount of mixing moves some of these nutrients into the surface waters of the gyres, but atmospheric inputs are often as large or larger. And as we noted earlier in this chapter, nitrogen fixation is a significant input of nitrogen to the subtropical gyres on a scale unmatched anywhere else in the oceans, and probably contributing roughly half the nitrogen inputs to the entire global biosphere from natural sources (Chapter 12). On the other hand, in upwelling ecosystems, the input of nutrients from the deep ocean is huge and dwarfs any other inputs.

> Deep-ocean nutrients and atmospheric deposition dominate nutrient sources in the ocean biomes away from shore

SUMMARY

The role of energy in ecology

All biological systems depend on a continuous input of energy to support their activities and structure. This is true for populations, communities, and ecosystems as well as for individual organisms.

Geographic patterns in primary productivity

The pattern of net primary productivity in both terrestrial ecosystems and the oceans varies dramatically across the Earth. Slightly more than half of all net primary productivity on the planet occurs in terrestrial ecosystems and slightly less than half in the oceans. Net primary productivity per area is at its highest in tropical rain forests, some wetlands, some estuaries, and giant kelp beds. Somewhat lower but still high rates occur in temperate forests and grasslands, tropical savannas, upwelling zones, and much of the North Atlantic Ocean north of 40° N. The lowest rates of net primary production occur in deserts and in the subtropical gyres of the world's oceans.

Factors limiting terrestrial primary productivity

Primary productivity on land is limited primarily by temperature and the availability of water. The highest rates of net primary productivity are found in tropical rain forests, where water is plentiful and temperatures favorable, and the lowest rates in deserts, where water availability is low. Nutrients are also an important control. Phosphorus is often limiting in tropical rain forests, while nitrogen is commonly limiting in terrestrial ecosystems at higher latitudes and in deserts.

Factors limiting aquatic primary productivity

Nutrient limitation is more important as a control on net primary productivity in aquatic ecosystems than on land, and patterns of net primary productivity are regulated by the interaction of nutrient availability and light penetration into the water column. The highest productivity is found in upwelling ecosystems and in nutrient-laden estuaries, while the lowest rates are found in the subtropical ocean gyres where nutrient availability is very low. Nitrogen and phosphorus tend to be equally limiting in these subtropical gyres, while nitrogen alone is often more limiting in estuaries and coastal marine ecosystems, and phosphorus alone is often more limiting in freshwater lakes.

The fate of primary productivity: grazing

In the oceans and lakes, secondary productivity by herbivores is approximately an order of magnitude less than the primary productivity on which it is based. That is, the trophic transfer efficiency is approximately 10%. Energy is lost at each feeding step because consumption efficiencies, assimilation efficiencies, and production efficiencies are all less than 100%. In terrestrial ecosystems, the trophic transfer efficiency from primary producers to herbivores is even lower than in phytoplankton-based ecosystems, since trees and grasses have far less protein, and far more hard-to-digest structural materials that are low in nitrogen, than do phytoplankton.

The process of decomposition

The decomposer system processes much more of the energy and matter than the live consumer system in terrestrial ecosystems. The energy pathways in the live consumer and decomposer systems are the same, with one critical exception—feces and dead bodies are lost to the grazer system (and enter the decomposer system), but feces and dead bodies from the decomposer system are simply sent back to the dead organic matter compartment at its base.

The flux of matter through ecosystems

While energy flows through an ecosystem, nutrients and other matter are cycled. For nutrients such as phosphorus, iron, potassium, and calcium, geological processes are important and the role of bacterial processes is relatively small (largely limited to mineralizing the nutrients during decomposition). On the other hand, geology plays a small role in the nitrogen cycle, which is dominated by a series of bacterially controlled processes such as nitrogen fixation, nitrification, and denitrification, in addition to mineralization.

In all ecosystems, nutrients regularly cycle between inorganic and organic forms. Inorganic forms of nitrogen and phosphorus, for instance, are assimilated by plants, algae, and cyanobacteria during primary production and converted into the organic tissues of these primary producers. As this material is consumed and decomposed by heterotrophic organisms, the nitrogen and phosphorus is mineralized and released back to the environment as inorganic forms. In ecosystems that are relatively closed to nutrient inputs, the rate of cycling of a limiting nutrient between inorganic and organic forms may be 50 to 100 times greater than the rate of new inputs of this limiting nutrient across the boundary of the ecosystem. In open systems, the external inputs of the limiting nutrient may equal or exceed the rate of cycling.

Nutrient budgets and cycling at the ecosystem scale

Nutrients both enter and leave the ecosystem across its boundaries and come from a variety of sources. In terrestrial ecosystems, weathering of soil and bedrock is often the dominant input of nutrients such as phosphorus, iron, and calcium in regions where the soils are less than a few tens of thousands of years old. Where the soils are older and more weathered, this weathering input is reduced, and the input of nutrients from atmospheric deposition becomes more critical. Weathering is seldom an important source of nitrogen, which comes largely from atmospheric deposition and from nitrogen fixation. Nutrients leave terrestrial ecosystems in flowing waters, dissolved in the water or as part of eroded material. Nitrogen is also lost as N_2 and N_2O gases, the products of denitrification and nitrification.

The nutrient losses that leave terrestrial ecosystems in flowing waters become the input to downstream aquatic ecosystems: streams, rivers, and lakes. Estuaries get nutrient inputs from rivers, but also from coastal ocean waters. Mixing of water from the deep ocean provides nutrients to many marine ecosystems, in addition to the river inputs for those near coastlines. Aquatic ecosystems also received nutrients from atmospheric deposition, which can be a major source in some of the more closed ecosystems such as the subtropical gyres. Nitrogen fixation is important in the subtropical gyres also and in many lakes, but it is seldom important elsewhere in the oceans.

REVIEW QUESTIONS

1   In terrestrial ecosystems, primary productivity tends to decrease as we move farther away from the Equator and towards higher and lower latitudes. Why don't we see this pattern in the oceans?

2   Compare and contrast the roles of nitrogen and phosphorus as limiting nutrients in lakes, estuaries, and subtropical ocean gyres.

3   Why is energy transfer from primary producers to herbivores less efficient in terrestrial ecosystems than in phytoplankton-based ones?

4   Account for the observation that in most terrestrial communities much more energy is processed through the decomposer system than through the live consumer system.

5   Why is production efficiency lower for warm-blooded animals than for insects?

6   Energy cannot be cycled and reused but matter can. Discuss this assertion and its significance for ecosystem functioning.

7   The phosphorus cycle is often considered to be more of a geological one, while the nitrogen cycle is more biologically mediated. What is the basis for this distinction?

Challenge Questions

1   Why is nutrient limitation a larger constraint on primary productivity in phytoplankton-based aquatic ecosystems than in terrestrial ecosystems?

2   Why is colimitation by nitrogen and phosphorus more likely in a closed ecosystem than in an open one?

3   Imagine a world in which denitrification no longer occurred. What would be the consequences on the global rates of net primary productivity, and on the controls on this productivity?

Part 5

Applied Issues in Ecology

Chapter 12

Global biogeochemical cycles and their alteration by humans

CHAPTER CONTENTS

KEY CONCEPTS

After reading this chapter, you will be able to:

- Describe how human activity has accelerated many biogeochemical cycles around the globe.

- Explain how the increase in atmospheric carbon dioxide, caused by fossil fuels and deforestation, is only temporarily reduced through uptake by the oceans and terrestrial biomes.

- Describe the natural and human-controlled sources of atmospheric methane and why it has increased by a larger percentage than carbon dioxide.

- Evaluate the causes and consequences of the global acceleration of the nitrogen cycle and how it varies among regions.

The extent of human activity has become so great that the some global cycles of elements such as carbon and nitrogen have become radically altered, particularly over the past few decades. Ecologists play a major role in understanding the global alteration of these cycles.

12.1 WHAT IS BIOGEOCHEMISTRY?

In this chapter, we formally define biogeochemistry and demonstrate how human activity has altered some critically important biogeochemical cycles at the global scale. **Biogeochemistry** is the science that addresses the "biotic controls on chemistry of the environment and the geochemical control of the structure and function of ecosystems" (Howarth, 1984). As we discussed in Chapter 11, the way phosphorus availability regulates net primary productivity in lakes is an example of a geochemical control of an ecosystem function, and the interaction of two biological processes (grazing and physiological assimilation of the trace metal molybdenum) to regulate nitrogen fixation (the creation of biologically available, reactive nitrogen) is an excellent example of a biotic control of the chemistry of the environment. Biogeochemistry as a science overlaps with much of ecological science at the ecosystem to global scale (see Figure 1.2), but it is not simply a subset of ecology because it also overlaps strongly with environmental chemistry and geochemistry.

In a provocative paper, Rockstrom et al. (2009) presented ten "planetary life-sustaining biophysical systems" they viewed as essential for life on Earth. They estimated how human activity has altered each of these compared to some level of change deemed safe (Figure 12.1). Their analysis, best viewed as a challenge for scientists to consider the safe levels of human alteration of global systems rather than as the final word on the topic, concludes that the most severe change has been a loss of biodiversity (see Chapter 13), followed closely by acceleration of the global nitrogen cycle. Three of the other life-sustaining systems they identify relate to biogeochemical cycles: the global phosphorus cycle, climate change, and ocean acidification, the latter two of which are tightly tied to the global carbon cycle.

In this chapter, we very briefly discuss the phosphorus cycle, introduce the sulfur cycle and its role in acid rain, and then focus on the global carbon and

biogeochemistry and planetary life-sustaining systems . . .

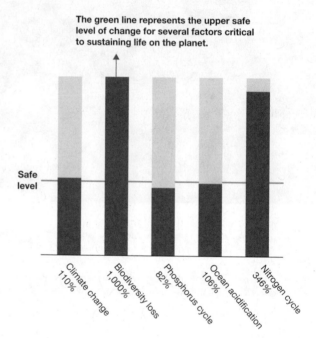

FIGURE 12.1 Life on the Earth is dependent upon several critical life-support systems, including major biogeochemical cycles and biodiversity. This figure illustrates parts of the assessment by Rockstrom and colleagues (2009) on how much humans have disturbed these key systems relative to a level (indicated by the green line) identified as providing some margin of safety. (Modified from Rockstrom et al. (2009) as reported in the Grist.)

nitrogen cycles, because these include many more ecological controls and feedbacks. We divide the carbon cycle into two separate parts—carbon dioxide and methane—since the ecological controls on these are quite different, and both are critical for understanding global change. For all these global cycles, we present both the natural cycle as it existed before the Industrial Revolution, and the modern cycle as influenced by human activities.

The rate of phosphorus flow through the global environment is now some three times greater than before the industrial and agricultural revolutions, due to human use of phosphorus fertilizer, use of phosphorus

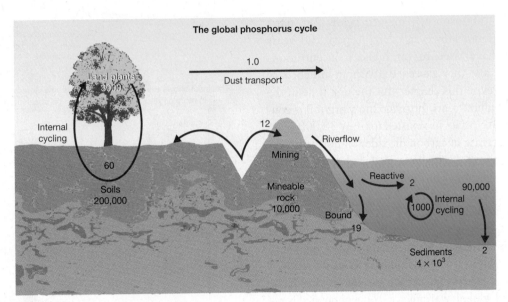

FIGURE 12.2   An illustration of the global phosphorus cycle (in units of Tg and Tg per year). Phosphorus cycles within ecosystems and flows downstream from terrestrial ecosystems in rivers to the ocean, largely bound in eroded sediments but also as dissolved inorganic phosphorus. Most of this phosphorus is stored in ocean sediments. Phosphorus also moves through the atmosphere in dust. Humans have accelerated the phosphorus cycle, primarily from mining phosphorus for fertilizer. (Schlesinger, 1997.)

compounds as cleaning agents, and increased flows of phosphorus-containing dust through the atmosphere due to desertification. This acceleration of the phosphorus cycle is of ecological importance, notably because it causes eutrophication in freshwater lakes (Chapters 1 and 11). However, the analysis of the global phosphorus cycle is largely a matter of estimating fluxes of sediment in rivers and dust in the air, since this is where the major phosphorus flows are (Figure 12.2; see also Figure 11.29). The ecological interactions and feedbacks within the phosphorus cycle lack the complexity of those for the carbon dioxide, methane, and nitrogen cycles we explore next.

The sulfur cycle has a great deal of complexity and interaction with ecological processes, which we touched on briefly when describing how sulfate can interfere with the assimilation of molybdenum by nitrogen-fixing cyanobacteria (Section 11.4). Sulfate is also very important in the decomposition processes in sediments in many coastal marine ecosystems, is essential for life in that it is a component of several amino acids and other biologically important compounds, and undergoes a vast array of geochemical and microbial-mediated reactions. However, while sulfur is essential, it is seldom if ever limiting to primary productivity, as it has a high availability relative to the requirements of plants and algae in almost all ecosystems. Human activity over the past century has doubled the flux of sulfur gases through the atmosphere globally, and increased the deposition of sulfur to terrestrial ecosystems by 30-fold in some regions. This sulfur, most of

which originates from burning fossil fuels such as coal, is a major component of acid rain (Figure 1.4), where it occurs as dilute sulfuric acid.

In the United States and much of Europe, aggressive regulation has reduced sulfur pollution since the 1970s, which has lessened the problem of acid rain. For example, at the Hubbard Brook Experimental Forest in the White Mountains of New Hampshire, where acid rain was first discovered in the 1960s, the acidity of rain fell by a factor of two between the 1960s and 2000 (Figure 1.4). In much of North America and Europe, sulfur pollution was once the major cause of acid rain, but nitrogen pollution (dilute nitric acid) is now more important. However, sulfur pollution and acid rain are growing problems elsewhere in the world, notably in China. Acid rain can cause many profound problems, including the loss of fish from lakes (Figure 1.8), the dieback of forests, and the loss of plant species richness generally (Figure 13.11). The acidification of lakes in the Adirondack Mountains of New York State and in Ontario led to many of them becoming fishless during the 1970s, and most of these still have no fish despite the decrease in the acidity of the rain over time.

12.2   THE GLOBAL CARBON DIOXIDE CYCLE

For at least 400,000 years before the start of the Industrial Revolution, the concentration of carbon dioxide in the atmosphere remained below 265 parts per million (ppm). However, over the past 200 years those levels have risen to far higher concentrations (Figure 12.3), and they exceeded 398 ppm in February 2014 (as we revise this chapter) for the first time in the past several million years (http://co2now.org/). For context, the human species has existed for only 30,000 years or so. This increase in carbon dioxide (and other greenhouse gases, as we discuss later in this chapter) has led to significant warming of the global environment over the past half-century (see Box 3.2). The global average temperature as of 2011 was 0.7°C above the long-term average from the early 20th century (see Figure 3.20 in Box 3.2), representing a warmer temperature than seen on the planet for at least one hundred thousand years.

We now know that carbon dioxide is the major driver behind global warming, a connection first proposed by the Swedish chemist Svante Arrhenius well over 100 years ago (Box 12.1). But how do scientists know what the atmosphere looked like hundreds and thousands of years ago? As ice was laid down in the Antarctic ice sheets, bubbles of atmospheric gases were trapped, preserving a historical record. With care, we can sample and assay the gases in these bubbles, and through a variety of techniques we can identify when the ice around them was formed, such as by looking for layers

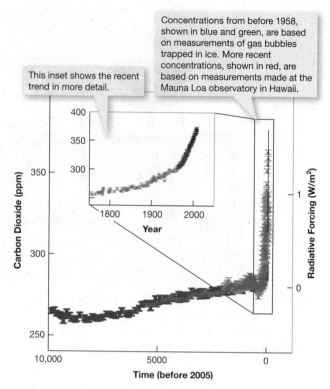

FIGURE 12.3 The concentration of carbon dioxide in the Earth's atmosphere was stable for most of the last 10,000 years but started to increase exponentially over the past 200 years. (Reprinted from IPCC, 2007.)

12.1 Historical Landmarks

Arrhenius and Keeling: Pioneers in understanding human alteration of the carbon dioxide cycle

With his pioneering work in 1896, the great Swedish chemist Svante Arrhenius became the first to understand that human activities could cause global warming by increasing the amount of carbon dioxide in the atmosphere (Figure 12.4a). Arrhenius estimated the extent to which carbon dioxide and water vapor absorb infrared radiation and calculated that a doubling of atmospheric carbon dioxide would result in an average increase in the global temperature of between 1.6°C and 6°C. In the subsequent 120 years, scientists have refined this understanding and included other factors, but Arrhenius's prediction is remarkably robust: in its most recent report (IPCC, 2013), the Intergovernmental Panel on Climate Change estimates that doubling of atmospheric carbon dioxide will increase the global temperature by between 2°C and 4.5°C.

Arrhenius also estimated the time he thought it would take for humans to cause a doubling of atmospheric carbon dioxide: 3,000 years. In this, he was way off target. Most modern scenarios predict 100 years instead. In part, Arrhenius's estimate was low because he overestimated the rate at which the world's oceans can absorb carbon dioxide. His bigger error, though, was his estimate of the rate at which humans would burn fossil fuels. He assumed that the release of carbon dioxide from fossil fuel combustion would continue into the future at 1900 values, while in fact emissions have already grown 20-fold.

At the time, Arrhenius's research was not widely accepted. Another famous chemist, Knut Ångström, criticized his conclusions and said increasing atmospheric carbon dioxide would have no effect on global warming. Other scientists believed the oceans would always take up carbon dioxide as quickly as human activity produced it. However, by the 1950s, fossil fuel consumption had increased dramatically, and many scientists began to suspect that the carbon dioxide level in the atmosphere was rising. One of these, Charles Keeling (Figure 12.4b) of the Scripps Institution of Oceanography in California, developed a sensitive method for measuring atmospheric carbon dioxide and began monitoring the air at a remote site in Hawaii in 1958 when he was 30 years old. Those measurements have been made continuously to the present day. Keeling became the first scientist to confirm that atmospheric carbon dioxide concentrations were indeed rising.

(a)

(b)

FIGURE 12.4 (a) Svante Arrhenius, the brilliant Swedish chemist, who, among other accomplishments, predicted the critical role of carbon dioxide in human-accelerated global climate change over a century ago. (b) Charles Keeling, a 20th century oceanographer who began the first monitoring station for atmospheric carbon dioxide in 1958, a station still in use today at the Mauna Loa Observatory in Hawaii. (Source: (a) http://en.wikipedia.org/wiki/Svante_Arrhenius; (b) Scripps Institution of Oceanography.)

of volcanic ash (or chemical signatures of the ash) that correspond to volcanic eruptions in the historical record.

Since 1958, carbon dioxide concentrations have been measured continuously at a site in Hawaii, at an elevation of 3,400 m on Mauna Loa (Box 12.1, and see Figure 3.21 in Box 3.2). The location was chosen to minimize short-term variation in carbon dioxide concentrations associated with metabolism, such as we see within a forest canopy (see Figure 3.22), and to obtain data more likely to reflect global, long-term trends. Nonetheless, the influence of metabolism is apparent in the "Keeling curve," driven by the seasonal balance of gross primary productivity (GPP) and ecosystem respiration in the temperate ecosystems of the Northern Hemisphere, as we noted in Section 3.3. The carbon dioxide concentrations are highest each year at the end

of winter in the Northern Hemisphere, reflecting the dominance of respiration over GPP in the winter. Monitoring stations in North America show a much more pronounced oscillation in carbon dioxide each year, while at the South Pole the oscillation is quite small—less than 1 ppm. At all monitoring sites, however, the trends from year to year show a steady and accelerating rate of increase in the carbon dioxide concentration: the global average increased by 1.7 ppm per year between 1993 and 2002, and then by 2.1 ppm per year from 2003 to 2013.

Since 1960, fossil fuel combustion has increasingly come to dominate the flux of carbon dioxide into the atmosphere, and today it represents more than 90% of the

carbon dioxide now comes largely from burning fossil fuels, but deforestation also contributes . . .

total flux (Figure 12.5). This reflects the incredible rate at which human consumption of coal, oil, and natural gas has grown (Figure 12.6). However, fossil fuels are not the only source of this atmospheric carbon dioxide: the cutting down of forests also contributes, and in fact this was the major source of atmospheric carbon dioxide before 1900. This deforestation releases carbon dioxide when wood is burned but also during decomposition and respiration of dead biomass and organic matter in forest soils. Forest clearing in both the temperate zone and tropics has been an important source of carbon dioxide since at least the mid 1800s, but since 1960, the tropics have grown and the

temperate zone has shrunk in importance (Figure 12.7). Ecologists do not measure these carbon dioxide fluxes in deforested regions directly, but rather infer them from knowledge of the change in organic carbon in forest soils after deforestation, and knowledge of the fate of woody biomass.

Not all the carbon dioxide emitted by human activity accumulates in the atmosphere. For instance, the increase of 2.1 ppm per year we observed in the first decade since 2000 corresponds to an accumulation rate of carbon dioxide mass in the atmosphere of 4.1 Pg C per year

less than half of emitted carbon dioxide stays in the atmosphere . . .

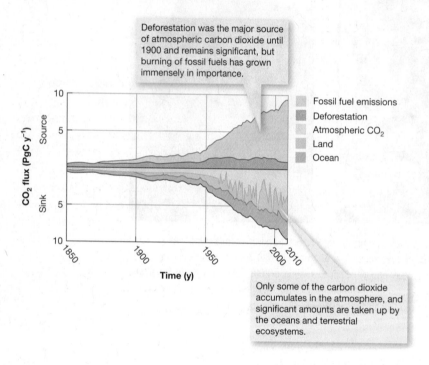

FIGURE 12.5 The global fluxes of carbon dioxide to the atmosphere (sources) and from the atmosphere (sinks) from 1850 to 2010. Sources include fossil fuel combustion and deforestation, while sinks include the accumulation of carbon dioxide in the atmosphere, the uptake by terrestrial ecosystems (land), and the assimilation by the oceans. (Source: Global Carbon Project (2010); updated from Le Quéré et al., 2009; Canadell et al., 2007.)

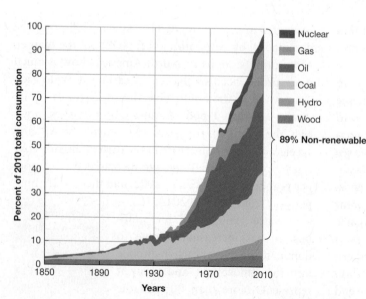

FIGURE 12.6 Global use of energy from 1850 to 2010. Wood has remained a major energy source and in fact has grown somewhat in importance since 1970. Hydropower, and to a lesser extent nuclear power, have grown since the 1960s. But these sources have been dwarfed by growth in the use of fossil fuels—coal, oil, and natural gas—over the past 150 years. (Source: David Hughes Lecture at Goldschmidt 2012 Geochemistry Conference.)

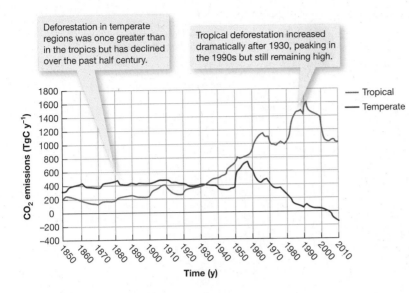

Deforestation in temperate regions was once greater than in the tropics but has declined over the past half century.

Tropical deforestation increased dramatically after 1930, peaking in the 1990s but still remaining high.

FIGURE 12.7 Release of carbon dioxide to the atmosphere from deforestation in tropical and temperate forests from 1850 to 2010. Deforestation in both the tropics and temperate zone was important until the 1960s. Since then, deforestation in the temperate zone has decreased, while tropical deforestation accelerated to high rates by the 1990s. The trend in the last decade has been downward. (Source: Richard Houghton.)

(a Pg, or petagram, is 10^{15} g). However, human activity was adding carbon dioxide to the atmosphere at a rate of 8.8 Pg C per year over this time period, with 7.7 from fossil fuel combustion and 1.1 from deforestation. Less than half this carbon dioxide stayed in the atmosphere. Where did the rest of it go? Ecologists, oceanographers, and other scientists who have studied this issue over the past few decades agree that the oceans and terrestrial ecosystems each have taken up and stored roughly equal portions of this missing carbon. They refer to the sites of such storage of an element, or to the mechanism for loss of a material (in this case atmospheric carbon dioxide), as **sinks**. On average, the ocean sink and the terrestrial sink for atmospheric carbon dioxide are estimated to be roughly equal: 2.3 to 2.4 Pg C per year respectively in the early 2000s. These sinks are critically important in reducing the impact of fossil fuel combustion. Without them, the concentration of carbon dioxide in the atmosphere would be going up much more quickly, and the planet would be warming much more quickly.

Understanding the carbon dioxide sinks

Will the sinks for atmospheric carbon dioxide continue to function in the same manner in the future as they have in the past? Or might either the ocean sink or the terrestrial sink become more or less effective in removing carbon dioxide from the atmosphere? The future trend in global warming is intimately tied to these sinks, and the trajectory of warming will change if their effectiveness changes. Below, we explore in some detail the mechanisms behind the sinks, the precision with which the sinks are known, and how the sinks may change in the future.

The terrestrial carbon dioxide sink is simply the balance between net primary productivity (NPP) and heterotrophic respiration, or net ecosystem production (NEP), summed for all the terrestrial ecosystems on Earth. Terrestrial NPP globally is an estimated 56.4 Pg C per year (see Table 11.1). The net average accumulation of carbon in terrestrial biomes (the global rate of terrestrial NEP) is 2.4 Pg C per year (as of the early 2000s) and is thus 4% of the rate of NPP. This implies that the rate of heterotrophic respiration in the terrestrial biosphere globally is 54.0 Pg C per year (Figure 12.8). Of course, we do not know the global rates of NPP and respiration with any precision: estimating NPP in any given ecosystem is itself subject to error, and scaling all the rates for all terrestrial ecosystems across the planet is difficult. We should not be surprised to learn that the uncertainty in the global rates may be 30% or even more. However, the accumulation of carbon in terrestrial NEP is not estimated by the difference between estimates of NPP and respiration globally; rather it is inferred from the difference between the known rates of input of carbon dioxide into the atmosphere (8.8 Pg C per year in the early 2000s) and the sum of the relatively well-known rate of accumulation in the atmosphere (4.1 Pg C per year in the early 2000s) and the estimate of uptake by the oceans (2.3 Pg C per year), which we will discuss next.

We know the ocean sink for carbon dioxide with reasonably high precision, thanks in part to the ability of oceanographers over several decades to track the fate of ^{14}C-radioactive carbon dioxide produced in the atmosphere during nuclear bomb tests in the 1960s. The

the net uptake of carbon dioxide by the oceans is known with moderate precision

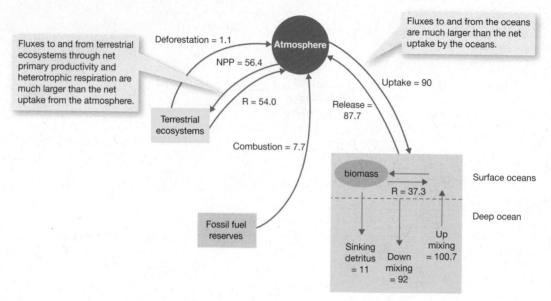

FIGURE 12.8 Major flows of the global carbon dioxide cycle as of the early 2000s, in Pg C per year.

mixing of waters makes the oceans far easier than terrestrial ecosystems to sample in a representative way. Carbon dioxide is quite soluble in seawater, but the solubility varies with water temperature; warm ocean waters release carbon dioxide to the atmosphere, while cold waters absorb it. Overall, there is an uptake of approximately 90 Pg C per year by the oceans and a release of 87.7 Pg per year, for a net uptake of 2.3 Pg C per year. See Figure 12.8.

The carbon cycle in the surface ocean is dynamic, with high rates of carbon dioxide uptake during NPP (48.3 Pg C per year, from Table 11.1) and a somewhat lower rate of carbon dioxide produced during heterotrophic respiration in the surface waters of the oceans. The difference between NPP and respiration of 11 Pg C per year represents organic carbon that sinks or is mixed into the deep oceans, where it is then largely respired to carbon dioxide. The exchange of dissolved carbon dioxide between the surface oceans and deep oceans is large, due to mixing and flows of water masses as we discussed in Chapter 4. On average, an estimated 100.7 Pg C per year is mixed up into the surface oceans and 92 Pg C per year is mixed down into the deep oceans (Figure 12.8). These flows, together with the sinking of organic detrital carbon to the deep oceans, represent a net flux to the deep oceans of 2.3 Pg C per year, the same value as for the net flux from the atmosphere to the surface oceans. It is the deep oceans that are the important sink for carbon dioxide, with the surface oceans simply serving as a vector.

The great oceanic conveyor belt (see Figure 4.7) is responsible for much of the flux of carbon

> the deep ocean sink is temporary . . .

dioxide and organic detrital carbon into the deep oceans, particularly due to the sinking of cold, carbon-dioxide-rich and organic-rich North Atlantic water. Remember this deep-ocean sink for carbon dioxide is temporary, lasting a few centuries on average; as deep-ocean water makes its way slowly to the Southern, Indian, and Pacific Oceans and then rises again to the surface, much of the carbon dioxide is released back to the atmosphere. In future centuries, uptake by the world's oceans will change to a net release of carbon dioxide back into the atmosphere. Even if the inputs from fossil fuel combustion are reduced, this will result in continuously high atmospheric levels for many hundreds of years to come.

As we discussed earlier, the rate of sinking is sensitive to climate change and may already be starting to slow as the North Atlantic Ocean waters become somewhat less salty

> the future of the ocean carbon dioxide sink is highly uncertain . . .

due to melting of Arctic Ocean ice and the ice sheets of Greenland (see Box 4.1). The slowing of the conveyor belt will result in less uptake of carbon dioxide by the world's oceans and therefore more global warming. This feedback of climate change on further global warming is one of many feedbacks of great concern to climate scientists.

In general, if there were no change in ocean physics and circulation, we would expect the oceans to take up more carbon dioxide as the concentration in the atmosphere increases, simply because the higher atmospheric concentration increases the diffusion of the gas. On the other hand, the surface oceans are warming as the global climate warms, and this lowers the solubility of carbon dioxide and therefore reduces the net rate of uptake by the oceans. The net effect of the increase in carbon dioxide and the higher temperatures is highly uncertain, as shown by the diverging predictions of 11 global change models in Figure 12.9 (Friedlingstein et al., 2006). All predict an increase in the net uptake of carbon dioxide beyond the current value of 2.3 Pg C per year, but they vary from 3.7 to 10 Pg C per year for the year 2100. None include the effect of the slowing of the oceanic conveyor belt we discussed above. As we will see below, the future uptake of carbon dioxide by terrestrial ecosystems is even more uncertain,

and these uncertainties are some of the greatest challenges to our ability to correctly predict future global climate change.

Effects of ocean acidification

As ocean waters take up carbon dioxide, they become more acidic; carbon dioxide dissolved in water is carbonic acid. Since the start of the Industrial Revolution, the oceans are estimated to have become 30% more acidic (Jacobson, 2005). From 1990 to 2006 alone, the average pH of the world's oceans dropped 0.04, a further increase in ocean acidity of more than 10% (remember that pH is a log scale) as a result of the additional carbon dioxide (Figure 12.10). Many marine organisms—including bivalves such as clams and oysters, corals, and some types of phytoplankton such as coccolithophores—have shells or other structures composed of carbonate minerals (Figure 12.11). With a drop in pH, it becomes more difficult or impossible for these organisms to secrete carbonates, which may result in serious ecological disruption of many marine ecosystems.

Will the terrestrial carbon dioxide sink change in the future?

Let's return to the net uptake of carbon dioxide by terrestrial ecosystems, estimated as 2.4 Pg C per year in recent years, but with significant year-to-year variability (see Figure 12.5). At least part of this sink results from net accumulation of organic matter in biomass and soils during forest succession on lands abandoned from agriculture in the temperate zone in the late 19th and 20th centuries (see Section 4.3). As these forests

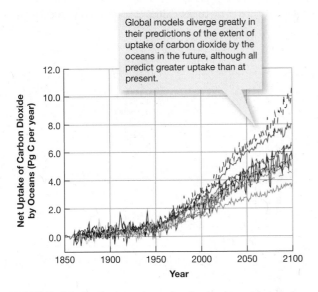

FIGURE 12.9   A comparison of estimates for uptake of atmospheric carbon dioxide between 1850 and 2100 by the oceans for several global carbon models. Several of the models do a good job of predicting the uptake observed through 2010, but the models increasingly diverge as they look into the future. (Friedlingstein et al., 2005.)

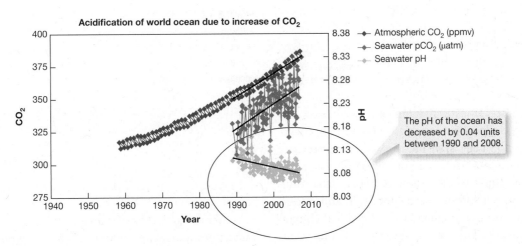

FIGURE 12.10   The global increase in atmospheric carbon dioxide has led to higher concentrations of dissolved carbon dioxide in the oceans, which has acidified the oceans since carbon dioxide dissolved in seawater forms carbonic acid. (Source: Intergovernmental Panel on Climate Change, Based on Feely et al., 2009.)

(a)

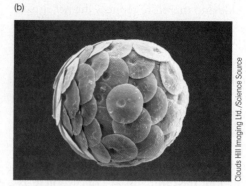
(b)

FIGURE 12.11 Increasing carbon dioxide levels are leading to acidification of the oceans, which makes it difficult for many marine organisms to form their carbonate shells and structures. (a) Corals are particularly sensitive to the effects of ocean acidification. (b) Coccolithophores, microscopic algae that are major primary producers in the subtropical oceanic gyres, have carbonate shells, as indicated in this scanning electron micrograph. Ocean acidification may result in their being replaced by totally different types of phytoplankton as the carbon dioxide concentration in the atmosphere continues to rise.

grow, GPP greatly exceeds ecosystem respiration for many decades, resulting in high rates of net ecosystem production and an accumulation of organic matter both in the tree biomass and in forest soils (Figure 12.12). Year-to-year variation in GPP and respiration, driven by variation in precipitation and temperature, may in part explain the large amount of variation we estimate for the terrestrial ecosystem sink. Eventually, as forest succession continues for 80 to 100 years or more, the rate of GPP levels off and may even fall. Respiration, however, continues to rise, and in a fully mature forest it may roughly equal GPP. At this point, net ecosystem production is zero and there is no further net uptake of atmospheric carbon dioxide; the amount of organic matter in biomass and soils remains high but no longer increases. We can expect the net rate of carbon dioxide uptake (net ecosystem production) in these temperate forests to slow in coming decades.

As we discussed in Section 3.3, an increasing amount of evidence from whole-ecosystem experiments in which carbon dioxide is added to forests indicates that higher concentrations of carbon dioxide increase rates of NPP and net ecosystem production. That is, we might expect the terrestrial carbon sink to increase as carbon dioxide levels in the atmosphere continue to rise, partially mitigating the problem. This response is not well understood at the global scale, though, and the 11 global climate models discussed above with regard to predicting ocean uptake of carbon dioxide also show dramatically different predictions for future rates of terrestrial NPP as carbon dioxide rises (Figure 12.13).

| the fertilizing effect of atmospheric carbon dioxide . . . |

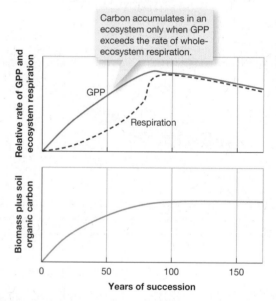

Carbon accumulates in an ecosystem only when GPP exceeds the rate of whole-ecosystem respiration.

FIGURE 12.12 The change in the rate of gross primary production and whole-ecosystem respiration (top) and in organic carbon in biomass plus soil organic carbon (bottom) over time during forest succession. For the first several decades, gross primary production increases rapidly, and organic carbon accumulates rapidly. Respiration increases over time, as the forest biomass increases and as more organic material is available to be decomposed; the rate of carbon storage in the ecosystem slows. After many decades (the exact timing varies across different ecosystems, based on climate and dominant species), rates of gross primary production slow and may even decrease. Respiration rates rise to equal the rate of gross primary production, and the ecosystem no longer continues to take up and store additional carbon. Vertical scales are purely relative, indicating general patterns.

Temperature also has an effect on net ecosystem production, both through effects on NPP and, even more, on ecosystem respiration (see Chapters 3 and 11). Since the effect of temperature is

| . . . and the influence of temperature |

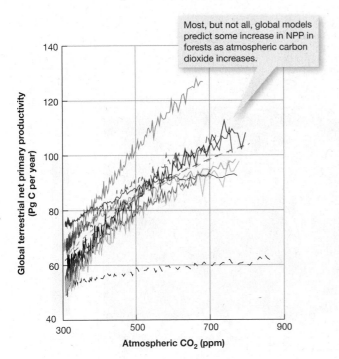

FIGURE 12.13 A comparison of model predictions for global rates of net primary productivity in terrestrial biomes as a function of the atmospheric concentration of carbon dioxide. All models predict some increase as carbon dioxide increases, but they vary greatly in their estimate of how big a response this is. (Friedlingstein et al., 2005.)

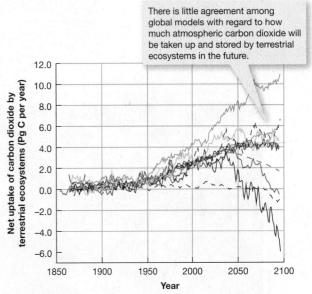

FIGURE 12.14 A comparison of global model predictions on rates of net carbon uptake by terrestrial biomes from the atmosphere from 1850 to 2100. Note the huge divergence of model predictions into the future, with some models estimating large net uptake and others showing large net releases to the atmosphere. (Friedlingstein et al., 2005.)

greater on respiration than NPP, we can expect that warmer temperatures will lead to lower rates of net ecosystem production (that is, less carbon storage). We would therefore also expect the future of terrestrial carbon uptake to be influenced both by increased carbon dioxide levels and by higher temperatures. With both factors included, though, the 11 global climate models again show widely divergent predictions for the future net uptake of carbon dioxide by terrestrial ecosystems. Compared to the current uptake rate of 2.4 Pg C per year, the models predict a range of values by the year 2100 from a net uptake of 10.8 Pg C per year to a net release to the atmosphere of 6 Pg C per year (Figure 12.14).

None of these global models consider other influences on NPP and carbon accumulation in terrestrial ecosystems, such as the fertilizing effects of nitrogen on both NPP and ecosystem respiration. As we discuss later in this chapter, human activity is altering the nitrogen cycle even more than the carbon dioxide cycle. Given the importance of nutrient limitation (see Chapter 11), it is no surprise that the altered nitrogen cycle may also have a major influence on the sinks of atmospheric carbon dioxide. Other feedbacks of climate change

other unknowns in the future of the global carbon cycle

on carbon storage also seem likely. One example is increased fires in the tundra biome caused by lightning strikes, burning vast stocks of organic matter in the soils. Because of global warming, convective thunderstorms (discussed in Chapter 4; Figure 4.20) and lightning strikes—once extremely rare in the tundra—have become 20 times more common in the Arctic over the past decade, although strikes are still far less frequent than in the temperate zone and tropics (Figure 12.15).

Given all this uncertainty about the future of the global carbon cycle, then, what *can* we conclude? The increase in carbon dioxide in the atmosphere—primarily from burning fossil fuels—is the main driver of the observed rate of global warming. This trend will continue for as long as we continue to burn fossil fuels. Scientists understand the current dynamics of the global carbon cycle with reasonable certainty, but our understanding of the future dynamics is very uncertain. As carbon dioxide increases and temperatures rise, both the ocean sink and the terrestrial ecosystem sink for atmospheric carbon dioxide will change, but we simply cannot predict with any accuracy how these changes will play out. Some model predictions give us reason to hope that the carbon sinks will grow and help mitigate global climate change, but others indicate the sinks may not grow sufficiently, and that the terrestrial biomes may actually shift and become a net source of carbon dioxide to the atmosphere, even further accelerating global warming.

Lightning strikes, once absent from the Arctic, are still far less frequent than at lower latitudes but are becoming more common.

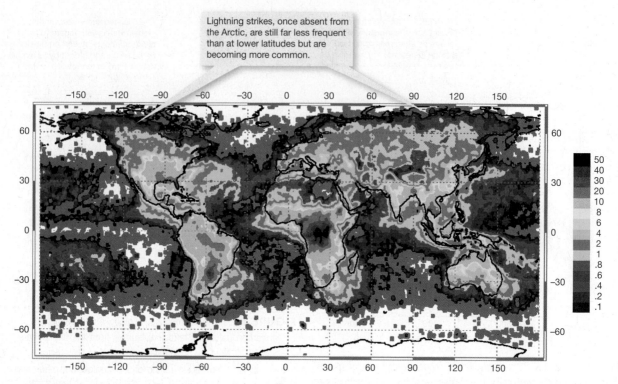

FIGURE 12.15 The frequency of lightning strikes across the planet, in strikes per square kilometer per year, as of late 2001. Lightning is much more common in the tropics. It was once absent in the tundra regions but is now becoming more common, leading to more tundra fires. (Source: http://science.nasa.gov/science-news/science-at-nasa/2001/ast05dec_1.)

Can we reduce carbon dioxide emissions?

Burning fossil fuels is the major cause of the increase in atmospheric carbon dioxide. Hence, any solution to the global warming problem requires a large reduction in these emissions from fossil fuels. An important step is to use energy more efficiently. Per capita, residents of the United States are the largest emitters of carbon dioxide (although total emissions from China, with its large population, have surpassed those for the entire United States since 2005) and also use energy far less efficiently than Europeans do. New technologies can help increase the efficiency of energy use; electric cars are more efficient than any gasoline-powered vehicles, and total emissions from driving an electric car are lower even if the electricity comes originally from burning fossil fuels (Jacobson & Delucchi, 2011).

Another approach is to trap the carbon dioxide released as fossil fuels are burned, perhaps storing it in water in deep caverns, although the water would become strongly acidic and corrosive, and leakage back to the surface would remain a concern. We could also continue to burn fossil fuels but rely more on natural gas and less on coal, since the carbon dioxide emissions to obtain the same amount of energy from natural gas are less. This idea, though, ignores the role of methane

as a greenhouse gas, as we discuss in the next section. Perhaps the best approach is to move as quickly as possible to implement truly renewable sources of energy, such as wind and solar power. Jacobson and Delucchi (2009) have proposed that such a radical change in energy use is obtainable, using currently available technologies, within 20 years.

12.3 THE GLOBAL METHANE CYCLE

Carbon dioxide is not the only greenhouse gas responsible for the present rate of global warming. Several other gases are also important, notably methane and nitrous oxide. We discuss nitrous oxide in the context of the global nitrogen cycle later in this chapter. In this section, we focus on methane.

Methane absorbs infrared radiation at wavelengths where carbon dioxide, water vapor, and most other greenhouse gases do not. As a result, per molecule it is far more potent than carbon dioxide as a greenhouse gas. On the other hand, the concentration of methane in the atmosphere is much lower than that of carbon dioxide, and the increase in methane concentrations over natural levels is less than for carbon dioxide in absolute terms (although greater as a proportional change, as we pick up on later in this section). The net consequence is that the global warming caused directly by methane today is equivalent to approximately 30% of the warming caused by carbon dioxide, if we look simply at the concentration of methane in the atmosphere (a **radiative forcing** of 0.48 Watts per square meter per year for the increase in methane, compared to 1.66 Watts per square meter for carbon dioxide; radiative forcing is the "imbalance between incoming solar radiation and outgoing infrared radiation" that results in global warming IPCC, 2007). This increases to 60% of the warming caused by carbon dioxide, however, if we also include the indirect effects of methane on other radiatively active materials in the atmosphere (which increases the radiative forcing for methane to 1.0 Watts per square meter per year; IPCC 2013). One example of such an indirect effect is the catalytic role of methane in producing ground-level ozone, another greenhouse gas.

As for carbon dioxide, we can examine gas bubbles laid down over past millennia to evaluate long-term changes in methane. Methane concentrations were reasonably low and constant for most of the previous 10,000 years, until 1800 or so (Figure 12.16). However, human activity has more than doubled the atmospheric concentration

> human alteration of the methane cycle is greater than for carbon dioxide . . .

of methane since the start of the Industrial Revolution, from 0.68 to 1.74 ppm. This is a far greater change than the 40% increase for carbon dioxide over the same time period (see Figure 12.3).

We saw that for atmospheric carbon dioxide, the sinks stored organic carbon (in biomass and soils) in terrestrial ecosystems and dissolved carbon dioxide in the oceans. For atmospheric methane, the sink is the chemical alteration of methane to other compounds. By far the largest methane sink is photooxidation in the atmosphere, which oxidizes the gas to carbon dioxide and water. The chemistry behind this reaction is reasonably well understood, and this allows us to conclude with some precision that the global sink for methane is approximately 570 Tg C (a Tg is 10^{12} g) per year (IPCC, 2007).

For carbon dioxide, the increase in the atmosphere is a significant proportion of the flux into the atmosphere. This accumulation is far less pronounced for methane, which has a shorter residence time in the atmosphere: approximately 12 years vs. a century for carbon dioxide. The input of methane to the atmosphere is roughly equal to the sink, or approximately 570 Tg C of methane per year in recent years. Since the change in concentration in the atmosphere over time is directly proportional to the inputs, we know from the time before the Industrial Revolution that the natural inputs then were approximately 220 Tg C of methane per year. Assume that these natural fluxes have remained constant over the past few centuries. We can then estimate the inputs that result from human activity as the difference between the current total and natural fluxes as 350 Tg C of methane per year (570 minus 220). See Table 12.1.

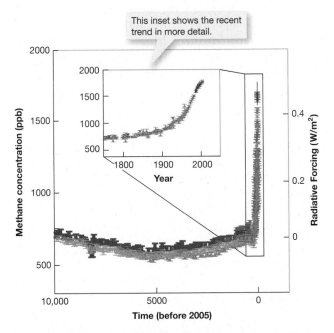

FIGURE 12.16 The concentration of methane in the Earth's atmosphere was stable for most of the last 10,000 years, but it started to increase exponentially over the past 200 years. The inset box shows the recent trend in more detail. Compare with Figure 12.3, which shows similar data for carbon dioxide. (Reprinted from IPCC, 2007.)

TABLE 12.1 Fluxes of methane to the atmosphere globally by source (Tg C of methane per year)

Natural sources	**220**
geological seeps	53
biological sources	167
Anthropogenic sources	**350**
fossil-fuel emissions	115
animal agriculture	90
rice cultivation	60
landfills and sewage	55
biomass burning	30
Total	**570**

Adapted from IPCC (2007), Howarth et al. (2011), and Kirschke et al. (2013).

The natural sources of methane

Although we can estimate the total magnitude of the natural methane flux with some precision, identifying its individual components is less easy. Methane is produced biologically in many different types of ecosystems, but some also seeps into the atmosphere from ancient geological formations. A global survey of such seeps concluded that this natural geological source represents a flux of 53 Tg C of methane per year, plus or minus 20% (Etiope et al., 2008). Accepting this estimate leaves us searching for another 167 Tg C of methane per year from natural sources (220 minus 53).

The natural biological sources are dominated by **methanogenesis**, the production of methane by bacteria in the absence of oxygen. As long as oxygen is present in an environment, bacteria will use oxygen in their respiration, decomposing organic matter and oxidizing the organic carbon to carbon dioxide. In the absence of oxygen, they use other strategies for their respiration and decomposition, one of which is methanogenesis. During methanogenesis, bacteria obtain only 100 KJ of energy per mole of carbon they decompose, as opposed to some 500 KJ per mole of carbon for bacteria using oxygen-based respiration. Bacteria making their living through methanogenesis could therefore never outcompete oxygen-respiring bacteria for organic substrates when oxygen is present. But when oxygen is absent, methanogenesis provides a mechanism for bacterial life and decomposition to continue, albeit with low energy gains. In freshwater wetlands and the sediments of freshwater lakes, methanogenesis is often the major form of decomposition. Little or no oxygen occurs in these environments, since the diffusion of oxygen is 10,000 times slower in waterlogged soils and sediments than in soils having air pores, and whatever oxygen starts to diffuse into the soils and sediments is rapidly consumed.

Much of the methane formed in these wetlands and lake sediments escapes to the atmosphere. The precise magnitude of these methane fluxes is unknown, since the fluxes are highly variable over both space and time, making measurement difficult. Our most reasonable estimates for the global flux from wetlands, developed from measurements from individual wetlands multiplied by the area of wetlands on the planet, vary from 80 to 280 Tg C of methane per year. Similar estimates for lakes vary from 10 to 73 Tg C of methane per year (IPCC 2007). Of course, at the higher end, these estimates exceed the value of 167 Tg C of methane per year that seems likely to be produced from all natural biological sources.

> most of the natural flux comes from oxygen-free decomposition

Some methane is also produced in the guts of animals, including termites, through methanogenesis by bacteria in their guts, which are oxygen-free (see Figure 1.3). Estimates for the atmospheric flux of methane from termites have ranged up to 150 Tg C of methane per year, based on methane production by individual termites and the global population of termites. However, recent studies have documented that most of the methane produced in termite mounds is oxidized by bacteria in the mound and never reaches the atmosphere (Jamali et al., 2011). The atmospheric methane source from termites is probably in the range of 1 Tg C of methane per year. The guts of cows and other ruminants are even more productive of methane, as we discuss below when we turn to anthropogenic sources.

> the oceans produce little methane . . .

The oceans are a source of methane to the atmosphere, but perhaps a surprisingly small one given the area of the planet covered by the oceans; they represent a flux of only about 10 Tg C of methane per year (IPCC, 2007). The concentration of methane in the oceans is above the saturation level relative to the atmosphere. It is therefore certain that the net flow of methane is from the oceans to the atmosphere. However, the flux is low. Also, as the salinity of wetlands increases, the methane flux to the atmosphere dramatically decreases (Figure 12.17). Hence, very little atmospheric methane comes from coastal marine wetlands compared to freshwater wetlands.

Why is the flux from oceans so much less than from lakes? And why is the flux from coastal wetlands so much less than from freshwater wetlands? The reason is simply that the concentration of sulfate is so much greater in seawater than in freshwaters. As we discussed above, methanogenesis dominates

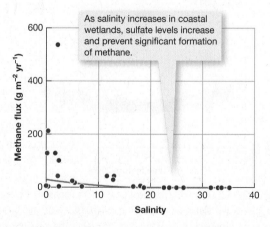

FIGURE 12.17 The flux of methane from coastal marshes to the atmosphere as a function of the salinity of the water flooding the marshes. (Poffenbarger et al., 2011.)

decomposition in many freshwater wetlands and lake sediments when oxygen is absent. But bacteria that use sulfate in their respiration to decompose organic matter obtain 130 KJ of energy per mole of carbon, which is far less than bacteria that can use oxygen, but nonetheless substantially more than obtained by methane-producing bacteria. Hence, bacteria using sulfate (reducing the sulfate to sulfides, while oxidizing the organic matter to carbon dioxide) will outcompete methane-producing bacteria for decomposable organic matter wherever sulfate is present. Sulfate reduction is therefore the dominant decomposition process in coastal marine wetlands whenever salinity is sufficiently high to supply sulfate (Figure 12.17) and in most coastal marine sediments.

Anthropogenic sources of methane

Emissions from developing and using fossil fuel use are one of the largest source of atmospheric methane from human activities, making up approximately one-third of the anthropogenic sources (see Table 12.1). Of the total estimated flux from fossil fuels of 115 Tg C per year, more than half comes from the natural gas industry (Howarth et al., 2011). Natural gas makes up only 20% of the global use of fossil fuels (see Figure 12.6), so we can see that the methane emissions per unit of production of natural gas are far greater than for coal and oil. This makes sense, since natural gas is composed largely of methane, and even small leakages and releases of natural gas add up to large methane emissions. In contrast, methane is a small contaminant of oil and coal.

As with all such global fluxes, our estimates for fossil fuel emissions are uncertain; the values in Table 12.1 may well be 20% to 30% lower or higher, and perhaps even higher (Howarth et al., 2011; Kirschke et al., 2013). However, the amount of radioactive ^{14}C in the methane in the atmosphere provides a check on these estimates. Methane formed from biological processes such as the decomposition of organic matter in recent years contains ^{14}C, because it is being continuously produced in the atmosphere from cosmic radiation and because nuclear bomb tests in the atmosphere in the 1960s contributed even more. On the other hand, methane from ancient geological formations contains no ^{14}C, because this radioactive isotope has a half-life of 5,730 years and will have decayed away over a time period of millions of years. From measurements of ^{14}C in present-day atmospheric methane, therefore, we know that roughly 30% of the methane comes from ancient sources, while the remainder comes from modern biologically mediated sources (Lassey et al., 2007; Etiope et al., 2008). Our estimates for the natural geological seeps (53 Tg C of methane per year) plus the anthropogenic fossil fuel emissions (115 Tg C of methane per year) equal a total estimate of 168 Tg C of methane per year from ancient sources. This, too, comes in at 30% of the total global flux, increasing our confidence in both estimates.

The other anthropogenic sources are many, all uncertain, although we can have some confidence that the sum of these is around 235 Tg C of methane per year (350 Tg C for all anthropogenic sources minus 115 Tg C for the fossil fuel emissions). Most of this is produced by bacterial methanogenesis, in flooded rice paddies (which are human-made wetlands, similar in function to natural freshwater wetlands), in the oxygen-free guts of cows and other livestock, and during bacterial decomposition of organic matter in landfills, sewage treatment facilities, and manure from livestock, all of which are often oxygen-free as well. Human activity is simply creating new environments for this natural process of methanogenesis on a massive scale. The single largest of these sources, after fossil fuels, is probably animal agriculture. Animal agriculture emissions have grown tremendously over the past half-century as the global production of meat has increased 5-fold and continues to rise. The drivers behind this increased consumption are both human population growth and an increase in meat consumption per person, which almost doubled since 1960. (See Chapter 14.)

agricultural sources are also large . . .

Methane and the global climate system

A critical difference between the effects of carbon dioxide and of methane on global warming is the time scale over which they operate. The climate system is far more responsive to methane, both because it has a shorter residence time in the atmosphere and because, molecule for molecule, it is a more potent greenhouse gas. Figure 12.18 illustrates this difference. By 2011, the average temperature of the planet had increased by 0.7°C compared to the baseline temperature of the early 20th century. All climate models predict further warming if greenhouse gas emissions continue unabated. In the absence of any major effort to reduce them, the planet is likely to warm by 1.5°C above baseline by 2030 and by 2°C before 2045. As Figure 12.18 shows, immediate efforts to reduce carbon dioxide emissions globally (starting in 2012) would do little or nothing to prevent this level of global warming by 2040 – because of the relatively long time that added carbon dioxide remains in the atmosphere—although warming would slow greatly in later years. On the other hand, efforts to reduce methane emissions would have more immediate benefits, greatly slowing the

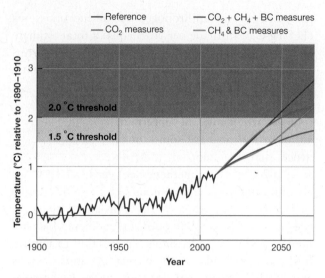

FIGURE 12.18 The average rate of warming of the global atmosphere from 1900 through 2010 and projections of future warming under four different scenarios. The 'reference' scenario assumes no major effort to mitigate greenhouse gas emissions. The 'CO$_2$' scenario assumes aggressive efforts to reduce carbon dioxide, but no efforts to reduce other greenhouse gas emissions. The 'CH$_4$ + BC' scenario assumes that emissions of methane and black carbon (BC, or soot) will be reduced, but carbon dioxide emissions will continue unabated. And the 'CO$_2$ + CH$_4$ + BC measures' assume that emissions of all three of these will be reduced. Because of lags in the climate system, controlling carbon dioxide emissions alone does little to slow the rate of global warming for several decades. Only reducing methane (and soot) emissions leads to some slowing of global warming over the critically important coming few decades. (Modified from Shindell et al., 2012.)

trend in global warming even without reductions in carbon dioxide emissions. Reducing both carbon dioxide and methane gives the best outcome, with not only immediate benefits in a slowed rate of warming but also continued slowing of warming after 2040 or so.

As global climate change accelerates, the risk increases that the world will reach some tipping points in the climate system, leading to runaway feedbacks that could further intensify global warming. We have already discussed some mechanisms for this, such as increased burning in the tundra biome and a slowing of the great oceanic conveyor belt. There is also much concern that warming will increase the flux of methane from natural ecosystems to the atmosphere, possibly at temperatures of only 1.8°C above the long-term baseline (Zimov et al., 2006; Hansen et al., 2007). One mechanism is that as the permanently frozen soils of the tundra biome melt, they will become wetlands, and huge amounts of methane will result from the decomposition of their soils, which are rich in organic carbon but lacking in oxygen (see Chapter 4). Another worry is that methane clathrates, forms of frozen methane or methane crystals that are abundant in sediments at intermediate depths in many coastal ocean areas, will melt and release huge quantities of gaseous methane. Reducing the anthropogenic fluxes of methane to the atmosphere is one of

tipping points in the global climate system . . .

12.2 **ECOncerns**

Hydraulic fracturing to produce natural gas from shale rock

Conventional sources of natural gas are being rapidly depleted around the world. Some, though, are optimistic that we will continue to have a large supply. Shale rock often holds large amounts of natural gas, but it is tightly held and was not available for commercial exploitation until the development of two technologies in the late1990s: high-volume hydraulic fracturing of rock ("fracking"), combined with precision drilling of gas wells along distances of 1 to 2 km or more into shale formations often only 50 m thick. Industry executives and many politicians have promoted this shale gas as a 'bridge fuel' to the future that would allow society to continue to use fossil fuels over the next few decades while reducing greenhouse gas emissions. This belief is based on the fact that far lower levels of carbon dioxide are released when we burn natural gas than when we consume oil or coal to obtain the same amount of energy.

Confidence in shale gas was challenged by new research in 2011, summarized in a commentary to the journal *Nature* by Howarth and Ingraffea (2011). Noting that carbon dioxide is not the only greenhouse gas of concern and that natural gas is composed largely of methane, they wrote: 'Methane is a potent greenhouse gas, so even small emissions matter. Over a 20-year time period, the greenhouse-gas footprint of shale gas is worse than that for coal or oil. The influence of methane is lessened over longer time scales,

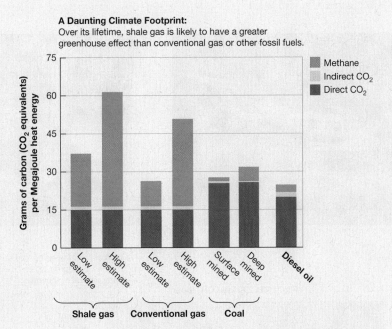

A Daunting Climate Footprint:
Over its lifetime, shale gas is likely to have a greater
greenhouse effect than conventional gas or other fossil fuels.

FIGURE 12.19   A comparison of the greenhouse gas footprints of several fossil fuels, based on the carbon dioxide emitted as the fuels are burned, the indirect emissions of carbon dioxide from developing and transporting the fuels, and methane emissions normalized to carbon dioxide equivalents, using an integrated assessment for a 20-year period after emission. High and low estimates of methane emissions are given for both shale gas and conventional natural gas. When we consider carbon dioxide emissions alone, natural gas has the lowest greenhouse gas emissions. Including methane emissions changes this picture and suggests that natural gas may have a very large greenhouse gas footprint. (Howarth & Ingraffea, 2011.)

because methane does not stay in the atmosphere as long as carbon dioxide. Still, over 100 years, the footprint of shale gas remains comparable to that of oil or coal' (Figure 12.19).

This argument assumes far larger emissions of methane from developing and using natural gas resources – especially shale gas, but also natural gas from conventional sources—than from oil or coal. Methane emissions from gas development are not known with certainty, and one of the recommendations from the commentary was to obtain better data about them. As we write, many new studies are coming out, some with surprising results (Figure 12.20). For instance, some new wells being drilled to develop shale gas emit very large quantities of methane to the atmosphere even before the shale formations are reached and hydraulically fractured to release gas, perhaps because more shallow pockets of gas are encountered on the way (Caulton et al., 2014).

A counterargument is that we must take the long-term view. Carbon dioxide added to the environment in our lifetimes will persist and influence the climate for centuries to come, while the time frame of influence for methane emissions is relatively short. In this alternate view, reducing carbon dioxide emissions is paramount to all other considerations.

Many approaches for reducing greenhouse gases do focus on carbon dioxide rather than on methane, and they have the perverse potential to increase methane emissions and therefore aggravate global warming over the coming few decades. For example, a 'carbon tax' on emissions of carbon dioxide may favor natural gas over coal or oil as a fuel because carbon dioxide emissions to obtain the same amount of energy are lower, even though total greenhouse gas emissions may be higher when we consider methane. Economic markets for trading of emission credits may act in a similar way if they focus only on carbon dioxide and do not consider other greenhouse gases. The time scale is paramount as well. If we compare the warming potential of methane to that of carbon dioxide only at time scales of 100 years rather than over the next few decades, we don't necessarily see the critical need to immediately reduce methane emissions.

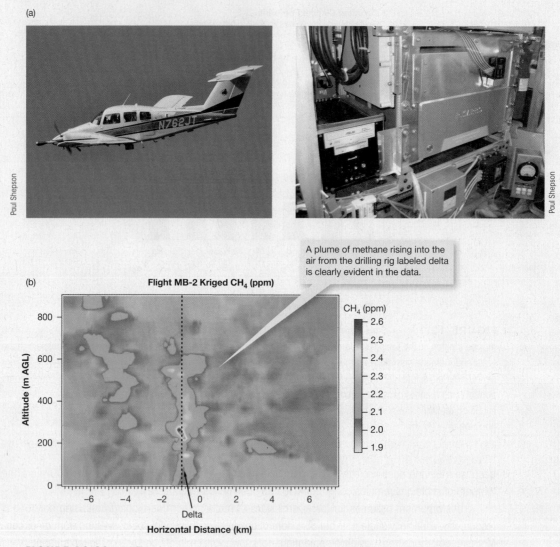

(a)

Paul Shepson

Paul Shepson

A plume of methane rising into the air from the drilling rig labeled delta is clearly evident in the data.

(b)

Flight MB-2 Kriged CH₄ (ppm)

CH₄ (ppm)

Delta

Horizontal Distance (km)

FIGURE 12.20 (a) The airplane and its measuring equipment used to estimate methane emissions from a shale-gas field in southwestern Pennsylvania in June of 2012. (Photos by Paul Shepson.) (b) The concentration of methane in the atmosphere as measured by the airplane, plotted as a function of height in the atmosphere and distance from a gas-drilling rig. (Cauton et al., 2014.)

How should we weigh the long-term consequences of carbon dioxide emissions against shorter-term concerns about methane? Currently, economic markets for carbon release and sequestration consider methane only on the time scale of 100 years. Is this approach protective enough to reduce global warming, if methane leads to excessive warming that surpasses critical tipping points in the global climate system? Does natural gas from shale make sense as a way to address global change, or should society work more aggressively to reduce use of all fossil fuels?

the best opportunities to slow global warming over the coming decades, and to lower the risk of encountering tipping points.

How do we reduce methane emissions?

Like carbon dioxide emissions, methane emissions can be greatly reduced if we move away from fossil fuels and towards sustainable wind and solar power sources. Doing so quickly buys the Earth's climate system some critically needed time (see Figure 12.18). Further reductions in methane emission can come from improved sewage and waste disposal (see Table 12.1) that does not place these organic-rich materials in oxygen-free environments. The other major sources of global methane are rice farming and animal agriculture, and ecologists and other biologists are hard at work to develop approaches for reducing these sources. For example, we can reduce the methane fluxes from rice paddies by managing the timing of flooding or by adding supplements such as iron. Fluxes from livestock can be managed by diet.

12.4 THE NITROGEN CYCLE AT GLOBAL AND REGIONAL SCALES

As we have discussed in earlier chapters, nitrogen is essential for all life, often limits primary productivity in both terrestrial and aquatic ecosystems, and plays a role in determining the diversity of ecological communities. Humans have had an immense effect on the global nitrogen cycle, particularly by using synthetic nitrogen fertilizer in agriculture. The rate of acceleration in the global nitrogen cycle due to human activity is staggering, with most of the growth in the use of synthetic nitrogen fertilizer occurring over just a few decades, compared to one or more centuries for carbon dioxide emissions from fossil fuels, deforestation, and human population growth (Figure 12.21). Before exploring this acceleration in more detail, we will examine the global nitrogen cycle as it functioned before the agricultural and industrial revolutions.

The vast majority of nitrogen on Earth is present as molecular N_2 gas, mostly in the atmosphere but also dissolved in the oceans. There is at least 10,000 times more nitrogen present as N_2 gas than in all the

> molecular N_2 dominates . . .

biomass, all the organic matter in soils and sediments, and all the plant-available nitrate and ammonium in the world combined. These scarcer forms of nitrogen are constantly being denitrified to N_2, either directly in the case of nitrate or indirectly in the case of ammonium (which is oxidized first to nitrate) and organic nitrogen (which is mineralized to ammonium, then oxidized to nitrate, then denitrified; see Figure 11.26). Biological nitrogen fixation is critical in counteracting this loss of biologically available nitrogen and providing new biologically available nitrogen to plants and other primary producers. At the global scale and over geological time (and before the agricultural and industrial revolutions), the loss through denitrification was balanced by nitrogen fixation.

Nitrogen fixation occurs widely in a variety of ecosystems (see Chapters 9 and 11), but in terms of the global fluxes, fixation in tropical and subtropical areas dominates. In terrestrial ecosystems, most of the nitrogen fixation is concentrated in tropical forests and savannas (see Figure 11.10). In the oceans, the subtropical gyres and tropical waters are the sites of almost all nitrogen fixation. Before humans' intervention in the global nitrogen cycle, the rate of nitrogen fixation globally was probably in the range of 240 Tg of nitrogen per year, half in terrestrial biomes and half in the oceans (Galloway et al., 2004).

> most nitrogen fixation occurs in tropical and subtropical biomes

Most of the nitrogen fixed in terrestrial biomes was also denitrified in these biomes, largely in wetlands and in soils that were seasonally waterlogged and therefore conducive to this process (see Chapter 11), and in the streams and rivers that drain the landscape. However, a significant amount of nitrogen, an estimated 30% of the quantity fixed in the landscape, flowed to the oceans (Galloway et al., 2004), where, in addition to the nitrogen fixed in the oceans themselves, it was largely denitrified. If we assume the global rate of denitrification was roughly equal to the 240 Tg of nitrogen per year that was fixed, then an estimated 160 Tg of was probably denitrified in the oceans and only 80 Tg in terrestrial and freshwater ecosystems. The oceans were thus twice as important globally in converting biologically available nitrogen to N_2.

> the oceans dominate denitrification . . .

Molecular N_2 is the major product of denitrification, but some nitrous oxide (N_2O) is also produced. From the analysis of gas bubbles in ice cores, we know that concentrations of nitrous oxide were relatively constant for most of the last 10,000 years, until 1900 or so (Figure 12.22). This provides us with some

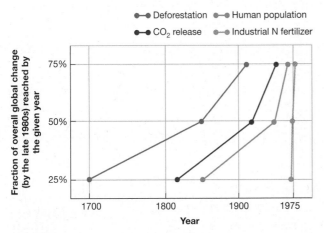

FIGURE 12.21 A comparison of the rate of change in global deforestation, emissions of carbon dioxide, world population growth, and the rate of production of synthetic nitrogen fertilizer since 1700, relative to the rates of each of these in the late 20th century. (Vitousek et al., 1997.)

confidence that the rate of denitrification globally was relatively constant over this time frame as well.

A majority of the denitrification in the world's oceans—both today and centuries ago—occurs in the sediments of the continental shelves. Although the sediments of the deep oceans cover a far greater area, they also receive much lower inputs of organic matter; the organic-rich sediments of the continental shelf provide a more favorable site for denitrification.

> denitrification in the geological past . . .

As noted by Falkowski (1997), the area of the continental shelves has varied over geological time; at the height of the last glaciation 22,000 years ago, sea levels were far lower since so much of the Earth's water was tied up in glacial ice. As a result, the area of the continental shelves was significantly less, and globally the rate of denitrification was probably less during this period of glaciation than during the past 10,000 years. The rate over the past 100 years has clearly increased, because human activity has created biologically available nitrogen at faster rates and much of it has flowed to coastal oceans.

Human acceleration of the nitrogen cycle

The global nitrogen cycle was forever changed in March 1909 when the German chemist Fritz Haber invented synthetic nitrogen fertilizer through a process that converts molecular N_2 to ammonia gas (Figure 12.23). Haber won the Nobel Prize for this discovery, which paved the way for the Green Revolution's huge increases in global food production over the past half-century (Chapter 14), greatly reducing world hunger and malnutrition. Remember that agro-ecosystems are very commonly limited by nitrogen (see Box 11.2). Before the invention of synthetic nitrogen fertilizer, most of the nitrogen needed to support agriculture came from nitrogen fixation by bacterial symbionts of crops. This was already a globally important source of nitrogen, with over 10 Tg of nitrogen per year fixed in 1900. Nitrogen fixation in agro-ecosystems then grew over the first half of the 20th century, but only slowly.

Then, particularly after the 1960s (Figure 12.23), synthetic nitrogen fertilizer became the dominant source of nitrogen for agriculture. Its use grew steadily over many decades, with only a slight hiccup in the late

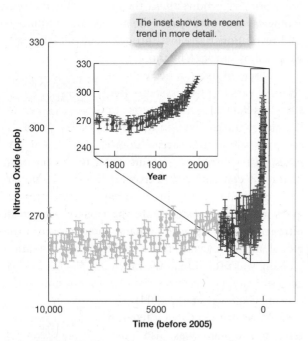

FIGURE 12.22 The concentration of nitrous oxide in the Earth's atmosphere was stable for most of the last 10,000 years but started to increase exponentially over the past 200 years. Compare with Figure 12.3 and Figure 12.16, which show comparable data for carbon dioxide and methane. (Reprinted from IPCC, 2007.)

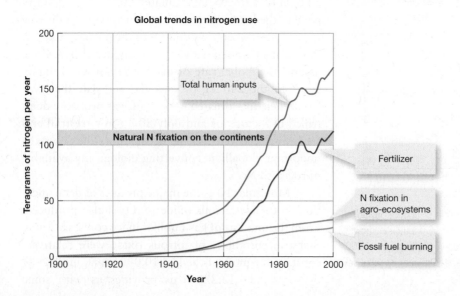

FIGURE 12.23 The rate of formation of new biologically available nitrogen on Earth since 1900 from combustion of fossil fuels, nitrogen fixation associated with agricultural crops, and manufacturing of synthetic fertilizer. Also shown is the total rate of formation from these three sources. For comparison, we also show the natural rate of nitrogen fixation in terrestrial biomes. (Howarth et al., 2005.)

1980s when the former Soviet Union collapsed and the production and use of fertilizer in Eastern Europe slowed dramatically. Global production and use quickly recovered, though, driven particularly by developing countries, which by 2007 were using more synthetic nitrogen fertilizer than the developed world (Figure 12.24). Use over the last 10 years has been particularly high in China, India, and Brazil, growing only slightly in the United States and falling in Europe. Today, 80% of the global nitrogen inputs for agriculture come from synthetic fertilizer, implying that on average 80% of the nitrogen in the proteins of the average human originate from this industrial process.

Some nitrogen is also converted from molecular N_2 to biologically available forms when fossil fuels are burned, although this is a less important process at the global scale. This conversion occurs because the molecular N_2 and oxygen that seem to coexist so peacefully in the atmosphere are in fact not chemically stable together. Kinetic barriers prevent them from reacting at the pressures and temperatures normally encountered on the Earth's surface, although the heat and energy from lightning blasts and volcanic eruptions do cause some formation of oxidized nitrogen

> the use of fossil fuels also fixes nitrogen, inadvertently

gases from reaction of oxygen and N_2. However, the temperatures and pressures created when fossil fuels are burned catalyze the reaction, producing nitrogen oxide gases that are released, for instance, in automobile exhaust. The creation of biologically available nitrogen from burning fossil fuels has risen steadily over the past 70 years, although it is still produced at a slower rate than nitrogen fixation by agricultural crops and far less than from synthetic fertilizer. These combustion-created nitrogen gases are a major contributor to acid rain and to the atmospheric deposition we have discussed in Chapters 1 and 11.

The rate of creation of biologically available nitrogen by human activity reached 120 Tg nitrogen per year—the amount created by all nitrogen fixation naturally in terrestrial biomes—back in the 1970s. If current trends continue, the total rate of natural fixation on the entire planet, both on land and in the oceans, will equal 240 Tg nitrogen per year by the early 2020s. Humans now dominate the nitrogen cycle, and although the use of fertilizer has helped improve human health and wellbeing immensely by reducing hunger and starvation, the cost to the environment—and to human health—may be high (Figure 12.25).

> human creation of nitrogen now exceeds the natural rate of nitrogen fixation on land . . .

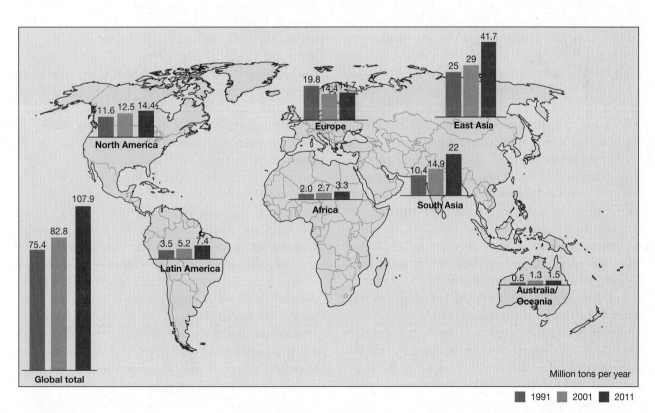

FIGURE 12.24   Use of synthetic nitrogen fertilizer by continent in 1991, 2001, and 2011 in millions of tons per year (data from the Food and Agricultural Organization).

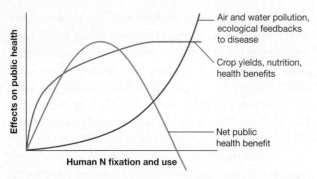

FIGURE 12.25 An illustration of how nitrogen use can affect human health. The blue line indicates a beneficial effect from using more fertilizer, with increasing agricultural productivity leading to less malnutrition. The red line indicates a rapid rate of increase in health-related problems as nitrogen use increases. The green line, which shows the net public health effect, is the sum of the other two lines. An optimal influence of human health occurs at an intermediate rate of use of nitrogen. (Townsend et al., 2003.)

The ecological and human-health costs of nitrogen

The huge increase in the rate of supply of biologically available nitrogen over the past half-century has led to numerous and diverse environmental problems including coastal eutrophication and dead zones (see Box 1.4) and loss of biotic diversity in both terrestrial and aquatic ecosystems (Figure 1.5). As we noted in Section 12.1, nitrogen is increasingly the dominant cause of acid rain in many regions, both because nitrogen pollution is growing and sulfur pollution is lessening in most developed countries. The acceleration of the nitrogen cycle has also led to increased fluxes of nitrous oxide, which is a potent greenhouse gas (see Figure 12.22). In fact, nitrous oxide is the third most important greenhouse gas after carbon dioxide and methane in the observed trends in global warming, and it is more potent per molecule in its global warming potential than either of these other two gases. In addition, nitrogen pollution contributes directly to human diseases, including cancer and heart disease, by catalyzing, along with methane and other hydrocarbon gases, the formation of toxic ozone in the lower atmosphere (which also damages forest and agricultural productivity), by forming into very fine particles, and by the reaction of nitrate with organic matter in the human gut to form potently carcinogenic compounds called nitrosamines. However, while the creation of more ozone in the lower atmosphere is harmful to us, nitrous oxide also contributes to the *destruction* of ozone in the upper atmosphere, the stratosphere, leading to "ozone holes," which allow excess ultraviolet radiation to reach the Earth's surface. These further effects on human health are taken up again in Section 14.3.

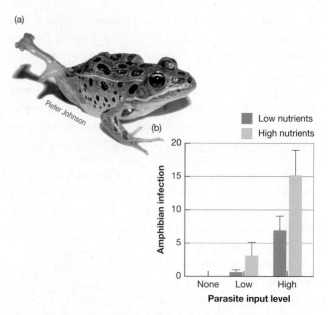

FIGURE 12.26 (a) Deformities in frogs and other amphibians are increasing globally. (b) In experimental mesocosms, the incidence of frogs with deformities increased as more parasites were added to the mesocosms and as more nutrients were added. The nutrients increased both the abundance of snails, the intermediate host of the parasite causing the deformities, and the number of parasites released per snail. (Johnson et al., 2010.)

Nitrogen pollution further affects the health of both animals and humans through indirect, ecologically mediated mechanisms. For instance, nitrogen deposition leads to increased pollen formation by many plants, increasing allergies. One of the best-studied effects of nitrogen pollution is on amphibian disease. Amphibian populations are in decline globally, in part due to an increase in deformities that reduce their fitness and survival (Figure 12.26a). Many of these deformities are caused by a parasite, a trematode (fluke worm) named *Ribeiroia ondatrae*, that has an intermediate snail host. That is, for part of its life cycle, the organism is a parasite of snails, and then later, a parasite of frogs and other amphibians. When snails and frogs are raised together in mesocosms, the frogs have no deformities if no parasites are present. When parasites are added, some deformities occur, and more occur when more parasites are added. The incidence of deformities increases further when nitrogen and phosphorus are added (Figure 12.26b). The mechanism is an increase in the parasite load in the snails, which are more productive in the fertilized mesocosms due to more algal production. *Ribeiroia ondatrae* is closely related to the parasite that causes schistosomiasis in humans, a deadly disease that also has a snail intermediate host. We can reasonably surmise that nitrogen pollution may also aggravate schistosomiasis.

Regional variation in nitrogen pollution

The acceleration of the nitrogen cycle is a global phenomenon, driven by global changes in agricultural practices (such as intensification of meat production and its segregation from sites of crop production, discussed in Chapter 14), energy use, food production, and other human activities. We have already noted the global effect of nitrous oxide gas on global warming and on the protective ozone layer in the stratosphere. However, the other effects of the change in nitrogen cycling are far from uniform across the planet.

Nitrogen pollution is much greater in some regions than in others: the hot spots include most of Europe, India, eastern China, the eastern portion of the United States, and southern Brazil (Figures 12.27a). These correspond to the major agricultural areas of the world, and are the regions seeing the greatest use of nitrogen fertilizer (Figure 12.24) and experiencing the highest rates of atmospheric deposition of nitrogen (Figure 12.27b). These areas of intense human use of nitrogen are generally not the regions with the greatest rates of natural nitrogen fixation. So while human activity has roughly doubled the rate of creation of nitrogen on the Earth, the rate of change is quite small in some areas such as the Amazon, but massive in regions such as Europe where the human use of nitrogen is high and the rate of natural nitrogen fixation is low. The regional variation in nitrogen fluxes and pollution is well illustrated by the amount of nitrogen that leaves the landscape and flows to estuaries and coastal oceans, leading to dead zones and other aspects of eutrophication (Box 12.3). Per area of watershed, these nitrogen fluxes vary across the planet from less than 100 to more than 5,000 kg per square kilometer per year. The fluxes are highly correlated with the rate of input of nitrogen to the landscape by human activity.

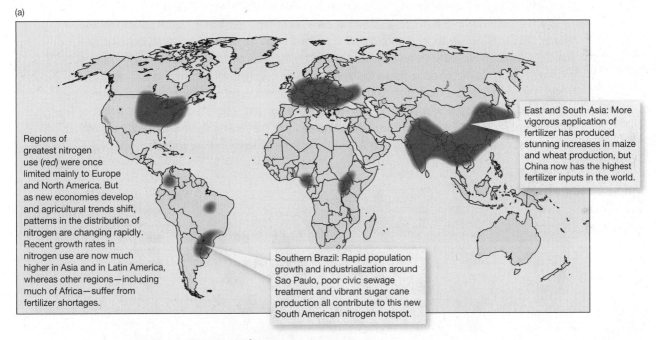

(a)

Regions of greatest nitrogen use (*red*) were once limited mainly to Europe and North America. But as new economies develop and agricultural trends shift, patterns in the distribution of nitrogen are changing rapidly. Recent growth rates in nitrogen use are now much higher in Asia and in Latin America, whereas other regions—including much of Africa—suffer from fertilizer shortages.

East and South Asia: More vigorous application of fertilizer has produced stunning increases in maize and wheat production, but China now has the highest fertilizer inputs in the world.

Southern Brazil: Rapid population growth and industrialization around Sao Paulo, poor civic sewage treatment and vibrant sugar cane production all contribute to this new South American nitrogen hotspot.

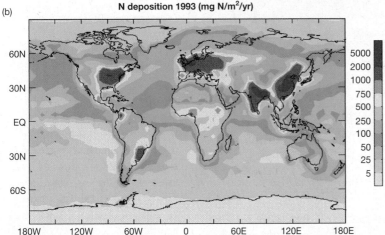

(b)

N deposition 1993 (mg N/m²/yr)

FIGURE 12.27   (a) The global distribution of hot spots of nitrogen pollution, from all sources. (Townsend & Howarth, 2010.) (b) The global pattern for deposition of nitrogen from the atmosphere onto the oceans and terrestrial landscape as of the mid 1990s. (Galloway et al., 2004.)

Quantitative Aspects

Nitrogen inputs to the landscape control nitrogen fluxes in rivers

In recent decades, human activity has dramatically increased nitrogen inputs to coastal marine ecosystems, with often devastating consequences (Box 1.4). This nitrogen pollution comes from a variety of sources in the landscape. Here we present a relatively simple tool for assessing these sources and determining how they relate to the flow of nitrogen in rivers to the coast. The approach is based on comparing the total mass of nitrogen inputs from all human-controlled sources to large watersheds with the measured mass flows of nitrogen in rivers, both normalized to the area of the watershed. Perhaps surprisingly, robust sources of data on nitrogen inputs such as fertilizer use and atmospheric deposition (see Section 11.8) are more reliable at larger spatial scales, such as a large river basin, than at the scale of a small forested catchment or plot of agricultural field.

Despite all the complex ecological processes that occur in terrestrial ecosystems and rivers – and despite the diversity of underlying soils and geologies, climates, and ways that humans interact with the landscape across regions—we can see a very strong relationship between the amount of nitrogen that humans put into the landscape and the amount that flows down rivers to coastal waters. This speaks strongly to the huge influence humans are having on the nitrogen cycle at regional scales.

The inputs of nitrogen from human activity at the scale of large river basins are the use of synthetic fertilizer, nitrogen fixation by bacterial symbionts of agricultural crops, atmospheric deposition that originates from fossil fuel combustion, and the net movement of nitrogen into or out of the river basin in food for humans and feeds for animals. We can obtain values for these inputs from governmental data, models of nitrogen flow through the atmosphere, and information about what humans and animals eat and how much nitrogen fixation is associated with particular crops. When we sum the inputs, we obtain an estimate of the total net anthropogenic nitrogen inputs (NANI) to a river basin or region. In many cases, there is a net export of food and feeds, and we treat these as negative input. Figure 12.28a illustrates what these net inputs look like across the United States.

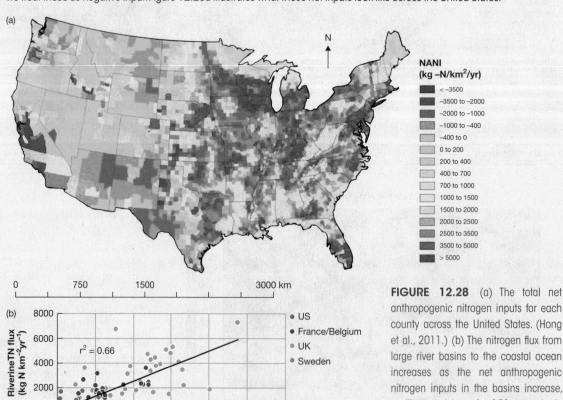

FIGURE 12.28 (a) The total net anthropogenic nitrogen inputs for each county across the United States. (Hong et al., 2011.) (b) The nitrogen flux from large river basins to the coastal ocean increases as the net anthropogenic nitrogen inputs in the basins increase, as illustrated here for 150 watersheds in Europe and North America. (Howarth et al., 2012.)

When the NANI sum is plotted against the average flux of nitrogen in rivers to the coast for 150 watersheds in Europe and the United States, a strong linear relationship results (Figure 12.28b). This relationship is highly significant, and the NANI term alone explains a high proportion of the variation in riverine nitrogen fluxes across these regions. Note that both the inputs and the riverine fluxes are expressed per area of watershed. The total mass flux of nitrogen to the coast is a function of both this area-specific flux and the area of the watershed. The Mississippi River exports a huge amount of nitrogen to the northern Gulf of Mexico, creating the 'dead zone' there. This arises from a moderately high nitrogen flux per area combined with a very large river basin consisting of more than 40% of the entire area of the lower 48 states of the United States.

In many estuaries, sewage is the major input of nitrogen. Why is sewage not included in this NANI-based analysis? In effect, it is. Sewage is part of the internal workings of NANI, with the nitrogen in sewage coming from the food humans eat, and this in turn originates with food grown in the river basin (based on fertilizer inputs and agricultural nitrogen fixation) or imported to the region. We can think of the sewage as a recycling of nitrogen. Double accounting would result if we considered it a separate input of nitrogen to the river basin.

On average, for the 150 watersheds shown in Figure 12.28b, 25% of the human-controlled nitrogen inputs (NANI) are exported to the coast in rivers. Some of the rest accumulates in the biomass and soil of the terrestrial ecosystems in the basins, but most is probably denitrified. Much of the variation in the regression shown in this figure is related to climatic variability. In river basins with more freshwater flowing out as a result of greater precipitation, a higher percentage of the nitrogen inputs are exported to the ocean. This suggests that less denitrification occurs in these wetter river basins. Given that denitrification occurs in wet places such as wetlands, seasonally flooded soils, and streams, this finding might surprise you. The likely answer is that in regions with more precipitation and greater flows, the water and associated nitrogen may pass through wetlands and streams so quickly that bacteria are less able to denitrify as much of it.

How can we reduce nitrogen pollution?

Many technical solutions exist for reducing nitrogen pollution. In fact, atmospheric nitrogen pollution in both Europe and the United States has decreased substantially over the past 20 years as power plants and vehicles have been required to use technologies that reduce their emissions of oxidized nitrogen gases. On the other hand, such pollution has increased in the developing world as energy use has increased. In many western cities, the use of nitrogen-removing technologies in sewage treatment facilities has greatly reduced water pollution. So far, efforts to reduce nitrogen pollution from large-scale agriculture have been less successful, although there are many proven approaches for doing so, as we will explore in Chapter 14. This suggests the need for better policy approaches for encouraging ways to reduce agricultural pollution.

There may be some reason to hope that improved lifestyles may reduce nitrogen pollution. For instance, if the average U.S. adult were to reduce consumption of meat to that of the average resident of southern Europe as of the 1980s, total fertilizer use would fall dramatically, since meat production and the crop production necessary to support it would be far less (Figure 12.29). Overconsumption of food, and particularly of meat, is increasingly recognized as a major health issue in many developed countries. An improvement in diet would be good for us individually. It could also help reduce nitrogen pollution from agriculture.

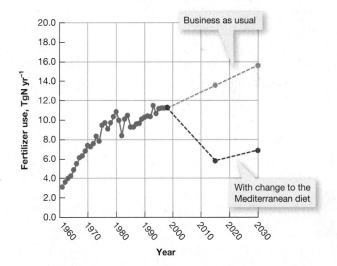

FIGURE 12.29   The total amount of synthetic nitrogen fertilizer used in the United States between 1960 and 2000, and two possible scenarios for future use into the future, presented in 2002. The top scenario shows 'business as usual,' with agricultural practices, grain exports from the United States, population growth, and diet remaining unchanged; this results in a large increase in fertilizer use, to support the growing population. The bottom scenario is the same, except it assumes a major shift in the diet of the average U.S. adult to the diet consumed by the average southern European as of the 1980s (the 'Mediterranean' diet). Far less fertilizer is used, since less meat is produced, and less crop production is required. (Modified from Howarth et al., 2002.)

SUMMARY

Human acceleration of global biogeochemical cycles

Human activity has greatly altered many biogeochemical cycles at the global scale, including those of phosphorus, carbon dioxide, methane, and nitrogen. The cycles of carbon dioxide, methane, and nitrogen have more ecologically important feedbacks than for phosphorus, and so we focus on those cycles in this chapter.

Global carbon dioxide cycle

Human activity is leading to a global warming of the planet. Carbon dioxide is the major greenhouse gas behind this warming, and concentrations are now at their highest levels during at least the last 400,000 years. Combustion of fossil fuels, which is the major source of atmospheric carbon dioxide, has increased exponentially over the past century. Deforestation also contributes to atmospheric carbon dioxide. Terrestrial biomes and the oceans both provide sinks for some atmospheric carbon dioxide, creating somewhat of a buffer against the increase from human activity. However, these sinks may be temporary.

Global methane cycle

Methane also contributes significantly to global warming. The concentration of methane in the atmosphere was relatively low until the Industrial Revolution but has risen exponentially over the past century. Expressed as a percentage change, the rate of methane increase is even greater than that for carbon dioxide. The climate system responds more rapidly to methane than to carbon dioxide, so immediate control of methane emissions provides one of the best ways to slow global warming and avoid possible tipping points in the climate over the coming decades. Methane is formed by bacterial decomposition of organic matter in the absence of oxygen, and the formation in wetlands and lakes from this mechanism is a large, natural source of methane. Humans have increased methane fluxes to the atmosphere by increasing sites for this bacterial formation of methane in the guts of livestock, in rice paddies, and in landfills. Humans have also increased the flux of methane to the atmosphere from fossil sources as natural gas, oil, and coal are developed for fuel.

Global nitrogen cycle

The rate of change in the global nitrogen cycle from human activity is even faster than that for the global carbon dioxide and methane cycles. Before the Industrial Revolution, the global nitrogen cycle was a balance between the bacterial fixation of nitrogen in the oceans and in terrestrial ecosystems and the bacterial conversion of nitrogen forms back to molecular N_2 through denitrification. The major driver of the acceleration of the nitrogen cycle is the intensification of agriculture based on synthetic nitrogen fertilizer, the use of which has increased exponentially over the past 50 years. The combustion of fossil fuels also contributes to nitrogen pollution. The human creation of biologically available nitrogen equaled the natural rate of nitrogen fixation in terrestrial biomes in the 1970s and is now approaching a rate equal to all natural nitrogen fixation on the planet, including that in the oceans. Nitrous oxide, the third most important greenhouse gas behind current global warming, has increased in the atmosphere as nitrogen use has increased. Nitrogen pollution has many other effects—both ecological and human health related—but these effects tend to play out at local to regional scales.

REVIEW QUESTIONS

1 How does deforestation result in more atmospheric carbon dioxide?

2 Why might the assimilation and storage of carbon dioxide in forests change in the future?

3 Why is the oceanic sink of atmospheric carbon dioxide described as a temporary one?

4 Why is so much methane released to the atmosphere from wetlands, and so little from the vastly larger oceans?

5 How can scientists distinguish between fossil methane and methane produced by bacteria in recent times?

6 How are methane and carbon dioxide similar as greenhouse gases? How are they different?

7   Explain how the global balance of nitrogen fixation and deni-
    trification may have changed over the past 50,000 years.

8   Compare and contrast the spatial patterns of human use of
    nitrogen over the planet with that for the natural process
    of nitrogen fixation.

9   Explain how nutrient pollution can increase the incidence of
    a disease such as schistosomiasis.

Challenge Questions

1   Why is the seasonal oscillation of the concentration of carbon
    dioxide controlled more by ecological processes in terrestrial
    biomes than by the uptake and release by the oceans?

2   Is it reasonable to assume that the natural flux of methane
    to the atmosphere has remained constant over hundreds
    and thousands of years? Why or why not?

3   On balance, was the invention of synthetic nitrogen fertilizer
    a good thing or a bad thing?

Peter Mcbride/Getty Images

Conservation ecology

CHAPTER CONTENTS

Chapter 13

KEY CONCEPTS

After reading this chapter you will be able to:

- explain what biodiversity is and why we wish to conserve it

- describe the demographic and genetic problems that rare species face because their populations are small

- identify the main threats currently facing biodiversity

- describe a range of approaches that can be taken to conserve individual species, communities, and habitats

- explain the purpose and value of an ecosystem services approach to conservation

onservation is the science concerned with increasing the probability that the Earth's species and communities (or, more generally, its biodiversity) will persist into the future. We need to appreciate the scale of the problem, understand the threats posed by human activities, and consider how our knowledge of ecology can be brought to bear to provide remedies.

13.1 THE NEED FOR CONSERVATION

We have discussed the processes by which new species are generated (Section 2.4) and by which they may migrate to new areas (Sections 5.4 and 10.4). We also know that species may become extinct, either locally or globally. At present, certainly globally and very often locally, rates of extinction greatly exceed those of species creation or immigration (see Figure 13.1). We are losing species, and as a consequence, we are witnessing declines in biodiversity. **Conservation** is the collective name we give to the various actions we can take to slow down or even reverse these losses of species and of biodiversity.

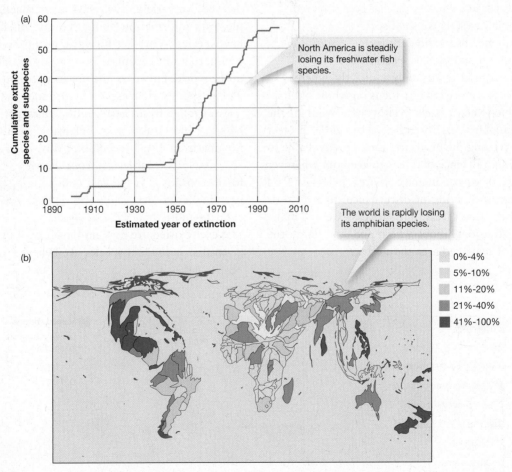

FIGURE 13.1 (a) The cumulative loss of species (39) and subspecies (18) of freshwater fish in North America since 1900. (After Burkhead, 2012.) (b) A map of the world in which countries are distorted in proportion to the number of amphibian species living there, relative to the country's size. Then, countries are coded according to the percentage of amphibian species threatened with extinction (either 'vulnerable,' 'endangered,' or 'critically endangered'—see Figure 13.4). (After Wake & Vredenburg, 2008.)

The term *biodiversity*, which we have mentioned throughout this book, also makes frequent appearances in both the popular media and the scientific literature. But what precisely does it mean? At its simplest, it means *species richness*, the number of species present in a defined geographic unit (see Chapter 10). But we can also view biodiversity at scales smaller and larger than the species. For example, we may include genetic diversity within species, perhaps seeking to conserve genetically distinct subpopulations and subspecies (see Chapter 8). Above the species level, we may wish to ensure that species without close relatives are afforded special protection, so that the overall evolutionary variety of the world's biota is maintained as large as possible. At a larger scale still, we may include in biodiversity the variety of community types present in a region—swamps, deserts, early and late stages in a woodland succession, and so on. Thus, biodiversity may itself have a diversity of meanings. Yet it is necessary to be specific if the term is to be of any practical use. Ecologists must define precisely what it is they mean to conserve in their particular circumstances, and how to measure whether this has been achieved.

> what is biodiversity?

In some cases, as we shall see in this chapter, conservation efforts are focused on individual species and hence on biodiversity only indirectly, whereas in others there is a more explicit focus on whole habitats and so on biodiversity itself. In either case, what are the options available to us? Broadly, we can either protect or restore. Among the protective measures available to us, we can simply fence off areas to keep out whatever the threat is, or we can manage an area (protect a food plant, eliminate hunting or disturbance) to reduce or even eliminate that threat. In either case we will make important decisions about where and how large the fenced-off or managed area should be.

> what is conservation?

Alternatively, we may relocate individuals from an area where the species is thriving (relatively speaking) to one where it is now extinct, or breed individuals in an artificial setting before reintroducing them, or even re-create whole ecosystems, especially in areas where all or most of the habitat has been destroyed (for example by logging or mining). We can think of these cases of restoration as conservation in a broader sense, since the aim is to reverse a past failure to conserve. We will meet examples of all these interventions later in the chapter.

Conservation biologists, then, are concerned with the loss of species in the face of human influence. To judge the scale of the problem, though, we need to know the total number of species currently in the world, the rate at which these are going extinct, and how this rate compares with that of pre-human times. Unfortunately, there are considerable uncertainties in our estimates of all three. About 1.8 million species have so far been named (Figure 13.2), but the real number must be much larger. Estimates have been derived in a variety of ways. One approach, for example, uses information about the rate of discovery of new species to project forward, group by taxonomic group, to a total estimate of up to 6–7 million species in the world. However, the uncertainties in estimating global species richness are profound, and our best guesses range from 5–10 million eukaryotes, plus the bacteria and viruses, with defensible numbers overall ranging from 3 to 100 million (May, 2010).

> how many species on Earth?

An important lesson from the fossil record is that the vast majority of species eventually become extinct—more than 99% of species that ever existed are now no longer with us. However, given that individual species are believed, on average,

> modern and historical extinction rates

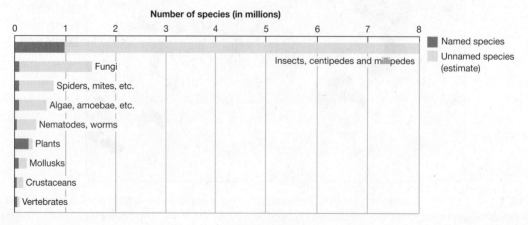

FIGURE 13.2 Numbers of species identified and named (blue histograms) and estimates of unnamed species that exist (green histograms). (Modified from Millennium Ecosystem Assessment, 2005.)

to have lasted about 1–10 million years, and if we estimate conservatively that the total number of species on Earth is 10 million, then we would predict that on average, only between 100 and 1,000 species (0.001–0.01%) would go extinct each century. The current observed rate of extinction of birds and mammals is about 1% per century, 100–1,000 times this 'natural' background rate (see also Figure 13.1).

The evidence, then, while based only on estimates, suggests that our children and grandchildren may live through a period of species extinction comparable to the 'natural' mass extinctions evident in the geological record, when, for example, many whole families of shallow-water invertebrates were lost towards the end of the Permian period, around 250 million years ago, and some estimate that around 96% of the species then living went extinct. But should we care? To most, the answer is an unhesitating Yes. It is nonetheless important to consider *why* we should care, that is, *why* biodiversity is valuable (Box 13.1), before we look more closely at the threats that confront us and how we may counter them.

13.1   ECOncerns

What is the value of biodiversity?

To most people, biological diversity is undeniably of value, but standard economics has generally failed to assign tangible value to ecological resources. Thus, the costs of environmental damage and the depletion of living resources have frequently been disregarded. A major challenge is the development of a new *ecological economics* (Daly & Farley, 2011) in which the worth of species, communities, and ecosystems can be assigned financial value to be set against the gains to be made in industrial and other human projects that may damage them. As we shall see in Section 13.5, for example, the value of biodiversity can be measured in terms of the *ecosystem services* it provides, and indeed, payments can be made to landowners to encourage them to continue to provide these (Figure 13.3).

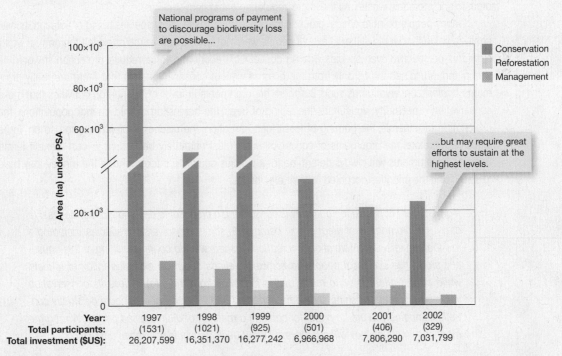

FIGURE 13.3   The economics of Costa Rica's Payment for Environmental Services program (*pagos por servicios ambientales –* PSA) in the years following its implementation. (After Sanchez-Azofeifa et al., 2007.)

Many species have direct value. Many more are likely to have a potential value that as yet remains untapped. For example, wild meat, fish, and plants remain vital resources in many parts of the world, while most of the world's food is derived from plants that were originally domesticated from wild plants in tropical and semiarid regions. In the future, wild strains of these species may be exploited for their genetic diversity, and quite different species of plants and animals may be found that are appropriate for domestication.

Further, as we shall see in Section 14.5, the potential benefits that might come from natural enemies if they could be used as biological control agents for pest species are enormous; most natural enemies of most pests remain unstudied and often unrecognized. Finally, about 40% of the prescription and nonprescription drugs used throughout the world have active ingredients extracted from plants and animals. Aspirin, probably the world's most widely used drug, was derived originally from the leaves of the tropical willow, *Salix alba*. The nine-banded armadillo (*Dasypus novemcinctus*) has been used to study leprosy and prepare a vaccine for the disease; the Florida manatee (*Trichechus manatus*), an endangered mammal, is being used to help understand hemophilia; while the rose periwinkle (*Catharanthus roseus*), a plant from Madagascar, has yielded two potent drugs effective in treating blood cancer. In all these cases, the species can be thought of as representing 'provisioning ecosystem services' (see Section 13.5).

Other species have indirect economic value. More ingenuity is required to find ways to measure this value. For example, many wild insects are responsible for pollinating crop plants. This is another provisioning service, and one that we are getting less benefit (value) from as many countries report a catastrophic decline in bees. In a different context, the monetary value of ecotourism, which depends on biodiversity, is becoming ever more considerable. Each year, around 200 million U.S. adults and children take part in nature recreation and spend about $4 billion on fees, travel, lodging, food, and equipment. Ecotourists, who visit a country wholly or partly to experience its biological diversity, spend approximately $12 billion a year worldwide on their enjoyment of the natural world (Primack, 1993).

On a smaller scale, a multitude of natural history films, books, and educational programs are consumed annually without harming the wildlife upon which they are based. In these contexts, biodiversity provides 'cultural ecosystem services.' Furthermore, biological communities can be of vital importance by maintaining the chemical quality of natural waters, buffering ecosystems against floods and droughts, protecting and maintaining soils, regulating local and even global climate, and breaking down or immobilizing organic and inorganic wastes. All these are 'regulating ecosystem services.'

Many people point to ethical grounds for conservation, with every species being of value in its own right—a value that would still exist even if people were not here to appreciate or exploit the natural world. From this perspective even species with no conceivable economic value require protection. It would be wrong, though, to see things only from the point of view of conservation. Not that there are really arguments *against* conservation as such. But there are arguments in favor of the human activities that make conservation a necessity: agriculture, the felling of trees, the harvesting of wild animal populations, the exploitation of minerals, the burning of fossil fuels, irrigation, the discharge of wastes, and so on. To be effective, therefore, the arguments of conservationists must ultimately be framed in cost–benefit terms, because governments will always determine their policies against a background of the money they have to spend and the priorities accepted by their electorates.

A government conservation authority is considering a proposal to designate a marine reserve at a rocky promontory of great scenic beauty. The site is very diverse in species, including a few that are rare. Commercial and recreational fishers wish to continue fishing at this unusually productive site, local people have mixed feelings about an expected influx of tourists, while conservationists (who mostly live a long way from the site) believe its conservation value is such that no fishing should be permitted and visitor numbers should be strictly controlled. Imagine you are an arbitrator chairing a meeting of all interested parties. What arguments do you think they will put forward? What decision would you reach and why?

13.2   SMALL POPULATIONS

The classification of risk

When conservation biologists are focused on individual species, those species are almost inevitably rare, and that rarity contributes to the risk of extinction. But how do we define the risk that a species faces? We can describe a species as:

- **critically endangered** if there is considered to be more than a 50% probability of extinction in 10 years or three generations, whichever is longer (Figure 13.4);

- **endangered** if there is more than a 20% chance of extinction in 20 years or five generations;

- **vulnerable** if there is a greater than 10% chance of extinction in 100 years;

- **near threatened** if the species is close to qualifying for a threat category or judged likely to qualify in the near future;

- or **of least concern** if it does not meet any of these threat categories (Rodrigues et al., 2006).

Based on the above criteria, for example, 12% of bird species, 20% of mammals, and 32% of amphibians are threatened with extinction (being critically endangered, endangered or vulnerable; Rodrigues et al.,

2006). We deal with some of the most important of these threats next.

Demographic risks associated with small populations

Much of conservation biology is a crisis discipline. For example, the residual population of giant pandas in China (or yellow-eyed penguins in New Zealand or spotted owls in North America) has become so small that if nothing is done, the species may become extinct within a few years or decades. There is thus a pressing need to understand the dynamics of small populations. These are governed by a high level of uncertainty, in contrast to those of large populations, which can be described as being governed by the law of averages (Caughley, 1994). Two different kinds of uncertainty are of particular importance.

1 *Demographic uncertainty*. Random variations in the number of individuals that are born male or female, or in the number that happen to die or reproduce in a given year, or in the genetic quality of the individuals in terms of survival/reproductive capacities can matter very much to the fate of small populations. Suppose a breeding pair produces a clutch consisting entirely of females. Such an event would go unnoticed in a large population but could be the last straw for a species down to its last pair.

2 *Environmental uncertainty*. Unpredictable changes in environmental factors, whether disasters (such as floods, storms, or droughts of a magnitude that occurs very rarely) or relatively minor events (year-to-year variation in average temperature or rainfall), can also seal the fate of a small population. A small population is more likely than a large one to be reduced by adverse conditions to zero or to numbers so low that recovery is impossible. And of course global climate change may greatly aggravate this influence.

To illustrate some of these ideas, take the demise in North America of the heath hen (*Tympanuchus cupido cupido*). This bird, once extremely common from Maine to Virginia, was highly edible, easy to shoot, susceptible to introduced cats, and affected by conversion of its grassland habitat to farmland. It is perhaps not surprising then that by 1830 it had disappeared from the mainland and was found only on the island of Martha's Vineyard. In 1908 a reserve was established for the remaining 50 birds, and by 1915 the population had increased to several thousand. But 1916 was a bad year. Fire (a disaster) eliminated much of the breeding ground, there was a particularly hard winter coupled with an influx of goshawks (environmental uncertainty), and finally poultry disease arrived

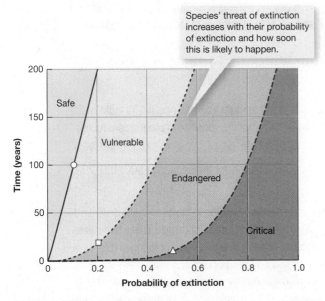

FIGURE 13.4   Levels of threat as a function of time and probability of extinction. The circle represents a 10% probability (0.1) of extinction in 100 years (minimum criterion for a population to be designated 'vulnerable'). The square represents a 20% probability of extinction in 20 years (minimum criterion for the designation 'endangered'). The triangle represents a 50% probability of extinction in 10 years (minimum criterion for the designation 'critically endangered'). (After Akçakaya, 1992.)

The heath hen

(c) Tom McHugh/Science Source

on the scene (another disaster). At this point, the remnant population was likely to have become subject to demographic uncertainty. For example, of the 13 birds remaining in 1928, only two were females. A single bird was left in 1930, and the species went extinct in 1932.

Genetic problems in small populations

Further problems may arise in small populations through loss of genetic variation (Box 13.2). The preservation of genetic diversity is important in the first place because of the long-term evolutionary potential it provides. Rare forms (*alleles*) of a gene, or combinations of alleles, may confer no immediate advantage but could turn out to be well suited to changed environmental

---

| 13.2 | **Quantitative Aspects** |

What determines genetic variation?

Genetic variation is determined primarily by the joint action of natural selection and genetic drift (where the frequency of genes in a population is determined by chance rather than evolutionary advantage). The relative importance of genetic drift is higher in small isolated populations that, as a consequence, are expected to lose genetic variation. The rate at which this happens depends on the **effective population size** (N_e), rather than simply on the number of individuals present (N). Effective population size is the size of the ideal population to which the actual population is equivalent in genetic terms, where 'ideal' means the sex ratio is 1:1, the distribution of numbers of offspring among parents is random, and the population size remains constant. N_e is usually less, often much less, than N, for a number of reasons [detailed formulas can be found in Lande and Barrowclough (1987)]:

1. If the sex ratio is not 1:1. For instance, with 100 breeding males and 400 breeding females, $N = 500$ but $N_e = 320$.

2. If the distribution of progeny from individual to individual is not random. For instance, if 500 individuals each produce one individual for the next generation on average but the variance in progeny production is five (with random variation this would be one), then $N_e = 100$.

3. If population size varies from generation to generation. For instance, for the sequence 500, 100, 200, 900, 800, mean $N = 500$ but $N_e = 258$.

Greater prairie chickens (*Tympanuchus cupido pinnatus*), closely related to the heath hens described above, provide a good example of how genetic diversity may be related to population size. These birds were once widespread throughout the prairies of North America, but with the loss and fragmentation of this habitat, many populations have become small and isolated. Prairie chicken populations were once linked by the gene flow provided by migrants, keeping genetic diversity high. But the current populations are not only small but also isolated in their habitat fragments. Johnson et al. (2003) used molecular biology techniques (see Section 8.2) to measure genetic diversity in prairie chicken populations both large (from 1,000 to more than 100,000 individuals) and small (fewer than 1,000 individuals). The mean number of alleles per gene was 7.7–10.3 in the large populations but only 5.1–7.0 in the small. Franklin and Frankham (1998) had previously suggested that an effective population size of 500–1,000 might be needed to maintain longer-term evolutionary potential.

conditions in the future. Small populations tend to have less variation and hence lower evolutionary potential.

A more immediate potential problem arises because when populations are small, related individuals tend to breed with one another. All populations carry recessive alleles that can be harmful, even lethal, to individuals when homozygous (when the alleles provided by the mother and father are identical). Individuals that breed with close relatives are more likely to produce offspring that receive harmful alleles from both parents. The deleterious effects that result are known as **inbreeding depression**. There are many examples of inbreeding depression—breeders of domesticated animals and plants, for example, have long been aware of reductions in fertility, survivorship, growth rates, and resistance to disease, and they commonly seek to prevent close relatives breeding with one another.

One study that demonstrated the importance of genetic effects for population persistence in rare species examined 23 local populations of the rare plant *Gentianella germanica* in grasslands in the Jura Mountains on the Swiss–German border. There was a negative correlation between reproductive performance and population size (Figure 13.5a–c). Seeds taken from small populations and grown in a common

garden produced fewer flowers than seeds from large populations grown under identical conditions. Furthermore, population size decreased between 1993 and 1995 in most of the studied populations, but it decreased most rapidly in the smallest populations (Figure 13.5d).

It is, though, possible to reduce and perhaps even solve the genetic problems in small populations. The pink pigeon (*Columba mayeri*) provides a good example. Once widespread on the island of Mauritius, it had declined to a population of only 9 or 10 birds by 1990. Other animals had been maintained in captivity, but that captive population was originally descended from just 11 founder individuals, augmented in 1989–1994 by adding 12 more (offspring of the remaining wild individuals). The aim, therefore, was to manage matings in captivity, so as to retain high levels of genetic diversity and minimize inbreeding, and then release offspring back into the wild. Between 1987 and 1997, scientists reintroduced 256 birds on Mauritius, wherever possible selecting birds with minimal inbreeding (based on records in breeding 'stud books') and releasing them in groups with good representation of the different founder ancestries. As a result, the population

> recovery of the pink pigeon

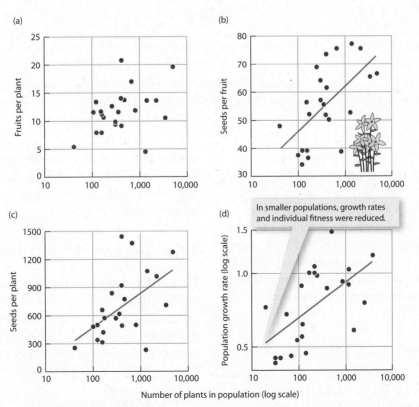

FIGURE 13.5 Relationships for 23 populations of *Gentianella germanica* between population size and (a) mean number of fruits per plant; (b) mean number of seeds per fruit and (c) mean number of seeds per plant; (d) the relationship between population growth rate from 1993 to 1995 (ratio of population sizes) and population size (in 1994). All regression lines are significant at $P < 0.05$; no line is shown in (a) because the regression is not significant. (From Fischer & Matthies, 1998.)

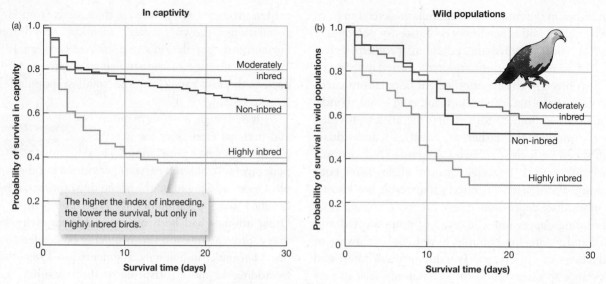

FIGURE 13.6   Effect of inbreeding on probability of survival to 30 days of age of pink pigeon nestlings (a) in captivity and (b) in the wild population. Inbreeding is expressed as an index derived from known ancestry in relationship to 23 founder individuals. (After Swinnerton et al., 2004.)

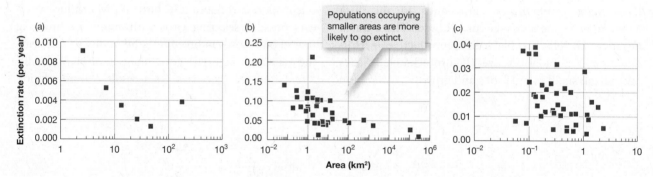

FIGURE 13.7   Percentage extinction rates as a function of habitat area for (a) zooplankton in lakes in northeastern USA, (b) birds on northern European islands, and (c) vascular plants in southern Sweden. (Data assembled by Pimm, 1991.)

had swelled to 355 free-living individuals by 2003 (plus more in captivity) and has remained at or above that level since.

Preoccupation with avoiding inbreeding has been justified by studies in which the genetics and ecological success of both captive and wild pink pigeon populations have been carefully monitored. Inbreeding reduced egg fertility and survival of nestlings (Figure 13.6), but effects were strongly marked only in the most inbred birds.

Habitat reduction

Both demographic and genetic factors, then, can make populations occupying smaller habitat patches more vulnerable to extinction. Figure 13.7 shows the negative relationships for a variety of taxa between annual extinction rate and area. No doubt the main reason for the vulnerability of populations in small areas is the fact that the populations themselves are small. This is illustrated in Figure 13.8 for bird species on islands and for bighorn sheep in various desert areas in the southwestern United States.

13.3 THREATS TO BIODIVERSITY

Clearly, populations on the way to extinction are small, whether they travel that road all the way to its destination or are stalled or even turned back. But what sets a population on this path? We turn next to some of the more important factors that threaten the future of species populations and hence of biodiversity more generally.

Overexploitation

The essence of overexploitation is that populations are harvested at a rate that is unsustainable, given their natural rates of mortality and capacities for reproduction.

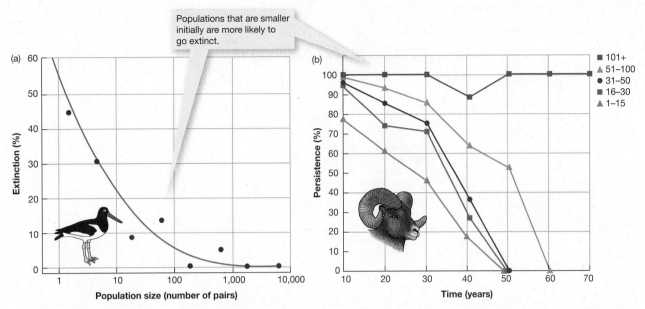

FIGURE 13.8   (a) The extinction rate of island birds is higher for small populations. The regression line is statistically significant. (b) The percentage of populations of bighorn sheep in North America that persists over a 70-year period for a range of initial population sizes. (After Berger, 1990.)

We shall examine the phenomenon in more detail in Section 14.7, but the basic idea is easy to grasp. In prehistoric times, for example, it seems likely that humans were responsible for the extinction of many large animals, the so-called megaherbivores, by overhunting them—removing them at a faster rate than the animals themselves could counter with their own reproduction. In much the same way, in more recent times, populations of the great whales and many commercial fish species are in danger of following a similar path.

In many cases, the endangered species is being exploited for food, often as a staple of the diet. But often the 'food' value of the species is exotic and imaginary (for instance, the supposed medicinal powers of rhino horn in some Asian cultures), or the species are exploited for 'aesthetic' reasons, whether for their body parts or as exotic pets. In these cases, their value to collectors goes up as they become rarer. Thus, instead of the normal safeguard of a density-dependent reduction in consumption rate at low density (see Section 7.5), increased rarity may actually increase the rate of exploitation, unless hunting and other forms of removal are strictly regulated.

Elephant ivory is a good example. African elephants have been exploited for ivory for many years, and as they have become rarer, the resource itself has become more valuable. For example, the estimated numbers of elephants in Africa was 1.3 million in 1979, reduced to 600,000 by 1989, whereas the value of uncarved ivory in Kenya was $2.5 a pound in 1969, $34 a pound in 1978, and more than $90 a pound in 1989 (Lemieux & Clarke, 2009). Between 1979 and 1989, the two countries suffering the greatest reduction in elephant numbers were the Democratic Republic of Congo and Tanzania (Figure 13.9a). The consequent rarity inevitably increased the value of what remained, so conservation of elephants would have been possible only with well-resourced and well-organized control of elephant hunting (poaching).

From the 1990s to 2007, Congo suffered more or less constant civil conflict, had one of the highest corruption perception indexes in Africa, and supported an unregulated ivory market. With no way to control elephant poaching, the increasing value of ivory drove elephant numbers still lower (Figure 13.9b). By contrast, Tanzania had no civil conflict, a low level of corruption, and no unregulated ivory market. It managed to jump from the second-largest fall to the second-largest rise in elephant numbers in the period leading to 2007 (Figure 13.9b). The rarity-value vicious circle is a problem, but one that can apparently be overcome.

Habitat disruption

Habitats can be adversely affected by human influence in three main ways. First, a proportion of the habitat available to a particular species may simply be *destroyed*, for urban and industrial development, or for the production of food and other natural resources such as timber. Of these, forest clearance has been, and still is, the most pervasive of the forces of habitat

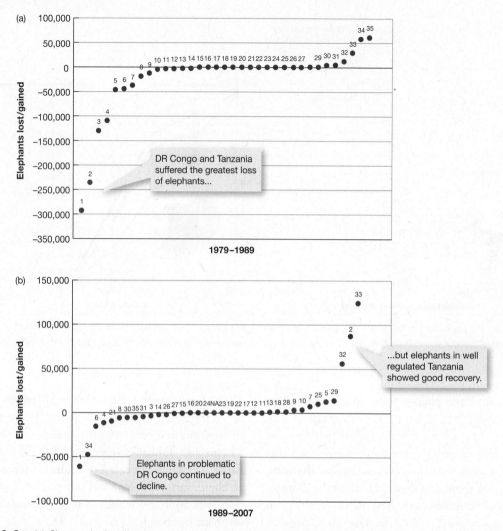

FIGURE 13.9 (a) Changes in the abundances of elephants in African countries for which data were available, 1979–1989, with countries ranked from greatest fall to greatest rise. '1' is the Democratic Republic of Congo; '2' is Tanzania. (b) Corresponding changes for 1989–2007, but with rank numbers retained from (a). (After Lemieux & Clarke, 2009.)

destruction (Figure 13.10a,b). Much of the native temperate forest in the developed world was destroyed long ago, but current rates of deforestation remain high in all forest biomes (Figure 13.10c), though it is likely that more of this forest loss currently is natural (for example, forest fires) in temperate zones, whereas more is human-induced in tropical zones. The process of habitat destruction also often makes the habitat available to a particular species more fragmented than in the past, with potentially serious effects for the populations concerned (see Section 13.4).

Second, habitat may be *degraded* by pollution to the extent that conditions become untenable for certain species. Degradation by pollution can take many forms, from the application of pesticides that harm nontarget organisms (discussed in Section 14.5), to acid rain (Sections 1.3 and 12.4) with its adverse effects on organisms as diverse as grasses (Figure 13.11),

forest trees, amphibians in ponds, and fish in lakes, to the runoff of fertilizers from agricultural land and the eutrophication to which it can give rise (Section 11.4).

Third, habitat may simply be *disturbed* by human activities to the detriment of some of its occupants. Habitat disturbance is not such a pervasive influence as destruction or degradation, but certain species are particularly sensitive. For example, diving and snorkeling on coral reefs, even in marine protected areas (see Section 13.4), can cause damage by means of direct physical contact with hands, body, equipment, and fins. Often individual disturbances are minor, but these can result in serious cumulative damage. In one analysis of 214 divers in a marine park on Australia's Great Barrier Reef, 15% of divers damaged or broke corals, mostly by fin flicks (Rouphael & Inglis, 2001). Specialist underwater photographers caused more damage on average (1.6 breaks per 10 minutes) than divers

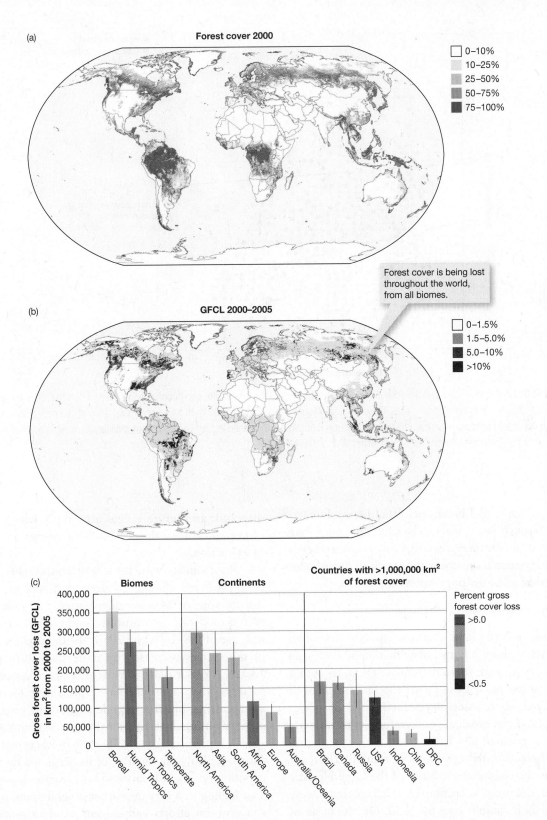

FIGURE 13.10 (a) The estimated percentage forest cover throughout the world in 2000. (b) The estimated gross forest cover loss (GFCL, %) throughout the world between 2000 and 2005. (c) GFCL 2000–2005 in absolute terms for different biomes, continents, and countries with large areas of forest cover. Bars are 95% confidence intervals. (After Hansen et al., 2010.)

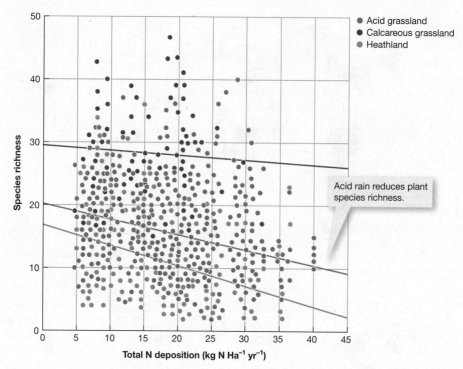

FIGURE 13.11 The relationship between plant species richness and the rate of nitrogen deposition in three habitats in Britain. For acid grasslands and heathlands, growing on acidic soils, there were significant declines ($P < 0.001$). In these communities, nitrogen deposition lowered pH further and the altered communities were indicative of increased acidity not increased fertility. For calcareous grasslands, growing on relatively neutral soils, there was no significant relationship. (After Maskell et al., 2010.)

without cameras (0.3 breaks per 10 minutes). Impacts were also much more likely to be caused by male than female divers. Nature recreation, ecotourism, and even ecological research are not without risk of disturbance and the decline of the populations concerned.

Global environmental change

When the environment changes, species may find themselves without a home, or with a home that is no longer large enough to accommodate them, or with a home that has moved too far or too fast for them to follow. One way to assess the extinction risk of species under global climate change, therefore, is to estimate the loss in area of key habitats on the basis of predicted changes to temperature and rainfall. Thus, for example, the characteristic biota of the Cape Floristic Province, discussed in Section 13.4, is expected to lose 65% of its habitable area by 2050. On the basis of the general pattern relating species richness to area (see Section 10.4), this represents a reduction of 24% in number of species (Thomas et al., 2004). However, this conclusion is based on the optimistic assumption that species are capable of dispersing to all currently uninhabited areas that become inhabitable. If we assume no

dispersal, and future ranges are simply what remains of current ranges, then 30–40% of species are at risk of extinction.

In a similar way, Beaumont and Hughes (2002) used predicted climate changes to model the future distributions of 24 Australian butterfly species. Under even a moderate set of future conditions (temperature increase of 0.8–1.4°C by 2050) the distributions of 13 of the species decreased by more than 20% (Figure 13.12). Most at risk are butterflies like *Hypochrysops halyetus*, that not only have specialized food plant requirements, but that also depend on the presence of ants for a mutualistic relationship (see Section 8.4). This species is predicted to lose 58–99% of its current range. Moreover, only one-quarter of its predicted future distribution occurs in locations that it currently occupies. This highlights a very important general point: regional conservation efforts and current nature reserves may turn out to be in the wrong place in a changing world.

Téllez-Valdés and Dávila-Aranda (2003) explored this issue for cacti, the dominant plant form in Mexico's Tehuacán-Cuicatlán Biosphere Reserve. From knowledge of the biophysical basis of current species distributions, and assuming one of three future climate

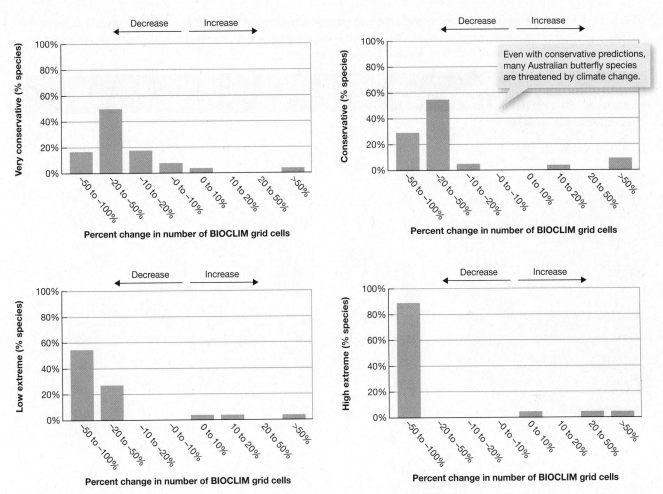

FIGURE 13.12   The effects of climate change to 2050, as predicted by the bioclimatic computer program BIOCLIM, for 24 species of Australian butterfly subjected to four climate change scenarios ranging from 'very conservative' [CO_2 increase to 479 ppm, Australian mean annual temperature (AMAT) increase 0.8–1.4 °C, Australian mean annual precipitation (AMAP) increase 0–18%] to 'high extreme' (CO_2 increase to 559 ppm, AMAT increase 2.1–3.9 °C, AMAP increase 0–59%). (After Beaumont & Hughes, 2002.)

scenarios, they predicted future species distributions in relationship to the location of the reserve. Table 13.1 shows how the potential ranges of species contracted or expanded in the various scenarios. In the most extreme scenario (an average temperature increase of 2.0°C and a 15% reduction in rainfall), more than half the species currently restricted to the reserve are predicted to go extinct. A second category of cacti, whose current ranges are almost equally within and outside the reserve, are expected to contract their ranges, but in such a way that their distributions become almost completely confined to the reserve. In the case of these cacti, then, the location of the reserve seems to cater adequately for potential range changes. But how many other nature reserves may turn out to be in the wrong place?

Introduced and invasive species

Invasions of exotic species into new geographic areas sometimes occur naturally and without human agency.

However, human actions have increased this trickle to a flood. Human-caused introductions may occur either accidentally as a consequence of human transport, or intentionally but illegally to serve some private purpose, or legitimately to procure some hoped-for public benefit by bringing a pest under control, producing new agricultural products, or providing novel recreational opportunities. Many introduced species are assimilated into communities without much obvious effect. However, some have been responsible for dramatic changes to native species and natural communities.

For example, the accidental introduction of the brown tree snake *Boiga irregularis* onto Guam, an island in the Pacific, has through nest predation reduced 10 endemic forest bird species to the point of extinction. The gradual spread of the snake from its bridgehead population in the center of the island has been paralleled by the loss of bird species to the north and south

TABLE 13.1 The distributions (km²) of cacti in Mexico under current conditions and as predicted for three climate change scenarios. Species in the first category of cacti are currently completely restricted to the 10,000 km² Tehuacán-Cuicatlán Biosphere Reserve. Those in the second category have a current range more or less equally distributed inside and outside the reserve.

Species Category	Current	+1.0° C −10% Rain	+2.0° C −10% Rain	+2.0° C −15% Rain
Restricted to the reserve				
Cephalocereus columna-trajani	138	27	0	0
Ferocactus flavovirens	317	532	100	55
Mammillaria huitzilopochtli	68	21	0	0
Mammillaria pectinifera	5,130	1,124	486	69
Pachycereus hollianus	175	87	0	0
Polaskia chende	157	83	76	41
Polaskia chichipe	387	106	10	0
Intermediate distribution				
Coryphantha pycnantha	1,367	2,881	1,088	807
Echinocactus platyacanthus f. grandis	1,285	1,046	230	1,148
Ferocactus haematacanthus	340	1,979	1,220	170
Pachycereus weberi	2,709	3,492	1,468	1,012

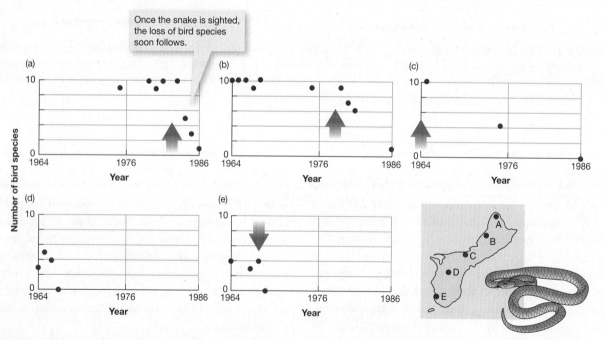

FIGURE 13.13 Decline in the number of forest bird species at five locations on the island of Guam. Large arrows indicate the first sightings of the brown tree snake at each location (in location d, the snake was first sighted in the early 1950s). (After Savidge, 1987.)

(Figure 13.13). Similarly, the introduction as a source of human food of the predaceous Nile perch (*Lates nilotica*) to the enormously species-rich Lake Victoria in East Africa has driven most of its 350 endemic species of fish to or near extinction (Kaufman, 1992).

Conservation biologists are particularly concerned about the effects of introduced species wherever there are communities of native organisms that are largely *endemic* (that is, that live nowhere else in the world). Indeed, one of the main reasons for the world's great biodiversity is the occurrence of centers of endemism, in which similar habitats in different parts of the world are occupied by different groups of species that happen to have evolved there (Section 2.6). If every species naturally had access to everywhere on the globe, we might expect a relatively small number of successful species to become dominant in each biome.

The extent to which this homogenization can happen naturally is restricted by the limited powers of dispersal of most species in the face of the physical barriers. But the transport opportunities offered by humans have allowed an ever-increasing number of species to breach these barriers. The resulting

homogenization is then partly a reflection of the same species being introduced repeatedly, and partly a result of the loss of local endemic species. For freshwater fish in the United States, at least, it seems that the former is more important than the latter (Figure 13.14a). But in more isolated communities, with a much higher proportion of endemics, species loss is bound to play an important part. Examples from Hawaii are shown in Figure 13.14b.

We can get some further sense of the scale of the problem of invasive or introduced species by noting that these rank second after habitat degradation among the factors threatening bird biodiversity (Figure 13.15), and that of the 958 species classified as

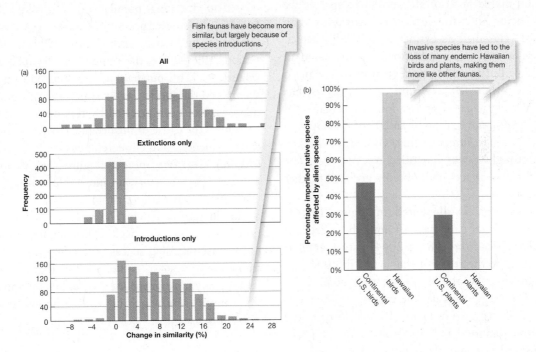

FIGURE 13.14   Biotic homogenization. (a) Changes in similarity among fish faunas for the 1,128 pairs of adjoining states in the United States, accounting for all changes (above), only for the local extinction of species (middle), and only for the introduction of species (below). State faunas were therefore classified around 1985 on the basis of whether species ever reported for the state were native and present, native but extinct, or introduced and present. (After Rahel, 2002.) (b) The impact of invasive species on the percentage of birds and plants of continental United States and Hawaii that are 'imperiled' (possibly extinct, critically endangered, or endangered). (After Wilcove et al., 1998.)

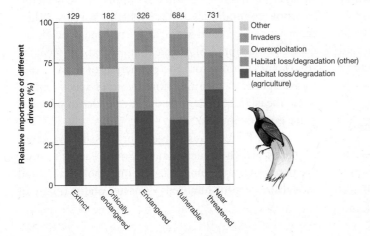

FIGURE 13.15   Relative importance of different drivers responsible for the loss or endangerment of bird biodiversity. Patterns are shown for five categories of extinction threat (see Section 14.2). The values above each histogram are the numbers of species in each threat category globally. Habitat loss/degradation poses a much bigger risk now than in the past (compare histograms for endangered and vulnerable categories with extinct birds), and this is set to increase in the future, in particular via agricultural expansion (histogram for near-threatened species). (Modified from Balmford & Bond, 2005.)

Infectious disease

We saw earlier, when we described the final years of the heath hen, how infectious disease may play its part in the decline of a species towards extinction. This integral role, embedded in a range of factors, is probably typical. Recently, however, the power of infectious disease in its own right in driving species to extinction has been emphasized by the plight of amphibian species worldwide (see Figure 13.1).

An estimated 50% of all amphibian species are currently facing an extinction crisis, with more than 100 already classified as extinct or possibly extinct. One cause is the worm infections we discussed in Section 12.5. But in addition, chytrid disease, caused by the fungus *Batrochochytrium dendrobatidis*, has recently been implicated as the immediate cause of catastrophic declines in more than 200 of the 350 species it is known to infect (Fisher, Garner, & Walker, 2009). Although it was identified for the first time only in 1997, its effects undoubtedly go back further than that.

One example is illustrated in Figure 13.16a, which shows the moving wave of amphibian declines in Central America from 1987 to 2004. A study at what turned out to be the front of that wave in 2004 had been monitoring the abundance of frogs going back to 1998. The precipitous decline in amphibian abundance in 2004 is shown in Figure 13.16b. Prior to the die-off, none of more than 1,500 frog samples from 43 species showed any sign of the infection. In 2004, the prevalence of infection in 879 samples from 48 species was around 50% (Lips et al., 2006).

In fact, the spread of the disease seems itself to be linked to global environmental change. Certainly in Costa Rica at least, the loss of species is closely linked to unusually warm conditions that favor its spread (Figure 13.16c). Overall, therefore, the threat to amphibians is grave. The infection is already found throughout the world (Figure 13.16d) and is likely to spread further still.

Combinations of risks and extinction vortices

Some species are at risk for a single reason, but often, as in the case of the heath hen discussed earlier, a combination of factors is at work, perhaps occurring sequentially. Among these may be the genetic effects described above, though no clear example of extinction due to genetic problems has so far come to light. On the other hand, it may be that inbreeding depression has occurred, though undetected, as part of the 'death rattle' of some dying populations (Caughley, 1994). Thus, a population may have been reduced to a very small size by one or more of the processes described above, and this may have led to an increased frequency of matings among relatives and the expression of deleterious recessive alleles in offspring, leading to reduced survivorship and fecundity and causing the population to become smaller still—a so-called **extinction vortex** (Figure 13.17a).

The declining population of a shore bird in Sweden, the southern dunlin, *Calidris alpina schinzii*, appears to be entering such a vortex. During the study period, 1993–2004, the population declined steadily (Figure 13.17b), but habitat loss that had previously threatened the species, and reduced the size of its populations, had been arrested by this time, so it cannot have been the cause of the decline. However, there was clear evidence that close relatives were increasingly breeding with one another (Figure 13.17c), and that reduced heterozygosity, an inevitable consequence of inbreeding, was leading to an increase in the failure of dunlin eggs to hatch (Figure 13.17d). The population has continued to decline, despite active measures to increase breeding success, including preventing the trampling of nests by cattle. It seems that earlier habitat loss may have propelled the dunlins into an extinction vortex from which inbreeding is now making it difficult to escape.

Chains of extinctions?

So far, we have looked at individual species, treating them as though they were largely independent entities. Conservation of biodiversity, however, also requires a broader perspective in which we apply our knowledge of whole communities and ecosystems. We can deduce from studies of food webs and keystone species, for example (Section 9.5), that a chain of extinctions may follow inexorably from the extinction of a particular native species, which therefore deserves special attention. The island of Guam in the Pacific Ocean provides some good examples.

First, flying foxes (large fruit bats) in the genus *Pteropus*, which occur on many South Pacific islands, are the major, and sometimes the only, pollinators and seed dispersers for hundreds of native plants (many of

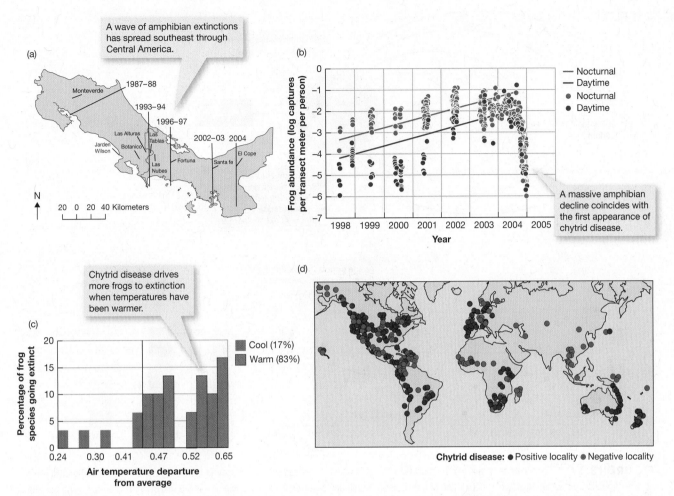

FIGURE 13.16 Amphibian extinctions and chytrid disease. (a) A map of Central America, with sites of reported major declines in amphibian populations, and lines showing the approximate location of a wave of declines as it moved southeast. (b) Frogs observed between 1998 and 2005 in daytime and nocturnal samples taken at El Copé [see map in (a)]. Statistical tests suggest a significant shift in the direction of the relationship ($P < 0.001$) on September 4, 2004. (c) The percentage of species of harlequin frog (*Atelopus* spp.) going extinct in Costa Rica at different departures in tropical air temperature from the 1950–1979 average. (d) The current distribution of chytrid disease in amphibians worldwide. ((a) and (b) After Lips et al., 2006; (c) After Pounds et al., 2006; (d) After Fisher et al., 2009)

which are of considerable economic importance, providing medicines, fiber, dyes, prized timber, and foods). Flying foxes are highly vulnerable to human hunters and there is widespread concern about declining numbers Cox et al., 1991. On Guam, the two indigenous flying fox species are either extinct or virtually so, and there are already indications of reductions in fruiting and dispersal. Guam is also now home to the brown tree snake, as we saw earlier (Figure 13.13), and indeed, the snake may have contributed to the decline of the flying foxes. Its most devastating effects, however, have been on indigenous birds, many species of which are also important pollinators, and here too, the

cascading effects from snake to pollinating birds to plants are apparent.

A comparison with the nearby but nearly snake-free island of Saipan for two important tree species showed, first, that visits by pollinating birds had ceased on Guam and pollinating visits overall were significantly down (Figure 13.18a,b), but also that recruitment of the trees (i.e., appearance of seedlings and saplings) was failing (Figure 13.18c). These studied species are only examples of tree species on Guam pollinated mainly by birds. Hence, if urgent conservation measures are not taken to protect the plants, this lowered recruitment may in due course be fatal.

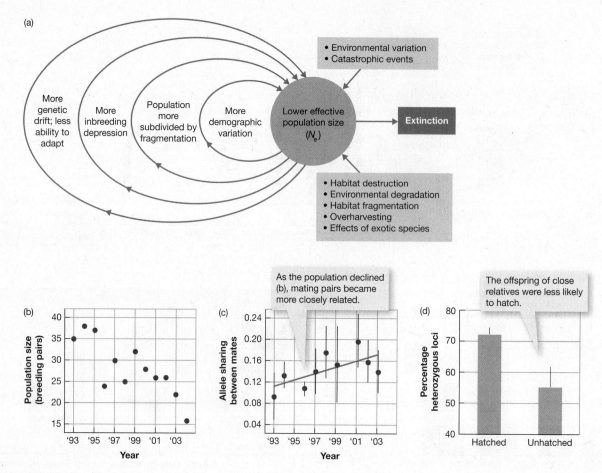

FIGURE 13.17 The extinction vortex. (a) The manner in which extinction vortices may progressively lower population sizes leading inexorably to extinction. (b) The declining size of a population of the southern dunlin in southwest Sweden between 1993 and 2004. (c) The increasing genetic similarity between mating pairs of the dunlins over the same period, as measured by the proportion of gene alleles they share (bars are SEs, $P < 0.05$). (d) The percentage of gene loci that are heterozygous among dunlin chicks that hatched or failed to do so (bars are SEs). ((a) from Primack, 1993; (b)–(d) After Blomqvist et al., 2010.)

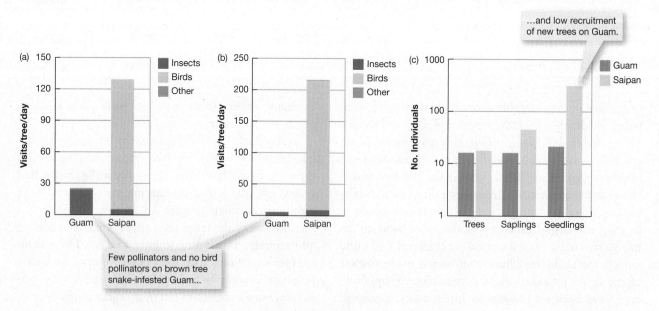

FIGURE 13.18 (a) Average daily number of pollinator visits per tree to the mangrove tree *Bruguiera gymnorrhiza* by insects, birds, and other visitors on the nearby islands of Guam and Saipan. (b) Equivalent results for the forest tree *Erythrina variegata*. (c) Mature trees, saplings, and seedlings of *B. gymnorrhiza* present in observation zones on Guam and Saipan. (After Mortensen et al., 2008.)

13.4 CONSERVATION IN PRACTICE

At the beginning of this chapter, we pointed out that conservation measures can be divided, broadly, into attempts to protect or to restore, and that often our most important decisions will be about where and how large fenced-off, managed, or restored areas should be.

We now return to these measures and to the decisions about where and when to apply them. Mainstream ecological scientists have not always applied fundamental principles to issues like conservation. A few far-sighted ecologists led the way for others to follow. The life and work of one of these, Tony Bradshaw, is described in Box 13.3.

13.3 Historical Landmarks

Tony Bradshaw: Evolutionary ecologist, restoration biologist

What follows is taken from an obituary of Tony Bradshaw, written by Alastair Fitter for *The Guardian* newspaper in the United Kingdom on September 11, 2008, with a few annotations to help an international audience.

In 2006, the Environment Agency [of the UK] published a poll of its top 100 eco-heroes of all time. Next to Charles Darwin was the name of Tony Bradshaw, a pioneer of restoration ecology, who has died aged 82. That juxtaposition was stunningly appropriate, for Tony made his name as an evolutionary biologist. His work on the evolution of tolerance to heavy metals in plants growing in contaminated soils remains the clearest and best example of evolution in action, and Darwin would have applauded it.

Where Tony left his academic colleagues in the shade was by applying that brilliant, fundamental research to the restoration of vegetation and functioning ecosystems to derelict land, first on mine sites in Wales and later in his adopted city of Liverpool and world-wide. His ability to move seamlessly between the worlds of fundamental and applied science may now be more common, but in the 1960s and 70s was unusual and even disdained by some.

Tony Bradshaw

The son of an architect, Tony was born in Kew, Surrey, and was interested in plants from an early age. As a 13-year-old at the outbreak of the Second World War, he dug for victory—literally. The entire front garden of the family house was turned over to vegetables under his care, despite being on heavy clay. Perhaps that experience seeded his lifelong interest in making soil in unlikely places. After graduating in botany at Jesus College, Cambridge, in 1947, he moved to the University College of Wales, first as a research student in Aberystwyth and then as a lecturer at Bangor. His work on metal tolerance was outstanding. He demonstrated the power of natural selection to bring about rapid evolutionary changes in natural grasses, even when the populations were separated by only very short distances, which conventional theory at the time regarded as impossible because gene flow between the populations was thought to swamp the disruptive effects of selection. Consequently, locally adapted races could evolve easily.

Tony quickly saw the potential of this discovery to help restore polluted soils. In 1968 he took the Chair of Botany at Liverpool University, and the range of degraded environments covered by his lab expanded quickly. Spoil heaps from coal, slate and china clay mines, limestone quarries, metal wastes, the bare margins of reservoirs and urban dereliction all fell under his gaze.

The key to his work was creating soil on these damaged sites without resorting to the then standard technique of importing topsoil at great expense (and at great damage to the source of the soil). Consequently, huge areas of land could be restored to health because the costs of doing so were so low.

In 1982 he was elected a Fellow of the Royal Society, and other honours followed, including the presidency of the British Ecological Society, of which he was made an honorary member in 1988. Tony was involved with many environmental organisations, including the first Groundwork Trust, set up in St Helens, Merseyside, in 1981 as a partnership between the then Countryside Commission and local authorities to work with the community to bring about environmental change and the consequent social and economic benefits.

The model was so successful that there are now more than 50 Groundwork Trusts in the UK, focused on communities where environmental dereliction goes hand-in-hand with social and economic deprivation, and has been replicated worldwide. In Britain, the movement has given rise to the Land Restoration Trust.

He published more than 250 works, not just in mainstream scientific journals. His book *The Restoration of Land* (1980), written jointly with Mike Chadwick, was the standard text on the subject, mixing science and practice fluently, and *Trees in the Urban Landscape* (1995) set the same standards for those much abused 'friends of the city.'

Tony married Betty Alliston in 1955, and they had three daughters, Jane, Penny and Sarah. When Betty died in 2000, Tony stayed in Liverpool and continued to devote his knowledge and experience to the city, and helped transform its oldest public park, St James Gardens. Shortly before he died, he was made Liverpool's first Citizen of Honour.

Population viability analysis

For the population of a rare species, one obvious question is: how great is the threat it faces? That is, given the environmental circumstances and its particular characteristics, what are the chances that it will go extinct in a specified period? How big does its population need to be to reduce the chance of extinction to an acceptable level? One simple approach to answering these questions is to determine the size of a **minimum viable population** (MVP). What defines an MVP is a matter of individual judgment, but we might suggest, for example, that an MVP will have at least a 95% probability of persistence for 100 years. We can then explore data sets such as those shown earlier for the bighorn sheep in Figure 13.8b to provide an approximate estimate of the MVP.

Bighorn populations of fewer than 50 individuals all went extinct within 50 years, while only 50% of populations of 51–100 sheep lasted for 50 years. Evidently, for an MVP here we require more than 100 individuals. Indeed, for these sheep, such populations

demonstrated close to 100% persistence over the maximum period studied of 70 years. However, the conservation value of studies like this is limited because they rely on long-term data sets and therefore deal with species that are generally not at risk.

Simulation models known as **population viability analyses** (PVAs) provide an alternative, more specific way of gauging viability. Usually, these start with estimates of survivorships and reproductive rates in age-structured populations (see Chapter 5) and then simulate random variations in these, or in carrying capacity (K), to represent the impacts of environmental variations, including those of disasters of specified frequency and intensity. The program is then run many times, each time giving a different population trajectory because of the random elements involved. For each set of parameters, the model's outputs include estimates of population size each year and the probability of extinction during the modeled period (the proportion of simulated populations that go extinct).

To take one example, koalas (*Phascolarctos cinereus*) are regarded as 'near threatened' in Australia, with populations in different parts of the country varying from secure to vulnerable or extinct. Penn et al. (2000) used a widely available PVA program known as VORTEX (Lacy, 1993) to model two populations in Queensland, one thought to be declining (Oakey) and the other secure (Springsure). Koala breeding commences at 2 years in females and 3 years in males. The other demographic values were derived from extensive knowledge of the two populations and are shown in Table 13.2. The Oakey population was modeled from 1971 and the Springsure population from 1976, when first estimates of density were available.

koalas—identifying populations at particular risk

We can see that the model trajectories were indeed declining for Oakey and stable for Springsure (Figure 13.19). Over the modeled period, therefore, the probability of extinction of the Oakey population was 0.380 (380 of 1,000 simulations went extinct), while that for Springsure was only 0.063. Conservation efforts should clearly be focused mostly on Oakey.

Managers concerned with critically endangered species do not usually have the luxury of monitoring populations to check the accuracy of their predictions. Penn et al. (2000), however, were able to compare the predictions of their PVAs with real population trajectories, because the koala populations have been continuously monitored since the 1970s (Figure 13.19). The predicted trajectories were close to the actual population trends, particularly for the Oakey population, and this gives added confidence to the modeling approach.

Related models use *population projection matrices*, which are particularly useful in acknowledging that most life cycles are made up of a sequence of distinct classes with different rates of fecundity and survival. The approach has been applied to the royal catchfly, *Silene regia*, a long-lived prairie perennial plant whose range has shrunk dramatically. Menges and Dolan (1998) collected demographic data for up to 7 years from 16 populations in the U.S. Midwest. The populations, whose total adult numbers ranged from 45 to 1,302, had been subject to different

TABLE 13.2 Values used as inputs for simulations of koala populations at Oakey (declining) and Springsure (secure), using the simulation program VORTEX. Values in brackets are standard deviations due to environmental variation; the model procedure involves the selection of values at random from the range. Catastrophes are assumed to occur with a certain probability; in years when the model 'selects' a catastrophe, reproduction and survival are reduced by the multipliers shown (e.g., in a year with a catastrophe, reproduction is reduced to 55% of what it would otherwise have been).

Variable	Oakey	Springsure
Maximum age	12	12
Sex ratio (proportion male)	0.575	0.533
Litter size of 0 (%)	57.00 (±17.85)	31.00 (±15.61)
Litter size of 1 (%)	43.00 (±17.85)	69.00 (±15.61)
Female mortality at age 0	32.50 (±3.25)	30.00 (±3.00)
Female mortality at age 1	17.27 (±1.73)	15.94 (±1.59)
Adult female mortality	9.17 (±0.92)	8.47 (±0.85)
Male mortality at age 0	20.00 (±2.00)	20.00 (±2.00)
Male mortality at age 1	22.96 (±2.30)	22.96 (±2.30)
Male mortality at age 2	22.96 (±2.30)	22.96 (±2.30)
Adult male mortality	26.36 (±2.64)	26.36 (±2.64)
Probability of catastrophe	0.05	0.05
Multiplier for reproduction	0.55	0.55
Multiplier for survival	0.63	0.63
% males in breeding pool	50	50
Initial population size	46	20
Carrying capacity, K	70 (±7)	60 (±6)

management regimes. A matrix for one of the populations of *S. regia* is illustrated in Table 13.3. Such matrices were produced for each population in each year. Multiple simulations, each lasting 1,000 years, were then run for every matrix to determine both the

probability of extinction and the population's projected rate of increase (where a value > 1 indicates an increase, a value < 1 a decrease, and hence a value of 1 an unchanging population size).

Figure 13.20 shows the median population rate of increase for the 16 populations, grouped into cases where particular management regimes were in place. This was done both for years when recruitment of seedlings occurred and when it did not. All sites with

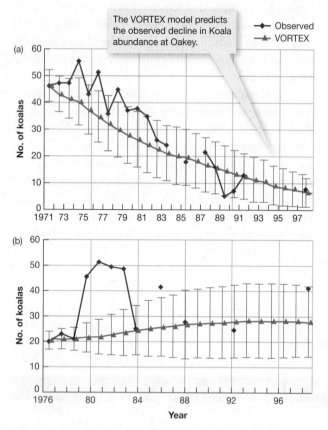

FIGURE 13.19 Observed koala population trends (red diamonds) compared with predicted population performance (blue triangles, ± 1 SD) based on 1,000 repeats of the VORTEX modeling procedure at (a) Oakey and (b) Springsure. Real population censuses were not performed every year. (After Penn et al., 2000.)

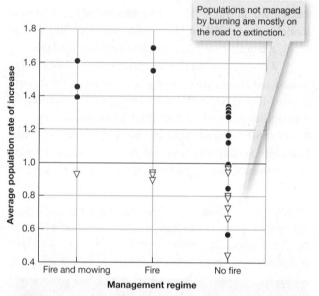

FIGURE 13.20 Median rates of population increase of *Silene regia* populations in relation to management regime, for years with seedling recruitment (red circles) and without (open triangles). Unburned management regimes include just mowing, herbicide use, or no management. All sites above the black solid line have values greater 1.0, indicating the capacity of their populations to grow in size. Those below the line are on paths to extinction. (After Menges & Dolan, 1998.)

TABLE 13.3   An example of a projection matrix (using the simulation modeling tool called RAMAS-STAGE) for a particular *Silene regia* population from 1990 to 1991, assuming successful germination of seedlings. The numbers represent the proportion changing from the stage in the column to the stage in the row (bold values represent plants remaining in the same stage). 'Alive undefined' represents individuals with no size or flowering data, usually as a result of mowing or herbivory. The numbers in the top row are seedlings produced by flowering plants. The projected rate of increase for this population is 1.67. The site is managed by regular burning.

	Seedling	Vegetative	Small Flowering	Medium Flowering	Large Flowering	Alive Undefined
Seedling	–	–	5.32	12.74	30.88	–
Vegetative	0.308	**0.111**	0	0	0	0
Small flowering	0	0.566	**0.506**	0.137	0.167	0.367
Medium flowering	0	0.111	0.210	**0.608**	0.167	0.300
Large flowering	0	0	0.012	0.039	**0.667**	0.167
Alive undefined	0	0.222	0.198	0.196	0	**0.133**

rates of increase greater than 1.35 when recruitment took place are managed by burning and some by mowing as well. None of these were predicted to go extinct during the modeled period. On the other hand, populations with no management regime, or whose management does not include fire, had lower rates of increase, and all except two had predicted extinction probabilities (over 1,000 years) from 0.10 to 1.00. The obvious management recommendation is to use prescribed burning to provide opportunities for seedling recruitment. Low establishment rates of seedlings may be due to rodents or ants eating fruits, or to competition for light with other plants—burnt areas probably reduce one or both of these negative effects.

Protected areas

Producing survival plans for individual species may be the best way to deal with species recognized to be in deep trouble and of special importance (e.g., keystone species described in Section 9.5, evolutionarily unique species, or charismatic large animals that are easy to 'sell' to the public). However, there is no possibility that all endangered species could be dealt with one at a time. Funds for conservation are simply too limited for this. We can, though, expect to conserve the greatest biodiversity if we protect whole communities by setting aside protected areas. In fact, protected areas of various kinds (national parks, nature reserves, sites of special scientific interest) grew in both number and area during the 20th century. Currently, about 13% of the world's land area is protected (and around 1% of the sea area; Mora & Sale, 2011).

A **protected area** has been defined by the International Union for the Conservation of Nature (IUCN, 2008) as 'a clearly defined geographical space, recognized, dedicated, and managed through legal or other effective means, to achieve the long term conservation of nature with associated ecosystem services and cultural values.' To better understand what this means, note that the IUCN has further defined several separate categories of protected areas, set out in Box 13.4.

13.4 ECOncerns

Protected areas

The IUCN, through one of its commissions, the World Commission on Protected Areas, has set out a categorization of protected areas summarized below

- **Strict Nature Reserves.** These are 'strictly protected areas set aside to protect biodiversity and also possibly geological/geomorphical features, where human visitation, use and impacts are strictly controlled and limited to ensure protection of the conservation values.'

- **Wilderness Areas** are 'large unmodified or slightly modified areas, retaining their natural character and influence without permanent or significant human habitation, which are protected and managed so as to preserve their natural condition.'

- **National Parks.** These are similar to wilderness areas but also add a more human perspective since they 'provide a foundation for environmentally and culturally compatible, spiritual, scientific, educational, recreational, and visitor opportunities.'

- **Natural Monuments or Features** are also 'more visitor-oriented and generally small, being areas set aside to protect a specific natural monument, which can be a landform, sea mount, submarine cavern, geological feature such as a cave or even a living feature such as an ancient grove.'

- **Habitat/Species Management Areas.** Here the focus is much more on particular species or habitats. These areas therefore typically 'need regular, active interventions to address the requirements of particular species or to maintain habitats.'

- **Protected Landscapes/ Seascapes.** In this case the focus is not only human but also aesthetic, since these are areas where the interaction of people and nature over time has produced an area of distinct character with 'significant, ecological, biological, cultural and scenic value: and where safeguarding the integrity of this interaction is vital to protecting and sustaining the area and its associated nature conservation and other values.'

- **Protected areas with sustainable use of natural resources.** Finally, here the human focus is much more practical. These areas 'conserve ecosystems and habitats together with associated cultural values and traditional natural resource management systems. They are generally large, with most of the area in a natural condition, where a proportion is under sustainable natural resource management and where low-level non-industrial use of natural resources compatible with nature conservation is seen as one of the main aims of the area.'

(http://cms.iucn.org/about/work/programmes/gpap_home/gpap_quality/gpap_pacategories/)

The IUCN website goes into more details, identifying for each category its objectives, distinguishing features, role in the land/seascape, what makes it unique, and issues for consideration. How would you weigh the relative value of conserving individual species, conserving whole biotas, and respecting human aesthetics and human practicalities? How much do you think the perspective is likely to change if the protected area is in a rich or a poor part of the world? Or if the person doing the weighing is from a rich or a poor part of the world? How useful do you believe the IUCN categorization is in balancing the perhaps conflicting purposes of protecting protected areas? Why?

Of course, areas may be protected without becoming part of the official IUCN portfolio, but we can see from the categories in Box 13.4 that any protected area must seek to satisfy a range of criteria that may often conflict with one another, and that different types of areas balance these criteria out in different ways. Notable among the criteria are conservation itself, access for humans for educational or recreational purposes, and the ability of humans to exploit the natural resources of the area, either directly for food, or more broadly to provide for their subsistence and well-being.

Deciding how best to resolve these conflicting requirements is itself an important issue in conservation. One example comes from the need to conserve butterflies in England but also for farmers to produce food from the same areas of land. Food can be produced conventionally, with the aid of artificial pesticides and fertilizers, or organically, without them. Yields tend to be lower on organic farms (although not by as much as is sometimes suggested; see Section 14.4), so more land is needed to produce the same crop yield. However, organic farms provide better habitat for butterflies, although they are not so good for butterfly conservation as nature reserves are (Figure 13.21a).

A choice therefore exists between 'land sharing'—farming organically and so producing food and protecting butterflies on the same land—and 'land sparing'—farming intensively to produce food in some areas while conserving intensively in others such as nature reserves. The right choice depends not only on

resolving conflicting requirements

the relative values for conservation of different land uses, shown in Figure 13.21a, but also on the relative yields of organic and conventional farms.

The search for the best solution is shown in Figure 13.21b, which combines the two conflicting requirements. First, contours represent the density of butterflies to be expected at different combinations of nature reserves and organic farmland in the habitat. Second, there are lines representing combinations of equal overall farmland yield for different assumptions about relative yield on organic and conventional farms. Our starting point is the current situation in terms of the percentages of reserve, organic, and conventional farmland, also shown in the figure. The question is: how can we conserve the most butterflies while maintaining current yields? Our aim, therefore, is to choose the line that most accurately represents the organic/conventional balance of yield, and then find the point on that line that takes us as high up the density contours as possible, that is, that conserves the most butterflies.

The analysis in Figure 13.21b indicates that if organic yields are lower than 87% of those achieved on conventional farms, then conventional farming and sparing land for nature reserves is the better conservation option. However, if we can achieve organic yields that are more than 87% of those on conventional farms, then sharing production and conservation on the same organic land is best. Clearly, decisions about whether and where to establish protected areas depend on more than simply what those areas are able to protect. And indeed, decisions about organic or

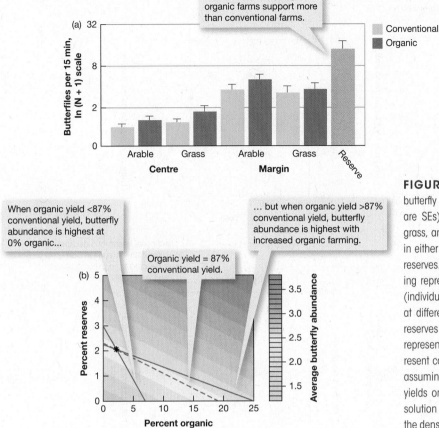

FIGURE 13.21 (a) The effect of land use on butterfly density at a range of sites in England (bars are SEs). Sites were classified as either arable or grass, and at either the center or margins of fields, in either conventional or organic farms, or as nature reserves. (b) Land sharing or land sparing? The shading represents contours in the density of butterflies (individuals seen per 15 minutes) to be expected at different combinations of percentages of nature reserves and organic farmland in the habitat. The star represents the estimated current situation. Lines represent combinations giving the same farmland yield, assuming a relationship (as indicated) between the yields on organic and conventional farms. The best solution for any line is found at the point highest on the density contours. (After Hodgson et al., 2010.)

conventional farming depend on more even than yields and conservation, as we discuss in Chapter 14.

The area covered by protected areas globally has been increasing steadily for at least the last 60 years (Figure 13.22). Throughout that period, marine protected areas have lagged far behind their terrestrial counterparts, but these are nonetheless useful to illustrate some important general points.

As the name suggests, the space included in a marine protected area (MPA) is all or mostly marine, although portions of adjacent land areas are often included, particularly when the land is intimately tied to the marine ecosystem in function or culturally. For instance, a tropical MPA may include an atoll as well as the associated fringing coral reefs and adjacent marine waters. MPAs were developed first in the 1960s as a way to better protect marine biodiversity and fishery resources, using as an analogy the way national parks on land had helped conserve some of the ecological functioning of terrestrial ecosystems. By 2010, MPAs around the world had grown to include almost 6,000 sites encompassing a little over 1% of the world's oceans (IUCN, 2010).

Most MPAs are in coastal areas, and since the coastal ecosystems make up a relatively small portion of the global oceans (Chapter 4), MPAs occupy a larger percentage of these coastal ecosystems: almost 3% of the waters lying within 320 km of coastlines (the "200-mile" limit of territorial waters claimed by most of the world's nations as part of their territories) and over 6% within 20 km of coastlines (the "12-mile" limit that many nations more intensively manage).

How effective are MPAs in conserving biodiversity and managing fishery resources? They vary tremendously in their effectiveness. Coral reefs are a particular focus for many MPAs, yet only a relatively few MPAs have been judged to be fully effective in protecting coral diversity and health (Figure 13.23).

Some MPAs are **no-take zones**: areas where no fishing is allowed. These areas are also well protected from land-based pollution sources and other human disturbances. In no-take zone MPAs, biodiversity is often maintained, and fishery resources grow in size and even help replenish fish populations in adjacent waters where fishing is still allowed. Hence, these MPAs are an overall benefit to commercial fishers despite being off limits to them. The Goat Island Marine Reserve in New Zealand, established in 1977, provides an example of such success, one that many subsequent MPAs have tried to emulate. Populations of snapper fish and crayfish that had been decimated by overfishing were

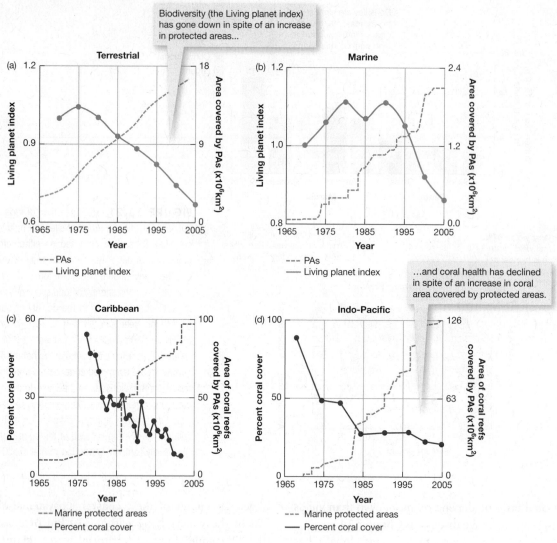

FIGURE 13.22 (a) Trends since 1965 in the global extent of area covered by terrestrial protected areas (PAs) (orange dashed line) and the 'living planet index' (green solid line), a measure of biodiversity based on the changing population size of 1,686 vertebrate species worldwide. (b) Equivalent plots for marine proteceted areas. (c) Trends since 1965 in the extent of area for coral reefs in the Caribbean Sea included in marine protected areas (blue dashed line) and the percent coral cover (a measure of coral health, red solid line). (d) Equivalent plots for the Indo-Pacific region. (After Mora and Sale, 2011.)

reestablished within ten years, and now these and the near-pristine coastal ecosystem support both scientific research and extensive ecotourism.

However, no-take zones do not always accomplish this goal. Sometimes the populations of predatory fish increase greatly, leading to a trophic cascade of fewer herbivores, an increase in macroalgae, and a decrease in biodiversity (see Section 9.5). This may be more likely to happen in areas under stress for other reasons, such as nutrient pollution, where the lack of predatory fish and the resulting high abundance of herbivores can mask some of the symptoms of nutrient pollution. Clearly, the ecological details matter. Most MPAs, however, are not no-take zones, and fishing and many other practices are commonly allowed. In the United States, less than 1% of the area of all MPAs consists of no-take zones.

The size of an MPA matters. Fish will of course swim freely in and out of the defined area of an MPA. To protect a fish population, therefore, the size of the MPA must be large relative to size of the home range of the fish of most concern. One calculation suggests that the MPA should be around 12.5 times the area of the species' home range, in order to keep fishing pressure on the population within the MPA at 2% or less of the pressure outside (Kramer & Chapman 1999). Since home range increases with body size, this means that for a typical species 20 cm in length, the MPA should be 1.8 km^2 in area. However, 30% of the world's MPAs are smaller than this, providing inadequate protection to typical fish species that grow to 20 cm in length or more.

> marine protected areas must be sufficiently large to be effective

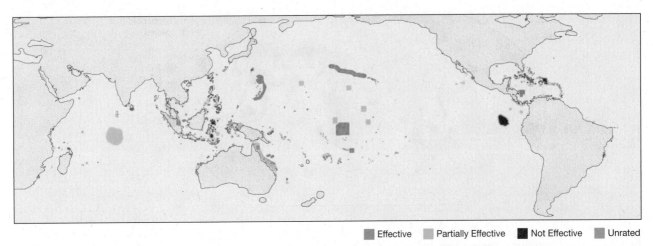

FIGURE 13.23   Marine protected areas have been established in many coral reef regions, but only a relatively few of these are judged to be fully effective in protecting the corals. (After World Resources Institute, 2011.)

While MPAs are an important step forward in the effort to protect marine ecosystems, they cannot do the job alone. They are focused on local protection of resources and provide little protection against global and regional threats. Global change poses high risk to many coastal ecosystems. Coral reefs, for example, are endangered by the increasing acidification of the world's oceans as levels of atmospheric carbon dioxide continue to rise (Chapter 12). Corals are damaged too by rising temperatures of surface-ocean waters, which can cause coral 'bleaching' (the loss of symbiotic algae in the corals, which causes the coral to lose color and become white).

> limits to the powers of MPAs

More generally, regional-scale nitrogen pollution can lead to eutrophication, dead zones, and loss of biodiversity and fishery resources at scales much larger than those of most MPAs. Hence, despite the growing area covered by MPAs, global biodiversity continues to fall, and the health of corals continues to decline (Figure 13.23). The trend globally for MPAs is similar to that for protected areas on land; for both, while protected areas have increased in size and number, they remain only a small percentage of the total, and biodiversity is clearly not adequately conserved.

Selecting conservation areas

We are bound to want to devise priorities so that the restricted number of new protected areas, in terrestrial and marine settings, can be evaluated systematically and chosen with care. We know that the biotas of different locations vary in species richness

> is conservation focused on biodiversity hotspots?

(with particular centers of diversity), in the extent to which the biota is unique (with centers of endemism), and in the extent to which the biota is endangered (with hotspots of extinction, for example, because of imminent habitat destruction). Locally, no doubt, these are important considerations when protected areas are chosen. But on a global scale, it is not so clear that protected areas are most concentrated where they are most needed.

For example, if we compare the global distribution of 'biodiversity hotspots' (as indicated by the number of globally threatened birds and amphibians in an area) with the global distribution of protected areas (Figure 13.24), we can see how the needs of conservation are being much better satisfied in the richer than in the poorer parts of the world. The reality is that protected areas will be placed not only where they are most needed but also where, simply, it is possible to place them.

Suggestions about how best to conserve species populations or biodiversity generally have appeared throughout this book. We have seen that small populations are particularly problematic (Section 13.3) and that larger areas, and especially larger habitat islands, contain not only larger populations but more species (Section 10.4). We have seen too that populations can sometimes be sustained by dispersal from outside (Section 5.4), and that for this reason, and also because the risk of extinction is spreading, connected metapopulations will be more likely to persist than their individual subpopulations (Section 9.3).

> designing nature reserves?

Of course, we have also seen throughout the book that species are found in habitats that provide their fundamental ecological niche (Section 6.2), and that species

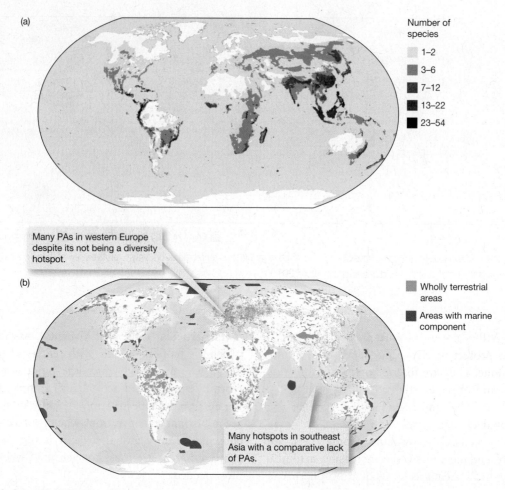

FIGURE 13.24 (a) Distribution of 'biodiversity hotspots' as measured by the numbers of species of globally threatened birds plus amphibians mapped on an equal area basis (each grid cell is 3113 km²). (After Rodrigues et al., 2006.) (b) Global distribution of the 177,547 nationally designated protected areas. Wholly terrestrial areas are green; those with at least a marine component are blue. (After Bertzky et al., 2012.)

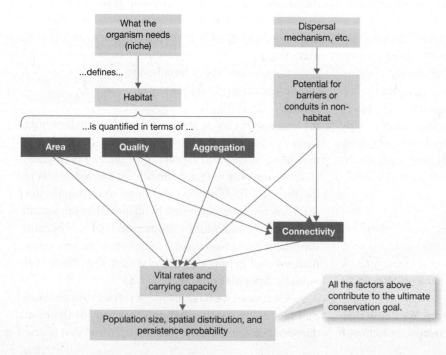

FIGURE 13.25 A simple recipe for species conservation. Arrows from A to B signify 'A determines B.' (After Hodgson et al., 2009.)

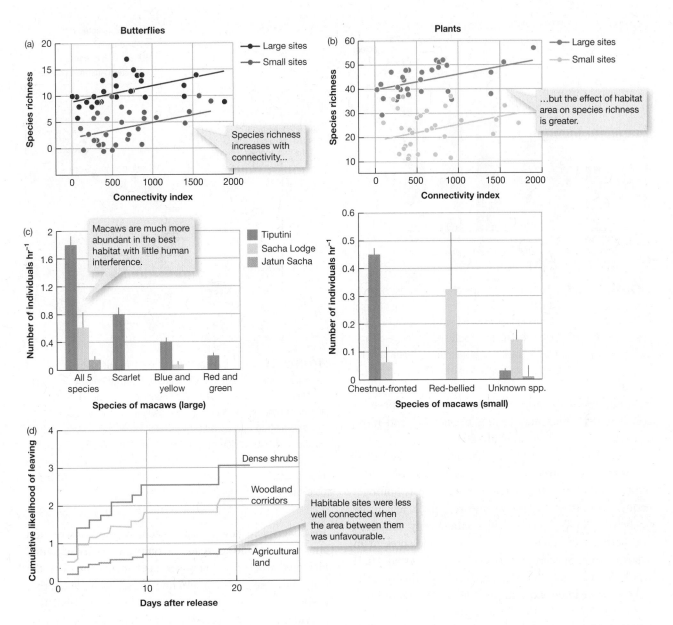

FIGURE 13.26   (a) The species richness of butterflies in different grassland patches in southern Germany in relation to a 'connectivity index' that combines edge-to-edge distances between a site and all other sites within a 2-km radius with the species' dispersal ability. Separate relationships are shown for large sites, c2.4 ha (red) and small sites, c0.12 ha (blue). Both relationships are significant ($P < 0.05$). (b) Similar relationships for plants in the same sites. (c) The abundance of various species of macaw, large above, small below, in the Ecuadorian Amazon region. Bars are SEs. Of the three sites, Tiputini is largely unaffected by human interference, Sacha Lodge is a tourist site with a human visitation rate twice that at Tiputini, surrounded by military and petroleum camps, and Jatun Sacha is another tourist site, close to a major highway and with an even higher visitation rate. (d) Wild caught birds, Chucao Tapaculos (*Scelorchilus rubecula*), were released from small patches of their favored woodland habitat in Chile and their dispersal from there monitored in landscapes where the release point was surrounded by agricultural land, woodland corridors, or dense shrubs. Dispersal was significantly lower in agricultural land than in the other two ($P < 0.05$), which were not significantly different from one another. ((a) and (b) After Bruckmann et al., 2010; (c) after Karubian et al., 2005; (d) after Castellon & Sieving, 2006.)

richness (biodiversity) is favored by a whole range of factors (Chapter 10). These ideas can be combined into a very basic recipe for successful conservation (Figure 13.25), which has three core ingredients: habitat quality, habitat area, and connectivity, the last of these being a combination of the spatial arrangement of suitable habitats and the provision of links between them. There

are examples showing the importance of each one of these (Figure 13.26). However, the argument has also been made that while, ideally, all should be considered when locating and designing nature reserves, in practice money and resources for conservation are likely to be limited. It may then be best to focus on habitat quality and area, where benefits are predictable and likely

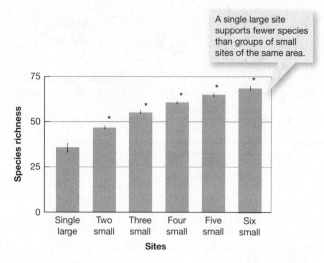

FIGURE 13.27 The species richness of plants in spruce (*Picea abies*) dominated mire (wetland) sites in Finland. With 24 sites in total, all combinations were examined in which a single large site could be compared with groups of between two and six sites of the same total area. Bars are SEs. Groups of small sites significantly different from the large one ($P < 0.05$) are marked *. (After Virolainen et al., 1998.)

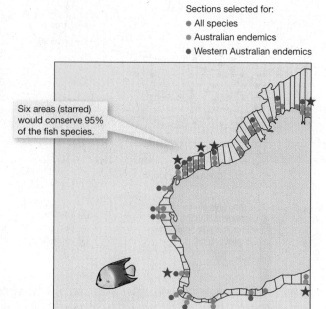

FIGURE 13.28 The coastline of Western Australia divided into 100 km lengths and showing the results of complementarity analysis to identify the minimum number of sites needed to include all the fish biodiversity for the region. Analyses were performed using all fish species, and separately for species endemic to Australia (found nowhere else) or those endemic to Western Australia. In the case of total fish biodiversity, 26 areas were needed if all 1,855 fish species were to be incorporated (green circles) but only 6 areas (stars) would be needed to incorporate more than 95% of the total. (After Fox & Beckley, 2005.)

to apply to a wide range of species, than on aspects of connectivity, which may be less predictable and more species-specific (Hodgson et al., 2011).

One particular question that may arise is whether to construct one large reserve or several small ones adding up to the same total area. If the region is homogeneous in terms of conditions and resources, it is quite likely that smaller areas will contain only a subset, and perhaps a similar subset, of the species present in a larger area. Species-area relationships (Section 10.4) would then suggest that it is preferable to construct the larger reserve in the expectation of conserving more species in total. On the other hand, if the region as a whole is heterogeneous, then each of the small reserves may support a different group of species and the total conserved might exceed that in one large reserve of the same size. In fact, collections of small islands tend to contain more species than a comparable area composed of one or a few large islands. The pattern is similar for habitat islands and, most significantly, for national parks. Thus, studies have tended to favor several small areas (see Figure 13.27). Similarly, sets of smaller parks contained more species than larger ones of the same area in studies of mammals and birds in East African parks, of mammals and lizards in Australian reserves, and of large mammals in national parks in the United States.

Collections of areas

Choosing the best areas for conservation is not limited to identifying the best individual sites. Conservation programs typically seek to identify whole sets of sites that

are collectively best suited to achieve the conservation aims in view. One approach is **complementarity selection**. Here, we proceed in a stepwise fashion, starting with the single best site, but then selecting at each step the site that is most complementary to those already selected, that provides most of what the existing sites lack. In the case of the coastal marine fishes around Western Australia, for example, the results of a complementarity analysis showed that more than 95% of the total of 1,855 species could be represented in just six appropriately located sections, each 100 km long (Figure 13.28).

We might also consider the **irreplaceability** of each potential area, defined as the likelihood of its being necessary if we are to achieve conservation targets or, conversely, the likelihood that one or more targets will not be achieved if the area is not included. Cowling et al., 2003 used irreplaceability analysis as part of their conservation plan for South Africa's Cape Floristic Province – a global hotspot with more than 9,000 plant species. A variety of conservation targets were identified, including, among others, the minimum acceptable number of species of *Protea* plants to be safeguarded (for which the region is famous), the minimum

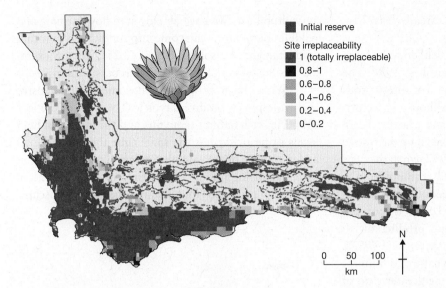

Initial reserve

Site irreplaceability
■ 1 (totally irreplaceable)
■ 0.8–1
▨ 0.6–0.8
▨ 0.4–0.6
▨ 0.2–0.4
□ 0–0.2

0   50   100
km

N

FIGURE 13.29 Map of South Africa's Cape Floristic Region showing site irreplaceability values for achieving a range of conservation targets in the 20-year conservation plan for the region. Irreplaceability is a measure, varying from 0 to 1, which indicates the relative importance of an area for the achievement of regional conservation targets. Existing reserves are shown in blue. (After Cowling et al., 2003.)

permissible number of ecosystem types, and even the minimum permissible number of individuals of large mammal species.

The researchers used an irreplaceability approach to guide the choice of areas to add to existing reserves that would best achieve the conservation targets (Figure 13.29), and they concluded that, in addition to areas that already have statutory protection, 42% of the Cape Floristic Province, comprising some 40,000 km², will need some level of protection. This includes all cases of high irreplaceability (> 0.8). It therefore includes some areas that are unimportant in terms of *Protea* and ecosystem types but that are critical to provide for the needs of large mammals in lowland areas.

13.5 ECOSYSTEM SERVICES

Throughout this chapter, we have seen examples where conservation efforts or concerns are concentrated on individual species, or on whole communities or habitats because of the collections of species they contain. The biodiversity that is the focus of attention in such cases has intrinsic value. But there is also a utilitarian view of nature that focuses on the value of what ecosystems provide for people to use and enjoy. The concept of ecosystem services, already touched upon in Box 13.1, formalizes this view. In this final section of the chapter, we examine this perspective further.

Ecosystem services are functions or attributes provided in support of human interests by ecosystems (natural or managed), generally saving a cost that would otherwise need to be paid. The concept of ecosystem services focuses on how ecosystems contribute to human well-being, providing a counterpoint to the purely market-driven economic reasons that often justify our degradation of nature through pollution, land

use, and other habitat destruction. We can divide ecosystem services into several components.

- *Provisioning services* include wild foods such as fish from the ocean and berries from the forest, medicinal herbs, fiber, fuel, and drinking water, the pollination of crops by bees, as well as the products of cultivation in agro-ecosystems.

- Nature also contributes the *cultural services* of aesthetic fulfillment and educational and recreational opportunities.

- *Regulating services* include the ecosystem's ability to break down or filter out pollutants, the moderation by forests and wetlands of disturbances such as floods, and the ecosystem's ability to regulate climate (via the capture or 'sequestration' by plants of the greenhouse gas carbon dioxide).

- Finally, and underlying all the others, there are *supporting services* such as primary production, the nutrient cycling upon which productivity is based, and soil formation.

Economists can put a value on nature in a variety of ways. A provisioning service for which there is a market is straightforward—values are easily ascribed to clean water for drinking or irrigation, to fish from the ocean, and to medicinal products from the forest. A more imaginative approach is required in other situations. Thus, the travel costs that tourists are willing to pay to visit a natural area provides at least a minimum value for the cultural service provided. We can also estimate how much we would need to spend to replace an ecosystem service with a man-made alternative, for example, by substituting the natural waste disposal capacity of a wetland by building a treatment works.

putting a value on Nature

And when an ecosystem service has already been lost, the real costs become apparent.

Take, for example, the largely deliberate burning of 50,000 km² of Indonesian vegetation in 1997. The economic cost was US$4.5 billion in lost forest products and agriculture, increased greenhouse gas emissions, reductions in tourism, and health care expenditure on 12 million people affected by the smoke (Balmford & Bond 2005). Even in 1997, when the concept of ecosystem services was first beginning to take hold, an aggregation of all the ecosystem services worldwide arrived at an estimate of US$38 trillion per year—more than the gross domestic product of all nations on the planet combined (Costanza et al., 1997). This view of the utilitarian value of nature to humans provides persuasive reasons for taking greater care of ecosystems and the biodiversity they contain.

In the case of three important provisioning services—production of crops, livestock, and aquaculture—human activities have had a positive effect. Likewise, the abandonment of agricultural land with regrowth of forest in many parts of the temperate zone over the past century has had a positive influence on the terrestrial sink for atmospheric carbon dioxide (Chapter 12). But we have degraded most of the other ecosystem services (Millennium Ecosystem Assessment, 2005). As we shall discuss in Chapter 14, many fisheries are now overexploited (a negative effect on this provisioning service), while intensive agriculture has worked against the ecosystem's ability to replace soil lost to erosion (a regulating service). The continuing loss of forest in tropical regions has negative effects on the ability of the terrestrial ecosystem to regulate river flow, because deforestation increases flow during flooding and decreases it during dry periods. And, as we saw in Chapters 1 and 11, deforestation (or even just the loss of riverside vegetation) can diminish the terrestrial ecosystem's capacity to hold and recycle nutrients (another regulating service), releasing large quantities of nitrate and other plant nutrients into waterways.

The modification of an ecosystem to enhance one service generally comes at a cost to other services previously provided. Thus, the intensification of agriculture to produce more crop per hectare – a provisioning service – is accompanied by the loss of regulating services such as nutrient uptake and of cultural services such as sacred sites, streamside walks, and valued biodiversity (Townsend, 2007).

We humans have appropriated much of global net primary production for ourselves, especially through

| positive and negative human effects on ecosystem services |
| farming and ecosystem services |

agricultural use. We have already seen how farming and wildlife conservation can come into conflict and how that conflict might be resolved (Figure 13.21). We now examine the interaction of farming with ecosystem services.

When farm production becomes too intensive and widespread, biodiversity is lost because of the loss of species-rich habitat remnants and the impact of high levels of pesticides. At the same time there may be an adverse effect on ecosystem services like the provision of water of high quality for drinking and recreation. Normally provided free from a healthy landscape,

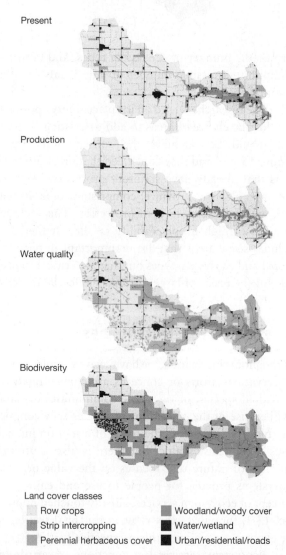

Present

Production

Water quality

Biodiversity

Land cover classes

▢ Row crops	▨ Woodland/woody cover
▨ Strip intercropping	■ Water/wetland
▨ Perennial herbaceous cover	■ Urban/residential/roads

FIGURE 13.30 Present landscape (top) and alternative future scenarios for the Walnut Creek catchment area in Iowa. In comparison to the current situation, row crops increase at the expense of perennial cover in the 'production' scenario. In the 'water quality' scenario, we see an increase in perennial cover (pasture and forage crops) and wider riparian buffers. In the 'biodiversity' scenario, there are increases in strip intercropping, the wide riparian buffers and the extensive prairie, forest, and wetland restoration reserves. (From Santelmann et al., 2004.)

these can be lost because of the input of large quantities of nitrogen and phosphorus, fine sediment from eroding land, and an increase in water-borne pathogens from farm animals. These impacts of agriculture depend on the proportion of the landscape that is used for production. One small farm – even if there is excessive use of plow, fertilizer and pesticide – will have little effect on biodiversity and water quality in the landscape as a whole. It is the cumulative effects of larger and larger areas of intensive agriculture that deplete the region's biodiversity and reduce the quality of water needed for other human activities. This suggests that management of agricultural landscapes needs to operate at a regional scale.

Santelmann et al. (2004) integrated the knowledge of experts in environmental, economic, and sociological disciplines into alternative visions of a particular landscape—the catchment area of Walnut Creek, in an intensively farmed part of Iowa. They mapped the present pattern of land use and then created three future management scenarios, assessing how farm income, water quality, and biodiversity would be expected to change under each. A *production* scenario imagines what the catchment will look like in 25 years if continued priority is given to corn and soybean production ('row crops'), following a policy that encourages extension of cultivation to all the highly productive soils available in the catchment. A *water quality* scenario envisions a new (hypothetical) federal policy that enforces chemical standards for river and ground water, and supports agricultural

practices that reduce soil erosion. Finally, a *biodiversity* scenario assumes a new (hypothetical) federal policy to increase the abundance and richness of native plants and animals—in this case a network of biodiversity reserves is established with connecting habitat corridors.

Figure 13.30 compares the distribution of agricultural and 'natural' habitats in 25 years' time for the three scenarios. Compared with the current situation, the production scenario produces the most homogeneous landscape, with an increase in row crops and a decrease in the less profitable pasture and forage crops. The water quality scenario leads to more extensive riverbank strips of natural vegetation cover and more perennial crop cover (pasture and forage crops), which are conducive to both higher water quality and biodiversity. Finally, the biodiversity scenario has even wider riverbank strips, together with prairie, forest and wetland reserves, and an increase in 'strip intercropping,' a farming practice that reduces soil erosion.

The percentage changes after 25 years in economic, water quality, and biodiversity terms are shown for each scenario in Figure 13.31. Not surprisingly, the biodiversity scenario ranks highest for improvements in plant and animal biodiversity. More unexpected is the finding that the land use and management practices required by the biodiversity scenario are nearly as profitable to farmers as current practices. The biodiversity scenario also ranks highest in terms of acceptability to farmers (based on farmer ratings of

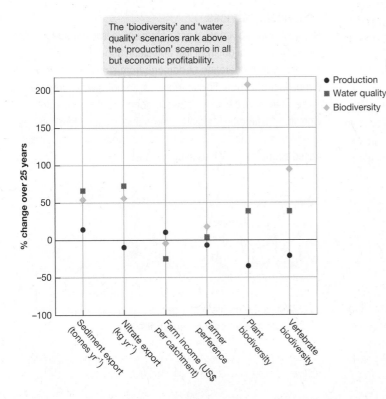

FIGURE 13.31 Percent change in the Walnut Creek catchment area for each scenario ('production,' 'water quality,' and 'biodiversity,' compared to the current situation) in water quality measures (sediment, nitrate concentration), an economic measure (farm income in the catchment as a whole), a measure of farmer preference for each scenario (based on farmer ratings of images of what the land cover would look like under each scenario) and two biodiversity measures (plant and vertebrate). (After Santelmann et al., 2004.)

images of land cover under each scenario), and it provides water quality improvements similar in magnitude to those in the water quality scenario. Despite the slightly higher profitability of the production scenario, it seems that the farmers would not be unhappy with a biodiversity strategy that provides the greatest benefits to the community at large in terms of biodiversity and ecosystem services.

This is an optimistic, and we hope not an overoptimistic, note on which to end. The problems of conservation are challenging, and those of combining conservation with other conflicting, and justified, demands on the land or water are even more so. But examples like this give us grounds for believing that, together, good science, public education, and consultation can help us in rising to these challenges.

SUMMARY

The need for conservation

Conservation is the science concerned with increasing the probability that the Earth's species and communities (or, more generally, its biodiversity) will persist into the future. Biodiversity is, at its most basic, the number of species present, but it can also be viewed at smaller scales (e.g., genetic variation within populations) and larger scales (e.g., the variety of community types present in a region). About 1.8 million species have so far been named, but the real number is probably between 3 and 100 million. The current observed rate of extinction may be as much as 100–1,000 times the background rate indicated by the fossil record.

Small populations

A population's risk of extinction can be classified in a graded scale combining the probability of going extinct and the timescale over which this might occur. Smaller populations are intrinsically most likely to go extinct. Both genetic and demographic problems contribute to this risk. Naturally, habitat reduction tends to make populations smaller and at greater risk of extinction.

Threats to biodiversity

Overexploitation, the disruption of habitats, for example by pollution or urban incursion, global environmental change, introduced or invasive species, and infectious disease may all pose serious threats to individual species and to biodiversity generally. Indeed, these may combine with one another and with genetic problems to drag threatened species into an extinction vortex. The extinction of one species may itself pose a threat to further species when chains of extinctions are generated.

Conservation in practice

Population viability analyses can be valuable in identifying species particularly at risk and in quantifying that risk. Most often, attempts to meet the aims of conservation involve the establishment of protected areas, though these must satisfy a range of criteria that may conflict, including conservation itself, human recreation, and the exploitation of natural resources. Selecting conservation areas is therefore never easy, and even satisfying the need to conserve requires that we consider the quality, size, and connectivity of protected areas, and also we consider areas in groups, not in isolation.

Ecosystem services

Nature provides a variety of ecosystem services to human society, including the availability of fish and clean drinking water, filtering of pollutants, moderation of damage from storms, and sinks for atmospheric carbon dioxide. Humans do not always appreciate these "free" services until they are gone, but they are valuable. One study estimated the global total value of ecosystem services as more than the total gross domestic product for all market-based economic activities in all nations.

REVIEW QUESTIONS

1 Explain the different meanings we can give to the word 'biodiversity' and describe the utility of each of these meanings in the context of conservation.

2 What is the nature of the genetic problems encountered by populations when they are small?

3   Explain the difference between the disturbance, destruction, and degradation of ecological communities. How can each be countered?

4   Explain, with examples, how the loss or introduction of a single species can have conservation consequences throughout a whole ecological community.

5   What are Population Viability Analyses and how are they carried out?

6   The IUCN has classified Protected Areas into a range of types. Discuss the rationale behind each.

7   Describe how ecosystems provide the service of clean drinking water to society.

Challenge Questions

1   Of the estimated 3–30 million species on Earth, only about 1.8 million have so far been named. How important is it for the conservation of biodiversity that we can name the species involved?

2   A famous ecologist of the early 20th century, A. G. Tansley, when asked what he meant by nature conservation, said it was maintaining the world in the state he knew as a child. From your perspective, in the early years of the 21st century, how would you define the aims of conservation biology?

3   Should all marine protected areas be no-take zones, with no fishing allowed? Fully justify your answer.

Peter Mcbride/Getty Images

Chapter 14

The ecology of human population growth, disease, and food supply

KEY CONCEPTS

After reading this chapter, you will be able to:

• explain how a growing human population and increased consumption put pressures on the environment

• identify the roots of and problems with global overcrowding

• describe how human health is affected by interactions with local and global ecosystems

• trace the effects of the increased use of synthetic nitrogen fertilizer over the past half century

• identify advantages and disadvantages of chemical and biological approaches to pest control, and describe why agricultural monocultures encourage pests

• explain the constraints on global food production

• describe how overfishing has depleted the world's fisheries' resources, and elaborate on the environmental consequences of aquaculture

As the human population has grown, new technologies have developed, and consumption per person has increased, we have had an ever-growing impact on the landscape and planet. Physical degradation and chemical pollution from agriculture, resource use, urban life, and industry have adversely affected human health and many ecosystem services that contributed greatly to human welfare. Our environmental problems have ecological, economic, and socio-political dimensions; we will need a multidisciplinary approach to find solutions.

14.1 HUMAN USE OF ECOLOGICAL RESOURCES

In this chapter, we explore how ecology can help us understand the issues of human population growth, disease, agriculture, and resource use. Human population growth itself has certainly affected the environment and the natural functioning of ecosystems, but equally important are the huge increases in consumption per person and the ever-growing new technologies that support it (Figure 14.1). During the first half of the 20th century, the global population increased by 40%, from 1.8 to 2.5 billion people, but since then the population has almost tripled to over 7 billion. The global gross domestic production also increased by 75% in the first half of the 20th century (from $3 trillion to $5.3 trillion U.S. dollars), but since then it has grown more than 10-fold to $55 trillion. What are the results of this affluence? Increases in consumption of food, energy, and material goods have been massive, with major local and global consequences. The science of ecology can help us understand these consequences and find more sustainable ways to move forward.

Not only has the world's population grown, the percentage of people living in cities has grown steadily as well (Figure 14.2), and by 2010 the number of humans living in cities equaled the number in rural environments for the first time in history. The United Nations predicts this trend will continue, with two-thirds of the global population living in cities by 2050. In some ways, this trend eases pressures on the environment: per capita greenhouse gas emissions for urban residents tend to be far lower than for the population as a whole, for example. But cities themselves take space, replacing rural landscapes of

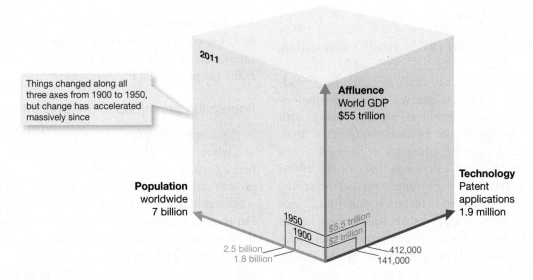

FIGURE 14.1 Humans are increasingly stressing the Earth's environment due to the interaction of population growth, growing affluence and consumption, and the development of new technologies. (Source: Kolbert, 2011.)

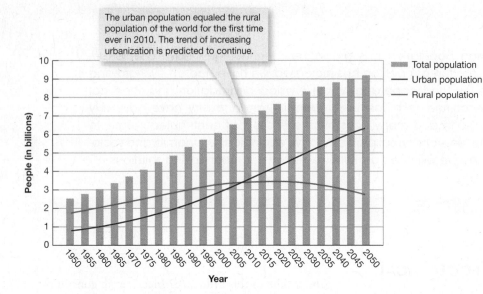

The urban population equaled the rural population of the world for the first time ever in 2010. The trend of increasing urbanization is predicted to continue.

FIGURE 14.2 The total population of the world, the rural population, and the urban population from 1950 to 2010 and projections from the United Nations for future trends through 2050. (Source: UNEP, 2014.)

agricultural and natural ecosystems, and they are sites of concentrated air and water pollution driven by the concentrated use of energy and resources and large quantities of human sewage. Increasing urbanization is made possible by an increasing intensification of agriculture. Fewer people are required to work the fields as agricultural productivity per area increases, fueled by inputs of fertilizer and fossil fuels (for tractors, irrigation, and other uses).

As we noted in Chapter 11, humans already appropriate a high proportion of the global rate of NPP for our own uses (see Figure 11.2). Humans will use even more of this NPP in the future, as food consumption increases and more crops are used to produce biofuel energy. Average food consumption per person has increased steadily over the past 50 years, from 2,360 calories per day in the mid-1960s to 2,940 calories today (World Health Organization, 2013). According to the U.S. National Institutes of Health, 2,250 calories per day is sufficient for a moderately active adult. Of course, hunger and malnutrition remain major problems in many areas, with perhaps 1 billion people receiving insufficient food. Yet even in developing countries, consumption has increased from 2,054 calories per day in the 1960s to 2,850 today. Why does hunger persist? Presently, hunger results not from inadequate global food production but from unequal distribution. In fact, health authorities are increasingly concerned about overconsumption of food: the global incidence of obesity has doubled over the past 30 years, and more than one-third of adults in the world are now overweight

(World Health Organization, 2013). Nonetheless, the global increase in food production is predicted to continue.

14.2 THE HUMAN POPULATION PROBLEM

The root of most, if not all, the environmental problems facing us is the Earth's large and growing population of humans. More people means an increased requirement for energy, nonrenewable resources like oil and minerals, renewable resources like fish and forests, and food production through agriculture. Clearly things cannot go on the way they are, though it is not so clear exactly what the problem is (Box 14.1). We examine first the size and growth rate of the global human population and how we reached our current state, then how accurately we can project the future. Finally we ask, 'How many people can the Earth support?'

Population growth up to the present

In exponential growth (see Chapter 5), the population as a whole grows at an accelerating rate (a plot of numbers against time sweeps upwards), simply because the growth rate is a product of the individual rate (which is constant) and the accelerating number of individuals. For thousands of years, as Figure 14.3 shows, growth, though it may have been exponential, was slow, despite a jump around 10,000 years ago at the dawn of agriculture. In brief periods, such as during the Black Death in Europe around 700 years ago, the world's population

14.1   ECOncerns

What *is* the human population problem?

What is 'the human population problem'? This is not an easy question, but here are some possible answers (Cohen, 1995, 2003, 2005). The real problem, of course, may be a combination of these—or of these and others. There is little doubt, though, that there *is* a problem, and that the problem is 'ours,' collectively.

• *The present size of the global human population is unsustainably high.* Around AD 200, when there were about a quarter of a billion people on Earth, Quintus Septimus Florens Tertullianus wrote that 'we are burdensome to the world, the resources are scarcely adequate to us.' By 2013 the population was estimated at more than 7 billion.

• *It is not the size of the population but its distribution over the Earth that is unsustainable.* The fraction of the population concentrated in urban environments has risen from around 3% in 1800 to more than 50% today. Each agricultural worker today has to feed her- or himself plus one city dweller; by 2050 that will have risen to two urbanites (Cohen, 2005).

• *The present rate of growth in size of the global human population is unsustainably high.* Before the agricultural revolution of the 18th century, the human population had taken roughly 1,000 years to double in size. The most recent doubling took just 39 years (Cohen 2001).

• *It is not the size but the age distribution of the global human population that is unsustainable.* In developed regions, the percentage of the population over 65 rose from 7.6% in 1950 to 12.1% in 1990. This proportion is now increasing faster still, as the large cohort born after World War II passes 65.

- *It is not the limits on resources but their uneven distribution among the global population that is unsustainable.* In 1992, the 830 million people of the world's richest countries enjoyed an average income equivalent to US$22,000 per annum. The 2.6 billion people in the middle-income countries received $1,600. But the 2 billion in the poorest countries got just $400. These averages themselves hide other enormous inequalities.

1 What role or responsibility does the individual, as opposed to government, have in responding to the human population problem?

2 We list five different variants of the human population problem, above. Which raise questions about the relationship between the developed and the developing parts of the world or between the 'haves' and the 'have nots'?

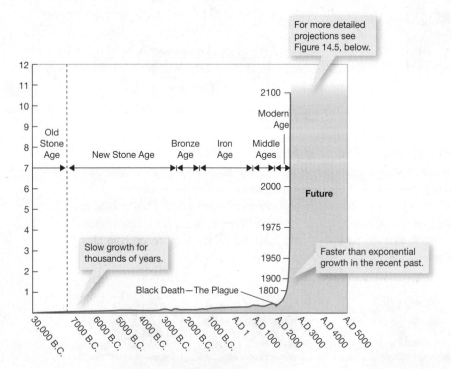

FIGURE 14.3 The estimated size of the global human population over the past 30,000 years and projected into the future. (After Population Reference Bureau, 2006.)

actually declined. Then, with growing urbanization and industrialization, growth accelerated and was faster than exponential for several centuries. Recently, however, it has again slowed, as we describe below.

Predicting the future

We can look at human populations of the past to predict future population sizes and rates of growth. But simply projecting forward would be making the almost certainly false assumption that the future will be just like the past. Prediction, in contrast, requires us to *understand* what has happened in the past, how the present differs, and how these differences might translate into future patterns of population growth. In particular, we must recognize that like all ecological populations, ours is heterogeneous.

We refer to a **demographic transition** as a switch from high birth and death rates to low birth and death rates. We can then recognize three human subpopulations: those that passed through the demographic transition 'early,' pre-1945 (Figure 14.4), 'late' (since 1945), and 'not yet.' The pattern is as follows. Initially, both the birth rate and the death rate are high, but the former is only slightly greater than the latter, so the overall rate of population increase is only moderate or small. (We assume this was the case in all human populations in much of the past.) Next, the death rate declines while the birth rate remains high, so the population growth rate increases. Next, however, the birth rate also declines until it is similar to or perhaps even lower than the death rate. The population growth rate eventually declines again (sometimes becoming negative, with death rate higher than birth rate), though with a far larger population than before the transition began.

The commonly proposed explanation is that transition is an inevitable consequence of industrialization, education, and general modernization, leading first, through medical advances, to the drop in death rates, and then, through the choices people make (such as delaying having children), to the drop in birth rates. Certainly, when we consider all the regional populations of the world together, there has been a dramatic decline from the peak population growth rate of about 2.1% per year in 1965–1970 to around 1.1–1.2% per year today (Figure 14.5a). And, as Cohen (2005) points out, while population growth rate has fallen in the past due to plague and war, never before the 20th century was such a fall 'voluntary.'

Two future inevitabilities

Even if it were possible to bring demographic transition to all countries of the world so birth rates equaled death rates and growth were zero, would the 'population problem' be solved? No, for at least two important reasons. First, there is a big difference in age structure between a population with equal but high birth and death rates and one in which both are low. We saw in Chapter 5 that the net reproductive rate of a population is a reflection of age-related patterns of survival and birth. A given net reproductive rate, though, can be arrived at through a literally infinite number of different birth and death patterns, and these combinations themselves give rise to different age structures within the population. If birth rates are high but survival rates low ('pre-transition'), there will be many young and relatively few old individuals in the population. But if

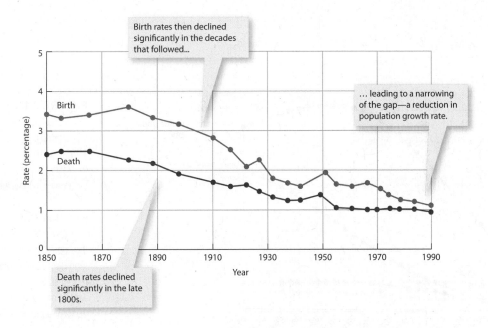

FIGURE 14.4 The birth and death rates in Europe since 1850. The annual net rate of population growth is given by the gap between the two. (After Cohen, 1995.)

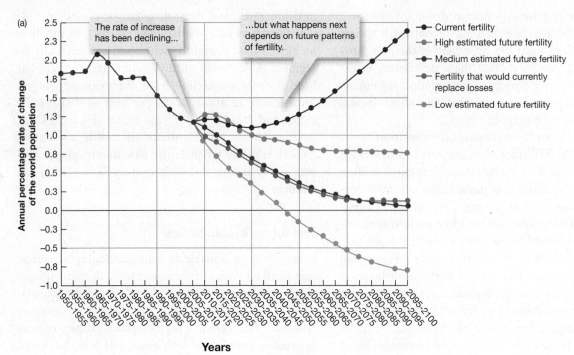

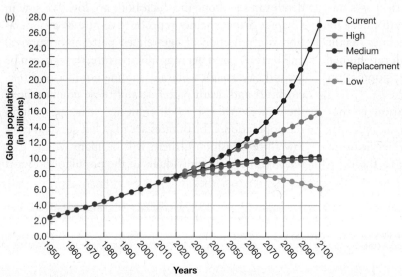

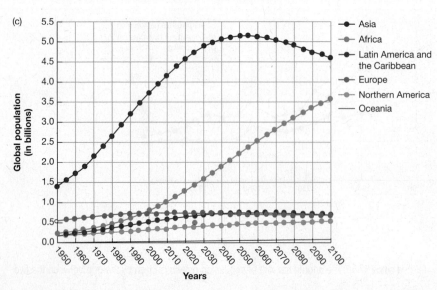

FIGURE 14.5 (a) The average annual percentage rate of change of the world population observed from 1950 to 2010, and projected forward to 2100 on the basis of various assumptions about future fertility rates. (b) The estimated size of the world's population from 1950 and 2010 and projected forward to 2100 on the basis of various assumptions regarding fertility rates. (c) The estimated size of the populations of the world's main regions from 1950 and 2010 and projected forward to 2100 assuming 'medium' fertility rates. (After United Nations, 2011.)

birth rates are low and survival rates high – the ideal to which we might aspire post-transition – relatively few young, productive individuals must support the many who are old, unproductive, and dependent. The size and growth rates of the human population are not the only problems: the age structure of a population adds yet another (Figure 14.6).

Suppose our understanding were so sophisticated, however, and our power so complete, that we could establish equal birth and death rates tomorrow.

Would the human population stop growing? The answer, once again, is 'No.' Population growth has its own momentum, and even with birth rate matched to death rate, it would take many years to establish a stable age structure, while considerable growth continued in the meantime. According to projections by the United Nations, even with low fertility the world's population will grow from slightly more than 7 billion today to more than 8 billion by 2050 (Figure 14.5b). There are many more babies in the world now than

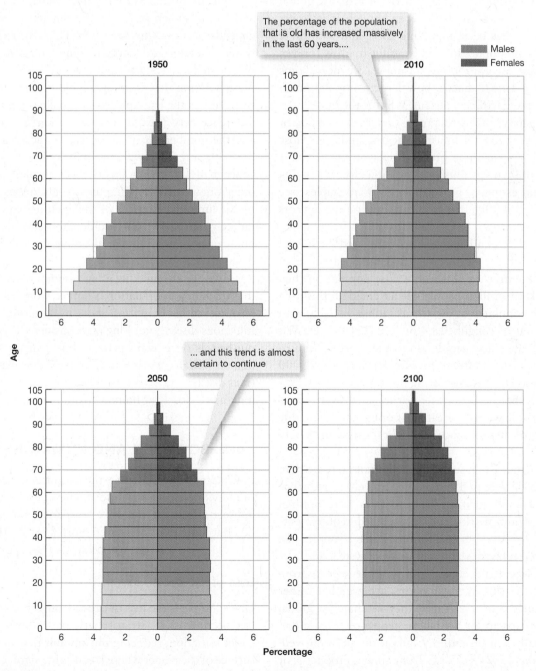

FIGURE 14.6 World 'population pyramids,' displayed as percentages in each age-sex class, for 1950 and 2010 and projected forward to 2050 and 2100 assuming 'medium' fertility rates (see Figure 14.5). (After United Nations, 2011.)

25 years ago, so even if birth rate per capita drops considerably now, there will still be many more births in 25 years' time than now, and these children, in turn, will continue the momentum before an approximately stable age structure is eventually established. As Figure 14.5c shows, it is the populations in the developing regions of the world, dominated by young individuals, that will provide most of the momentum for further population growth.

A global carrying capacity?

The current rate of increase in the size of the global population is unsustainable even though it is lower now than it has been; in a finite space and with finite resources, no population can continue to grow forever. To suggest an appropriate response, we need some sense of a target, that is, how large a population of humans the Earth can sustain. What is the global carrying capacity? Estimates proposed over the last 300 or so years vary to an astonishing degree, and even those since 1970 span three orders of magnitude—from 1 to 1000 billion. To illustrate the difficulty of arriving at a good estimate, we look at a few examples (see Cohen, 1995, 2005, for further details on the authors mentioned below).

In 1679, van Leeuwenhoek estimated the inhabited area of the Earth as 13,385 times larger than his home nation of Holland, whose population then was about 1 million people. He assumed all this area could be populated as densely as Holland, yielding an upper limit of roughly 13.4 billion. In 1967, De Wit asked, 'How many people can live on Earth if photosynthesis is the limiting process?' His answer—1,000 billion—assumed the growing season varies with latitude, but that neither water nor minerals were limiting. If people wanted to eat meat or have a reasonable amount of living space, De Wit's estimate would be lower.

By contrast, in suggesting a limit of no more than 1 billion, Hulett in 1970 assumed levels of affluence and consumption in the United States were optimal for the whole world, and he included requirements not only for food but also for renewable resources like wood and nonrenewable resources like steel and aluminum. Kates and others, in a series of reports from 1988, made similar assumptions using global rather than United States averages; they estimated a global carrying capacity of 5.9 billion people on a basic diet (principally vegetarian), 3.9 billion on an 'improved' diet (about 15% of calories from animal products), or 2.9 billion on a diet with 25% of calories from animal products.

More recently, Wackernagel and his colleagues in 2002 sought to quantify the amount of land humans use to supply resources and to absorb wastes (embodied in their 'ecological footprint' concept). Their preliminary assessment was that people were using 70% of the biosphere's capacity in 1961 and 120% by 1999. They reasoned, in other words, that global carrying capacity had been exceeded before the turn of the millennium—when our population was about 6 billion.

As Cohen (2005) has pointed out, many estimates have been based or rely heavily on a single dimension – biologically productive land area, water, energy, food, and so on—when in reality the impact of one factor depends on the value of others. Thus, for example, if water is scarce and energy is abundant, water can be desalinated and transported to where it is in short supply, a solution that is not available if energy is expensive. And as the examples above make clear, there is a difference between the number the Earth can support and the number it can support at an acceptable standard of living. The higher estimates come closer to the concept of a carrying capacity we normally apply to other organisms (see Chapter 5)—a number imposed by the limiting resources of the environment. But it is unlikely that many of us would choose to live crushed up against an environmental ceiling or wish it on our descendants.

We are also making a big assumption in implying that the human population is limited from below by its resources rather than from above by its natural enemies. Infectious disease, not long ago considered largely vanquished, is once again perceived as a major threat to human welfare. We saw in Chapter 7 that many infectious diseases thrive best in the densest populations. We look more closely at human health in the next section.

14.3 ECOLOGY AND HUMAN HEALTH

In Chapter 13 we saw that the loss of biodiversity and of individual species is unfortunate in itself and for its indirect effects on our own prosperity and quality of life. But human health is even more directly affected by the ecological changes we see around us.

Loss of the ozone layer

One example is the loss of ozone from the stratosphere (see Section 12.4). Stratospheric ozone absorbs ultraviolet radiation, reducing the amount that reaches the surface of the planet where it can cause significant environmental and human-health consequences. Ozone in the stratosphere began to decrease in the 1970s at a rate of 4% per decade. The most dramatic effect was the growing hole in the ozone layer centered over the Antarctic (Shanklin 2010; Figure 14.7), but levels have declined globally. Naturally occurring chemicals such as nitrous oxide (see Section 12.4) help cause the decline, but a breakthrough discovery revealed the major cause was a buildup in the atmosphere of chlorofluorocarbons (CFCs), manufactured as refrigerants and particularly powerful in catalyzing the destruction of the ozone. The Montreal Protocol of 1987 banned chlorofluorocarbons globally as a result and has been hugely helpful in slowing the rate of ozone depletion, but nitrous oxide, which can play a similar role in destroying stratospheric ozone, has been having an increasing effect as its concentrations rise—one of many consequences of the human acceleration of the global nitrogen cycle—and ozone depletion remains a critical issue.

Among the medical consequences of increased ultraviolet radiation penetrating the depleted ozone layer is a rise in skin cancers. Data for Northern Ireland are shown in Figure 14.8. We are not helpless in the face of such increased threats, however. In Australia, for example, where the threat has been taken particularly seriously and educational campaigns have been thorough, the incidence rate of cutaneous melanomas (a type of skin cancer) has stabilized in men and actually declined in women, despite the increased risk.

Extreme events

A variety of atmospheric changes have led to a change in global climate patterns (Chapter 12). Average values are changing—for example, mean global temperatures are rising—but there are also more frequent extreme climate events, as defined by the international Expert Team on Climate Change Detection Monitoring and Indices (ETCCDMI): heat waves, floods, droughts, high winds, and so on. Past patterns and forward projections are shown for one extreme event in Figure 14.9: the annual number of 'tropical nights' in Europe, when temperatures remain above 20°C (68°F), indicative of heat wave conditions. To validate the forward projections, the output of the chosen climate model (shown in Figure 14.9) was first compared with observed values up to the present. The correspondence was good. The same model was then used to project forward on the basis of two contrasting scenarios proposed by the International Panel on Climate Change. Model B1 imagines a rapid switch to a less production-intensive economy and the introduction of clean and resource-efficient technologies; model A1B stays much closer to

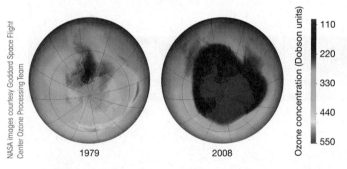

FIGURE 14.7 The 'hole' in the ozone layer centered over the Antarctic soon after it was first identified in 1979 and in 2008. (After Shanklin, 2010.)

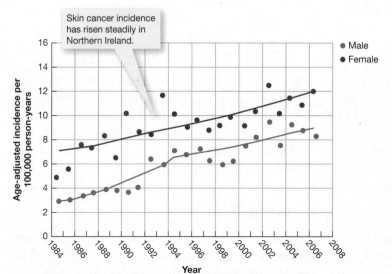

FIGURE 14.8 The incidence of cutaneous melanomas (a type of skin cancer) in Northern Ireland since 1984. Trend lines are significant ($P < 0.05$). That for males shows a lower rate of increase after 1995 (similar to women) than before. (After Montella et al., 2009.)

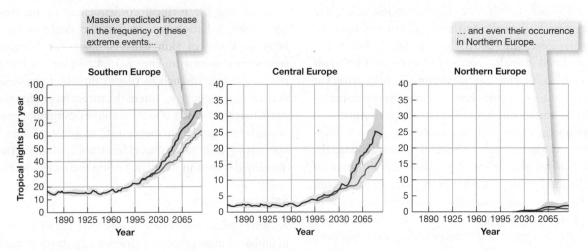

FIGURE 14.9 The number of 'tropical nights' in Europe (when temperature remained above 20°C) in the past and projected into the future. The black line uses a climatic simulation model but is validated by direct observations. The red and blue lines take that model and project into the future on the basis of two scenarios, model B1 (optimistic; better control of greenhouse gas emissions: blue) and model A1B (less optimistic; emissions more status quo; red). The lines are the average outputs from different runs of the models; the shading covers the whole range of outputs. (After Sillmann & Roekner, 2008.)

TABLE 14.1 The numbers of extreme weather events, and people killed and severely affected by them, for each region of the world in the 1980s and 1990s.

	1980s			1990s		
	Events	Killed (thousands)	Affected (millions)	Events	Killed (thousands)	Affected (millions)
Africa	243	417	137.8	247	10	104.3
Eastern Europe	66	2	0.1	150	5	12.4
Eastern Mediterranean	94	162	17.8	139	14	36.1
Latin America and Caribbean	265	12	54.1	298	59	30.7
South East Asia	242	54	850.5	286	458	427.4
Western Pacific	375	36	273.1	381	48	1,199.8
Developed	563	10	2.8	577	6	40.8
Total	1,848	692	1,336	2,078	601	1,851

After World Health Organization, 2013a.

the current status quo. But both scenarios predict still further acceleration in the number of tropical nights in southern and central Europe, and even in northern Europe by the end of the century.

The seriousness of the human health consequences is evident from Table 14.1. The World Health Organization estimates, for example, that more than 600,000 people were killed worldwide as a result of extreme climate events in the 1990s, and more than three times as many suffered severe health effects. A July 1995 heat wave in Chicago caused 514 heat-related deaths and 3,300 emergency hospital admissions. And the excess summer mortality attributable to climate change is expected to reach 500–1,000 per year in New York by 2050 if the population adjusts to that change (by changing their behavior, the city's infrastructure and so on), and even more if it does not (Kalkstein et al., 1997).

Changing global patterns of infection

Infectious diseases are also affected by the changing climate, both in their local and global distribution among humans and in their intensity and prevalence.

Even in rich countries with sophisticated sanitation like the United States, fatalities from food-borne illness are in the thousands annually, and hospitalizations in the hundreds of thousands. Incidence increases sharply with temperature, and particularly above a threshold value (Figure 14.10)—a pattern that is likely to become increasingly common if extreme climate events also increase.

Most likely to be affected are vector-borne diseases like malaria, since the vectors themselves, often insects, are especially at the mercy of environmental temperatures because they are ectotherms (see Section 3.2).

They will be affected not only by average temperatures but by the even more uncertain daily and yearly patterns. The effects of climate on vectors are complex, and our knowledge of even the most important ones is far from complete. Clearly we must be cautious, especially as earlier suggestions of massive range expansions have tended to be superseded by more modest predictions, as shown, for example, for malaria in Figure 14.11. On the other hand, we must not be complacent. What we need is better predictions, and for that we need improved ecological understanding.

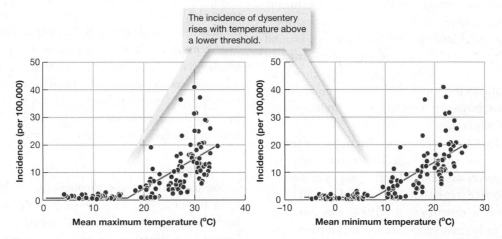

FIGURE 14.10 The relationships between mean monthly maximum and minimum temperatures and the incidence of dysentery caused by *Shigella* bacteria in the city of Jinan in northeast China between 1996 and 2003. Fitted models are significant at $P < 0.01$. (After Zhang et al., 2007.)

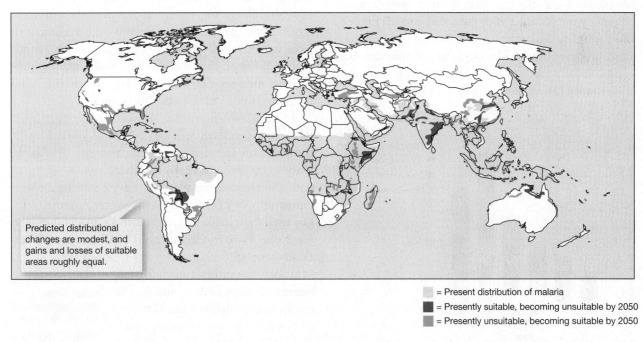

FIGURE 14.11 The present distribution of *falciparum* malaria (the most serious form, transmitted by anopheline mosquitoes), and the changes predicted by 2050 to the regions in the world that are suitable for *falciparum* malaria. (Malaria is not necessarily present wherever conditions are suitable.) Predictions are based on a widely used future climate scenario, with a 1% annual increase in overall greenhouse gas concentrations and relative high sensitivity of climate to those concentrations. (After Rogers & Randolph, 2000.)

Emerging infectious diseases

Recent years have also brought a rise in the number of **emerging infectious diseases**: diseases appearing in the human population for the first time or expanding rapidly in incidence or geographic range (much faster than the rate of environmental change). The most notable is HIV, but many others have acquired a high public profile, for example Ebola virus, severe acute respiratory syndrome (SARS), and Lyme disease in the United States. The rising numbers of new emerging infections since the 1940s is shown in Figure 14.12, which also identifies their likely origin. Around 60% are **zoonoses**, infections that circulate naturally in nonhuman vertebrate hosts but can be transmitted from these to humans. Of these zoonoses, around 70% originate in wildlife (as opposed to domesticated animals). In some cases, for example HIV, they may become established enough to become human infections. In most, however, there is little or no human-to-human transmission.

As ecologists, we might ask what accounts for the sudden success of any biological invasion—increased migration? increased survival following migration? a new evolutionary step? There are examples of all these. The peak of new diseases in the 1980s was driven by an increase in human susceptibility to infection associated with the HIV/AIDS pandemic. The emergence of Lyme disease in the 1960s was a case of increased migration of the pathogen *Borrelia burgdorferi*, following land use changes that brought people more frequently into contact with ticks that had fed on deer and squirrels, the natural hosts. And many emerging infections have been driven by the use of antimicrobial agents, usually antibiotics, and the resulting evolution of resistance (as we discuss in Box 14.2).

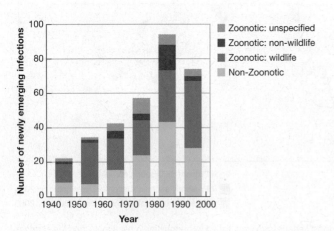

FIGURE 14.12 The number of newly emerging infections since the 1940s, classified according to whether they were zoonotic or not in their origin. (After Jones et al., 2008.)

14.4 SYNTHETIC FERTILIZER AND THE INTENSIFICATION OF AGRICULTURE

The huge increase in the global population would not have been possible without an equally huge intensification in agricultural production. For most of the history of our species—since humans evolved 300,000 years ago—all people were hunter-gatherers, obtaining their food by harvesting natural populations of animals and plants. The first global increase in human population could occur only with the development of agriculture, beginning in many parts of the world independently some 10,000 years ago. Plants and animals were domesticated and, through gradual experimentation over centuries and millennia, humans increased their productivity, allowing our population densities to rise.

For much of the past 10,000 years, the global human population has been regulated by the supply of food, by disease, and by war. The rate of food production was central, since shortages contribute greatly to disease and war. Providing a larger and more secure food supply has been critical for all civilizations, and a major focus of science throughout the modern era. With the discovery by 19th-century German chemist Justus von Liebig (see Box 11.2) that nitrogen often limited agricultural productivity, the search was on for ways to find more nitrogen to support increased production. Until the 20th century, much of this came from nitrogen fixation by bacterial symbionts of plants such as alfalfa, clover, soybeans, and peanuts (all classified as leguminous). The mining of guano—the accumulated feces and urine of birds in rich deposits—on islands such as those off the shores of Peru and Chile also provided a nitrogen-rich fertilizer that was widely used from the 1700s well into the 20th century.

The 1909 invention of synthetic nitrogen fertilizer from N_2 in the air by the German chemist Fritz Haber (see Section 12.4), together with the engineering work of Carl Bosch to make it industrially practical, is now the source of most nitrogen used in agriculture (see Figure 12.24). Haber won the Nobel Prize in Chemistry in 1919 for this 'Haber-Bosch process.'

> the industrial production of nitrogen fertilizer . . .

Before synthetic nitrogen fertilizer became readily available, the return to crop fields of the nitrogen-rich manure wastes of the livestock was critical. Manure is wet and heavy, and therefore difficult and expensive to transport, mandating that the livestock were raised near the crop fields or allowed to graze directly in pastures. But

> . . . led to the intensification of animal agriculture far from the site of crop production

with relatively cheap synthetic fertilizer, farmers no longer needed animal wastes. It became more economically efficient to feed animals in concentrated feedlots, often far from the crop fields, and simply dispose of the manure as waste. These feedlot operations became a source of nitrogen pollution of both water and air (ammonia gas volatilizes to the atmosphere), in addition to nitrogen pollution from crop fields that produce grain fed to the animals.

The trend toward a spatial segregation of crop production and animal production began in the United States after the Second World War and accelerated rapidly during the 1950s and 1960s. This allowed an intensification of animal agriculture that supported a large increase in the consumption of meat. Per capita meat consumption in the United States rose by 60% to a little over 70 kg per year between 1950 and 1970 and has remained fairly constant since then. Of course, total consumption for the nation has continued to grow as the population has grown.

Over the past several decades, many other regions of the world have intensified animal agriculture and begun segregating crop and animal production. Per capita meat consumption has risen globally, with that in Latin America rapidly approaching the level of the United States (Figure 14.13a). This naturally demands an increase in crop production, with accompanying inputs of fertilizer, to support the increased animal production. The intensification of animal production leads to many other concerns about human health, such as the increase in antibiotic resistance (Box 14.2). The

> the intensification of animal agriculture goes global . . .

percentage of protein consumption met by meat is strongly related to a country's affluence (Figure 14.13b). We can expect meat consumption and intensification of animal agriculture to increase as the wealth of developing countries continues to grow.

Recognizing that meat consumption and the intensification of animal agriculture cause significant environmental damage, a group of ecologists and biogeochemists in 2009 passed the "Barsac Declaration," urging everyone in the world to obtain no more than 35 to 40% of their protein from meat (Billen et al., 2012). This "demitarian" diet would reduce by half the meat now consumed by citizens of the United States and western Europe, is in line with what physicians recommend as an upper limit, and roughly equals the global average, although regional variations exist (Figure 14.13a). As an example of how much this could reduce nitrogen pollution, consider the case of Paris: if every Parisian were to convert to the demitarian diet and primarily eat organic food, nitrogen pollution in the Seine River would decrease by more than half. Although the reduction in meat consumption is a critical part of this reduction, replacing food grown with synthetic nitrogen fertilizer with organic food helps, as less nitrogen leaches from the organic agricultural fields per unit of crop harvested (Billen et al., 2012).

> diet matters . . .

Nitrogen losses from conventional agriculture can also be significantly reduced simply by using less fertilizer. In the "corn belt" of the upper Midwestern United States—the source both of most of the corn grown in the United

> other approaches to reduce nitrogen pollution . . .

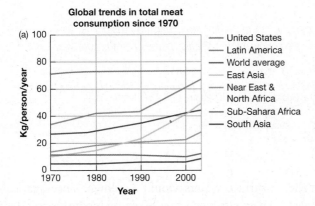

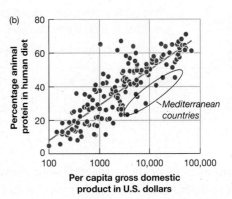

FIGURE 14.13 (a) The per capita rate of meat consumption has risen globally and in many regions over the past few decades, but has remained steady at a high level in the United States and at low levels in South Asia and Sub-Sahara Africa. The global average now of 43 kg per person per year matches the per capita meat consumption rate in the United States in 1952. (b) Share of animal products in human total protein consumption of the world's countries in 2003, plotted against their gross domestic product (GDP). (Source: (a) Modified from http://www.icsu-scope.org/unesco/USUPB06_LIVESTOCK.pdf, with Additional data from UNEP (2014); (b) Billen et al., 2012.)

14.2 ECOncerns

Scientists: overuse of antibiotics in animal agriculture endangers humans*

The overuse of antibiotics in animal agriculture and medicine is putting human lives at unnecessary risk and driving up medical costs, according to a group of 150 scientists that includes a former head of the Food and Drug Administration (FDA). Along with 50 U.S. farmers and ranchers who have opted out of using non-therapeutic antibiotics in their animal feed, the scientists are calling on the FDA and Congress to work together to regulate unnecessary use of antibiotics in animal agriculture. In twin statements released on Wednesday, the scientists and farmers said that a growing body of research supported the conclusion that overuse of antibiotics in animal agriculture is fueling a health crisis. One statement cited a study which estimated that antibiotic-resistant infections cost $20bn annually to hospitals alone.

Donald Kennedy, former FDA commissioner and president emeritus at Stanford University, said: 'There's no question that routinely administering non-therapeutic doses of antibiotics to food animals contributes to antibiotic resistance.' Kennedy said the FDA's current voluntary approach, which asks the animal drug industry to stop selling antibiotics medically important to human disease as growth promoters in animal feed, was not enough. Kennedy, who was also former editor-in-chief of *Science* magazine for eight years, said: 'Unless it reaches the industry as a regulatory requirement it will not be taken seriously.' Three decades after the FDA determined that growth-promoting uses of penicillin and tetracycline in agriculture were threatening human health, its own data show that 80% of all antimicrobial drugs sold nationally are used in animal agriculture.

Louise Slaughter, a Democratic Representative for New York who is a microbiologist, joined the scientists and others in calling for regulation of animal agriculture's use of non-therapeutic antibiotics. At a press conference organised by the Union of Concerned Scientists to launch the statement, she compared agriculture's use of antibiotics in animal feed to mothers sprinkling them on their children's cereal every morning. Antibiotic-resistant diseases now kill more Americans than HIV/Aids, Slaughter said. 'Every year, more than 100,000 Americans die from bacterial infections acquired in hospitals, and seventy percent of these infections are resistant to drugs commonly used to treat them. This abuse and overuse must stop.'

The scientists said that while the medical community had 'stepped up to the plate' by educating doctors and reducing prescriptions, the agriculture industry was lagging behind. They said that while the principal linking antibiotic resistance and non-therapeutic uses of antibiotics was widely accepted, antibiotics are still routinely added in massive quantities to animal feed, not to treat disease but to promote faster growth and to stave off diseases caused by poor diets and raising animals in overcrowded unsanitary living conditions.

The scientific community has understood the risk of feeding antibiotics to livestock for well over 40 years. Why has the practice persisted? How might scientists better communicate their concerns to the public and politicians?

*This article by Karen McVeigh was published in *The Guardian* (United Kingdom) on 19 September 2012.

States and of most of the nitrogen flowing down the Mississippi River to fuel the large hypoxic "dead zone" in the northern Gulf of Mexico—farmers routinely use more fertilizer than makes economic sense. They would save money by buying and using less fertilizer, since their crop yields and gross sales would remain as high (Figure 14.14), and the nitrogen leaching to surface waters would go down, since again this surplus nitrogen is not taken up by the crop and is readily leached. Why do farmers waste their money on this overfertilization? The answer seems to be that economic cost (if one does not include the environmental damage from the nitrogen pollution!) is relatively low, and so farmers use the extra fertilizer hoping that an

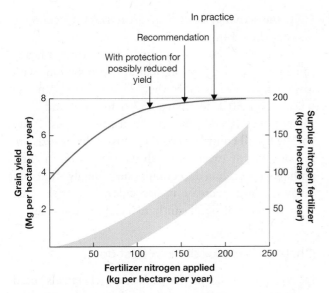

FIGURE 14.14   The yield of corn and soybeans for the average farm in the corn belt of the United States (blue curve) plotted as a function of the input of nitrogen fertilizer, and the surplus nitrogen fertilizer not used by the crop that is left in the soil to be leached away to surface and groundwaters (wide green curve). The arrows indicate the rate at which farmers actually fertilize, the rate at which farm extension agents recommend that farmers fertilize in order to maximize their profit, and an indicated level of fertilization where the farm yield and profit would go down slightly but far less nitrogen pollution would result. (Source: Howarth et al., 2005.)

above-average year climatically (perfect temperatures and rainfall) may lead to an extremely bountiful harvest. If farmers used even a little less synthetic nitrogen fertilizer than recommended for their maximum economic return, the resulting nitrogen pollution could be reduced greatly with only slight reductions in crop yields (Figure 14.14). Farmers could be reimbursed for any loss in yield in exchange for better water quality.

Agricultural scientists have developed many other tools for reducing downstream nitrogen pollution from agriculture. For instance, a winter cover crop is a secondary crop (for example rye) planted in the fall and harvested in the following spring before again planting the primary crop of economic value (for example corn). In the temperate zone, soil nitrate concentrations in the spring are often very high due to continuous low-level bacterial mineralization and lack of plant assimilation over the winter. This soil nitrate can be assimilated, though, by cover crops in early spring; otherwise, the snowmelt and rains of spring carry the nitrates away downstream. Another approach is to treat soils with substances that inhibit nitrification (see Figure 11.26); ammonium is readily adsorbed onto soil particles and is far less likely to be leached to downstream waters than nitrate, so preventing the conversion to nitrate reduces nitrogen pollution.

14.5   MONOCULTURES, PESTS, AND PESTICIDES IN AGRICULTURE

At its simplest, a **pest species** is one that humans consider undesirable, usually because it either harms us directly or damages or competes with our crops. Agriculture is an industry whose practices must make economic sense. Therefore, a pest is any species that causes economic damage.

what is a pest?

Is the aim of pest control, then, to totally eradicate the pest? Not usually. Rather, it is to reduce the population to a level at which further reduction would cost more than we would gain in improved yield. This is called the **economic injury level** or EIL. The EIL for a hypothetical pest is illustrated in Figure 14.15a. It is

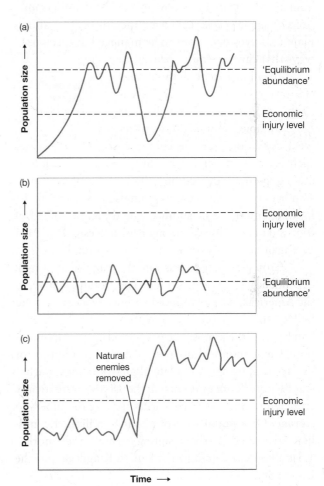

FIGURE 14.15   (a) The population fluctuations of a hypothetical pest. Abundance fluctuates around an 'equilibrium abundance' set by the pest's interactions with its food, predators, etc. It makes economic sense to control the pest when its abundance exceeds the economic injury level (EIL). Being a pest, its abundance exceeds the EIL most of the time (assuming it is not being controlled). (b) By contrast, a species that always fluctuates below its EIL cannot be a pest. (c) 'Potential' pests fluctuate normally below their EIL but rise above it in the absence of one or more of their natural enemies.

greater than zero abundance, because total eradication is not profitable. But it is also below the average abundance of the species, and it is this that makes the species a pest. If the species were naturally self-limited to a density below the EIL, then it would never make economic sense to apply control measures, and the species could not, by definition, be considered a pest (Figure 14.15b). Some species have a carrying capacity (Section 5.5) in excess of their EIL but a typical abundance that is kept below the EIL by their natural enemies (Figure 14.15c). These are *potential pests*. As we shall see below, they can become pests if their enemies are removed.

Farmers have always dealt with pests, but the intensification of agriculture has worsened the problem, since the great bulk of the human food resource is now farmed as dense populations of single species called **monocultures**. This allows plant and livestock crops to be managed in specialized ways that can maximize their productivity.

> monocultures aggravate pest problems

To what extent is the monoculture method sustainable? There is abundant evidence of its high price. For example, monocultures offer ideal conditions for the epidemic spread of diseases such as swine fever among livestock and coccidiosis among poultry. In addition, high rates of transmission between herds occur as animals are sold from one farming enterprise to another. The dramatic spread of foot and mouth disease among British livestock in 2001 and again in 2007 provides a graphic example.

> but disease spreads in monocultures

Crop plants, too, illustrate the fragility of human dependence on monocultures. The potato, native to the Americas, was introduced to Europe in the second half of the 16th century, and three centuries later it had become the almost exclusive food crop of Ireland's poor. Dense monoculture provided ideal conditions for the devastating spread of late blight (the fungal pathogen *Phytophthora infestans*) in the 1840s, dramatically reducing potato yields and decomposing the tubers in storage. Of a population of about 8 million, 1.1 million Irish died in the resulting famine and another 1.5 million emigrated to the United Kingdom and the United States.

An outbreak of southern corn leaf blight (caused again by a fungus, *Helminthosporium maydis*) developed in the southeastern United States in the late 1960s and spread rapidly after 1970. Most of the corn grown in the area had been derived from the same stock and was genetically almost uniform. This extreme monoculture allowed one specialized race of the pathogen to have devastating consequences. The damage was estimated as at least $1 billion and had repercussions on grain prices worldwide. A favorite

fruit, the banana, is also at great risk of economic disaster (Box 14.3).

The best available information indicates a heavy toll by pests on global agricultural production, with losses of somewhat more than 25% for soybean and wheat, over 30% for maize, 37% for rice, and 40% for potatoes (Oerke, 2006). Weeds, insect pests, and pathogens all contribute to these losses, but weeds are responsible for twice as much damage as insects and pathogens combined. Perhaps surprisingly (but see below), a large increase in pesticide use over the past 40 years has not significantly reduced this damage.

Chemical approaches to pest control

Heavy use of chemical **pesticides**, chemicals used to control pests, developed simultaneously with the increased use of synthetic nitrogen fertilizer after the Second World War. These pesticides include insecticides (which target insects), herbicides (which target weeds), and fungicides (which target fungi). Here we consider the sustainability of insect pest control in agriculture to illustrate the problems that arise. We could equally well have chosen the control of weeds or fungi.

Chemical insecticides are intended to control particular target pests at particular places and times. Problems arise when they are toxic to other species, and particularly when they drift beyond the target areas and persist in the environment beyond the target time. Many pesticides have become environmental pollutants. The potential for disaster was clear when massive doses of the insecticide dieldrin were applied to large areas of Illinois farmland from 1954 to 1958 in efforts to eradicate the Japanese beetle. Cattle and sheep were poisoned, 90% of farm cats and a number of dogs were killed, and 12 species of wildlife mammals and 19 species of birds suffered losses (Luckman & Decker, 1960).

The organochlorine insecticides have caused particularly severe problems because they are **biomagnified** (Figure 14.16). Biomagnification occurs when a pesticide is present in an organism that becomes the prey of another, and the predator fails to excrete or metabolize the pesticide. It then accumulates in the body of the predator. The predator may itself be eaten by a further predator, and the insecticide becomes ever more concentrated as it passes up the food chain. Top predators, never intended as targets, can then accumulate extraordinarily high doses, sometimes with lethal effects.

> pesticides are most polluting when they are unselective and persistent, and when they biomagnify in food chains

The peregrine falcon (*Falco peregrinus*) is a particularly distinctive and beautiful bird of prey with an

14.3   ECOncerns

Can this fruit be saved? The banana as we know it is on a crash course toward extinction

In June 2005, Dan Koeppel filed the story below.

For nearly everyone in the US, Canada and Europe, a banana is a banana: yellow and sweet, uniformly sized, firmly textured, always seedless.

The Cavendish banana—as the slogan of Chiquita, the globe's largest banana producer, declares – is 'quite possibly the world's perfect food'. . . . It also turns out that the 100 billion Cavendish bananas consumed annually worldwide are perfect from a genetic standpoint, every single one a duplicate of every other. It doesn't matter if it comes from Honduras or Thailand, Jamaica or the Canary Islands – each Cavendish is an identical twin to one first found in Southeast Asia, brought to a Caribbean botanic garden in the early part of the 20th century, and put into commercial production about 50 years ago.

That sameness is the banana's paradox. After 15,000 years of human cultivation, the banana is too perfect, lacking the genetic diversity that is key to species health. What can ail one banana can ail all. A fungus or bacterial disease that infects one plantation could march around the globe and destroy millions of bunches, leaving supermarket shelves empty.

A wild scenario? Not when you consider that there's already been one banana apocalypse. Until the early 1960s, American cereal bowls and ice cream dishes were filled with the Gros Michel, a banana that was larger and, by all accounts, tastier than the fruit we now eat. Like the Cavendish, the Gros Michel, or 'Big Mike', accounted for nearly all the sales of sweet bananas in the Americas and Europe.

But starting in the early part of the last century, a fungus called Panama disease began infecting the Big Mike harvest. It appeared first in Suriname, then plowed through the Caribbean, finally reaching Honduras in the 1920s. By 1960, the major importers were nearly bankrupt, and the future of the fruit was in jeopardy.

The Cavendish was eventually accepted as Big Mike's replacement after billions of dollars in infrastructure changes were made to accommodate different growing and ripening needs. Its advantage was its resistance to Panama disease. But in 1992, a new strain of the fungus - one that can affect the Cavendish - was discovered in Asia. Since then, Panama disease Race 4 has wiped out plantations in Indonesia, Malaysia, Australia and Taiwan, and it is now spreading through much of Southeast Asia. It has yet to hit Africa or Latin America, but most experts agree that it is coming.

A global effort is now under way to save the fruit - an effort defined by two opposing visions of how best to address the looming crisis. On one side are traditional banana growers, who raise experimental breeds in the fields, trying to create a replacement plant that looks and tastes so similar to the Cavendish that consumers won't notice the difference. On the other side are bioengineers, who, armed with a largely decoded banana genome, are manipulating the plant's chromosomes, sometimes crossing them with DNA from other species, with the goal of inventing a tougher Cavendish that will resist Panama disease and other ailments.

Currently, there is no way to effectively combat Panama disease and no Cavendish replacement in sight. And so traditional scientists and geneticists are in a race - against one another, for certain, but mostly against time.

Use a web search to discover the options that might be used to safeguard the banana industry. How farfetched do you consider the risk of global economic terrorism by deliberate spread of a banana disease?

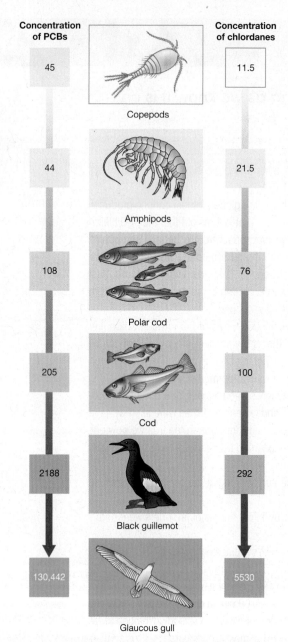

Concentration of PCBs

45

44

108

205

2188

130,442

Concentration of chlordanes

11.5

21.5

76

100

292

5530

Copepods

Amphipods

Polar cod

Cod

Black guillemot

Glaucous gull

FIGURE 14.16 Organochlorines, applied as pesticides on land, are transported to the Arctic through river runoff and oceanic and atmospheric circulation. A study in the Barents Sea showed how two classes of pesticide are biomagnified during passage through the marine food chain. Concentrations in sea water are very low. Herbivorous copepods (that feed on phytoplankton) have higher concentrations (measured in nanograms per gram of lipid in the organisms), and predatory amphipods higher concentrations still. The polar cod (*Boreogadus saida*), which feeds on the invertebrates, and cod (*Gadus morhua*), which also includes polar cod in its diet, show further evidence of biomagnification. However, it is the higher steps in the food chain where biomagnification is most marked, because the sea birds that feed on the fish (black guillemots, *Cepphus grylle*) or on fish and other sea birds (glaucous gull, *Larus hyperboreus*) have much less ability to eliminate the chemicals than fish or invertebrates. Note how chlordanes are biomagnified to a lesser extent than polychlorinated biphenyls (PCBs). This results from the birds' greater ability to metabolize and excrete the former class of pesticide. (Based on data in Borga et al., 2001.)

almost worldwide range. Until the 1940s, about 500 pairs bred regularly in the eastern states of the United States and about 1,000 pairs in the west and in Mexico. In the late 1940s their numbers started a rapid decline, and by the mid-1970s the bird had disappeared from almost all the eastern states and its numbers had fallen by 80 to 90% in the west. Similar dramatic declines were occurring in Europe. Peregrine falcons were listed as an *endangered species*, at risk of extinction. The decline, which occurred in many other birds of prey, was traced to failure to hatch normal broods and very high breakage of eggs in the nest.

The cause was eventually identified as the accumulation of DDT (dichlorodiphenyltrichloroethane) in the parent birds. The pesticide had apparently contaminated seeds and insects eaten by small birds and had accumulated in their tissues. In turn these had been caught and eaten by birds of prey, and the pesticide interfered with their reproduction, causing the eggs to have thin shells and be more likely to break (Figure 14.17a).

The use of DDT was banned in the United States in 1972, at least in part because of Rachel Carson's book *Silent Spring*, published a decade earlier (Box 14.4). Programs were developed to breed peregrines in captivity, and at least 4,000 birds were later released to the wild. Peregrines are now breeding successfully over much of the United States and are no longer considered an endangered species. In Britain, recovery has been so successful that the peregrine is regarded as a pest by pigeon fanciers and lovers of smaller songbirds. The use of DDT as an agricultural pesticide continued in some developing countries until it was banned globally in 2001 by the Stockholm Convention, but it is still allowed for controlling populations of insects that carry diseases such as malaria.

It was possible to identify DDT pollution as a cause of eggshell thinning because eggshells had been collected as dated specimens in museums and private collections. A measure of eggshell thickness in collections of eggs of the sparrowhawk (*Accipiter nisus*) showed a sudden stepwise fall of 17% in 1947, when DDT began to be used widely in agriculture, and a steady increase in thickness after DDT was banned (Figure 14.17b).

A pesticide gets a bad name if, as is often the case, it kills more species than the one at which it is aimed, and especially if it kills the pest's natural enemies and so contributes to undoing what it was employed to do. **Target pest resurgence** occurs when treatment kills large numbers of the pest *and* its natural enemies. Pest individuals that survive the pesticide or migrate into the area

target pest resurgence and secondary pest outbreaks

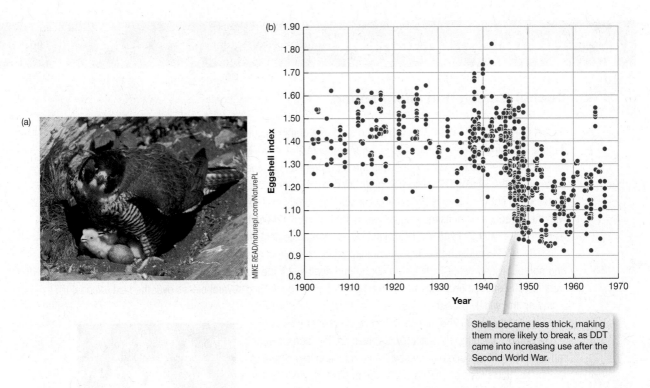

MIKE READ/naturepl.com/NaturePL

Shells became less thick, making them more likely to break, as DDT came into increasing use after the Second World War.

FIGURE 14.17   (a) A peregrine falcon sits on an egg and a recently hatched chick. (b) Graph showing the changes in sparrowhawk eggshell thickness (museum specimens) in Britain. Ratcliffe, 1970.)

later therefore find themselves with a plentiful food resource and few natural enemies. The pest population may then explode.

A pesticide can cause even more subtle reactions by destroying the enemies of potential pest species that had otherwise been kept below their EILs. These now become real, "secondary," pests released from natural control (see Figure 14.15c). In 1950, when mass dissemination of organic insecticides began, there were two primary pests on cotton in Central America: the Alabama leafworm and the boll weevil (Smith, 1998). Organochlorine and organophosphate insecticides were applied fewer than five times a year and initially had apparently miraculous results – crop yields soared. By 1955, however, three new types of pests had emerged: the cotton bollworm, the cotton aphid, and the false pink bollworm. In response, farmers increased pesticide applications to 8 to 10 times per year. This reduced the aphid and the false pink bollworm, but it led to the emergence of another five pests. By the 1960s, the original two pest species had become eight, and there were, on average, 28 applications of insecticide per year.

Chemical pesticides lose their value if the targeted pests evolve resistance. This natural selection in action (see Chapter 2) is almost certain to occur when vast numbers of a genetically variable population are killed. One or a few individuals may be unusually resistant (perhaps because they possess an enzyme that can detoxify the pesticide). If the pesticide is applied repeatedly, each successive generation of the pest will contain a larger proportion of resistant individuals. Pests typically have a high intrinsic rate of reproduction, so a few individuals in one generation may give rise to hundreds or thousands in the next, and resistance spreads very rapidly in a population. The same processes are at work when bacteria develop resistance to antibiotics, something greatly favored by the routine feeding of antibiotics to healthy animals in feedlot operations (Box 14.2).

The first case of insecticide resistance was reported as early as 1946 (housefly resistance to DDT in Sweden). The current scale of the problem is illustrated in Figure 14.19. Resistance has been recorded in every family of arthropod pest (including flies, beetles, moths, wasps, fleas, lice, moths, and mites) as well as in weeds and plant pathogens. The Alabama leafworm (see above), a moth pest of cotton, has developed resistance in one or more regions of the world to aldrin, DDT, dieldrin, endrin, lindane, and toxaphene.

evolved resistance . . .

14.4   Historical Landmarks

Rachel Carson and *Silent Spring*

Rachel Carson (1907–1964) was a U.S. ecologist and author (Figure 14.18). After completing a master's degree in zoology, she was forced to quit her graduate studies in 1935 to support her mother when her father died. She took a temporary job with the U.S. Bureau of Fisheries, writing the script for a weekly radio show on marine ecology. She excelled at this and in 1936 landed a regular position with the bureau as a professional aquatic biologist, becoming the second woman ever to have such a job there. Carson stayed with the bureau through its reconfiguration as the U.S. Fish and Wildlife Service following World War II, and as part of her work, in 1945 she began to look into the ecological effects of the then-new insecticide DDT.

By 1949, Carson had become the chief editor of publications for the U.S. Fish and Wildlife Service and had published several articles on nature and ecology for the general public. Finding her government work tedious, she left and became an independent writer, publishing several pieces over the next decade including the widely praised books *The Sea Around Us* (1950) and *The Edge of the Sea* (1955). These still make excellent reads.

Carson became increasingly involved in the conservation movement and was quite alarmed by the introduction into nature of exotic chemicals such as DDT. In 1962, she published *Silent Spring*, documenting the destructive potential of carelessly used pesticides on natural ecosystems and human health. Carson solicited expert review of her chapters before publication, and her work was very well received by environmental scientists. However, she and the book were savagely attacked by the chemical industry and industry-connected scientists.

Silent Spring was a game changer. Public attention to the use of synthetic pesticides was galvanized, leading to the banning of DDT, first in the United States in the early 1970s and eventually throughout the world as of 2001. The book also contributed to the birth of the modern environmental movement in the late 1960s and early 1970s. More than a decade after her death, Carson was awarded the Presidential Medal of Freedom by President Jimmy Carter.

Rachel Carson Collection. Chatham University Archives, Pittsburgh, PA

FIGURE 14.18   Rachel Carson in 1940 at age 33. (Source: http://en .wikipedia.org/wiki/Rachel_Carson)

If chemical pesticides brought nothing but problems, they would already have fallen out of widespread use. Instead, their rate of production has increased rapidly. Farmers find the cost of buying and spraying the pesticides to be worth the savings in reduced damage to crops. In many poorer countries, the prospect of mass starvation or epidemic disease is so frightening that the social and health costs of pesticides have to be ignored. Their use is generally justified by objective measures such as lives saved, economic efficiency of food production, and total food produced. In this very fundamental sense, their use may be described as sustainable. In practice, sustainability depends on

. . . but pesticides work

continually developing new pesticides that keep at least one step ahead of the pests—and that are less persistent, biodegradable, and more accurately targeted.

Biological control

Biologists can sometimes replace chemicals with other tools that do the same job and cost a great deal less, both economically and environmentally. **Biological control** manipulates the natural enemies of a pest to control it. One form of biological control, often called classical or **importation biological control**, imports a known natural enemy from another geographic area—often where the pest originated. The objective is for the

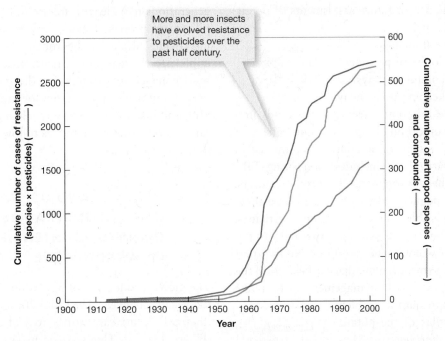

FIGURE 14.19   Global increases in the number of arthropod pest species reported to have evolved pesticide resistance and in the number of pesticide compounds against which resistance has developed. Each pest, on average, has evolved resistance to more than one pesticide, so there are now more than 2,500 cases of evolution of resistance (pests × compounds). (From Michigan State University's database of arthropods resistant to pesticides, www.pesticideresistance.org/DB/; © Patrick Bills, David Mota-Sanchez & Mark Whalon.)

control agent to persist and maintain the pest population at low levels, creating a desirable invasion of an exotic species or, put differently, restoration of the status quo between the species.

The classic example is the story of the cottony cushion scale insect (*Icerya purchasi*), discovered as a pest of Californian citrus orchards in 1868. By 1886 it had brought the citrus industry to its knees. A search for natural enemies led to the importation from Australia and New Zealand of two candidate species. The first was a parasitoid, a two-winged fly (*Cryptochaetum* sp.) that laid its eggs on the scale insect, giving rise to a larva that consumed the pest. The other was a predatory ladybird beetle (*Rodolia cardinalis*). Initially, the parasitoids seemed to have disappeared, but the predatory beetles underwent such a population explosion that, amazingly, all scale insect infestations in California were controlled by the end of 1890. Although the beetles have usually taken the credit, the long-term outcome has been that the beetles keep the scale insects in check inland, but *Cryptochaetum* is the main control near the coast (Flint & van den Bosch, 1981). The economic return on investment in biological control was very high in California, and ladybird beetles have since been transferred to 50 countries.

Conservation biological control, in contrast, aims to increase the equilibrium density of natural

conservation biological control

enemies already native to the region where the pest occurs. Predators of the aphid pests of wheat include ladybirds and other beetles, heteropteran bugs, lacewings, fly larvae, and spiders. Many of these spend the winter in grassy boundaries at the edge of wheat fields, from where they disperse to reduce aphid populations around field edges. Farmers can protect grass habitat around their fields and even plant grassy strips in the interior to enhance these natural populations and the scale of their impact on the pests. Insects have been the main agents of biological control against both insect pests and weeds.

Biological control may appear to be a particularly environmentally friendly approach, but even carefully chosen, and apparently successful, introductions of biological control agents have affected nontarget species (Pearson & Callaway, 2003). *Cactoblastis* moths were introduced to Australia and succeeded dramatically in controlling exotic cactuses. Accidentally introduced to Florida, however, they have been attacking several native cacti (Cory & Myers, 2000). A seed-feeding weevil (*Rhinocyllus conicus*) introduced to North America to control exotic *Carduus* thistles attacks several native thistles and harms populations of a native picture-winged fly (*Paracantha culta*) that feed on the thistle seeds (Louda et al., 1997). Such ecological effects need

... but sometimes nontarget organisms are affected

to be better evaluated in future assessments of potential biocontrol agents.

Integrated pest management is a practical philosophy of pest management that combines physical control (simply keeping pests away from crops), cultural control (rotating the crops in a field so pests cannot build up their numbers over years), biological and chemical control, and the use of resistant crop varieties. It includes monitoring EILs and also relies heavily on natural mortality factors, such as natural enemies and weather, disrupting them as little as possible. Broad-spectrum pesticides, although not excluded, are used only sparingly and in ways that minimize costs and quantities. No two pest problems are the same, even in adjacent fields, and the essence of integrated pest management is to make the control measures fit the problem.

> combining chemical and biological controls . . .

The caterpillar of the potato tuber moth (*Phthorimaea operculella*) commonly damages crops in New Zealand. An invader from a warm, temperate subtropical country, it is most devastating when conditions are warm and dry. There can be as many as 6 to 8 generations per year, and different

> integrated pest management for the potato tuber moth

generations mine leaves, stems, and tubers. The caterpillars are protected from both natural enemies and insecticides when in the tuber, so control must be applied to the leaf-mining generations. The integrated pest management strategy for the potato tuber moth uses monitoring (via pheromone traps), soil cultivation, and insecticides, but only when absolutely necessary (Herman, 2000). Farmers follow the decision tree in Figure 14.20.

14.6 GLOBAL LAND USE AND OTHER CONSTRAINTS ON CONTINUED INTENSIFICATION OF AGRICULTURE

The global production of agricultural crops tripled in the decades between 1961 and 2008, though with less than a 10% increase in the area of cropland farmed (Figure 14.21a). This statistic masks some important regional differences: the cropland area of Eastern Europe and Russia fell, as did that in the United States to a lesser extent, while Latin America, Asia, Oceania, and Africa all saw increases in the area of land farmed for crops. Still, thanks to high population growth, the area of cropland farmed per capita globally fell by 50% (Figure 14.21b). The increase in food production was driven by agricultural intensification, particularly the use of nitrogen fertilizer, but also by increased use of fossil fuels for tractors and combines, greater efforts to control pests, increased irrigation, and breeding of inherently more productive crops. Can production continue to increase?

Water is one constraint. On a global scale, agriculture is the largest consumer of fresh water, taking more than 70% of available supplies. Irrigation is particularly important in developing countries, and more than 90% of available fresh water is used by agriculture in parts of South America, central Asia, and Africa. Already, the Nile in Africa, the Yellow River in China, and the Colorado River in North America dry up for at least part of the year, with dramatic consequences for the downstream coastal waters. Clearly there are limits to how much more irrigation we can employ. And global climate change seems certain to aggravate issues of water supply and use, as many parts of the Earth become not only warmer but also much drier (Figure 14.22).

> water is a finite global resource

Around 300 million hectares of soil around the world are now severely degraded, unable to continue to support productive agriculture because of nutrient exhaustion,

> soil degradation provides another constraint on agriculture

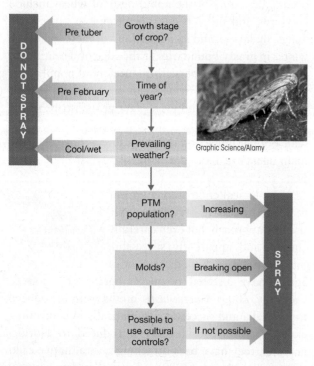

FIGURE 14.20 Decision flowchart for the integrated pest management of potato tuber moths (PTM) in New Zealand. Boxed phrases are questions ('Growth stage of crop?'), words in arrows are the farmer's answers to the questions ('Pre tuber'), and the recommended action is shown in the vertical boxes ('Do not spray'). Note that February is late summer in New Zealand. (After Herman, 2000.)

Graphic Science/Alamy

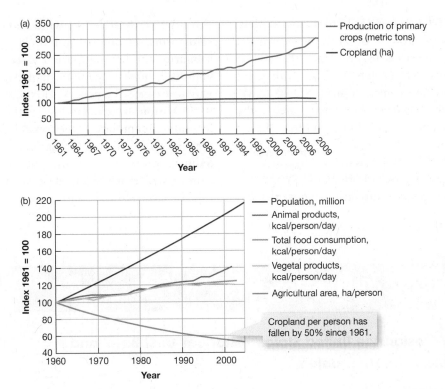

FIGURE 14.21 (a) Increase in the global production of crops and in the area of cropland since 1961, expressed relative to the values in 1961; and (b) change in the global population, per capita consumption of meat, per capita total food consumption, per capita consumption of vegetable matter, and cropland per person globally since 1961, expressed relative to the values in 1961. (Source: (a) UNEP, 2014; and (b) Bringezu et al., 2009.)

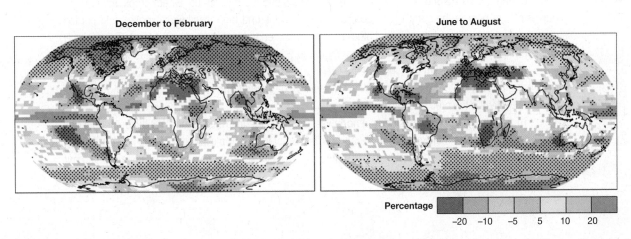

FIGURE 14.22 Predicted change in precipitation by the period 2090–2099 relative to late 20th century, based on output from multiple global models. White areas indicates less than 66% agreement among the models. Stippled areas indicate greater than 90% agreement among models. (Source: IPCC, 2007.)

salt buildup, or other problems. A further 1.2 billion hectares—10% of the Earth's vegetated surface—are moderately degraded. Clearly much of agricultural practice has not been sustainable. Land without soil can support only very small primitive organisms, such as lichens and mosses that can cling onto a rock surface. The rest of the world's terrestrial vegetation has to be rooted in soil, which gives it physical support and

supplies roots with essential mineral nutrients and water during plant growth. Soil develops by the accumulation of finely divided mineral products of rock weathering and decomposing organic residues from previous vegetation. It can be lost by being washed or blown away, perhaps to be redeposited as fine-textured *loess* somewhere else. Soil is best protected when it contains organic matter, is always wholly covered with vegetation, is finely interwoven with roots and rootlets, and is on horizontal ground. Generally, when natural ecosystems are converted to cropland, the soil is degraded to at least some extent. Dramatic evidence of unsustainable land use comes from the "Dust Bowl" disaster in the Great Plains of the United States, and a similar disaster happening now in China (Box 14.5).

Agricultural land is particularly susceptible to degradation in arid and semiarid regions. Both overgrazing and excessive cultivation expose the soil

> desertification and salinization

directly to erosion by the wind and to rare but fierce rainstorms. In **desertification,** land that is arid or semiarid but has supported natural ecosystems— and often subsistence or nomadic agriculture—gives way to desert. The process can be slowed for a time by irrigation, which also lowers the water table and lets salts accumulate in the topsoil. Such **salinization** makes continued use for agriculture very difficult. Salinization tends to spread and leads to an expansion of sterile, white salt deserts. It has been a particular hazard in irrigated areas of Pakistan and

---

14.5 Historical Landmarks

Soil erosion, the United States' historical 'Dust Bowl' and China's current problem

Large areas of southeastern Colorado, southwestern Kansas, and parts of Texas, Oklahoma, and northeastern New Mexico once supported rangeland management of livestock. The vegetation was native perennial grasses and the land had been neither ploughed nor sown with seed.

During the First World War, much of the land was ploughed and annual wheat crops were grown. Harvests were poor in the early 1930s due to severe drought, and the topsoil was exposed and carried away by the wind in black blizzards that blocked the sun and ended in drifts (Figure 14.23a). Occasionally dust storms swept completely across the country to the East Coast. Thousands of families were forced to leave the Dust Bowl region at the height of the Great Depression in the early and mid-1930s. The wind erosion was gradually halted with federal aid; windbreaks were planted and much of the grassland restored. By the early 1940s the area had largely recovered.

This story is being repeated today in northwest China, where the need to feed 1.3 billion people has led to the raising of too many cattle and sheep, and the use of too many plows. About 2300 km² are turning to desert each year. A huge 2011 dust storm (Figure 14.23b) blanketed areas not only in eastern China but in western North America as well.

(a)

(b)

FIGURE 14.23 (a) A dust storm during the Dust Bowl years of the 1930s in the southwestern United States; and (b) a dust storm in China in 2011.

some Asian countries that were part of the former Soviet Union.

Although the productivity per unit area for many major agricultural crops continues to increase globally, the rate of increase in productivity is slowing for all crops, and for some it is now static (Figure 14.24a). If present trends continue, our ability to grow more crops on the same amount of land may come to an end within the next decade, at least at the global scale. There is room for increased productivity for some crops in some regions, such as maize (corn) in Brazil, Mexico, and India, and for palm oil in many countries, particularly Nigeria (Figure 14.24b). But globally, we appear to be reaching the limits of what may be possible, in part due to the constraints of water availability and climatic variability. In fact, scientists are concerned that global climate change will *decrease* agricultural productivity, and evidence indicates air pollution is already constraining crop production. Recent model results suggest the current level of ozone pollution (mostly from nitrogen, methane, and hydrocarbon pollution from producing and using fossil fuels; see Section 12.4) is already depressing the global production of staples like wheat (by 3.9 to 15%), soybeans (8.5 to 14%) , and corn (2.2 to 5.5%). This negative effect may further decrease crop production markedly over the coming decades, if current air pollution trends continue (Avnerya et al., 2011).

> the trend of increased agricultural productivity is slowing . . .

Further increases in global crop productivity are therefore likely to require more land. How much more? As we write, UN estimates suggest an additional 70 to 300 million hectares will be needed by 2050 (Table 14.2), or up to a 20% increase over the 1,530 million currently farmed (UNEP, 2014). The growth of cities and suburbs, however, is predicted to cause a loss of 107 to 129 million hectares, and continued soil degradation seems likely to cause the abandonment of between 90 and 225 million by 2050. Further, many nations have aggressive policies to grow *biofuels* for energy. Meeting their goals, and the need for agriculturally produced biomaterials for industry (for example, biodegradable plates and forks made from corn), is estimated to require yet another 52 to 195 million hectares (UNEP, 2014). Summing all these potential needs, we find that up to 859 million hectares of new land may be required by 2050 (Table 14.2), an increase of more than 50% of the land currently under cultivation worldwide.

> further increases in global crop production likely require more land . . .

Biofuels are sources of energy developed from biomass derived from agriculture or natural ecosystems. **Liquid biofuels** are substances such as ethanol or hydrocarbons that can substitute for gasoline and diesel oil. To many, growing crops for liquid biofuels seemed a win-win scenario, and over the last decade a global explosion in the production of liquid biofuels has occurred, particularly for ethanol (Figure 14.25). Most of this increased production was in the United States, with a huge increase in the production of ethanol from corn.

> Liquid biofuels may be a bad idea . . .

Perhaps this vision was too good to be true. A 2010 letter to the U.S. President from the Council of Scientific Society Presidents, an umbrella organization representing 1.4 million scientists in the United States, argued that while addressing global warming was critical, some "solutions" had received insufficient scientific scrutiny, noting particularly the production of liquid biofuels from corn, and the development of natural gas from shale using high-volume hydraulic fracturing (see Box 12.2). Both solutions made superficial sense and could work politically, but the environmental consequences could be very high, aggravating rather than reducing global warming.

Biofuels supply 13% of global energy, but less than 1% is from liquid biofuels. The global use of biomass is dominated by combustion of solid forms, wood, and animal waste; dried dung makes up 5% of all global energy use. Virtually all the ethanol in the world is now produced from crops that can also be used for food; corn in the United States is the largest source globally, followed by sugar cane in Brazil. In 2007, the United States used 24% of its corn harvest to produce 1.3% of the total liquid fuel supply for the country (gasoline, diesel, and other petroleum products made up the rest). Ethanol production has grown since then, consuming 40% of the U.S. corn harvest in 2012 but still producing only a very small percentage of the nation's energy.

> a closer look at ethanol

The process of making ethanol from these crops is inherently inefficient, particularly for corn. Corn itself is an energy-intensive crop to grow, and half the energy content of the harvest is lost during fermentation to produce ethanol. Large amounts of energy from coal or natural gas are needed to distill the ethanol into an anhydrous (water-free) form that can be mixed with gasoline. Per unit of land, we can generate nine times more energy by instead growing perennial grasses and burning them to coproduce heat and electricity. The land requirement is of course critical, given an impending shortage of global land for agriculture.

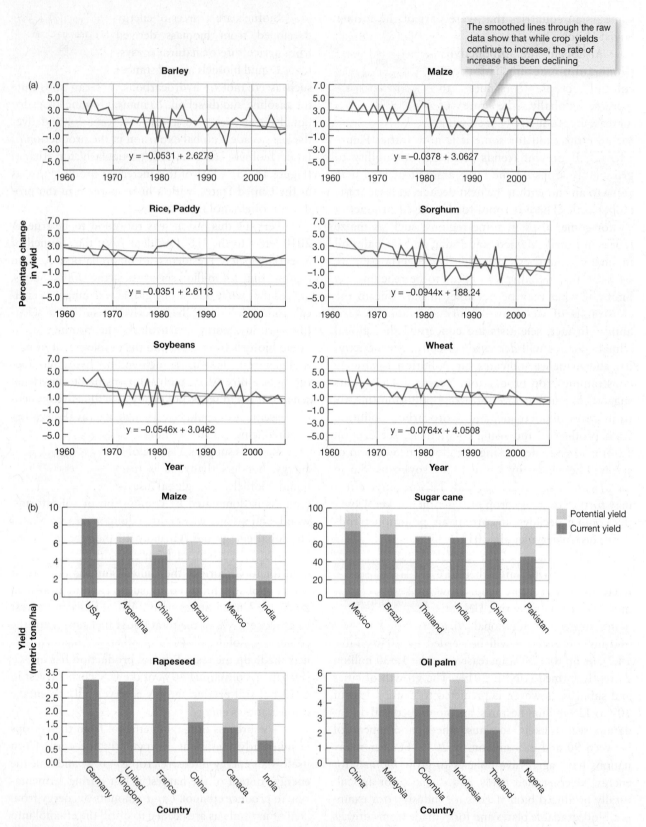

The smoothed lines through the raw data show that while crop yields continue to increase, the rate of increase has been declining

FIGURE 14.24 (a) Trends in the global increase in yields of various major crops since 1960. A value of 0% indicates no change in yield. (b) Actual yield and potential maximum yield for four major crops. For each crop, six of the major producing countries for that crop are shown. (Source: UNEP 2014.)

Ethanol causes more air pollution than gasoline and also leads to high levels of water pollution, in Brazil from the wastes of fermentation, and in the United States from growing the corn. The United States has a national policy to reduce the nitrogen flows down the Mississippi River by 45% and to reduce the size of the dead zone in the northern Gulf of Mexico (Box 1.4, and Chapter 11), yet several studies conclude that the national policy to produce more ethanol from corn will actually increase this nitrogen pollution by 30% or more.

Ethanol as a fuel may also aggravate greenhouse gas emissions more than fossil fuels; it certainly has a larger greenhouse gas footprint than renewable energy sources such as wind and solar power. It also has major indirect land-use consequences that can greatly increase greenhouse gas emissions (Searchinger et al., 2008). As more of the U.S. corn harvest is used to produce ethanol, less corn is available for food or livestock feed. In response, the demand for animal feeds grown elsewhere has increased, in particular for soybeans grown in the Cerrado region of Brazil (a large area of tropical savanna; see Chapter 4). In turn, some of the low-intensity grazing of beef cattle in the Cerrado is displaced to the Amazon, where rain forest is cut down to create pastures (Figure 14.26). Cattle grazing is one of the largest drivers of deforestation in the Amazon, and tropical deforestation is a major source of atmospheric carbon dioxide (see Figure 12.8). Tracing the environmental consequences of local actions demands a determinedly global perspective.

TABLE 14.2 Required areas of new cropland that will be required globally by 2050 under low-end and high-end estimates for business-as-usual expansion.

	Low-end estimate (millions of hectares)	High-end esiimate (millions of hectares)
Food supply	71	300
Biofuel supply	48	80
Biomaterial supply	4	115
Net expansion	**123**	**495**
Compensation for built environment	107	129
Compensation for soil degradation	90	225
Gross expansion	**320**	**859**

Source: UNEP, 2014.

Global production of liquid biofuels

- ● Ethanol
- ● Biodiesel equivalent

(Graph: Peta Joules vs Year, 1975 to 2007)

FIGURE 14.25 Global production of ethanol and biodiesel from 1975 to 2007. (Source: Howarth & Bringezu, 2009.)

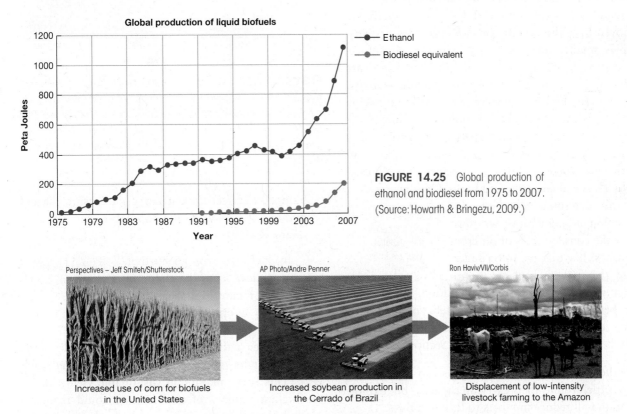

Perspectives – Jeff Smiteh/Shutterstock

AP Photo/Andre Penner

Ron Haviv/VII/Corbis

Increased use of corn for biofuels in the United States

Increased soybean production in the Cerrado of Brazil

Displacement of low-intensity livestock farming to the Amazon

FIGURE 14.26 An example of the indirect land effects that result from producing more ethanol for corn. As more corn in the United States is used for ethanol, production of soybeans in the Cerrado region of Brazil increases to make up for the animal feeding that had been based on corn. The soybean expansion displaces low-intensity livestock grazing, which instead moves to the Amazon on land newly deforested for this purpose.

14.7 FOOD FROM FISHERIES AND AQUACULTURE

Is there a difference between obtaining food by farming and by fishing? Farms and herds tend to be owned and managed by a farmer or organization. In contrast, most of the ocean waters that are fished have at one time been common property, leaving them open to potentially unsustainable looting by all comers. Over the past 50 years, though, fishing has come under increasing national and international regulation and national claims to '"ownership", though this has not necessarily led to good management, as we shall see. The global catch of wild fish has remained steady for most of the past 25 years, perhaps even declining in the past decade (Figure 14.27). At the same time, aquaculture—the true farming of fish—has increased.

Whenever a natural popula-
tion is exploited, there is a risk of | aiming for the narrow path between over- and underexploitation
overexploitation, in which too
many individuals are removed and
the population is driven into bio-
logical jeopardy or economic insignificance – perhaps even extinction. Global catches of marine fish rose fivefold between 1950 and 1989, and many of the world's fish stocks are now beyond the point of overexploitation (Figure 14.28). But harvesters also want to avoid underexploitation; if fewer individuals are removed than the population can bear, the harvested crop is smaller than necessary, potential consumers are deprived, and those who do the harvesting are underemployed. It is not easy to tread the narrow path between under- and overexploitation. It is asking a great deal of a management policy to combine the well-being of the exploited species, the profitability of the harvesting enterprise, continuing employment for the workforce, and the maintenance of traditional lifestyles, social customs, and natural biodiversity.

To determine the best way
to exploit a population, we must | population dynamics in the absence of exploitation – humped net recruitment curves
know the consequences of different
strategies. For that we must understand the dynamics of the popu-
lation without, or before, exploitation. We assume that, before it is exploited, a harvestable population is crowded and intraspecific competition is intense. We base these broad generalizations on Section 5.5:

- without exploitation, populations can be expected to settle around their carrying capacity, but exploitation will reduce numbers to less than this;

- exploitation, by reducing the intensity of competition, moves the population leftward along the

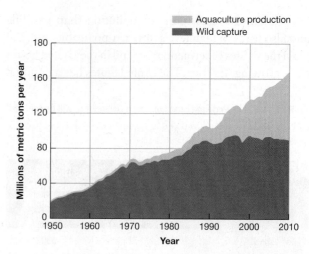

FIGURE 14.27 Global trends in the fisheries catch of wild fish and of aquaculture since 1950. (Source: FAOSTAT Data from The Food and Agricultural Organization [downloaded from Wikipedia].)

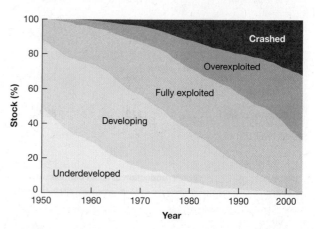

FIGURE 14.28 Changes in the contribution to global marine fish production made by fisheries in different phases of their exploitation. In the 1950s most of the catches were from undeveloped fisheries, but by 2000 most fisheries were fully exploited (near their maximum sustainable yield), or overexploited or had already collapsed. (After Khan et al., 2006.)

humped net recruitment curve, increasing the net number of recruits to the population per unit time (Figure 14.29).

In fact, we can go further,
since it is clear from the shape of | maximum sustainable yield – the narrow path?
the curve in Figure 14.29 that there
must be an 'intermediate' popula-
tion size at which the rate of net recruitment is highest. Consider a time scale of years. The peak of the curve might be '10 million new fish recruited each year.' This is the **maximum sustainable yield (MSY)**, the highest number of new fish we could regularly remove from the population each year that the population itself could replenish indefinitely. It looks as though a fishery could tread the narrow path between

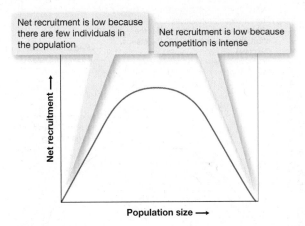

FIGURE 14.29   The humped relationship between the net recruitment into a population (births minus deaths) and the size of that population, resulting from the effects of intraspecific competition (see Chapter 5). Population size increases from left to right, but increasing rates of exploitation take the population from right to left.

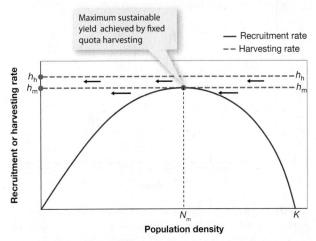

FIGURE 14.30   Fixed-quota harvesting. The figure shows a single recruitment curve (solid line; recruitment in relation to density, N) and two fixed-quota harvesting curves (dashed lines): high quota (h_h), and MSY quota (h_m). The arrows in the figure refer to changes to be expected in abundance under the influence of the harvesting rate to which the arrows are closest. The blue dots indicate equilibria. At h_h the only 'equilibrium' is when the population is driven to extinction. The MSY is obtained at h_m because it just touches the peak of the recruitment curve (at a density N_m): populations greater than N_m are reduced to N_m, but populations smaller than N_m are driven to extinction. K is the carrying capacity, the density where the population is expected to settle in the absence of exploitation.

under- and overexploitation if it could find a way to achieve this MSY.

The MSY concept has been the guiding principle in resource management for many years in fisheries (as well as in forestry and wildlife exploitation and hunting), but it is far from a perfect answer for several reasons.

the MSY concept has shortcomings

1   By treating the population as a number of similar individuals, it ignores all aspects of population structure such as size or age classes and their differential rates of growth, survival, and reproduction.

2   Based on a single recruitment curve, it treats the environment as unvarying.

3   In practice, it may be impossible to obtain a reliable estimate of the MSY.

4   Achieving an MSY is by no means the only, nor necessarily the best, criterion for successful management of a harvesting operation. (It may be more important to provide stable, long-term employment for the fisher workforce, for instance, or to conserve biodiversity in an area.)

There are two simple ways of obtaining an MSY on a regular basis: through a fixed quota and a fixed effort. With a **fixed quota** (Figure 14.30), we remove the same amount (the MSY) from the population every year. If the population stayed exactly at the peak of its net recruitment curve (and that is a big if), this could work: each year the population, through growth and reproduction, would add exactly what harvesting removed. But if by chance numbers fell even slightly below those at which the

the fragility of fixed-quota harvesting . . .

curve peaked (say because a year's bad weather led to lower recruitment), then the numbers harvested would exceed those recruited. Population size would decline below the peak of the curve, and if we maintained a fixed quota at the MSY level, the population would decline until it was extinct. If we even slightly overestimated the MSY (and reliable estimates are hard to come by), the harvesting rate would always exceed the recruitment rate and extinction would again follow. In short, a fixed quota at the MSY level might be desirable and reasonable in a wholly predictable world about which we had perfect knowledge. But in the real world of fluctuating environments and imperfect data sets, these fixed quotas are open invitations to disaster.

Nevertheless, the fixed-quota strategy has frequently been used. A typical example is the Peruvian anchovy (*Engraulis ringens*) fishery (Figure 14.31) in the rich coastal upwelling biome along the west coast of South America. From 1960 to 1972 this was the world's largest single fishery, and a major sector of the Peruvian economy. Fisheries experts advised that the MSY was around 10 million tons annually, and catches were limited accordingly. But the fishing capacity of the fleet expanded, and in 1972 the catch crashed. Overfishing seems, at the least, to have been a major cause, although

. . . borne out in practice

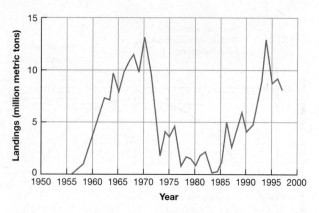

FIGURE 14.31 Landings of the Peruvian anchovy since 1950. Note the dramatic crash that resulted mainly as a result of overfishing. The stock took 20 years to rebuild. (After Jennings et al., 2001.)

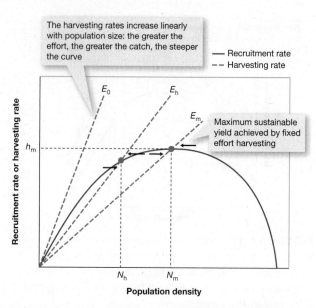

The harvesting rates increase linearly with population size: the greater the effort, the greater the catch, the steeper the curve

— Recruitment rate
-- Harvesting rate

Maximum sustainable yield achieved by fixed effort harvesting

FIGURE 14.32 Fixed-effort harvesting. Curves, arrows, and dots as in Figure 14.31. The MSY is obtained with an effort of E_m, leading to a stable equilibrium at a density of N_m with a yield of h_m. At a somewhat higher effort (E_h), the equilibrium density and the yield are both lower than with E_m, but the equilibrium is still stable. Only at a much higher effort (E_0) is the population driven to extinction.

its effects were compounded by profound environmental fluctuations, discussed below. A moratorium on fishing would have been an ecologically sensible step, but this was not politically feasible: 20,000 people were dependent on the anchovy industry for employment. The Peruvian government therefore allowed fishing to continue. The stock took more than 20 years to recover.

The second method of obtaining an MSY is a **fixed effort**, such as the number of trawler-days, in which the amount harvested increases with the size of the population being harvested (Figure 14.32). Here, if density drops below the peak, new recruitment exceeds the amount harvested and the population recovers. The risk of extinction is much reduced. The disadvantages are that the yield varies with population size (there are good and bad years), and that nobody can be allowed to make a greater effort than they are supposed to. Harvesting of the Pacific halibut (*Hippoglossus stenolepis*) is limited by seasonal closures and protected areas with no-take zones (see Chapter 13), though heavy investment in fisheries protection vessels is needed to control law breakers.

relative robustness of fixed-effort harvesting

Fishing often exerts a great strain on populations. But the collapse of fish stocks in a given year is often the result of unusually unfavorable environmental conditions, rather than simply overfishing. Harvests of the Peruvian anchovy (see Figure 14.31) collapsed from 1972 to 1973, but the steady rise in catches had already dipped in the mid-1960s as a result of an El Niño event, in which warm tropical water from the north reduces the upwelling, and hence the productivity, of the Peruvian coastal ecosystem. By 1973, however, commercial fishing had so

environmental fluctuations – the anchovy and El Niño

greatly increased that the next El Niño event had even more severe consequences. There were some signs of recovery from 1973 to 1982, but a further collapse occurred in 1983 with yet another El Niño event. While it is unlikely the events would have had such severe effects if the anchovy had been only lightly fished, it is clear that the history of the Peruvian anchovy fishery cannot be explained simply in terms of overfishing.

So far, we have ignored population structure of the exploited species. But most harvesting practices are primarily interested in only a portion of the harvested population (for example, fish that are large enough to be saleable). Also, recruitment in practice is a complex process incorporating adult survival, adult fecundity, juvenile survival, juvenile growth, and so on, each of which may respond in its own way to changes in density and harvesting strategy. One model that takes some of these variables into account was developed for the Arcto-Norwegian cod fishery, the most northerly fish stock in the Atlantic Ocean. The numbers of fish in different age classes were known for the late 1960s and this information was used to predict the tonnage of fish likely to be caught with different intensities of harvesting and net mesh sizes. The model predicted that the long-term prospects for the fishery were best ensured

population structure and the Arctic cod (Gadus morhua)

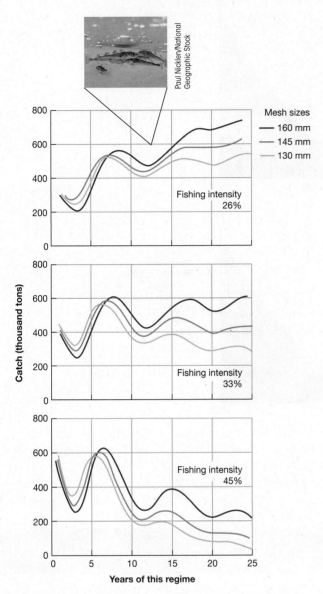

FIGURE 14.33 Predictions for the stock of Arctic cod under three intensities of fishing and three different sizes of mesh in the nets. Larger meshes allow more and larger fish to escape capture. The largest effort (45%, bottom panel) is clearly unsustainable, regardless of the mesh size used. The largest sustainable catches are achieved with a low fishing effort (26%, upper panel) and a large mesh size. (After Pitcher & Hart, 1982.)

with a low intensity of fishing (less than 30%) and a large mesh size. These gave the fish more opportunity to grow and reproduce before they were caught (Figure 14.33). The model's recommendations were ignored, and the stocks of cod fell disastrously.

As global fisheries have stagnated, the loss of the fish resource for food has been made up in part by a growing trend in aquaculture (Figure 14.27).

> the growing trend of aquaculture . . .

Aquaculture is the farming of domesticated fish, shellfish, and algae and occurs in both coastal marine ecosystems and freshwaters. The roots of aquaculture run deep. In 475 BC, Fan Lai published a book entitled *The Classic of Fish Culture* on the production of domesticated carp in fish ponds in China, characterizing an enterprise that probably started at least 4,000 years ago (FAO, 2012). However, aquaculture on the scale practiced today began in the second half of the 20th century, essentially the same time as the intensification of agriculture. Fish, particularly freshwater species such as carp and tilapia, but also marine species such as salmon, are the largest products of aquaculture (Figure 14.34a). The production of algae (used in many food products, as agar in desserts, and as a thickening agent, as well as the rolled algae in sushi), of molluscs (mussels, oysters, clams), and of crustaceans (shrimp) has been on the rise. Most aquaculture occurs in Asia, particularly China (Figure 14.34b).

Aquaculture now provides a globally important source of food and protein, but it has also attracted attention for adverse environmental consequences. One problem is the loss of coastal mangrove swamps: the mangroves are frequently cut and the swamps converted to aquaculture ponds, particularly for shrimp culture, creating a leading cause of wetland loss globally. Although the operators of shrimp ponds profit, the external costs of wetland loss are greater and are borne by society, for instance through losses of native fish species (Gunawardena & Rowan, 2005). Some aquaculture species, particularly salmon, are commonly fed wild-caught food as a major part of their diet. This is essential to produce salmon that are healthy for humans to consume (Box 14.6), but the result is a net loss of fish protein, since the biomass of the salmon produced is less than the biomass of the fish they consume, a necessary result of food-chain efficiencies (see Section 11.5). Because fish and shellfish are highly concentrated in aquaculture operations, high levels of nutrient pollution often result. Further, as in animal feeding operations (see Box 14.2), fish and shellfish in aquaculture are commonly fed antibiotics to encourage growth and reduce bacterial infections.

We see here, therefore, what we have seen throughout the final chapters of this book. The world around us presents us with urgent and complex problems. An understanding of ecology will help us and arguably gives us some of our best hopes. But none of these problems can be viewed in isolation from either its social and economic context or other processes elsewhere in the global ecosystem.

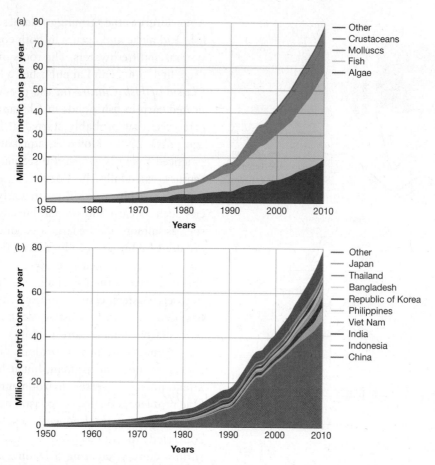

FIGURE 14.34 (a) Global increases in production of different types of products in aquaculture since 1950; and (b) increases in aquaculture production in the major producer nations of the world since 1950. (Source: FAOSTAT Data from The Food and Agricultural Organization.)

14.6 ECOncerns

Aquaculture, fatty acids, and human health

A 2008 report in the online Science Daily led off with the headline 'Popular Fish, Tilapia, Contains Potentially Dangerous Fatty Acid Combination' (Wake Forest University, 2008). The article went on: "Farm-raised tilapia, one of the most highly consumed fish in America, has very low levels of beneficial omega-3 fatty acids and, perhaps worse, very high levels of omega-6 fatty acids, according to new research from Wake Forest University School of Medicine. The researchers say the combination could be a potentially dangerous food source for some patients with heart disease, arthritis, asthma and other allergic and auto-immune diseases that are particularly vulnerable to an 'exaggerated inflammatory response.' Inflammation is known to cause damage to blood vessels, the heart, lung and joint tissues, skin, and the digestive tract."

The article received widespread attention, since one of the major reasons many people eat fish is to consume omega-3 fatty acids for their health benefits.

Why do the tilapia have more of the adverse omega-6 fatty acids, and less of the beneficial omega-3 fatty acids? Fish do not produce their own fatty acids but rather biomagnify the fatty acids in their foods, just as animals biomagnify pesticides and other toxins (Section 14.5). Algae are high in the omega-3 acids, so tilapia feeding on algae (as they do in the wild) are very high in these as well. Tilapia fed on corn or soybean, which are low in the omega-3 fatty acids and high in the omega-6 ones, are the problem.

A similar problem results when salmon are provided with foods with the 'wrong' fatty acids. This is one reason that cultured salmon are commonly fed wild-caught fish, in addition to feeds that contain corn and soybeans. Further, more than half the world's production of fish oil from wild-caught fish is fed to salmon in aquaculture.

Tilapia can eat practically any plant or algae species. Should those who grow tilapia in aquaculture be encouraged to use algae as a food, so the fish will have more of the beneficial fatty acids, and less of the undesirable ones? Should fish foods based on corn or soybean be augmented with fish oils, to increase the desirable fatty acids? What are the benefits and problems of this approach?

SUMMARY

Human use of ecological resources

Over the past half century, the world has experienced explosive growth in affluence and consumption, and the population has become more urban. In 2010, the global urban population matched the rural population for the first time. These trends place unprecedented stresses on the environment.

The human population problem

The problems of human overcrowding are various; total numbers, growth rates, and distributions, especially relative to resources, are all unsustainable. Projections suggest a continued slowing down of growth rates, but momentum from past growth and an increasingly aged population will inevitably cause further problems. Identifying an acceptable global carrying capacity is a challenge with no easy answers.

Ecology and human health

Biogeochemical changes to the global ecosystem cause a variety of health problems. Loss of protective ozone from the stratosphere, caused by man-made chemicals such as CFCs or increased concentrations of natural products such as nitrous oxide, is leading to increased penetration of UV radiation and higher incidence of skin cancers. Increasingly frequent extreme climate effects accompanying global warming account for many additional deaths and even more hospitalizations each year. Global climate change is heightening the incidence and distribution of existing human infectious diseases; these and other human influences are leading to a rise in the number of new infections, many of them zoonotic.

Synthetic fertilizer and the intensification of agriculture

Increased use of synthetic nitrogen fertilizer has intensified agriculture and allowed the spatial segregation of concentrated animal agriculture from the fields, which grow the crops to feed the animals. Meat consumption has risen globally and in many countries, requiring more crop production and creating large amounts of animal wastes. Diets less rich in meat can reduce pollution from agriculture and are already recommended in many Western countries, where current consumption levels adversely affect health. Excessive fertilization of croplands also contributes significantly to nitrogen pollution.

Monocultures, pests, and pesticides in agriculture

The intensification of agriculture is based on crop monocultures susceptible to pest outbreaks. Heavy use of chemical pesticides after the Second World War sometimes killed predators, allowing rapid growth of novel pests. Some pesticides such as DDT can be biomagnified, leading to damage in nontarget species, including birds. Another problem is the ability of pests to rapidly evolve resistance to chemical pesticides. Biological approaches to pest control offer many advantages, as does integrated pest management, a pragmatic approach that combines some use of chemicals with biological controls.

Global land use and other constraints on continued intensification of agriculture

The huge increase in agricultural production over the past 50 years was based on a continuous increase in yield per area farmed, but further increases will be limited by constraints such as inadequate water for irrigation and adverse effects of global climate change and air pollution. Further, some cropland becomes degraded through salinization or is lost to urban encroachment. The increased production of liquid biofuels also requires land. There may not be sufficient land available globally to meet the future demand for agricultural production of food and biofuels.

Food from fisheries and aquaculture

Most fishery resources have been poorly managed for decades, often through use of the maximum sustainable yield model, which does not allow for environmental variation and complex ecological interactions. More appropriate models have often been ignored for political reasons.

The global catch of fish has remained stagnant for several decades. Aquaculture, on the other hand, has expanded greatly over the past two decades and is an increasingly important source of protein globally. Like land-based agriculture, however, it poses significant environmental challenges.

REVIEW QUESTIONS

1 To what extent are global environmental problems driven by human population growth? Compare and contrast this growth with other drivers of change.

2 Explain how the age structure of a population affects human population growth in different regions of the world.

3 How does global climate change affect disease and human health?

4 Explain how the use of synthetic nitrogen fertilizer has allowed intensive animal feedlot operations to develop far from the site of the production of the crops that feed them, and why this creates pollution problems.

5 How does human diet affect nitrogen pollution?

6 What is meant by biomagnification, and why is this process of concern?

7 Why do agricultural monocultures aggravate pest problems?

8 How was the global production of crops able to increase so greatly over recent decades without an accompanying increase in cropland? Why might this trend be coming to an end?

9 How does climate change undermine the effectiveness of the maximum sustainable yield approach for managing fisheries?

Challenge Questions

1 Why does organic agriculture tend to result in less nitrogen pollution than conventional agriculture?

2 What use, if any, is appropriate for antibiotics in animal agriculture and aquaculture? Why? What drives the current high use of antibiotics in these industries?

3 The precautionary principle calls for caution in taking any action when the results are uncertain but the potential dangers highly serious. How would you apply this principle to planning for the future needs of land for agriculture?

Abramsky, Z. & Rosenzweig, M.L. (1983) Tilman's predicted productivity–diversity relationship shown by desert rodents. *Nature*, 309, 150–151.

Agrawal, A.A. (1998) Induced responses to herbivory and increased plant performance. *Science*, 279, 1201–1202.

Ainsworth, E.A. & Long, S.P. (2005) What have we learned from 15 years of free-air CO_2 enrichment (FACE)?: A meta-analytical review of the responses of photosynthesis, canopy properties, and plant production to rising CO_2. *New Phytologist*, 165, 351–372.

Akçakaya, H.R. (1992) Population viability analysis and risk assessment. In: *Proceedings of Wildlife 2001: Populations* (D.R. McCullough, ed.), pp. 148–157. Elsevier, Amsterdam.

Al-Hiyaly, S.A., McNeilly, T. & Bradshaw, A.D. (1988) The effects of zinc contamination from electricity pylons—evolution in a replicated situation. *New Phytologist*, 110, 571–580.

Allan, J.D. & Flecker, A.S. (1993) Biodiversity conservation in running waters. *Bioscience*, 43, 32–43.

Alliende, M.C. & Harper, J.L. (1989) Demographic studies of a dioecious tree. I. Colonization, sex and age-structure of a population of *Salix cinerea*. *Journal of Ecology*, 77, 1029–1047.

Ames, S., Pischke, K., Schoenfuss, N., Snobl, Z., Soine, J., Weiher, E. & Wellnitz, T. (2012) Biogeographic patterns of lichens and trees on islands of the Boundary Waters Canoe Area Wilderness. *BIOS*, 83, 145–154.

Anderson, R.M. & May, R.M. (1991) *Infectious Diseases of Humans: Dynamics and Control*. Oxford University Press, Oxford.

Anderson, R.M. (1982) Epidemiology. In: *Modern Parasitology* (F.E.G. Cox, ed.), pp. 205–251. Blackwell Scientific Publications, Oxford.

Andrewartha, H.G. (1961) *Introduction to the Study of Animal Populations*. Methuen, London.

Angel, M.V. (1994) Spatial distribution of marine organisms: patterns and processes. In: *Large Scale Ecology and Conservation Biology* (P.J. Edwards, R.M. May & N.R. Webb, eds), pp. 59–109. Blackwell Science, Oxford.

Antoniadou, C., Voultsiadou, E. & Chintiroglou, C. (2010) Benthic colonization and succession on temperate sublittoral rocky cliffs. *Journal of Experimental Marine Biology and Ecology*, 382, 145–153.

Antonovics, J. & Bradshaw, A.D. (1970) Evolution in closely adjacent plant populations. VIII. Clinical patterns at a mine boundary. *Heredity*, 25, 349–362.

Antonovics, J. (2006) Evolution in closely adjacent plant populations. X. Long term persistence of prereproductive isolation at a mine boundary. *Heredity*, 97, 33–37.

Arendt, J.D. & Reznick, D.N. (2005) Evolution of juvenile growth rates in female guppies (Poecilia reticulata): predator regime or resource level? *Proceedings of the Royal Society of London, Series B*, 272, 333–337.

Arheimer, B. & Wittgren, H.B. (2002) Modelling nitrogen retention in potential wetlands at the catchment scale. *Ecological Engineering*, 19, 63–80.

Aston, J.L. & Bradshaw, A.D. (1966) Evolution in closely adjacent plant populations. II. *Agrostis stolonifera* in maritime habitats. *Heredity*, 21, 649–664.

Atkinson, D., Ciotti, B.J. & Montagnes, D.J.S. (2003) Protists decrease in size linearly with temperature: *ca*. 2.5%°C⁻¹. *Proceedings of the Royal Society of London, Series B*, 270, 2605–2611.

Audesirk, T. & Audesirk, G. (1996) *Biology: Life on Earth*. Prentice Hall, Upper Saddle River, NJ.

Avnerya, S., Mauzella, D.L., Liu, J. & Horowitz, L.W. (2011) Global crop yield reductions due to surface ozone exposure: 2. Year 2030 potential crop production losses and economic damage under two scenarios of O_3 pollution. *Atmospheric Environment*, 45, 2297–2309.

Ayre, D.J. (1985) Localized adaptation of clones of the sea anemone *Actinia tenebrosa*. *Evolution*, 39, 1250–1260.

Ayre, D.J. (1995) Localized adaptation of sea anemone clones: evidence from transplantation over two spatial scales. *Journal of Animal Ecology*, 64, 186–196.

Bach, C.E. (1994) Effects of herbivory and genotype on growth and survivorship of sand-dune willow (*Salix cordata*). *Ecological Entomology*, 19, 303–309.

Baker, A.J.M. (2002) The use of tolerant plants and hyperaccumulators. In: *The Restoration and Management of Derelict Land: Modern Approaches* (M.H. Wong & A.D. Bradshaw, eds), pp. 138–148. World Scientific Publishing, Singapore.

Balmford, A. & Bond, W. (2005) Trends in the state of nature and their implications for human well-being. *Ecology Letters*, 8, 1218–1234.

Balmford, A., Bruner, A., Cooper, P. et al. (2002) Economic reasons for conserving wild nature. *Science*, 297, 950–953.

Barluenga, M., Stolting, K.N., Salzburger, W., Muschick, M. & Meyer, A. (2006) Sympatric speciation in Nicaraguan crater lake cichlid fish. *Nature*, 439, 719–723.

Barnett, M.A. (1983) Species structure and temporal stability of mesopelagic fish assemblages in the Central Gyres of the North and South Pacific Ocean. *Marine Biology*, 74, 245–256.

Barrios, L. & Rodriguez, A. (2004) Behavioural and environmental correlates of soaring-bird mortality at on-shore wind turbines. *Journal of Applied Ecology*, 41, 72–81.

Bauerfeind, S.S., Theisen, A. & Fischer, K. (2009) Patch occupancy in the endangered butterfly Lycaena helle in a fragmented landscape: effects of habitat quality, patch size and isolation. *Journal of Insect Conservation*, 13, 271–277.

Bayliss, P. (1987) Kangaroo dynamics. In: *Kangaroos, their Ecology and Management in the Sheep Rangelands of Australia* (G. Caughley, N. Shepherd & J. Short, eds), pp. 119–134. Cambridge University Press, Cambridge.

Bazzaz, F.A. & Williams, W.E. (1991) Atmospheric CO_2 concentrations within a mixed forest: implications for seedling growth. *Ecology*, 72, 12–16.

Bazzaz, F.A. (1979) The physiological ecology of plant succession. *Annual Review of Ecology and Systematics*, 10, 351–371.

Bazzaz, F.A. (1996) *Plants in Changing Environments*. Cambridge University Press, Cambridge.

Bazzaz, F.A., Miao, S.L. & Wayne, P.M. (1993) CO_2-induced growth enhancements of co-occurring tree species decline at different rates. *Oecologia*, 96, 478–482.

Beaumont, L.J. & Hughes, L. (2002) Potential changes in the distributions of latitudinally restricted Australian butterfly species in response to climate change. *Global Change Biology*, 8, 954–971.

Becker, P. (1992) Colonization of islands by carnivorous and herbivorous Heteroptera and Coleoptera: effects of island area, plant species richness, and 'extinction' rates. *Journal of Biogeography*, 19, 163–171.

Begon, M., Sait, S.M. & Thompson, D.J. (1995) Persistence of a predator–prey system: refuges and generation cycles?

Proceedings of the Royal Society of London, Series B, 260, 131–137.

Begon, M., Townsend, C.R. & Harper, J.L. (2006) *Ecology: from Individuals to Ecosystems*, 4th edn. Blackwell Publishing, Oxford.

Belmaker, J. & Jetz, W. (2011) Cross-scale variation in species richness–environment associations. *Global Ecology & Biogeography*, 20, 464–474.

Belt, T. (1874) *The Naturalist in Nicaragua*. J.M. Dent, London.

Berger, J. (1990) Persistence of different-sized populations: an empirical assessment of rapid extinctions in bighorn sheep. *Conservation Biology*, 4, 91–98.

Berner, E.K. & Berner, R.A. (1987) *The Global Water Cycle: Geochemistry and Environment*. Prentice Hall, Englewood Cliffs, NJ.

Bertzky, B., Corrigan, C., Kemsey, J., Kenney, S., Ravilious, C., Besançon, C. & Burgess, N. (2012) Protected Planet Report 2012: Tracking progress towards global targets for protected areas. IUCN, Gland, Switzerland and UNEP-WCMC, Cambridge, UK.

Berven, K.A. (1995) Population regulation in the wood frog, *Rana sylvatica*, from three diverse geographic localities. *Australian Journal of Ecology*, 20, 385–392.

Bilby, R.E., Fransen, B.R. & Bisson, P.A. (1996) Incorporation of nitrogen and carbon from spawning coho salmon into the trophic system of small streams: Evidence from stable isotopes. *Canadian Journal of Fisheries and Aquatic Sciences*, 53, 164–173.

Billen, G., Garnier, J., Thieu, V., Silvetre, M., Barles, S. & Chatzimpiros, P. (2012) Localising the nitrogen imprint of the Paris food supply: potential of organic farming and change in human diet. *Biogeosciences*, 9, 607–616.

Blaustein, A.R., Wake, D.B. & Sousa, W.P. (1994) Amphibian declines: judging stability, persistence, and susceptibility of populations to local and global extinctions. *Conservation Biology*, 8, 60–71.

Blomqvist, D., Pauliny, A., Larsson, M. & Flodin, L-A. (2010) Trapped in the extinction vortex? Strong genetic effects in a declining vertebrate population. *BMC Evolutionary Biology*, 10, 33.

Bobbink, R., Hicks, K., Galloway, J., Spranger, T., Alkemade, R., Ashmore, M., Bustamante, M., Cinderby, S., Davidson, E., Dentener, F., Emmett, B., Erisman, J.W., Fenn, M., Gilliam, F., Nordin, A., Pardo, L. & De Vries, W. (2010) Global assessment of nitrogen deposition effects on terrestrial plant diversity: a synthesis. *Ecological Applications*, 20, 30–59.

Bobko, S.J. & Berkeley, S.A. (2004) Maturity, ovarian cycle, fecundity, and age-specific parturition of black rockfish (*Sebastes melanops*). *Fisheries Bulletin*, 102, 418–429.

Bodmer, R.E., Eisenberg, J.F. & Redford, K.H. (1997) Hunting and the likelihood of extinction of Amazonian mammals. *Conservation Biology*, 11, 460–466.

Bonnet, X., Lourdais, O., Shine, R. & Naulleau, G. (2002) Reproduction in a typical capital breeder: costs, currencies and complications in the aspic viper. *Ecology*, 83, 2124–2135.

Bonsall, M.B., French, D.R. & Hassell, M.P. (2002) Metapopulation structure affects persistence of predator–prey interactions. *Journal of Animal Ecology*, 71, 1075–1084.

Borer, E.T., Halpern, B.S. & Seabloom, E.W. (2006) Asymmetry in community regulation: effects of predators and productivity. *Ecology*, 87, 2813–2820.

Borga, K., Gabrielsen, G.W. & Skaare, J.U. (2001) Biomagnification of organochlorines along a Barents Sea food chain. *Environmental Pollution*, 113, 187–198.

Bouzille, J.B., Bonis, A., Clement, B. & Godeau, M. (1997) Growth patterns of *Juncus gerardi* clonal populations in a coastal habitat. *Plant Ecology*, 132, 39–48.

Bowles, K.C., Apte, S.C., Maher, W.A., Kawei, M. & Smith, R. (2001) Bioaccumulation and biomagnification of mercury in Lake Murray, Papua New Guinea. *Canadian Journal of Fisheries and Aquatic Science*, 58, 888–897.

Bradshaw, A.D. (2002) Introduction—an ecological perspective. In: *The Restoration and Management of Derelict Land: Modern Approaches* (M.H. Wong & A.D. Bradshaw, eds), pp. 1–6. World Scientific Publishing, Singapore.

Breznak, J.A. (1975) Symbiotic relationships between termites and their intestinal biota. In: *Symbiosis* (D.H. Jennings & D.L. Lee, eds.), pp. 559–580. Symposium 29, Society for Experimental Biology. Cambridge University Press, Cambridge.

Briand, F. (1983) Environmental control of food web structure. *Ecology*, 64, 253–263.

Bringezu, S., Schutz, H., O'Brien, M., Kauppi, L., Howarth, R. & McNeely, J. (2009) Towards Sustainable Production and Use of Resources: Assessing Biofuels. International Panel for Sustainable Resource Management, United Nations Environment Program, Paris, France (http://www.unep.fr/scp/rpanel/biofuels.htm).

Brockhurst, M., Morgan, A.D., Rainey, P.B. & Buckling, A. (2003) Population mixing accelerates coevolution. *Ecology Letters*, 6, 975–979.

Brook, B.W., O'Grady, J.J., Chapman, A.P., Burgman, M.A., Akçakaya, H.R. & Frankham, R. (2000) Predictive accuracy of population viability analysis in conservation biology. *Nature*, 404, 385–387.

Brookes, M. (1998) The species enigma. *New Scientist*, June 13, 1998.

Brown, J.H. & Davidson, D.W. (1977) Competition between seed-eating rodents and ants in desert ecosystems. *Science*, 196, 880–882.

Brown, P.H., Miller, D.M., Brewster, C.C. & Fell, R.D. (2013) Biodiversity of ant species along a disturbance gradient in Puerto Rico ref - Urban Ecosystems, 16, 175–192.

Brown, V.K. & Southwood, T.R.E. (1983) Trophic diversity, niche breadth, and generation times of exopterygote insects in a secondary succession. *Oecologia*, 56, 220–225.

Brown, W.P. & Roth, R.R. (2009) Age-specific reproduction and survival of individually marked Wood Thrushes, *Hylocichla mustelina*. *Ecology*, 90, 218–229.

Bruckmann, S.V., Krauss, J. & Steffan-Dewenter, I. (2010) Butterfly and plant specialists suffer from reduced connectivity in fragmented landscapes. *Journal of Applied Ecology*, 47, 799–809.

Brunet, A.K. & Medellín, R.A. (2001) The species–area relationship in bat assemblages of tropical caves. *Journal of Mammalogy*, 82, 1114–1122.

Bryden, H.L., Hannah, R., Longworth, H.R. & Cunningham, S.A. (2005) Slowing of the Atlantic meridional overturning circulation at 25° N. *Nature*, 438: 655–657.

Brylinski, M. & Mann, K.H. (1973) An analysis of factors governing productivity in lakes and reservoirs. *Limnology and Oceanography*, 18, 1–14.

Buckling, A. & Rainey, P.B. (2002) Antagonistic coevolution between a bacterium and a bacteriophage. *Proceedings of the Royal Society of London, Series B*, 269, 931–936.

Buckling, A., McLean, R.C., Brockhurst, M.A & Colgrave, N. (2009) The *Beagle* in a bottle. *Nature*, 457, 824–829.

Bullock, J.M., Moy, I.L., Pywell, R.F., Coulson, S.J., Nolan, A.M. & Caswell, H. (2002) Plant dispersal and colonization processes at local and landscape scales. In: *Dispersal Ecology* (J.M. Bullock, R.E. Kenward & R.S. Hails, eds.), pp. 279–302. Blackwell Publishing, Oxford.

Burdon, J.J. (1987) *Diseases and Plant Population Biology*. Cambridge University Press, Cambridge.

Burdon-Sanderson, J.S. (1893) Inaugural address. *Nature*, 48, 464–472.

Burg, T.M. & Croxall, J.P. (2001) Global relationships amongst black-browed and grey-headed albatrosses: analysis of population structure using mitochondrial DNA and microsatellites. *Molecular Ecology*, 10, 2647–2660.

Burgman, M.A., Ferson, S. & Akçakaya, H.R. (1993) *Risk Assessment in Conservation Biology*. Chapman & Hall, London.

Burkhead, N.M. (2012) Extinction rates of North American freshwater fishes, 1900–2010. *BioScience*, 62, 798–808.

Buscot, F., Munch, J.C., Charcosset, J.Y., Gardes, M., Nehls, U. & Hampp, R. (2000) Recent advances in exploring physiology and biodiversity of ectomycorrhizas highlight the functioning of these symbioses in ecosystems. *FEMS Microbiology Reviews*, 24, 601–614.

Butet, A. & Leroux, A.B.A. (2001) Effects of agricultural development on vole dynamics and conservation of Montagu's harrier in western French wetlands. *Biological Conservation*, 100, 289–295.

Cain, M.L., Pacala, S.W., Silander, J.A. & Fortin, M.-J. (1995) Neighbourhood models of clonal growth in the white clover *Trifolium repens*. *American Naturalist*, 145, 888–917.

Canadell, J.G., Le Quéré, C., Raupach, M.R., Field, C.B., Buitenhuis, E.T., Ciais, P., Conway, T.J., Gillet, N.P., Houghton, R.A. & Marland. G. (2007) Contributions to accelerating atmospheric CO_2 growth from economic activity, carbon intensity, and efficiency of natural sinks. *Proceedings of the National Academy of Sciences*, 104, 18866–18870.

Carignan, R., Planas, D. & Vis, C. (2000) Planktonic production and respiration in oligotrophic shield lakes. *Limnology and Oceanography*, 45, 189–199.

Carpenter, S.R., Christensen, D.L., Cole, J.J., Kottingham, K.L., He, X., Hodgson, J.R., Hitchell, J.F., Knight, S.E., Pace, M.L., Post, D.M., Schindler, D.E. & Voichick, N. (1995) Biological control of eutrophication in lakes. *Environmental Science and Technology*, 29, 784–786.

Carruthers, R.I., Larkin, T.S., Firstencel, H. & Feng, Z. (1992) Influence of thermal ecology on the mycosis of a rangeland grasshopper. *Ecology*, 73, 190–204.

Castellon, T.D. & Sieving, K.E. (2006) An experimental test of matrix permeability and corridor use by an endemic understory bird. *Conservation Biology*, 20, 135–145.

Caughley, G. (1994) Directions in conservation biology. *Journal of Animal Ecology*, 63, 215–244.

Caulton, D.R., Shepson, P. B., Santoro, R.L., Sparks, J.P., Howarth, R.W., Ingraffea, A., Cambaliza, M.O., Sweeney, C., Karion, A., Davis, K.J., Lauvaux, T., Stirm, B., Belmecheri, S. & Sarmiento D. (2014) Quantifying methane emissions from an area of shale gas development in Pennsylvania. *Proceedings of the National Academy of Sciences*, 111, 6237-6242.

Cebrian, J. (1999) Patterns in the fate of production in plant communities. *American Naturalist*, 154, 449–468.

Charnov, E.L. (1976) Optimal foraging: attack strategy of a mantid. *American Naturalist*, 110, 141–151.

Chase, J.M. & Leibold, M.A. (2002) Spatial scale dictates the productivity-biodiversity relationship. *Nature*, 416, 427–430.

Chase, J.M. (2003) Experimental evidence for alternative stable equilibria in a benthic pond food web. *Ecology Letters*, 6, 733–741.

Choquenot, D. (1998) Testing the relative influence of intrinsic and extrinsic variation in food availability on feral pig populations in Australia's rangelands. *Journal of Animal Ecology*, 67, 887–907.

Clements, F.L. (1905) *Research Methods in Ecology*. University of Nevada Press, Lincoln, NV.

Cleveland, C.C., Townsend, A.R., Schimel, D.S., Fisher, H., Howarth, R.W., Hedin, L.O., Perakis, S.S., Latty, E.F., von Fischer, J., Elseroad, C.A. & Wasson, M.F. (1999) Global patterns of terrestrial biological nitrogen (N_2) fixation in natural systems. *Global Biogeochemical Cycles*, 13, 623–645.

Cohen, J.E. (1995) *How Many People Can the Earth Support?* W.W. Norton & Co., New York.

Cohen, J.E. (2001) World population in 2050: assessing the projections. In: *Seismic Shifts: the Economic Impact of Demographic Change* (J.S. Little & R.K. Triest, eds.), pp. 83–113. Conference Series No. 48. Federal Reserve Bank of Boston, Boston.

Cohen, J.E. (2003) Human population: the next half century. *Science*, 302, 1172–1175.

Cohen, J.E. (2005) Human population grows up. *Scientific American*, 293(3), 48–55.

Cole, J.J., Findlay, S. & Pace, M.L. (1988) Bacterial production in fresh and salt water ecosystems: a cross-system overview. *Marine Ecology Progress Series*, 4, 1–10.

Connell, J.H. (1978) Diversity in tropical rainforests and coral reefs. *Science*, 199, 1302–1310.

Connell, J.H. (1983) On the prevalence and relative importance of interspecific competition: evidence from field experiments. *American Naturalist*, 122, 661–696.

Cook, L.M., Dennis, R.L.H. & Mani, G.S. (1999) Melanic morph frequency in the peppered moth in the Manchester area. *Proceedings of the Royal Society of London, Series B*, 266, 293–297.

Coomes, D.A., Rees, M., Turnbull, L. & Ratcliffe, S. (2002) On the mechanisms of coexistence among annual-plant species, using neighbourhood techniques and simulation models. *Plant Ecology*, 163, 23–38.

Cornell, H.V. & Hawkins, B.A. (2003) Herbivore responses to plant secondary compounds: a test of phytochemical coevolution theory. *American Naturalist*, 161, 507–522.

Cortes, E. (2002) Incorporating uncertainty into demographic modeling: application to shark populations and their conservation. *Conservation Biology*, 16, 1048–1062.

Cory, J.S. & Myers, J.H. (2000) Direct and indirect ecological effects of biological control. *Trends in Ecology and Evolution*, 15, 137–139.

Costanza, R., D'Arge, R., de Groot, R. et al. (1997) The value of the world's ecosystem services and natural capital. *Nature* 387, 253–260.

Cotrufo, M.F., Ineson, P., Scott, A. et al. (1998) Elevated CO_2 reduces the nitrogen concentration of plant tissues. *Global Change Biology*, 4, 43–54.

Cottingham, K.L., Brown, B.L. & Lennon, J.T. (2001) Biodiversity may regulate the temporal variability of ecological systems. *Ecology Letters*, 4, 72–85.

Coulson, T., Gaillard, J-M. & Festa-Bianchet, M. (2005) Decomposing the variation in population growth into contributions from multiple demographic rates. *Journal of Animal Ecology*, 74, 789–901.

Courant, S. & Fortin, D. (2011) Time allocation of bison in meadow patches driven by potential energy gains and group size dynamics. *Oikos*, 121, 1163–1173.

Courchamp, F., Clutton-Brock, T. & Grenfell, B. (1999) Inverse density dependence and the Allee effect. *Trends in Ecology and Evolution*, 14, 405–410.

Cowling, R.M., Pressey, R.L., Rouget, M. & Lombard, A.T. (2003) A conservation plan for a global biodiversity hotspot—the Cape Floristic Region, South Africa. *Biological Conservation*, 112, 191–216.

Cox, P.A., Elmquist, T., Pierson, E.D. & Rainey, W.E. (1991) Flying foxes as strong interactors in South Pacific island ecosystems: a conservation hypothesis. *Conservation Biology*, 5, 448–454.

Crisp, M.D. & Lange, R.T. (1976) Age structure distribution and survival under grazing of the arid zone shrub *Acacia burkitii*. *Oikos*, 27, 86–92.

Currie, D.J. & Paquin, V. (1987) Large-scale biogeographical patterns of species richness in trees. *Nature*, 39, 326–327.

Currie, D.J. (1991) Energy and large-scale patterns of animal and plant species richness. *American Naturalist*, 137, 27–49.

Dabbagh, A., Gacic-Dobo, M., Simons, E., Featherstone, D., Strebel, P., Okwo-Bele, J. M., Hoekstra, E., Chopra, M., Uzicanin, A. & Cochi, S. (2009) Global measles mortality, 2000–2009. *Morbidity and Mortality Weekly Report 2009*, 58, 1321–1326.

Daly, H.E. & Farley, J. (2011) *Ecological Economics: principles and Applications. Second Edition.* Island Press, Washington, D.C.

Darwin, C. (1859) *On the Origin of Species by Means of Natural Selection*, 1st edn. John Murray, London.

Davidson, D.W. (1977) Species diversity and community organization in desert seed-eating ants. *Ecology*, 58, 711–724.

Davidson, J. & Andrewartha, H.G. (1948) The influence of rainfall, evaporation and atmospheric temperature on fluctuations in the size of a natural population of *Thrips imaginis* (Thysanoptera). *Journal of Animal Ecology*, 17, 200–222.

Davies, S.J., Palmiotto, P.A., Ashton, P.S., Lee, H.S. & Lafrankie, J.V. (1998) Comparative ecology of 11 sympatric species of *Macaranga* in Borneo: tree distribution in relation to horizontal and vertical resource heterogeneity. *Journal of Ecology*, 86, 662–673.

Davis, M.B. & Shaw, R.G. (2001) Range shifts and adaptive responses to quarternary climate change. *Science*, 292, 673–679.

Davis, M.B. (1976) Pleistocene biogeography of temperate deciduous forest. *Geoscience and Management*, 13, 13–26.

De Barro, P.J., Liu, S-S., Boykin, L.M. & Dinsdale, A.B. (2011) *Bemisia tabaci*: a statement of species status. *Annual Review of Entomology*, 56, 1–19.

de Wit, C.T. (1965) Photosynthesis of leaf canopies. *Verslagen van Landbouwkundige Onderzoekingen*, 663, 1–57.

de Wit, C.T., Tow, P.G. & Ennik, G.C. (1966) Competition between legumes and grasses. *Verslagen van Landbouwkundige Onderzoekingen*, 112, 1017–1045.

Deevey, E.S. (1947) Life tables for natural populations of animals. *Quarterly Review of Biology*, 22, 283–314.

Denno, R.F., McClure, M.S. & Ott, J.R. (1995) Interspecific interactions in phytophagous insects: competition reexamined and resurrected. *Annual Review of Entomology*, 40, 297–331.

Descamps, S., Boutin, S., McAdam, A.G., Berteaux, D. & Gaillard, J-M. (2009) Survival costs of reproduction vary with age in North American red squirrels. *Proceedings of the Royal Society of London, Series B*, 276, 1129–1135.

Detwiler, R.P. & Hall, C.A.S. (1988) Tropical forests and the global carbon cycle. *Science*, 239, 42–47.

Diamond, J.M. (1972) Biogeographic kinetics: estimation of relaxation times for avifaunas of south-west Pacific islands. *Proceedings of the National Academy of Science of the USA*, 69, 3199–3203.

Dickie, I.A., Xu, B. & Koide, R.T. (2002) Vertical niche differentiation of ectomycorrhizal hyphae in soil as shown by T-RFLP analysis. *New Phytologist*, 156, 527–535.

Dinsdale, A., Cook, L., Riginos, C., Buckley, Y.M. & De Barro, P. (2010) Refined global analysis of *Bemisia tabaci* (Hemiptera: Sternorrhyncha: Aleyrodoidea: Aleyrodidae) mitochondrial cytochrome oxidase 1 to identify species level genetic boundaries. *Annals of the Entomological Society of America*, 103, 196–208.

Disma, G., Sokolowski, M.B.C. & Tonneau, F. (2011) Children's competition in a natural setting: evidence for the ideal free distribution. *Evolution and Human Behaviour*, 32,373–379.

Dobson, A.P. & Carper, E.R. (1996) Infectious diseases and human population history. *Bioscience*, 46, 115–126.

Dodson, S.I., Arnott, S.E. & Cottingham, K.L. (2000) The relationship in lake communities between primary productivity and species richness. *Ecology*, 81, 2662–2679.

Dölle, M., Bernhardt-Römmermann, M., Parth, A. & Schmidt, W. (2008) Changes in life history trait composition during undisturbed old-field succession. *Flora*, 203, 508–522.

Doube, B.M., Macqueen, A., Ridsdill-Smith, T.J. & Weir, T.A. (1991) Native and introduced dung beetles in Australia. In: *Dung Beetle Ecology* (I. Hanski & Y. Cambefort, eds.), pp. 255–278. Princeton University Press, Princeton, NJ.

Drès, M. & Mallet, J. (2001) Host races in plant-feeding insects and their importance in sympatric speciation. *Philosophical Transactions of the Royal Society of London, Series B*, 357, 471–492.

Dunne, J.A., Williams, R.J. & Martinez, N.J. (2002) Network structure and biodiversity loss in food webs: robustness increases with connectance. *Ecology Letters*, 5, 558–567.

Eamus, D. (1999) Ecophysiological traits of deciduous and evergreen woody species in the seasonally dry tropics. *Trends in Ecology and Evolution*, 14, 11–16.

Eberhardt, L.L. & Breiwick, J.M. (2010) Trend of the Yellowstone grizzly bear population. *International Journal of Ecology*, 2010, 924197.

Ebert, D., Zschokke-Rohringer, C.D. & Carius, H.J. (2000) Dose effects and density-dependent regulation in two microparasites of *Daphnia magna*. *Oecologia*, 122, 200–209.

Ehleringer, J.R. & Cerling, T.E. (2002) C3 and C4 photosynthesis. Pages 186–190 in H.A., Mooney & J.C. Canadell, (editors), Volume 3, *the Earth System: Biological and Ecological Dimensions of Global Environmental Change. Encyclopedia of Global Environmental Change.* Wiley, Chichester.

Ehrlen, J. (2003) Fitness components versus total demographic effects: evaluating herbivore impacts on a perennial herb. *American Naturalist*, 162, 796–810.

Ehrlich, P. & Raven, P.H. (1964) Butterflies and plants: a study in coevolution. *Evolution*, 18, 586–608.

Eis, S., Garman, E.H. & Ebel, L.F. (1965) Relation between cone production and diameter increment of douglas fir (*Pseudotsuga menziesii* (Mirb). Franco), grand fir (*Abies grandis* Dougl.) and western white pine (*Pinus monticola* Dougl.). *Canadian Journal of Botany*, 43, 1553–1559.

Elliott, J.K. & Mariscal, R.N. (2001) Coexistence of nine anemonefish species: differential host and habitat utilization, size and recruitment. *Marine Biology*, 138, 23–36.

Elliott, J.M. (1994) *Quantitative Ecology and the Brown Trout.* Oxford University Press, Oxford.

Elton, C. (1927) *Animal Ecology.* Sidgwick & Jackson, London.

Elton, C.S. (1958) *The Ecology of Invasions by Animals and Plants.* Methuen, London.

Endler, J.A. (1980) Natural selection on color patterns in *Poecilia reticulata*. *Evolution*, 34, 76–91.

Erickson, G.M., Makovicky, P.J., Inouye, B.D., Zhou, C-F & Gao, K-Q. (2009) A life table for Psittacosaurus lujiatunensis: initial insights into Ornithischian dinosaur population biology. *The Anatomical Record*, 292, 1514–1521.

Erwin, T.L. (1982) Tropical forests: their richness in Coleoptera and other arthropod species. *Coleopterists Bulletin*, 36, 74–75.

Etiope, G., Lassey, K.R., Klusman, R.W. & Boschi, E. (2008) Reappraisal of the fossil methane budget and related emission from geologic sources, *Geophysical Research Letters*, 35, L09307.

European Environment Agency (2002) Europe's environment: The third assessment. *Environmental assessment report No 10.* 242pp. Office for Official Publications of the European Communities, Luxembourg.

Falge, E., Baldocchi, D., Tenhunen, J. et al. (2002) Seasonality of ecosystem respiration and gross primary production as derived from FLUXNET measurements. *Agricultural and Forest Meteorology*, 113, 53–74.

Falkowski, P.G. (1997) Evolution of the nitrogen cycle and its influence on the biological sequestration of CO_2 in the oceans. *Nature*, 387, 272–275.

FAO (2012) Milestones in Aquaculture Development. Food and Agricultural Organization, United Nations, Rome, Italy. (http://www.fao.org/docrep/field/009/ag158e/AG158E02.htm).

Fasham, M.J.R., Balino, B.M. & Bowles, M.C. (2001) A new vision of ocean biogeochemistry after a decade of the Joint Global Ocean Flux Study (JGOFS). *Ambio Special Report*, 10, 4–31.

Feely, R.A., Doney, S.C. & Cooley S.R. (2009) Ocean acidification: present conditions and future changes in a high CO_2 world. *Oceanography*, 22, 36–47.

Fenner, F. (1983) Biological control, as exemplified by smallpox eradication and myxomatosis. *Proceedings of the Royal Society of London, Series B*, 218, 259–285.

Ferguson, R.G. (1933) The Indian tuberculosis problem and some preventative measures. *National Tuberculosis Association Transactions*, 29, 93–106.

Field, C.B., Behrenfeld, M.J., Randerson, J.T. & Falkowski. P. (1998) Primary production of the biosphere: integrating terrestrial and oceanic components. *Science*, 281, 237–240.

Fischer, M. & Matthies, D. (1998) Effects of population size on performance in the rare plant *Gentianella germanica*. *Journal of Ecology*, 86, 195–204.

Fisher, M.C., Garner, T.W.J. & Walker, S.F. (2009) Global emergence of *Batrochochytrium dendrobatidis* and amphibian chytridiomycosis is space, time and host. *Annual Review of Microbiology*, 63, 291–310.

FitzGibbon, C.D. & Fanshawe, J. (1989) The condition and age of Thomson's gazelles killed by cheetahs and wild dogs. *Journal of Zoology*, 218, 99–107.

FitzGibbon, C.D. (1990) Anti-predator strategies of immature Thomson's gazelles: hiding and the prone response. *Animal Behaviour*, 40, 846–855.

Flecker, A.S. & Townsend, C.R. (1994) Community-wide consequences of trout introduction in New Zealand streams. *Ecological Applications*, 4, 798–807.

Fleischer, R.C., Perry, E.A., Muralidharan, K., Stevens, E.E. & Wemmer, C.M. (2001) Phylogeography of the Asian elephant (*Elephus maximus*) based on mitochondrial DNA. *Evolution*, 55, 1882–1892.

Flessa, K.W. & Jablonski, D. (1995) Biogeography of recent marine bivalve mollusks and its implications of paleobiogeography and the geography of extinction: a progress report. *Historical Biology*, 10, 25–47.

Flett, R.J., Schinder, D.W., Hamilton, R.D. & Campbell, E.R. (1980) Nitrogen fixation in Canadian Precambrian Shield lakes. *Canadian Journal of Fisheries and Aquatic Sciences*, 37, 494–505.

Flint, M.L. & van den Bosch, R. (1981) *Introduction to Integrated Pest Management*. Plenum Press, New York.

Flower, R.J., Rippey, B., Rose, N.L., Appleby, P.G. & Battarbee, R.W. (1994) Palaeolimnological evidence for the acidification and contamination of lakes by atmospheric pollution in western Ireland. *Journal of Ecology*, 82, 581–596.

Fonseca, C.R. (1994) Herbivory and the long-lived leaves of an Amazonian ant-tree. *Journal of Ecology*, 82, 833–842.

Fonseca, D.M. & Hart, D.D. (1996) Density-dependent dispersal of black fly neonates is mediated by flow. *Oikos*, 75, 49–58.

Ford, E.B. (1975) *Ecological Genetics*, 4th edn. Chapman & Hall, London.

Ford, M.J. (1982) *The Changing Climate: Responses of the Natural Fauna and Flora*. George Allen & Unwin, London.

Fowler, S.V. (2004) Biological control of an exotic scale, *Orthezia insignis* Browne (Homoptera: Orthexiidae), saves the endemic gumwood tree, *Commidendrum robustum* (Roxb.) DC (Asteraceae) on the island of St Helena. *Biological Control*, 29, 367–374.

Fox, C.J. (2001) Recent trends in stock-recruitment of blackwater herring (*Clupea harengus* L.) in relation to larval production. *ICES Journal of Marine Science*, 58, 750–762.

Fox, N.J. & Beckley, L.E. (2005) Priority areas for conservation of Western Australian coastal fishes: a comparison of hotspot, biogeographical and complementarity approaches. *Biological Conservation*, 125, 399–410.

Frank, K.T., Petrie, B., Choi, J.S. & Leggett, W.C. (2005) Trophic cascades in a formerly cod-dominated ecosystem. *Science*, 308, 1621–1623.

Franklin, I.R. & Frankham, R. (1998) How large must populations be to retain evolutionary potential? *Animal Conservation*, 1, 69–73.

Franzen, M. & Nilsson, M. (2010) Both population size and patch quality affect local extinctions and colonizations. *Proceedings of the Royal Society of London, B* 277, 79–85.

Fredrickson, M.E. (2005) Ant species confer different partner benefits on two neotropical myrmecophytes. *Oecologia*, 143, 387–395.

Fredrickson, M.E., Greene, M.J. & Gordon, D.M. (2005) 'Devil's gardens' bedevilled by ants. *Nature*, 437, 495–496.

Fredrickson, R.J. & Hedrick, P.W. (2006) Dynamics of hybridization and introgression in red wolves and coyotes. *Conservation Biology*, 20, 1272–1283.

Friedlingstein, P. et al. (2006) Climate-carbon cycle feedback analysis: Results from the C4 MIP model intercomparison. *Journal of Climate*, 19, 3337–3353.

Gadgil, M. (1971) Dispersal: population consequences and evolution. *Ecology*, 52, 253–261.

Galloway, J.N., Dentener, F.J., Capone, D.G., Boyer, E.W., Howarth, R.W., Seitzinger, S.P., Asner, G.P., Cleveland, C., Green, P.A., Holland, E., Karl, D.M., Michaels, A., Porter, J.H., Townsend A. & Vorosmarty, C. (2004) Nitrogen cycles: past, present, and future. *Biogeochemistry*, 70, 153–226.

Galloway, L.F. & Fenster, C.B. (2000) Population differentiation in an annual legume: local adaptation. *Evolution*, 54, 1173–1181.

Garthe, S. & Huppop, O. (2004) Scaling possible adverse effects of marine wind farms on seabirds: developing and applying a vulnerability index. *Journal of Applied Ecology*, 41, 724–734.

Gaston, K.J. (1998) *Biodiversity*. Blackwell Science, Oxford.

Geange, S.W. & Stier, A.C. (2009) Order of arrival affects competition in two reef fishes. *Ecology*, 90, 2868–2878.

Geider, R.J., Delucia, E.H., Falkowski, P.G. et al. (2001) Primary productivity of planet earth: biological determinants and physical constraints in terrestrial and aquatic habitats. *Global Change Biology*, 7, 849–882.

Gende, S.M., Quinn, T.P. & Willson, M.F. (2001) Consumption choice by bears feeding on salmon. *Oecologia*, 127, 372–382.

Gilman, M.P. & Crawley, M.J. (1990) The cost of sexual reproduction in ragwort (*Senecio jacobaea* L.). *Functional Ecology*, 4, 585–589.

Godfray, H.C.J. & Crawley, M.J. (1998) Introductions. In: *Conservation Science and Action* (W.J. Sutherland, ed.), pp. 39–65. Blackwell Science, Oxford.

Gotelli, N.J. & McCabe, D.J. (2002) Species co-occurrence: a meta-analysis of J.M. Diamond's assembly rules model. *Ecology*, 83, 2091–2096.

Gotthard, K., Nylin, S. & Wiklund, C. (1999) Seasonal plasticity in two satyrine butterflies: state-dependent decision making in relation to daylength. *Oikos*, 84, 453–462.

Gould, W.A. & Walker, M.D. (1997) Landscape-scale patterns in plant species richness along an arctic river. *Canadian Journal of Botany*, 75, 1748–1765.

Graham, C.H., Moritz, C. & Williams, S.E. (2006) Habitat history improves prediction of biodiversity in rainforest fauna. *Proceedings of the National Academy of Sciences of the USA*, 103, 632–636.

Grant, B.S., Cook, A.D., Clarke, C.A. & Owen, D.F. (1998) Geographic and temporal variation in the incidence of melanism in peppered moth populations in America and Britain. *Journal of Heredity*, 89, 465–471.

Grant, P.R., Grant, B.R., Keller, L.F. & Petren, K. (2000) Effects of El Niño events on Darwin's finch productivity. *Ecology*, 81, 2442–2457.

Grant, P.R. & Grant, B.R. (2010) Conspecific versus heterospecific gene exchange between populations of Darwin's finches. *Philosophical Transactions of the Royal Society of London, Series B*, 365, 1065–1076.

Gray, M.E., Sappington, T.W., Miller, N.J., Moeser, J. & Bohn, M.O. (2009) Adaptation and invasiveness of western corn rootworm: intensifying research on a worsening pest. *Annual Review of Entomology*, 54, 303–321.

Gray, S.M. & Robinson, B.W. (2001) Experimental evidence that competition between stickleback species favours adaptive character divergence. *Ecology Letters*, 5, 264–272.

Green, R.E. (1998) Long-term decline in the thickness of egg-shells of thrushes, *Turdus* spp., in Britain. *Proceedings of the Royal Society of London, Series B*, 265, 679–684.

Green, R.E., Newton, I., Shultz, S., Cunningham, A.A., Gilbert, M., Pain, D.J. & Prakash, V. (2004) Diclofenac poisoning as a cause of vulture population declines across the Indian subcontinent. *Journal of Applied Ecology*, 41, 793–800.

Greenwood, P.J., Harvey, P.H. & Perrins, C.M. (1978) Inbreeding and dispersal in the great tit. *Nature*, 271, 52–54.

Grutter, A.S. (1999) Cleaner fish really do clean. *Nature*, 398, 672–673.

Grytnes, J.A. & Vetaas, O.R. (2002) Species richness and altitude: a comparison between null models and interpolated plant species richness along the Himalayan altitudinal gradient, Nepal. *American Naturalist*, 159, 294–304.

Gunawardena, M. & Rowan, J.D. (2005) Economic valuation of a mangrove ecosystem threatened by shrimp aquaculture in Sri Lanka. *Journal of Environmental Management*, 36, 535–550.

Gurevitch, J., Morrow, L.L., Wallace, A. & Walsh, J.S. (1992) A meta-analysis of competition in field experiments. *American Naturalist*, 140, 539–572.

Haberl, H. et al. (2007). Quantifying and mapping the human appropriation of net primary production in Earth's terrestrial ecosystems. *Proceedings of the National Academy of Sciences of the USA*, 104, 12942–12947.

Hairston, N.G., Smith, F.E. & Slobodkin, L.B. (1960) Community structure, population control, and competition. *American Naturalist*, 44, 421–425.

Halaj, J., Ross, D.W. & Moldenke, A.R. (2000) Importance of habitat structure to the arthropod food-web in Douglas-fir canopies. *Oikos*, 90, 139–152.

Hall, S.J. & Raffaelli, D.G. (1993) Food webs: theory and reality. *Advances in Ecological Research*, 24, 187–239.

Hall, S.J. (1998) Closed areas for fisheries management—the case consolidates. *Trends in Ecology and Evolution*, 13, 297–298.

Hamel, S., Côté, S.D., Gaillard, J-M. & Festa-Bianchet, M. (2009) Individual variation in reproductive costs of reproduction: high-quality females always do better. *Journal of Animal Ecology*, 78, 143–151.

Hansen, J., Sato, M., Kharecha, P., Russell, G., Lea, DW. & Siddall, M. (2007) Climate change and trace gases. *Philosophical Transactions of the Royal Society of London, Series A*, 365, 1925–1954.

Hansen, M.C., Stehman, S.V. & Potapov, P.V. (2010) Quantification of global gross forest cover loss. *Proceedings of the National Academy of Sciences of the USA*, 107, 8650–8655.

Hanski, I. (1999) *Metapopulation Ecology*. Oxford University Press, Oxford.

Hanski, I., Pakkala, T., Kuussaari, M. & Lei, G. (1995) Metapopulation persistence of an endangered butterfly in a fragmented landscape. *Oikos*, 72, 21–28.

Harcourt, D.G. (1971) Population dynamics of *Leptinotarsa decemlineata* (Say) in eastern Ontario. III. Major population processes. *Canadian Entomologist*, 103, 1049–1061.

Harper, J.L. & White, J. (1974) The demography of plants. *Annual Review of Ecology and Systematics*, 5, 419–463.

Harper, J.L. (1977) *The Population Biology of Plants*. Academic Press, London.

Hart, A.J., Bale, J.S., Tullett, A.G., Worland, M.R. & Walters, K.F.A. (2002) Effects of temperature on the establishment potential of the predatory mite *Amblyseius californicus* McGregor (Acari: Phytoseiidae) in the UK. *Journal of Insect Physiology*, 48, 593–599.

Hassell, M.P., Latto, J. & May, R.M. (1989) Seeing the wood for the trees: detecting density dependence from existing life-table studies. *Journal of Animal Ecology*, 58, 883–892.

Hautier, Y., Niklaus, P.A. & Hector A. (2009) Competition for light causes plant biodiversity loss following eutrophication. *Science*, 324, 636–638.

Hawkins, B.A., Field, R., Cornell, H.V., Currie, D.J., Guegan, J-F., Kaufman, D.M., Kerr, J.T., Mittelbach, G.G., Oberdorff, T., O'Brien, E.M., Porter, E.E. & Turner, J.R.G. (2003) Energy, water, and broad-scale geographic patterns of species richness. *Ecology*, 84, 3105–3117.

Hay, M.E. & Taylor, P.R (1985) Competition between herbivorous fishes and urchins on Caribbean reefs. *Oecologia*, 65, 591–598.

Hebert, P.D.N., Penton, E.H., Burns, J.M., Janzen, D.H. & Hallwachs, W. (2004) Ten species in one: DNA barcoding reveals cryptic species in the neotropical skipper butterfly *Astraptes fulgerator*. *Proceedings of the National Academy of Sciences of the USA*, 101, 14812–14817.

Hereford, J. (2009). A quantitative survey of local adaptation and fitness trade-offs. *American Naturalist*, 173, 579–588.

Herman, T.J.B. (2000) Developing IPM for potato tuber moth. *Commercial Grower*, 55, 26–28.

Hermannsson, S. (2000) *Surtsey Research Report No. XI*. Museum of Natural History, Reykjavik, Iceland.

Hermoyian, C.S., Leighton, L.R. & Kaplan, P. (2002) Testing the role of competition in fossil communities using limiting similarity. *Geology*, 30, 15–18.

Herre, E.A. & West, S.A. (1997) Conflict of interest in a mutualism: documenting the elusive fig wasp–seed trade-off. *Proceedings of the Royal Society of London, Series B*, 264, 1501–1507.

Hilborn, R. & Walters, C.J. (1992) *Quantitative Fisheries Stock Assessment*. Chapman & Hall, New York.

Hodgson, J.A., Thomas, C.D., Wintle, B.A. & Moilanen, A. (2009) Climate change, connectivity and conservation decision making: back to basics. *Journal of Applied Ecology*, 46, 964–969.

Hodgson, J.A., Kunin, W.E., Thomas, C.D., Benton, T.G. & Gabriel, D. (2010) Comparing organic farming and land sparing: optimizing yield and butterfly populations at a landscape scale. *Ecology Letters*, 13, 1358–1367.

Hodgson, J.A., Moilanen, A., Wintle, B.A. & Thomas, C.D. (2011) Habitat area, quality and connectivity: striking the balance for efficient conservation. *Journal of Applied Ecology*, 48, 148–152.

Holloway, J.D. (1977) *The Lepidoptera of Norfolk Island, their Biogeography and Ecology*. Junk, The Hague.

Holloway, J.M., Dahlgren, R.A., Hansen, B. & Casey, W.H. (1998) Contribution of bedrock nitrogen to high nitrate concentrations in stream water. *Nature*, 395, 785–788.

Holyoak, M. & Lawler, S.P. (1996) Persistence of an extinction-prone predator–prey interaction through metapopulation dynamics. *Ecology*, 77, 1867–1879.

Hong, B., Swaney, D. & Howarth, R.W. (2011) A toolbox for calculating net anthropogenic nitrogen inputs (NANI). *Environmental Modeling and Software*, 26, 623–33.

Hooper, D.U., Chapin, F.S., Ewel, J.J. et al. (2005) Effects of biodiversity on ecosystem functioning: a consensus of current knowledge. *Ecological Monographs*, 75, 3–35.

Howarth, R.W. & S. Bringezu (eds.). (2009) Biofuels: Environmental Consequences and Interactions with Changing Land Use. Proceedings of the International SCOPE Biofuels Project Rapid Assessment, 22–25 September 2008, Gummersbach Germany. (http://cip.cornell.edu/biofuels/).

Howarth, R. W., Swaney, D., Billen, G., Garnier, J., Hong, B., Humborg, C., Johnes, P., Morth C. & Marino, R. (2012) Nitrogen fluxes from large watershed to coastal ecosystems controlled by net anthropogenic nitrogen inputs and climate. *Frontiers in Ecology & Environment*, 10, 37–43.

Howarth, R.W., Boyer, E.W., Pabich, W.J. & Galloway, J.N. (2002) Nitrogen use in the United States from 1961–200 and potential future trends. *Ambio*, 31, 88–96.

Howarth, R. W., Billen, G., Chan, F., Conley, D., Doney, S.C., Garnier, J. & Marino, R. (2011) Coupled biogeochemical cycles: eutrophication and hypoxia in coastal marine ecosystems. *Frontiers in Ecology & Environment*, 9, 18–26.

Howarth, R. W., Ramakrishna, K., Choi, E., Elmgren, R., Martinelli, L., Mendoza, A., Moomaw, W., Palm, C., Boy, R., Scholes, M. & Zhu Zhao-Liang (2005) Chapter 9: Nutrient Management, Responses Assessment. Pages 295–311 in *Ecosystems and Human Well-being*, Volume 3, *Policy Responses, the Millennium Ecosystem Assessment*. Island Press, Washington, DC.

Howarth, R.W. & Ingraffea, A. (2011) Should fracking stop? Yes, it is too high risk. *Nature*, 477, 271–273.

Howarth, R.W., Santoro, R. & Ingraffea, A. (2011) Methane and the greenhouse gas footprint of natural gas from shale formations. *Climatic Change Letters*, 106, 679–690.

Hoyer, M.V. & Canfield, D.E. (1994) Bird abundance and species richness on Florida lakes: influence of trophic status, lake morphology and aquatic macrophytes. *Hydrobiologia*, 297, 107–119.

Hudson, P.J., Dobson, A.P. & Newborn, D. (1992) Do parasites make prey vulnerable to predation? Red grouse and parasites. *Journal of Animal Ecology*, 61, 681–692.

Hudson, P.J., Dobson, A.P. & Newborn, D. (1998) Prevention of population cycles by parasite removal. *Science*, 282, 2256–2258.

Huffaker, C.B. (1958) Experimental studies on predation: dispersion factors and predator–prey oscillations. *Hilgardia*, 27, 343–383.

Hughes, A.R., Byrnes, J.E., Kimbro, D.L. & Stachowicz, J.J. (2007) Reciprocal relationships and potential feedbacks between biodiversity and disturbance. *Ecology Letters*, 10, 849–864.

Hughes, L. (2000) Biological consequences of global warming: is the signal already apparent? *Trends in Ecology and Evolution*, 15, 56–61.

Hunter, M.L. & Yonzon, P. (1992) Altitudinal distributions of birds, mammals, people, forests, and parks in Nepal. *Conservation Biology*, 7, 420–423.

Hurd, L.E. & Eisenberg, R.M. (1990) Experimentally synchronized phenology and interspecific competition in mantids. *American Midland Naturalist*, 124, 390–394.

Huryn, A.D. (1998) Ecosystem-level evidence for top-down and bottom-up control of production in a grassland stream system. *Oecologia*, 115, 173–183.

Husband, B.C. & Barrett, S.C.H. (1996) A metapopulation perspective in plant population biology. *Journal of Ecology*, 84, 461–469.

Huston, M.A. (1997) Hidden treatments in ecological experiments: Re-evaluating the ecosystem function of biodiversity. *Oecologia*, 110, 449–460.

Hut, R.A., Barnes, B.M. & Daan, S. (2002) Body temperature patterns before, during and after semi-natural hibernation in the European ground squirrel. *Journal of Comparative Physiology B*, 172, 47–58.

Hutchinson, G.E. (1957) Concluding remarks. *Cold Spring Harbour Symposium on Quantitative Biology*, 22, 415–427.

Inouye, R.S. & Tilman, D. (1995) Convergence and divergence of old-field vegetation after 11 yr of nitrogen addition. *Ecology*, 76, 1872–1877.

Inouye, R.S., Huntly, N.J., Tilman, D., Tester, J.R., Stillwell, M. & Zinnel, K.C. (1987) Old-field succession on a Minnesota sand plain. *Ecology*, 68, 12–26.

Intergovernmental Panel on Climate Change, (2007). *Climate Change 2007: The Physical Science Basis. Contribution of Working Group I to the Fourth Assessment Report of the Intergovernmental Panel on Climate Change*. Cambridge University Press.

Intergovernmental Panel on Climate Change, (2013). *Climate Change 2013: The Physical Science Basis. Contribution of Working Group I to the Fifth Assessment Report of the Intergovernmental Panel on Climate Change*. Cambridge University Press.

Interlandi, S.J. & Kilham, S.S. (2001) Limiting resources and the regulation of diversity in phytoplankton communities. *Ecology*, 82, 1270–1282.

International Organisation for Biological Control (1989) *Current Status of Integrated Farming Systems Research in Western Europe* (P. Vereijken & D.J. Royle, eds.). IOBC West Palaearctic Regional Service Bulletin No. 12(5). IOBC, Zurich.

Irvine, R.J., Stien, A., Dallas, J.F., Halvorsen, O., Langvatn, R. & Albon, S.D. (2001) Contrasting regulation of fecundity in two abomasal nematodes of Svarlbard reindeer (*Rangifer tarandus platyrynchus*). *Parasitology*, 122, 673–681.

Irwin, D.E., Irwin, J.H. & Price, T.D. (2001) Ring species as bridges between microevolution and speciation. *Genetica*, 112–113, 223–243.

IUCN/UNEP/WWF (1991) *Caring for the Earth. A Strategy for Sustainable Living*. World Conservation Union/United Nations Environmental Program/World Wide Fund, Gland, Switzerland.

Jackson, S.T. & Weng, C. (1999) Late quaternary extinction of a tree species in eastern North America. *Proceedings of the National Academy of Sciences of the USA*, 96, 13847–13852.

Jacobson, M.A. (2005) Studying ocean acidification with conservative, stable numerical schemes for nonequilibrium air-ocean exchange and ocean equilibrium chemistry. *Journal of Geophysical Research*, 110, D07302.

Jacobson, M.Z. & Delucchi, M.A. (2011) Providing all global energy with wind, water, and solar power, Part I: technologies, energy resources, quantities and areas of infrastructure, and materials. *Energy Policy*, 39, 1154–1169.

Jacobson, M.Z. & Delucchi, M.A. (2009) A path to sustainable energy by 2030. *Scientific American*. November 2009, 58–65.

Jain, S.K. & Bradshaw, A.D. (1966) Evolutionary divergence among adjacent plant populations. I. The evidence and its theoretical analysis. *Heredity*, 21, 407–411.

Jamali, H., Livesley, S.J., Dawes, T.Z. et al. (2011) Termite mound emissions of CH_4 and CO_2 are primarily determined by seasonal changes in termite biomass and behaviour. *Oecologia*, 167, 525–534.

Janis, C.M. (1993) Tertiary mammal evolution in the context of changing climates, vegetation and tectonic events. *Annual Review of Ecology and Systematics*, 24, 467–500.

Janssen, A., van Gool, E., Lingeman, R., Jacas, J. & van de Klashorst, G. (1997) Metapopulation dynamics of a persisting predator-prey system in the laboratory: time series analysis. *Experimental and Applied Acarology*, 21, 415–430.

Jedrzejewska, B., Sidorovich, V., Greco, C., Randi, E., Musiani, M., Kays, R., Bustamante, C.D., Ostrander, E.A., Novembre, J. & Wayne, R.K. (2011) A genome-wide perspective on the evolutionary history of the enigmatic wolf-like canids. *Genome Research*, 21, 1294–1305.

Jennings, S., Kaiser, M.J. & Reynolds, J.D. (2001) *Marine Fisheries Ecology*. Blackwell Publishing, Oxford.

Jeppesen, E., Sondergaard, M., Jensen, J.P. et al. (2005) Lake responses to reduced nutrient loading—an analysis of contemporary long-term data from 35 case studies. *Freshwater Biology*, 50, 1747–1771.

Johannes, R.E. (1998) Government-supported village-based management of marine resources in Vanuatu. *Ocean Coastal Management*, 40, 165–186.

Johnson, C.G. (1967) International dispersal of insects and insect-borne viruses. *Netherlands Journal of Plant Pathology*, 73 (Suppl. 1), 21–43.

Johnson, J.A., Toepfer, J.E. & Dunn, P.O. (2003) Contrasting patterns of mitochondrial and microsatellite population structure in fragmented populations of greater prairie-chickens. *Molecular Ecology*, 12, 3335–3347.

Johnson, P.T. J., Townsend, A.R., Cleveland, C.C., Glibert, P.M., Howarth, R.W., McKenzie, V.J., Rejmankova, E., & Ward, M.H. (2010) Linking environmental nutrient enrichment and disease emergence in humans and wildlife. *Ecological Applications*, 20, 16–29.

Jones, K.E., Patel, N.G., Levy, M.A., Storeygard, A., Balk, D., Gittleman, J.L. & Daszak, P. (2008). Global trends in emerging infectious diseases. *Nature*, 451, 990–994.

Jones, M. & Harper, J.L. (1987) The influence of neighbours on the growth of trees. I. The demography of buds in *Betula pendula*. *Proceedings of the Royal Society of London, Series B*, 232, 1–18.

Jonsson, M. & Malmqvist, B. (2000) Ecosystem process rate increases with animal species richness: evidence from leaf-eating, aquatic insects. *Oikos*, 89, 519–523.

Jutila, H.M. (2003) Germination in Baltic coastal wetland meadows: similarities and differences between vegetation and seed bank. *Plant Ecology*, 166, 275–293.

Kaiser, J. (2000) Rift over biodiversity divides ecologists. *Science*, 89, 1282–1283.

Kalkstein, L.S. & Greene, J.S. (1997) An evaluation of climate/mortality relationships in large US cities and the possible impacts of climate change. *Environmental Health Perspectives*, 105, 84–93.

Kamijo, T., Kitayama, K., Sugawara, A., Urushimichi, S. & Sasai, K. (2002) Primary succession of the warm-temperate broad-leaved forest on a volcanic island, Miyake-jima, Japan. *Folia Geobotanica*, 37, 71–91.

Kaplan, I. & Denno, R.F. (2007) Interspecific interactions in phytophagous insects revisited: a quantitative assessment of competition theory. *Ecology Letters*, 10, 977–994.

Karban, R., Agrawal, A.A., Thaler, J.S. & Adler, L.S. (1999) Induced plant responses and information content about risk of herbivory. *Trends in Ecology and Evolution*, 14, 443–447.

Karels, T.J. & Boonstra, R. (2000) Concurrent density dependence and independence in populations of arctic ground squirrels. *Nature*, 408, 460–463.

Karl, B.J. & Best, H.A. (1982) Feral cats on Stewart Island: their foods, and their effects on kakapo. *New Zealand Journal of Zoology*, 9, 287–294.

Karlsson, P.S. & Jacobson, A. (2001) Onset of reproduction in *Rhododendron lapponicum* shoots: the effect of shoot size, age, and nutrient status at two subarctic sites. *Oikos*, 94, 279–286.

Karubian, J., Fabara, J., Yunes, D., Jorgeson, J.P., Romo, D. & Smith, T.B. (2005) Temporal and spatial patterns of macaw abundance in the Ecuadorian Amazon. *The Condor*, 107, 617–626.

Keeling, M.J., Rohani, P. & Grenfell, B.T. (2001) Seasonally-forced disease dynamics explored as switching between attractors. *Physica D*, 148, 317–335.

Kennedy, P.G., Peay, K.G. & Bruns, T.D. (2009) Root tip competition among ectomycorrhizal fungi: are priority effects a rule or an exception? *Ecology*, 90, 2098–2107.

Kerbes, R.H., Kotanen, P.M. & Jefferies, R.L. (1990) Destruction of wetland habitats by lesser snow geese: a keystone species on the west coast of Hudson Bay. *Journal of Applied Ecology*, 27, 242–258.

Kersch-Becker, M.F. & Lewinsohn, T.M. (2012) Bottom-up multitrophic effects in resprouting plants. *Ecology*, 93, 9–16.

Kettlewell, H.B.D. (1955) Selection experiments on industrial melanism in the Lepidoptera. *Heredity*, 9, 323–342.

Khan, A.S., Sumaila, U.R., Watson, R., Munro, G. & Pauly, D. (2006) The nature and magnitude of global non-fuel fisheries subsidies. In: *Catching More Bait: a Bottom-up Re-estimation of Global Fisheries Subsidies* (U.R. Sumaila & D. Pauly, eds.), pp. 5–37. Fisheries Centre Research Reports Vol. 14, No. 6. Fisheries Centre, University of British Columbia, Vancouver.

Kicklighter, D.W., Bruno, M., Donges, S. et al. (1999) A first-order analysis of the potential role of CO_2 fertilization to affect the global carbon budget: a comparison of four terrestrial biosphere models. *Tellus*, 51B, 343–366.

Kirschke, S., Bousquet, P., Ciais, P., Saunois, M., Canadell, J.G., Dlugokencky, E.J., Bergamaschi, P., Bergmann, D., Blake, D.R., Bruhwiler, L., Cameron-Smith, P., Castaldi, S., Chevallier, F., Feng, L., Fraser, A., Heimann, M., Hodson, E.L., Houweling, S., Josse, B., Fraser, P.J., Krummel, P.B., Lamarque, J.-F., Langenfelds, R.L., Le Quéré, C., Naik, V., O'Doherty, S., Palmer, P.I., Pison, I., Plummer, D., Poulter, B., Prinn, R.G., Rigby, M., Ringeval, B., Santini, M., Schmidt, M., Shindell, D.T., Simpson, I.J., Spahni, R., Steele, L.P., Strode, S.A., Sudo, K., Szopa, S., van der Werf, G.R., Voulgarakis, A., van Weele, M., Weiss, R.F., Williams, J.E. and Zeng, G. (2013) Three decades of global methane sources and sinks. *Nature Geoscience*, 6, 813–823.

Kingsland, S.E. (1985) *Modeling Nature*. The University of Chicago Press, Chicago.

Kirk, J.T.O. (1994) *Light and Photosynthesis in Aquatic Ecosystems*. Cambridge University Press, Cambridge, UK.

Klemola, T., Koivula, M., Korpimaki, E. & Norrdahl, K. (2000) Experimental tests of predation and food hypotheses for population cycles of voles. *Proceedings of the Royal Society of London, Series B*, 267, 351–356.

Koch, H. & Schmid-Hempel, P. (2012) Gut microbiota instead of host genotype drive the specificity in the interaction of a natural host-parasite system. *Ecology Letters*, 15, 1095–1103.

Kodric-Brown, A. & Brown, J.M. (1993) Highly structured fish communities in Australian desert springs. *Ecology*, 74, 1847–1855.

Kolbert, E. (2011) Enter the anthropocene: Age of man. *National Geographic*, 219, 60–85.

Koop, J.A.H., Huber, S.K., Laverty, S.M. & Clayton, D.H. (2011) Experimental demonstration of the fitness consequences of an introduced parasite of Darwin's finches. *PLoS ONE*, 6, e19706.

Kramer, D.L. & Chapman, M.R. (1999) Implications of fish home range size and relocation for marine reserve function. *Environmental Biology of Fishes*, 55, 65–79.

Krebs, C.J. (1972) *Ecology*. Harper & Row, New York.

Krebs, C.J., Boonstra, R., Boutin, S. & Sinclair, A.R.E. (2001) What drives the 10-year cycle of snowshoe hares? *Bioscience*, 51, 25–35.

Krebs, C.J., Sinclair, A.R.E., Boonstra, R., Boutin, S., Martin, K. & Smith, J.N.M. (1999) Community dynamics of vertebrate herbivores: how can we untangle the web? In: *Herbivores: between Plants and Predators* (H. Olff, V.K. Brown & R.H. Drent, eds.), pp. 447–473. Blackwell Science, Oxford.

Kremen, C., Williams, N.M., Bugg, R.L., Fay, J.P. & Thorp, R.W. (2004) The area requirements of an ecosystem service: crop pollination by native bee communities in California. *Ecology Letters*, 7, 1109–1119.

Kullberg, C. & Ekman, J. (2000) Does predation maintain tit community diversity? *Oikos*, 89, 41–45.

Lacy, R.C. (1993) VORTEX: a computer simulation for use in population viability analysis. *Wildlife Research*, 20, 45–65.

Lalli, C.M. & Parsons, T.R. (2004) *Biological Oceanography: An Introduction, 2nd Ed.*, Elsevier Butterworth Heinemann, Burlington, MA.

Lande, R. & Barrowclough, G.F. (1987) Effective population size, genetic variation, and their use in population management. In: *Viable Populations for Conservation* (M.E. Soulé, ed.), pp. 87–123. Cambridge University Press, Cambridge.

Larcher, W. (1980) *Physiological Plant Ecology*, 2nd edn. Springer-Verlag, Berlin.

Lassey, K. R., Lowe, D.C. & Smith, A.M. (2007) The atmospheric cycling of radiomethane and the "fossil fraction" of the methane source, *Atmospheric Chemistry and Physics*, 7, 2141–2149.

Lathrop, R.C., Johnson, B.M., Johnson, T.B. et al. (2002) Stocking piscivores to improve fishing and water clarity: a synthesis of the Lake Mendota biomanipulation project. *Freshwater Biology*, 47, 2410–2424.

Laugen, A.T., Laurila, A., Rasanen, K. & Merila, J. (2003) Latitudinal countergradient variation in the common frog (*Rana temporaria*) development rates—evidence for local adaptation. *Journal of Evolutionary Biology*, 16, 996–1005.

Laurance, W.F. (2001) Future shock: forecasting a grim fate for the Earth. *Trends in Ecology and Evolution*, 16, 531–533.

Law, B.E., Thornton, P.E., Irvine, J., Anthoni, P.M. & van Tuyl, S. (2001) Carbon storage and fluxes in ponderosa pine forests at different developmental stages. *Global Climate Change*, 7, 755–777.

Lawlor, L.R. (1980) Structure and stability in natural and randomly constructed competitive communities. *American Naturalist*, 116, 394–408.

Lawrence, W.H. & Rediske, J.H. (1962) Fate of sown douglas-fir seed. *Forest Science*, 8, 211–218.

Lawton, J.H. & May, R.M. (1984) The birds of Selborne. *Nature*, 306, 732–733.

Le Cren, E.D. (1973) Some examples of the mechanisms that control the population dynamics of salmonid fish. In: *The Mathematical Theory of the Dynamics of Biological Populations* (M.S. Bartlett & R.W. Hiorns, eds.), pp. 125–135. Academic Press, London.

Le Quéré, C. et. al. (2009) Trends in the sources and sinks of carbon dioxide. *Nature Geoscience*, 2, 831–836.

Lehmann, N., Eisenhawer, A., Hansen, K., Mech, L.D., Peterson, R.O., Gogan, P.J.P. & Wayne, R.K. (1991) Introgression of mitochondrial DNA into sympatric North American grey wolf population. *Evolution*, 45, 104–119.

Lemieux, A.M. & Clarke, R.V. (2009) The international ban on ivory sales and its effects on elephant poaching in Africa. *British Journal of Criminology*, 49, 451–471.

Lennartsson, T., Nilsson, P. & Tuomi, J. (1998) Induction of overcompensation in the field gentian, *Gentianella campestris*. *Ecology*, 79, 1061–1072.

Leroy, F. & de Vuyst, L. (2001) Growth of the bacteriocin-producing *Lactobacillus sakei* strain CTC 494 in MRS broth is strongly reduced due to nutrient exhaustion: a nutrient depletion model for the growth of lactic acid bacteria. *Applied and Environmental Microbiology*, 67, 4407–4413.

Letourneau, D.K. & Dyer, L.A. (1998a) Density patterns of *Piper* ant-plants and associated arthropods: top-predator trophic cascades in a terrestrial system? *Biotropica*, 30, 162–169.

Letourneau, D.K. & Dyer, L.A. (1998b) Experimental test in a lowland tropical forest shows top-down effects through four trophic levels. *Ecology*, 79, 1678–1687.

Levins, R. (1969) Some demographic and genetic consequences of environmental heterogeneity for biological control. *Bulletin of the Entomological Society of America*, 15, 237–240.

Lichter, J. (2000) Colonization constraints during primary succession on coastal Lake Michigan sand dunes. *Journal of Ecology*, 88, 825–839.

Likens, G.E. & Bormann, F.G. (1975) An experimental approach to New England landscapes. In: *Coupling of Land and Water Systems* (A.D. Hasler, ed.), pp. 7–30. Springer-Verlag, New York.

Likens, G.E. & Bormann, F.H. (1994) *Biogeochemistry of a Forested Ecosystem*, 2nd edn. Springer-Verlag, New York.

Likens, G.E. (1989) Some aspects of air pollutant effects on terrestrial ecosystems and prospects for the future. *Ambio*, 18, 172–178.

Likens, G.E. (1992) *The Ecosystem Approach: its Use and Abuse*. Excellence in Ecology, Book 3. Ecology Institute, Oldendorf-Luhe, Germany.

Likens, G.E. (2004) Some perspectives on long-term biogeochemical research from the Hubbard Brook Ecosystem Study. *Ecology*, 85, 2355–2362.

Likens, G.E., Bormann, F.H., Pierce, R.S. & Fisher, D.W. (1971) Nutrient–hydrologic cycle interaction in small forested watershed ecosystems. In: *Productivity of Forest Ecosystems* (P. Duvogneaud, ed.), pp. 553–563. UNESCO, Paris.

Likens, G.E., Driscoll, C.T. & Buso, D.C. (1996) Long-term effects of acid rain: response and recovery of a forest ecosystem. *Science*, 272, 244–245.

Lindeman, R.L. (1942) The trophic–dynamic aspect of ecology. *Ecology*, 23, 399–418.

Lipowsky, A., Roscher, C., Schumacher, J. & Schmid, B. (2012) Density-independent mortality and increasing plant diversity are associated with differentiation of *Taraxacum officinale* into r- and K-strategists. *PLoS ONE* 7, e28121.

Lips, K.R., Brem, F., Brenes, R., Reeve, J.D., Alford, R.A., Voyles, J., Carey, C., Livo, L., Pessier, A.P. & Collins, J.P. (2006) Emerging infectious disease and the loss of biodiversity in a Neotropical amphibian community. *Proceedings of the National Academy of Sciences of the USA*, 103, 3165–3170.

Lofgren, A. & Jerling, L. (2002) Species richness, extinction and immigration rates of vascular plants on islands in the Stockholm Archipelago, Sweden, during a century of ceasing management. *Folia Geobotanica*, 37, 297–308.

Lotka, A.J. (1932) The growth of mixed population: two species competing for a common food supply. *Journal of the Washington Academy of Sciences*, 22, 461–469.

Louda, S.M. (1982) Distributional ecology: variation in plant recruitment over a gradient in relation to insect seed predation. *Ecological Monographs*, 52, 25–41.

Louda, S.M. (1983) Seed predation and seedling mortality in the recruitment of a shrub, *Haplopappus venetus* (Asteraceae), along a climatic gradient. *Ecology*, 64, 511–521.

Louda, S.M., Kendall, D., Connor, J. & Simberloff, D. (1997) Ecological effects of an insect introduced for the biological control of weeds. *Science*, 277, 1088–1090.

Lövei, G.L. (1997) Global change through invasion. *Nature*, 388, 627–628.

Lubchenco, J. (1978) Plant species diversity in a marine intertidal community: importance of herbivore food preference and algal competitive abilities. *American Naturalist*, 112, 23–39.

Lubchenco, J., Olson, A.M., Brubaker, L.B. et al. (1991) The sustainable biosphere initiative: an ecological research agenda. *Ecology*, 72, 371–412.

Luckman, W.H. & Decker, G.C. (1960) A 5-year report on observations in the Japanese beetle control area of Sheldon, Illinois. *Journal of Economic Entomology*, 53, 821–827.

Luo, Z., Tang, S., Li, C., Fang, H., Hu. H., Yang, J., Ding. J. & Jiang, Z. (2012) Environmental effects on vertebrate species richness: testing the energy, environmental stability and habitat heterogeneity hypotheses. *PLoS ONE*, 7, e35514.

Lussenhop, J. (1992) Mechanisms of microarthropod–microbial interactions in soil. *Advances in Ecological Research*, 23, 1–33.

MacArthur, J.W. (1975) Environmental fluctuations and species diversity. In: *Ecology and Evolution of Communities* (M.L. Cody & J.M. Diamond, eds.), pp. 74–80. Belknap, Cambridge, MA.

MacArthur, R.H. & Pianka, E.R. (1966) On optimal use of a patchy environment. *American Naturalist*, 100, 603–609.

MacArthur, R.H. & Wilson, E.O. (1967) *The Theory of Island Biogeography*. Princeton University Press, Princeton, NJ.

MacArthur, R.H. (1955) Fluctuations of animal populations and a measure of community stability. *Ecology*, 36, 533–536.

MacArthur, R.H. (1972) *Geographical Ecology*. Harper & Row, New York.

MacLulick, D.A. (1937) Fluctuations in numbers of the varying hare (*Lepus americanus*). *University of Toronto Studies, Biology Series*, 43, 1–136.

Mahowald, N., Jickells, T.D., Baker, A.R., Artaxo, P., Benitez-Nelson, C., Bergametti, R.G., Bond, T. Chen, C.Y., Cohen, D.D., Herut, B., Kubilay, N., Losno, R., Luo, C., Maenhaut, W., McGee, K.S., Okin, G.S., Siefert, R.L. & Tsukuda, S. (2008) Global distribution of atmospheric phosphorus sources, concentrations and deposition rates, and anthropogenic impacts. *Global Biogeochemical Cycles*, 22, GB4026.

Malmqvist, B., Wotton, R.S. & Zhang, Y. (2001) Suspension feeders transform massive amounts of seston in large northern rivers. *Oikos*, 92, 35–43.

Malthus, T. (1798) *An Essay on the Principle of Population*. J. Johnson, London.

Marino, R., Chan, F., Howarth, R.W., Pace, M.L. & Likens, G.E. (2006) Experimental tests of ecological constraints on planktonic nitrogen fixation in saline estuaries: 1. Nutrient and trophic controls. *Marine Ecology Progress Series*, 309, 25–39.

Marino, R., Chan, F., Howarth, R., Pace M. & Likens, G. (2002) Ecological and biogeochemical interactions constrain planktonic nitrogen fixation in estuaries. *Ecosystems*, 5, 719–725.

Maron, J.L. & Kauffman, M.J. (2006) Habitat-specific impacts of multiple consumers on plant population dynamics. *Ecology*, 8, 113–124.

Martin, J.H. & Fitzwater, S.E. (1988) Iron deficiency limits phytoplankton growth in the Northeast Pacific Subarctic. *Nature*, 331, 341–343.

Martin, J-L. & Thibault, J-C. (1996) Coexistence in Mediterranean warblers: ecological differences or interspecific territoriality? *Journal of Biogeography*, 13, 169–178.

Martin, P.R. & Martin, T.E. (2001) Ecological and fitness consequences of species coexistence: a removal experiment with wood warblers. *Ecology*, 82, 189–206.

Martin, P.S. (1984) Prehistoric overkill: the global model. In: *Quaternary Extinctions: a Prehistoric Revolution* (P.S. Martin & R.G. Klein, eds.), pp. 354–403. University of Arizona Press, Tuscon, AZ.

Marzusch, K. (1952) Untersuchungen über di Temperaturabhängigkeit von Lebensprozessen bei Insekten unter besonderer Berücksichtigung winter-schlantender Kartoffelkäfer. *Zeitschrift für vergleicherde Physiologie*, 34, 75–92.

Maskell, L.C., Smart, S.M., Bullock, J.M., Thomson. K. & Stevens, C. (2010) Nitrogen deposition causes widespread loss of species richness in British habitats. *Global Change Biology*, 16, 671–679.

May, R.M. (1981) Patterns in multi-species communities. In: *Theoretical Ecology: Principles and Applications*, 2nd edn (R.M. May, ed.), pp. 197–227. Blackwell Scientific Publications, Oxford.

May, R.M. (2010) Tropical arthropod species, more or less? *Science*, 329, 41–42.

McGrady-Steed, J., Harris, P.M. & Morin, P.J. (1997) Biodiversity regulates ecosystem predictability. *Nature*, 390, 162–165.

McIntosh, A.R. & Townsend, C.R. (1994) Interpopulation variation in mayfly antipredator tactics: differential effects of contrasting predatory fish. *Ecology*, 75, 2078–2090.

McIntosh, A.R. & Townsend, C.R. (1996) Interactions between fish, grazing invertebrates and algae in a New Zealand stream: a trophic cascade mediated by fish-induced changes to grazer behavior. *Oecologia*, 108, 174–181.

McKane, R.B., Johnson, L.C., Shaver, G.R. et al. (2002) Resource-based niches provide a basis for plant species diversity and dominance in arctic tundra. *Nature*, 415, 68–71.

McKay, J.K., Bishop, J.G., Lin, J.-Z., Richards, J.H., Sala, A. & Mitchell-Olds, T. (2001) Local adaptation across a climatic gradient despite small effective population size in the rare sapphire rockcress. *Proceedings of the Royal Society of London, Series B*, 268, 1715–1721.

McKey, D. (1979) The distribution of secondary compounds within plants. In: *Herbivores: their Interaction with Secondary Plant Metabolites* (G.A. Rosenthal & D.H. Janzen, eds.), pp. 56–134. Academic Press, New York.

Mduma, S.A.R., Sinclair, A.R.E. & Hilborn, R. (1999) Food regulates the Serengeti wildebeest: a 40 year record. *Journal of Animal Ecology*, 68, 1101–1122.

Menges, E.S. & Dolan, R.W. (1998) Demographic viability of populations of *Silene regia* in midwestern prairies: relationships with fire management, genetic variation, geographic location, population size and isolation. *Journal of Ecology*, 86, 63–78.

Menges, E.S. (2000) Population viability analyses in plants: challenges and opportunities. *Trends in Ecology and Evolution*, 15, 51–56.

Merryweather, J.W. & Fitter, A.H. (1995) Phosphorus and carbon budgets: mycorrhizal contribution in *Hyacinthoides nonscripta* (L.) Chouard ex Rothm. under natural conditions. *New Phytologist*, 129, 619–627.

Millennium Ecosystem Assessment (2005) *Ecosystems and Human Well-being: Biodiversity Synthesis*. World Resources Institute, Washington, DC.

Miller, D.D. & Mariani, S. (2010) Smoke, mirrors, and mislabeled cod: poor transparency in the European seafood industry. *Frontiers in Ecology and the Environment*, 8, 517–521.

Milner-Gulland, E.J. & Mace, R. (1998) *Conservation of Biological Resources*. Blackwell Science, Oxford.

Mittelbach, G.G., Steiner, C.F., Scheiner, S.M. et al. (2001) What is the observed relationship between species richness and productivity? *Ecology*, 82, 2381–2396.

Moilanen, A., Smith, A.T. & Hanski, I. (1998) Long-term dynamics in a metapopulation of the American pika. *American Naturalist*, 152, 530–542.

Monks, A. & Kelly, D. (2006) Testing the resource-matching hypothesis in the mast seeding tree Nothofagus truncata (Fagaceae). *Austral Ecology* 31: 366–375.

Montagnes, D.J.S., Kimmance, S.A. & Atkinson, D. (2003) Using Q_{10}: can growth rates increase linearly with temperature? *Aquatic Microbial Ecology*, 32, 307–313.

Montella, A., Gavin, A., Middleton, R., Autier, P. & Boniol. M. (2009) Cutaneous melanoma mortality starting to change: a study of trends in Northern Ireland. *European Journal of Cancer*, 45, 2360–2366.

Mora, C. & Sale, P.F. (2011) Ongoing global biodiversity loss and the need to move beyond protected areas: a review of the technical and practical shortcomings of protected areas on land and sea. *Marine Ecology Progress Series*, 434, 251–266.

Morin, P.J., Lawler, S.P. & Johnson, E.A (1988) Competition between aquatic insects and vertebrates: interaction strength and higher order interactions. *Ecology*, 69, 1401–1409.

Mortensen, H.K., Dupont, Y.L. & Olesen, J.M. (2008) A snake in paradise: disturbance of plant reproduction following extirpation of bird flower-visitors on Guam. *Biological Conservation*, 141, 2146–2154.

Mosier, A.R., Bleken, M.A., Chaiwanakupt, P. et al. (2002) Policy implications of human-accelerated nitrogen cycling. *Biogeochemistry*, 57/58, 477–516.

Mueller, N.D., J.S. Berber, M. Johnston, D.K. Ray, N. Ramankutty, & J.A. Foley (2012) Closing yield gaps through nutrient and water management. *Nature*, 490, 254–257.

Murdoch, W.W. & Stewart-Oaten, A. (1975) Predation and population stability. *Advances in Ecological Research*, 9, 1–131.

Murdoch, W.W. (1966) Community structure, population control and competition—a critique. *American Naturalist*, 100, 219–226.

Murray, D.L., Cary, J.R. & Keith, L.B. (1997) Interactive effects of sublethal nematodes and nutritional status on snowshoe hare vulnerability to predation. *Journal of Animal Ecology*, 66, 250–264.

Mwendera, E.J., Saleem, M.A.M. & Woldu, Z. (1997) Vegetation response to cattle grazing in the Ethiopian Highlands. *Agriculture, Ecosystems and Environment*, 64, 43–51.

Myers, R.A. (2001) Stock and recruitment: generalizations about maximum reproductive rate, density dependence, and variability using meta-analytic approaches. *ICES Journal of Marine Science*, 58, 937–951.

Nagy, E.S & Rice, K.J. (1997) Local adaptation in two subspecies of an annual plant: implications for migration and gene flow. *Evolution*, 51, 1079–1089.

National Research Council (1990) *Alternative Agriculture*. National Academy of Sciences, Academy Press, Washington, DC.

Navas, M-L., Roumet, C., Bellmann, A., Laurent, G. & Garnier, E. (2010) Suites of plant traits in species from different stages of a Mediterranean secondary succession. *Plant Biology*, 12, 183–196.

Neilson, R.P., Prentice, I.C., Smith, B., Kittel, T. & Viner, D. (1998) Simulated changes in vegetation distribution under global warming. Available as Annex C at www.epa.gov/global warming/reports/pubs/ipcc/annex/index.html.

NERC (1990) *Our Changing Environment*. Natural Environment Research Council, London.

Newsham, K.K., Fitter, A.H. & Watkinson, A.R. (1994) Root pathogenic and arbuscular mycorrhizal fungi determine fecundity of asymptomatic plants in the field. *Journal of Ecology*, 82, 805–814.

Newsham, K.K., Fitter, A.H. & Watkinson, A.R. (1995) Multifunctionality and biodiversity in arbuscular mycorrhizas. *Trends in Ecology and Evolution*, 10, 407–411.

Niklas, K.J., Tiffney, B.H. & Knoll, A.H. (1983) Patterns in vascular land plant diversification. *Nature*, 303, 614–616.

Nilsson, L.A. (1988) The evolution of flowers with deep corolla tubes. *Nature*, 334, 147–149.

Nixon, S.W. et al. (1996) The fate of nitrogen and phosphorus at the land-sea margin of the North Atlantic Ocean. *Biogeochemistry*, 35, 141–180.

Norton, I.O. & Sclater, J.G. (1979) A model for the evolution of the Indian Ocean and the breakup of Gondwanaland. *Journal of Geophysical Research*, 84, 6803–6830.

Nowak, R.M. (1979) *North American Quaternary Canis*. Monograph No. 6, Museum of Natural History. University of Kansas, Lawrence, KS.

O'Brien, E.M. (1993) Climatic gradients in woody plant species richness: towards an explanation based on an analysis of southern Africa's woody flora. *Journal of Biogeography*, 20, 181–198.

Oaks, J.L., Gilbert, M., Virani, M.Z. et al. (2004) Diclofenac residues as the cause of vulture population decline in Pakistan. *Nature*, 427, 629–633.

Odum, E.P. (1953) *Fundamentals of Ecology, 1st ed*. Saunders, Philadelphia.

Oedekoven, M.A. & Joern, A. (2000) Plant quality and spider predation affects grasshoppers (Acrididae): food-quality-dependent compensatory mortality. *Ecology*, 81, 66–77.

Oerke, E.C. (2006) Crop losses to pests. *Journal of Agricultural Science*, 144, 31–43.

Ogden, J. (1968) *Studies on reproductive strategy with particular reference to selected composites*. PhD thesis, University of Wales, Bangor.

Osborne, C.P. & D.J. Beerling (2006) Nature's green revolution: the remarkable evolutionary rise of C4 plants. *Philosophical Transactions of the Royal Society of London, Series B: Biological Sciences*, 361, 173–194.

Osmundson, D.B., Ryel, R.J., Lamarra, V.L. & Pitlick, J. (2002) Flow–sediment–biota relations: implications for river regulation effects on native fish abundance. *Ecological Applications*, 12, 1719–1739.

Oviatt, C.A., Doering, P., Nowicki, B., Reed, L., Cole, J. & Frithsen, J. (1995) An ecosystem level experiment on nutrient limitation in temperate coastal marine environments. *Marine Ecology Progress Series*, 116, 171–179.

Owen-Smith, N. (1987) Pleistocene extinctions: the pivotal role of megaherbivores. *Paleobiology*, 13, 351–362.

Pace, M.L., Cole, J.J., Carpenter, S.R. & Kitchell, J.F. (1999) Trophic cascades revealed in diverse ecosystems. *Trends in Ecology and Evolution*, 14, 483–488.

Paine, R.T. (1966) Food web complexity and species diversity. *American Naturalist*, 100, 65–75.

Paine, R.T. (1979) Disaster, catastrophe and local persistence of the sea palm *Postelsia palmaeformis*. *Science*, 205, 685–687.

Palmer, T.M., Young, T.P., Stanton, M.L. & Wenk, E. (2000) Short term dynamics of an acacia ant community in Laikipia, Kenya. *Oecologia*, 123, 425–435.

Passarge, J., Hol, S., Escher, M. & Huisman, J. (2006) Competition for nutrients and light: stable coexistence, alternative stable states, or competitive exclusion? *Ecological Monographs*, 76, 57–72.

Paterson, S. & Viney, M.E. (2002) Host immune responses are necessary for density dependence in nematode infections. *Parasitology*, 125, 283–292.

Pauly, D. & Christensen, V. (1995) Primary production required to sustain global fisheries. *Nature*, 374, 255–257.

Pavia, H. & Toth, G.B. (2000) Inducible chemical resistance to herbivory in the brown seaweed *Ascophyllum nodosum*. *Ecology*, 81, 3212–3225.

Pearl, R. (1927) The growth of populations. *Quarterly Review of Biology*, 2, 532–548.

Pearl, R. (1928) *The Rate of Living*. Knopf, New York.

Pearson, D.E. & Callaway, R.M. (2003) Indirect effects of host-specific biological control agents. *Trends in Ecology and Evolution*, 18, 456–461.

Peay, K.G., Garbelotto, M. & Bruns, T.D. (2010) Evidence of dispersal limitation in soil microorganisms: isolation reduces species richness on mycorrhizal tree islands. *Ecology*, 91, 3631–3640.

Penn, A.M., Sherwin, W.B., Gordon, G., Lunney, D., Melzer, A. & Lacy, R.C. (2000) Demographic forecasting in koala conservation. *Conservation Biology*, 14, 629–638.

Pennings, S.C. & Callaway, R.M. (2002) Parasitic plants: parallels and contrasts with herbivores. *Oecologia*, 131, 479–489.

Perrins, C.M. (1965) Population fluctuations and clutch size in the great tit, *Parus major* L. *Journal of Animal Ecology*, 34, 601–647.

Petren, K. & Case, T.J. (1996) An experimental demonstration of exploitation competition in an ongoing invasion. *Ecology*, 77, 118–132.

Petren, K., Grant, B.R. & Grant, P.R. (1999) A phylogeny of Darwin's finches based on microsatellite DNA variation. *Proceedings of the Royal Society of London, Series B*, 266, 321–329.

Pimentel, D. (1993) Cultural controls for insect pest management. In: *Pest Control and Sustainable Agriculture* (S. Corey, D. Dall & W. Milne, eds.), pp. 35–38. Commonwealth Scientific and Research Organisation, East Melbourne, New South Wales.

Pimentel, D., Krummel, J., Gallahan, D. et al. (1978) Benefits and costs of pesticide use in U.S. food production. *Bioscience*, 28, 777–784.

Pimentel, D., Lach, L., Zuniga, R. & Morrison, D. (2000) Environmental and economic costs of nonindigenous species in the United States. *BioScience*, 50, 53–65.

Pimm, S.L. (1991) *The Balance of Nature: Ecological Issues in the Conservation of Species and Communities*. University of Chicago Press, Chicago.

Pitcher, T.J. & Hart, P.J.B. (1982) *Fisheries Ecology*. Croom Helm, London.

Poffenbarger, H.J., Needelman, B.A. & Megonical, J.P. (2011) Salinity influences on methane emissions from tidal marshes. *Wetlands*, 31, 831–842.

Pope, S.E., Fahrig, L. & Merriam, H.G. (2000) Landscape complementation and metapopulation effects on leopard frog populations. *Ecology*, 81, 2498–2508.

Population Reference Bureau (2006) http://www.prb.org/Publications/GraphicsBank/PopulationTrends.aspx.

Pounds, A.P., Bustamante, M.R., Coloma, L.A., Consuegra, J.A., Fogden, M.P.L., Foster, P.N., Marca, E., Masters, K.L., Merino-Viteri, A., Puschendorf, R., Ron, S.R., Sanchez-Azofeifa, G.A., Still, C.J. & Young, B.E. (2006) Widespread amphibian extinctions from epidemic disease driven by global warming. *Nature*, 439, 161–167.

Power, M.E., Tilman, D., Estes, J.A. et al. (1996) Challenges in the quest for keystones. *Bioscience*, 46, 609–620.

Primack, R.B. (1993) *Essentials of Conservation Biology*. Sinauer Associates, Sunderland, MA.

Pywell, R.F., Bullock, J.M., Walker, K.J., Coulson, S.J., Gregory, S.J. & Stevenson, M.J. (2004) Facilitating grassland diversification using the hemiparasitic plant *Rhinanthus minor*. *Journal of Applied Ecology*, 41, 880–887.

Raffaelli, D. & Hawkins, S. (1996) *Intertidal Ecology*. Kluwer, Dordrecht.

Rahbek, C. (1995) The elevational gradient of species richness: a uniform pattern? *Ecography*, 18, 200–205.

Rahel, F.J. (2002) Homogenization of freshwater faunas. *Annual Review of Ecology and Systematics*, 33, 291–315.

Rainey, P.B. & Trevisano, M. (1998) Adaptive radiation in a heterogeneous environment. *Nature*, 394, 69–72.

Ramankutty, N., Evan, A.T., Monfreda, C. & Foley, J.A. (2008) Farming the planet: 1. Geographic distribution of global agricultural lands in the year 2000. *Global Biogeochemical Cycles*, **22**, GB1003.

Randall, M.G.M. (1982) The dynamics of an insect population throughout its altitudinal distribution: *Coleophora alticolella* (Lepidoptera) in northern England. *Journal of Animal Ecology*, 51, 993–1016.

Ratcliffe, D.A. (1970) Changes attributable to pesticides in egg breakage frequency and eggshell thickness in some British birds. *Journal of Applied Ecology*, 7, 67–107.

Rätti, O., Dufva, R. & Alatalo, R.V. (1993) Blood parasites and male fitness in the pied flycatcher. *Oecologia*, 96, 410–414.

Reganold, J.P., Glover, J.D., Andrews, P.K. & Hinman, H.R. (2001) Sustainability of three apple production systems. *Nature*, 410, 926–929.

Rezende, E.L., Albert, E.M., Fortuna, M.A. & Bascompte, J. (2009) Compartments in a marine food web associated with phylogeny, body mass, and habitat structure. *Ecology Letters*, 12, 779–788.

Reznick, D. & Endler, J.A. (1982) The impact of predation on life history evolution in Trinidadian guppies (*Poecilia reticulata*). *Evolution*, 36, 160–177.

Ribas, C.R., Schoereder, J.H., Pic, M. & Soares, S.M. (2003) Tree heterogeneity, resource availability, and larger scale processes regulating arboreal ant species richness. *Austral Ecology*, 28, 305–314.

Ricklefs, R.E. & Lovette, I.J. (1999) The role of island area *per se* and habitat diversity in the species–area relationships of four Lesser Antillean faunal groups. *Journal of Animal Ecology*, 68, 1142–1160.

Ricklefs, R.E. (1973) *Ecology*. Nelson, London.

Ridley, M. (1993) *Evolution*. Blackwell Science, Boston.

Riis, T. & Sand-Jensen, K. (1997) Growth reconstruction and photosynthesis of aquatic mosses: influence of light, temperature and carbon dioxide at depth. *Journal of Ecology*, 85, 359–372.

Risebrough, R. (2004) Fatal medicine for vultures. *Nature*, 427, 596–598.

Rockström, J. *et al.* (2009) A safe operating space for humanity. *Nature*, 461, 472–475.

Rodrigues, A.S.L., Pilgrim, J.D., Lamoreux, J.F., Hoffmann, M. & Brooks, T.M. (2006) The value of the IUCN Red List for conservation. *Trends in Ecology and Evolution*, 21, 71–76.

Rogers, D.J. & Randolph, S.E. (2000) The global spread of malaria in a future, warmer world. *Science*, 489, 1763–1766.

Rohr, D.H. (2001) Reproductive trade-offs in the elapid snakes *Austrelap superbus* and *Austrelap ramsayi*. *Canadian Journal of Zoology*, 79, 1030–1037.

Root, R. (1967) The niche exploitation pattern of the blue-grey gnatcatcher. *Ecological Monographs*, 37, 317–350.

Rosenthal, G.A., Dahlman, D.L. & Janzen, D.H. (1976) A novel means for dealing with L-canavanine, a toxic metabolite. *Science*, 192, 256–258.

Rosenthal, G.A., Hughes, C.G. & Janzen, D.H. (1982) L-Canavanine, a dietary nitrogen source for the seed predator *Caryedes brasiliensis* (Bruchidae). *Science*, 217, 353–355.

Rosenzweig, M.L. & Sandlin, E.A. (1997) Species diversity and latitudes: listening to area's signal. *Oikos*, 80, 172–176.

Rosenzweig, M.L. (1971) Paradox of enrichment: destabilization of exploitation ecosystems in ecological time. *Science*, 171, 385–387.

Rouphael, A.B. & Inglis, G.J. (2001) 'Take only photographs and leave only footprints'? An experimental study of the impacts of underwater photographers on coral reef dive sites. *Biological Conservation*, 100, 281–287.

Roura-Pascual, N., Suarez, A.V., Gomez, C., Pons, P., Touyama, Y., Wild, A.L. & Townsend Peterson, A. (2004) Geographical potential of Argentine ants (*Linepithema humile* Mayr) in the face of global climate change. *Proceedings of the Royal Society of London, Series B*, 271, 2527–2534.

Rowe, C.L. (2002) Differences in maintenance energy expenditure by two estuarine shrimp (*Palaemonetes pugio* and *P. vulgaris*) that may permit partitioning of habitats by salinity. *Comparative Biochemistry and Physiology A*, 132, 341–351.

Roy, M.S., Geffen, E., Smith, D., Ostrander, E.A. & Wayne, R.K. (1994) Patterns of differentiation and hybridization in North American wolflike canids, revealed by analysis of microsatellite loci. *Molecular Biology and Evolution*, 11, 553–570.

Ruiters, C. & McKenzie, B. (1994) Seasonal allocation and efficiency patterns of biomass and resources in the perennial geophyte *Sparaxis grandiflora* subspecies *fimbriata* (Iridaceae) in lowland coastal Fynbos, South Africa. *Annals of Botany*, 74, 633–646.

Rundle, H.D. & Nosil, P. (2005). Ecological speciation, *Ecology Letters*, 8, 336–352.

Saccheri, I.J., Rousset, F., Watts, P.C., Brakefield, P.M. & Cook, L.M. (2008). Selection and gene flow on a diminishing cline of melanic peppered moths. *Proceedings of the National Academy of Sciences of the USA*, 105, 16212–16217.

Sale, P.F. & Douglas, W.A. (1984) Temporal variability in the community structure of fish on coral patch reefs and the relation of community structure to reef structure. *Ecology*, 65, 409–422.

Sale, P.F. (1979) Recruitment, loss and coexistence in a guild of territorial coral reef fishes. *Oecologia*, 42, 159–177.

Salek, M. (2012) Spontaneous succession on opencast mining sites: implications for bird biodiversity. *Journal of Applied Ecology*, 49, 1417–1425.

Salisbury, E.J. (1942) *The Reproductive Capacity of Plants*. Bell, London.

Sanchez-Azofeifa, G.A., Pfaff, A., Robalino, J.A. & Boomhower, J. (2007) Costa Rica's Payment for Environmental Services program: intention, implementation and impact. *Conservation Biology*, 21, 1165–1173.

Sanders, N.J., Moss, J. & Wagner, D. (2003) Patterns of ant species richness along elevational gradients in an arid ecosystem. *Global Ecology and Biogeography*, 12, 93–102.

Santelmann, M.V., White, D., Freemark, K. et al. (2004) Assessing alternative futures for agriculture in Iowa, USA. *Landscape Ecology*, 19, 357–374.

Savidge, J.A. (1987) Extinction of an island forest avifauna by an introduced snake. *Ecology*, 68, 660–668.

Sax, D.F. & Gaines, S.D. (2003) Species diversity: from global decreases to local increases. *Trends in Ecology and Evolution*, 18, 561–566.

Schindler, D.W. (1971) Carbon, nitrogen, and phosphorus and the eutrophication of freshwater lakes. *Journal of Phycology*, 7, 321–329.

Schindler, D.W. (1974) Eutrophication and recovery in experimental lakes: implications for lake management. *Science*, 184, 897–899.

Schindler, D.W. (1977) Evolution of phosphorus limitation in lakes. *Science*, 195(4275), 260–262.

Schindler, D.W. (2009) A personal history of the Experimental Lakes Area project. *Canadian Journal of Fisheries an Aquatic Sciences*, 66, 1837–1847.

Schindler, D.W. (1998) Replication versus realism: The need for ecosystem-scale experiments. *Ecosystems*, 1, 323–334.

Schindler, D.W., Mills, K.H., Malley, D.F., Findlay, D.L., Shearer, J.A., Davies, I.J., Turner, M.A., Linsey, G.A. & Cruikshank, D.R. (1985) Long-term ecosystem stress: the effects of years of experimental acidification on a small lake. *Science*, 228, 1395–401.

Schlesinger, W.H. (1997) *Biogeochemistry: An Analysis of Global Change, 2nd edition*. Academic Press.

Schluter, D. (2001) Ecology and the origin of species. *Trends in Ecology and Evolution*, 16, 372–380.

Schoener, T.W. (1983) Field experiments on interspecific competition. *American Naturalist*, 122, 240–285.

Schoenly, K., Beaver, R.A. & Heumier, T.A. (1991) On the trophic relations of insects: a food-web approach. *American Naturalist*, 137, 597–638.

Schroeder, M.S. & Janos, D.P. (2004) Phosphorus and intraspecific density alter plant responses to arbuscular mycorrhizas. *Plant and Soil*, 264, 335–348.

Schulze, E.D. (1970) Dre CO_2-Gaswechsel de Buche (*Fagus sylvatica* L.) in Abhäbgigkeit von den Klimafaktoren in Freiland. *Flora, Jena*, 159, 177–232.

Schulze, E.D., Fuchs, M.I. & Fuchs, M. (1977a) Spatial distribution of photosynthetic capacity and performance in a mountain spruce forest in northern Germany. I. Biomass distribution and daily CO_2 uptake in different crown layers. *Oecologia*, 29, 43–61.

Schulze, E.D., Fuchs, M.I. & Fuchs, M. (1977b) Spatial distribution of photosynthetic capacity and performance in a mountain spruce forest in northern Germany. III. The significance of the evergreen habit. *Oecologia*, 30, 239–249.

Schwartz, O.A., Armitage, K.B. & Van Vuren, D. (1998) A 32-year demography of yellow-bellied marmots (*Marmota flaviventris*). *Journal of Zoology*, 246, 337–346.

Searchinger et al. (2008) Use of U.S. croplands for biofuels increases greenhouse gases through emissions from land use change. *Science*, 5867, 1238–1240.

Shankar Raman, T., Rawat, G.S. & Johnsingh, A.J.T. (1998) Recovery of tropical rainforest avifauna in relation to vegetation succession following shifting cultivation in Mizoram, north-east India. *Journal of Applied Ecology*, 35, 214–231.

Shanklin, J. (2010) Reflections on the ozone hole. *Nature*, 465, 34–35.

Sherman, C.D.H & Ayre, D.L (2008). Fine-scale adaptation in a clonal sea anemone. *Evolution*, 62, 1373–1380.

Shindell, D., Kuylenstierna, J.C.I., Vignati, E., van Dingenen, R., Armann, M., Klimont, Z., Anenberg, S.C., Muller, N., Janssents-Maenhout, G., Rase, F., Schwartsz, J., Faluvegi, G., Pozzoli, L, Kupiainen, K., Hoglund-Isaksson, L, Emberson, L, Streets, D., Ramanathan, V., Hicks, K., Kim Oanh, N.T., Milly, G., Wiklliams, M., Demkine, V. & Fowler, D. (2012) Simultaneously mitigating near-term climate change and improving human health and food security. *Science*, 335, 183–189.

Shindell, D.T., Faluvegi, G., Koch, D.M., Schmidt, G.A., Unger, N. & Bauer, S.E. (2009) Improved attribution of climate forcing to emissions. *Science*, 326, 716–718.

Sibly, R.M. & Hone, J. (2002) Population growth rate and its determinants: an overview. *Philosophical Transactions of the Royal Society of London, Series B*, 357, 1153–1170.

Sieman, E., Haarstad, J. & Tilman, D. (1999) Dynamics of plant and arthropod diversity during old field succession. *Ecography*, 22, 406–414.

Sillmann, J. & Roekner E. (2008) Indices for extreme events in projections of anthropogenic climate change. *Climatic Change*, 86, 83–104.

Simberloff, D.S. (1976) Experimental zoogeography of islands: effects of island size. *Ecology*, 57, 629–648.

Simberloff, D.S., Dayan T., Jones, C. & Ogura, G. (2000) Character displacement and release in the small Indian mongoose, *Herpestes javanicus*. *Ecology*, 91, 2086–2099.

Simon, K.S., Townsend, C.R., Biggs, B.J.F., Bowden, W.B. & Frew, R.D. (2004) Habitat-specific nitrogen dynamics in New Zealand streams containing native or invasive fish. *Ecosystems*, 7, 777–792.

Sinclair, B.J. & Sjursen, H. (2001) Cold tolerance of the Antarctic springtail *Gomphiocephalus hodgsoni* (Collembola, Hypogastruridae). *Antarctic Science*, 13: 277–279.

Singleton, G., Krebs, C.J., Davis, S., Chambers, L. & Brown, P. (2001) Reproductive changes in fluctuating house mouse populations in southeastern Australia. *Proceedings of the Royal Society of London, Series B*, 268, 1741–1748.

Slobodkin, L.B., Smith, F.E. & Hairston, N.G. (1967) Regulation in terrestrial ecosystems, and the implied balance of nature. *American Naturalist*, 101, 109–124.

Smith, C.R., De Leo, F.C., Bernardino, A.F., Sweetman, A.K. & Martinez Arbizu, P. (2008) Abyssal food limitation, ecosystem structure and climate change. *Trends in Ecology and Evolution*, 23, 518–528.

Smith, J.W. (1998) Boll weevil eradication: area-wide pest management. *Annals of the Entomological Society of America*, 91, 239–247.

Smith, S.V. (1984) Phosphorus versus nitrogen limitation in the marine environment. *Limnology and Oceanography*, 29, 1149–1160

Sousa, M.E. (1979a) Experimental investigation of disturbance and ecological succession in a rocky intertidal algal community. *Ecological Monographs*, 49, 227–254.

Sousa, M.E. (1979b) Disturbance in marine intertidal boulder fields: the nonequilibrium maintenance of species diversity. *Ecology*, 60, 1225–1239.

Spiller, D.A. & Schoener, T.W. (1990) A terrestrial field experiment showing the impact of eliminating predators on foliage damage. *Nature*, 347, 469–472.

Stanton, M.L., Palmer, T.M. & Young, T.P. (2002) Competition-colonization trade-offs in a guild of African acacia ants. *Ecological Monographs*, 72, 347–363.

Stenseth, N.C., Falck, W., Bjornstad, O.N. & Krebs, C.J. (1997) Population regulation in snowshoe hare and lynx populations: asymmetric food web configurations between the snowshoe hare and the lynx. *Proceedings of the National Academy of Science of the USA*, 94, 5147–5152.

Stevens, C. J., Dupre, C., Dorland, E., Gaudnik, C., Gowing, D.J.G., Bleeker, A., Diekmann, M., Alard, D., Bobbink, R., Fowler, D., Corcket, D., Mountford, J.O., Vandvick, V., Aarrestad, P.A., Muller, S. & Dise, N.B. (2011) The impact of nitrogen deposition on acid grasslands in the Atlantic region of Europe. *Environmental Pollution*, 159, 2243–2250.

Stevens, C.E. & Hume, I.D. (1998) Contributions of microbes in vertebrate gastrointestinal tract to production and conservation of nutrients. *Physiological Reviews*, 78, 393–426.

Stoll, P. & Prati, D. (2001) Intraspecific aggregation alters competitive interactions in experimental plant communities. *Ecology*, 82, 319–327.

Stouffer, D.B. & Bascompte, J. (2011) Compartmentalization increases food web persistence. *Proceedings of the National Academy of Sciences of the USA*, 108, 3648–3652.

Strauss, S.Y. & Agrawal, A.A. (1999) The ecology and evolution of plant tolerance to herbivory. *Trends in Ecology and Evolution*, 14, 179–185.

Strauss, S.Y., Irwin, R.E. & Lambrix, V.M. (2004) Optimal defence theory and flower petal colour predict variation in the secondary chemistry of wild radish. *Journal of Ecology*, 92, 132–141.

Strong, D.R. Jr., Lawton, J.H. & Southwood, T.R.E. (1984) *Insects on Plants: Community Patterns and Mechanisms*. Blackwell Scientific Publications, Oxford.

Susarla, S., Medina, V.F. & McCutcheon, S.C. (2002) Phytoremediation: an ecological solution to organic chemical contamination. *Ecological Engineering*, 18, 647–658.

Sutherland, W.J., Gill, J.A. & Norris, K. (2002) Density-dependent dispersal in animals: concepts, evidence, mechanisms and consequences. In: *Dispersal Ecology* (J.M. Bullock, R.E. Kenward & R.S. Hails, eds.), pp. 134–151. Blackwell Publishing, Oxford.

Sutton, S.L. & Collins, N.M. (1991) Insects and tropical forest conservation. In: *The Conservation of Insects and their Habitats* (N.M. Collins & J.A. Thomas, eds.), pp. 405–424. Academic Press, London.

Swan, G., Naidoo, V., Cuthbert, R. et al. (2006) Removing the threat of diclofenac to critically endangered Asian vultures. *Public Library of Science Biology*, 4(3), e66. doi: 10.1371/journal.pbio.0040066.

Swift, M.J., Heal, O.W. & Anderson, J.M. (1979) *Decomposition in Terrestrial Ecosystems*. Blackwell Scientific Publications, Oxford.

Swinnerton, K.J., Groombridge, J.J., Jones, C.G., Burn, R.W. & Mungroo, Y. (2004) Inbreeding depression and founder diversity among captive and free-living populations of the endangered pink pigeon *Columba mayeri*. *Animal Conservation*, 7, 353–364.

Symonides, E. (1979) The structure and population dynamics of psammophytes on inland dunes. II. Loose-sod populations. *Ekologia Polska*, 27, 191–234.

Symonides, E. (1983) Population size regulation as a result of intra-population interactions. I. The effect of density on the survival and development of individuals of *Erophila verna* (L.). *Ekologia Polska*, 31, 839–881.

Tanaka, M.O. & Magalhaes, C.A. (2002) Edge effects and succession dynamics in *Brachidontes* mussel beds. *Marine Ecology Progress Series*, 237, 151–158.

Taniguchi, Y. & Nakano, S. (2000) Condition-specific competition: implications for the altitudinal distribution of stream fishes. *Ecology*, 81, 2027–2039.

Tansley, A.G. (1904) The problems of ecology. *New Phytologist*, 3, 191–200.

Taylor, I. (1994) *Barn Owls. Predator–Prey Relationships and Conservation*. Cambridge University Press, Cambridge.

Téllez-Valdés, O. & Dávila-Aranda, P. (2003) Protected areas and climate change: a case study of the cacti in the Tehuacán-Cuicatlán Biosphere Reserve, Mexico. *Conservation Biology*, 17, 846–853.

terHorst, C.P. (2011) Experimental evolution of protozoan traits in response to interspecific competition. *Journal of Evolutionary Biology*, 24, 36–46.

Thomas, C.D. & Harrison, S. (1992) Spatial dynamics of a patchily distributed butterfly species. *Journal of Applied Ecology*, 61, 437–446.

Thomas, C.D. & Jones, T.M. (1993) Partial recovery of a skipper butterfly (*Hesperia comma*) from population refuges: lessons for conservation in a fragmented landscape. *Journal of Animal Ecology*, 62, 472–481.

Thomas, C.D., Cameron, A., Green, R.E. et al. (2004) Extinction risk from climate change. *Nature*, 427, 145–148.

Thomas, C.D., Thomas, J.A. & Warren, M.S. (1992) Distributions of occupied and vacant butterfly habitats in fragmented landscapes. *Oecologia*, 92, 563–567.

Thompson, R.M., Townsend, C.R., Craw, D., Frew, R. & Riley, R. (2001) (Further) links from rocks to plants. *Trends in Ecology and Evolution*, 16, 543.

Tilman, D. (1982) *Resource Competition and Community Structure*. Princeton University Press, Princeton, NJ.

Tilman, D. (1986) Resources, competition and the dynamics of plant communities. In: *Plant Ecology* (M.J. Crawley, ed.), pp. 51–74. Blackwell Scientific Publications, Oxford.

Tilman, D. (1996) Biodiversity: population versus ecosystem stability. *Ecology*, 77, 350–363.

Tilman, D. (1999) The ecological consequences of changes in biodiversity: a search for general principles. *Ecology*, 80, 1455–1474.

Tilman, D., Fargione, J., Wolff, B. D'Antonio, C., Dobson, A., Howarth, R.W., Schindler, D., Schlesinger, W., Simberloff D. & Swackhamer, D. (2001) Forecasting agriculturally driven global environmental change. *Science*, 292, 281–284.

Tilman, D., Mattson, M. & Langer, S. (1981) Competition and nutrient kinetics along a temperature gradient: an experimental test of a mechanistic approach to niche theory. *Limnology and Oceanography*, 26, 1020–1033.

Tittensor, D.P., Rex, M.A., Stuart, C.T., McClain, C.R. & Smith, C.R. (2011) Species-energy relationships in deep-sea molluscs. *Biology Letters*, 7, 718–722.

Tokeshi, M. (1993) Species abundance patterns and community structure. *Advances in Ecological Research*, 24, 112–186.

Tonn, W.M. & Magnuson, J.J. (1982) Patterns in the species composition and richness of fish assemblages in northern Wisconsin lakes. *Ecology*, 63, 137–154.

Townsend, A. & Howarth, R.W. (2010) Human acceleration of the global nitrogen cycle. *Scientific American*, 302, 32–39.

Townsend, A.R., Howarth, R., Bazzaz, F.A., Booth, M.S., Cleveland, C.C., Collinge, S.K., Dobson, A.P., Epstein, P.R., Holland, E.A., Keeney, D.R., Mallin, M.A., Rogers, C.A., Wayne P. & Wolfe, A.H. (2003) Human health effects of a changing global nitrogen cycle. *Frontiers in Ecology & Environment*, 1, 240–246.

Townsend, C.R. & Crowl, T.A. (1991) Fragmented population structure in a native New Zealand fish: an effect of introduced brown trout? *Oikos*, 61, 348–354.

Townsend, C.R. (2007) *Ecological Applications: Toward a Sustainable World*. Blackwell Publishing, Oxford.

Townsend, C.R., Hildrew, A.G. & Francis, J.E. (1983) Community structure in some southern English streams: the influence of physiochemical factors. *Freshwater Biology*, 13, 521–544.

Townsend, C.R., Scarsbrook, M.R. & Dolédec, S. (1997) The intermediate disturbance hypothesis, refugia and bio-diversity in streams. *Limnology and Oceanography*, 42, 938–949.

Townsend, C.R., Thompson, R.M., McIntosh, A.R. et al. (1998) Disturbance, resource supply, and food-web architecture in streams. *Ecology Letters*, 1, 200–209.

Turkington, R. & Harper, J.L. (1979) The growth, distribution and neighbour relationships of *Trifolium repens* in a permanent pasture. IV. Fine scale biotic differentiation. *Journal of Ecology*, 67, 245–254.

Turner, J.R.G., Lennon, J.J. & Greenwood, J.J.D. (1996) Does climate cause the global biodiversity gradient? In: *Aspects of the Genesis and Maintenance of Biological Diversity* (M. Hochberg, J. Claubert & R. Barbault, eds.), pp. 199–220. Oxford University Press, London.

Uchida, Y. & Inoue, M. (2010) Fish species richness in spring-fed ponds: effects of habitat size versus isolation in temporally variable environments. *Freshwater Biology*, 55, 983–994.

UNEP (2003) *Global Environmental Outlook Year Book 2003*. United Nations Environmental Program (UNEP), GEO Section, Nairobi, Kenya.

UNEP (2014) *Assessing global land use: Balancing consumption with sustainable supply*. United Nations Environment Programme, Paris, France.

United Nations (1998) *Global Change and Sustainable Development: Critical Trends*. Report of the Secretary General, United Nations, New York. (Also available on the world wide web at www.un.org/esa/sustdev/trends.html.)

United Nations (2002) *Global Environmental Outlook 3*. Report of the United Nations Environmental Program (UNEP). UNEP, www.unep.org/GEO/geo3.

United Nations (2005) *The World Population Prospects: the 2004 Revision*. Analytical Report Vol. III. Department of Economic and Social Affairs, Population Division, United Nations. United Nations, New York.

United Nations, Department of Economic and Social Affairs, Population Division (2011) *World Population Prospects: The 2010 Revision*. United Nations, New York.

Valdivia, N., Hiedemann, A., Thiel, M., Molis, M. & Wahl, M. (2005) Effects of disturbance on the diversity of hard-bottom macrobenthic communities on the coast of Chile. *Marine Ecology Progress Series*, 299, 45–54.

Valeix, M., Loveridge, A.J., Chamaille-Jammes, S., Davidson, Z., Murindagomo, F., Fritz, H. & McDonald, D.W. (2009) Behavioral adjustments of African herbivores to predation risk by lions: spatiotemporal variations influence habitat use. *Ecology*, 90, 23–30.

Valentine, J.W. (1970) How many marine invertebrate fossil species? A new approximation. *Journal of Paleontology*, 44, 410–415.

Valladares, V.F. & Pearcy, R.W. (1998) The functional ecology of shoot architecture in sun and shade plants of *Heteromeles arbutifolia* M. Roem., a Californian chaparral shrub. *Oecologia*, 114, 1–10.

van der Jeugd, H.P. (1999) *Life history decisions in a changing environment: a long-term study of a temperate barnacle goose population*. PhD thesis, Uppsala University, Uppsala.

Vannotte, R.L., Minshall, G.W., Cummins, K.W., Sedell, J.R. & Cushing, C.E. (1980) The river continuum concept. *Canadian Journal of Fisheries and Aquatic Sciences*, 37, 130–137.

Vázquez, G.J.A. & Givnish, T.J. (1998) Altitudinal gradients in tropical forest composition, structure, and diversity in the Sierra de Manantlán. *Journal of Ecology*, 86, 999–1020.

Verhoeven, J.T.A., Arheimer, B., Yin, C. & Hefting, M.M. (2006) Regional and global concerns over wetlands and water quality. *Trends in Ecology and Evolution*, 21, 96–103.

Virolainen, K.M., Suomi, T., Suhonen, J. & Kuitunen, M. (1996) Conservation of vascular plants in single large and several small mires: species richness, rarity and taxonomic diversity. *Journal of Applied Ecology*, 35, 700–707.

Vitousek, P.M. & Farrington, H. (1997) Nutrient limitation and soil development: Experimental test of a biogeochemical theory. *Biogeochemistry*, 37, 63–75.

Vitousek, P.M., Mooney, H.A., Lubchenco, J. & Melillo, J.M. (1997) Human domination of Earth's ecosystems. *Science*, 277, 494–499.

Volterra, V. (1926) Variations and fluctuations of the numbers of individuals in animal species living together. (Reprinted in 1931. In: *Animal Ecology* (R.N. Chapman, ed.), pp. 409–448. McGraw Hill, New York.)

vonHoldt, B.M., Pollinger, J.P., Earl, D.A., Knowles, J.C., Boyko, A.R., Parker, H., Geffen, E., Pilot, M., Jedrzejewski W., Jedrzejewska, B., Sidorovich, V., Greco, C., Randi, E., Musiani, M., Kays, R., Bustamante, C.D., Ostrander, E.A., Novembre, J. & Wayne, R.K. (2011) A genome-wide perspective on the evolutionary history of enigmatic wolf-like canids. *Genome Research*, 21, 1294–305.

Waage, J.K. & Greathead, D.J. (1988) Biological control: challeges and opportunities. *Philosophical Transactions of the Royal Society of London, Series B*, 318, 111–128.

Wackernagel, M., Schulz, N.B., Deumling, D., Callejas Linares, A., Jenkins, M., Kapos, V., Monfreda, C., Loh, J., Myers, M., Norgaard, R. & Randers, J. (2002) Tracking the ecological overshoot of the human economy. *Proceedings of the National Academy of the USA*, 99, 9266–9271

Wake, D.B. & Vredenburg, V.T. (2008) Are we in the midst of the sixth mass extinction? A view from the world of amphibians. *Proceedings of the National Academy of Sciences of the USA*, 105, 11466–11473.

Walsh, J.A. (1983) Selective primary health care: strategies for control of disease in the developing world. IV. Measles. *Reviews of Infectious Diseases*, 5, 330–340.

Wang, G.-H. (2002) Plant traits and soil chemical variables during a secondary vegetation succession in abandoned fields on the Loess Plateau. *Acta Botanica Sinica*, 44, 990–998.

Wardle, D.A., Bonner, K.I. & Barker, G.M. (2000) Stability of ecosystem properties in response to above-ground functional group richness and composition. *Oikos*, 89, 11–23.

Warren, P.H. (1989) Spatial and temporal variation in the structure of a freshwater food web. *Oikos*, 55, 299–311.

Watkinson, A.R. & Harper, J.L. (1978) The demography of a sand dune annual: *Vulpia fasciculata*. I. The natural regulation of populations. *Journal of Ecology*, 66, 15–33.

Wayne, R.K. & Jenks, S.M. (1991) Mitochondrial DNA analysis implying extensive hybridization of the endangered red wolf *Canis rufus*. *Nature*, 351, 565–568.

Wayne, R.K. (1996) Conservation genetics in the Canidae. In: *Conservation Genetics* (J.C. Avise & J.L. Hamrick, eds.), pp. 75–118. Chapman & Hall, New York.

Webb, S.D. (1987) Community patterns in extinct terrestrial invertebrates. In: *Organization of Communities: Past and Present* (J.H.R. Gee & P.S. Giller, eds.), pp. 439–468. Blackwell Scientific Publications, Oxford.

Webb, W.L., Lauenroth, W.K., Szarek, S.R. & Kinerson, R.S. (1983) Primary production and abiotic controls in forests, grasslands and desert ecosystems in the United States. *Ecology*, 64, 134–151.

Wegener, A. (1915) *Entstehung der Kontinenter und Ozeaner*. Samml. Viewig, Braunschweig. (English translation 1924. *The Origins of Continents and Oceans*, translated by J.G.A. Skerl. Methuen, London.)

Wheeler, B.D. & Shaw, S.C. (1991) Above-ground crop mass and species richness of the principle types of rich-fen vegetation of lowland England and Wales. *Journal of Ecology*, 79, 285–301.

White, G. (1789) *The Natural History and Antiquities of Selborne*. (Reprinted in 1977 as *The Natural History of Selborne* (G. White and R. Mabey). Penguin, London.)

Whitehead, A.N. (1953) *Science and the Modern World*. Cambridge University Press, Cambridge.

Whittaker, R.J. (2010) Meta-analyses and mega-mistakes: calling time on meta-analysis of the species richness-productivity relationship. *Ecology*, 91, 2522–2533.

Whittaker, R.J., Willis, K.J. & Field, R. (2003) Climatic–energetic explanations of diversity: a macroscopic perspective. In: *Macroecology: Concepts and Consequences* (T.M. Blackburn & K.J. Gaston, eds), pp. 107–129. Blackwell Publishing, Oxford.

Wilcove, D.S., Rothstein, D., Dubow, J., Phillips, A. & Losos, E. (1998) Quantifying threats to imperiled species in the United States. *Bioscience*, 48, 607–615.

Williams, W.D. (1988) Limnological imbalances: an antipodean viewpoint. *Freshwater Biology*, 20, 407–420.

Winemiller, K.O. (1990) Spatial and temporal variation in tropical fish trophic networks. *Ecological Monographs*, 60, 331–367.

Withler, R.E., Candy, J.R., Beacham, T.D. & Miller, K.M. (2004) Forensic DNA analysis of Pacific salmonid samples for species and stock identification. *Environmental Biology of Fishes*, 69, 275–285.

Woiwod, I.P. & Hanski, I. (1992) Patterns of density dependence in moths and aphids. *Journal of Animal Ecology*, 61, 619–629.

Wolff, J.O., Schauber, E.M. & Edge, W.D. (1997) Effects of habitat loss and fragmentation on the behavior and demography of gray-tailed voles. *Conservation Biology*, 11, 945–956.

Wootton, J.T. (1992) Indirect effects, prey susceptibility, and habitat selection: impacts of birds on limpets and algae. *Ecology*, 73, 981–991.

Worland, M.R. & Convey, P. (2001) Rapid cold hardening in Antarctic microarthropods. *Functional Ecology*, 15, 515–524.

World Health Organization (2013a) http://www.who.int/globalchange/summary/en/index4.html.

World Health Organization (2013b) *Obesity and overweight, fact sheet #311*. United Nations, Geneva, Switzerland. http://www.who.int/nutrition/topics/3_foodconsumption/en/

World Resources Institute (2011) *Reefs at Risk Revisited*. The World Resources Institute.

Wright, S., Keeling, J. & Gillman, L. (2006) The road from Santa Rosalia: a faster tempo of evolution in tropical climates. *Proceedings of the National Academy of Sciences of the USA*, 103, 7718–7722.

Yao, I., Shibao, H. & Akimoto, S. (2000) Costs and benefits of ant attendance to the drepanosiphid aphid *Tuberculatus quercicola*. *Oikos*, 89, 3–10.

Yodzis, P. (1986) Competiton, mortality and community structure. In: *Community Ecology* (J. Diamond & T.J. Case, eds.), pp. 480–491. Harper & Row, New York.

Yoshida, T., Jones, L.E., Ellner, S.P., Fussmann, G.F. & Hairston, N.G., Jr. (2003) Rapid evolution drives ecological dynamics in a predator–prey system. *Nature*, 424, 303–306.

Zhang, Y., Bi, P., Hiller, J.E., Sun, Y. & Ryan, P. (2007) Climate variations and bacillary dysentery in northern and southern cities of China. *Journal of Infection*, 55, 194–200.

Zimmer, M. & Topp, W. (2002) The role of coprophagy in nutrient release from feces of phytophagous insects. *Soil Biology and Biochemistry*, 34, 1093–1099.

Zimov, S.A., Schuur, E.A.G. & Chapin, F.S. (2006) Permafrost and the global carbon budget. *Science*, 312, 1612–1613.